Ken Ring's

Australia

Biennial Weather Almanac

2025 - 2026

By The Same Author

Predict Weather Almanacs for New Zealand 1999-Present)
Predict Weather Almanacs for Australia (2006- Present)
Predict Weather Almanac for Ireland (2010-Present)
Predict Weather Almanac for the Unites States of America (2025)
How to Get Your Kid to Like Maths
Predicting Weather by The Moon
New Meteorological Techniques
We Used to Call It Summer
The Romany's Apprentice
Moon and WeatherLore
Secrets of the Moon
Notes From the Pit
The Lunar Code
Fingermaths
Supertramp
A Fiery Tale

Paperback ISBN: 978-1-0670445-2-7
Hardback ISBN: 978-1-0670445-3-4
eBook ISBN: 978-1-0670445-4-1

Email: enquiries@predictweather.com
Web: www.predictweather.com

INDEX

INTRODUCTION

Cycles rule the planet's weather. This book has brought together the relevant cycles of the past, and reproduces the maps and conditions of the those years, as a collection that it is calculated will be relevant to 2025 and 2026. This is therefore weather guide for the next two years and should answer all questions that growers, producers, distributors and planners may have. It replaces the Australian Almanacs of the past, which were limited to one year when it was conveyed to us that a minimum of two years was necessary to properly plan ahead.

We cover 81 towns and area in the 7 states. These are population centers or growing areas of importance. They also reflect the industry in their area sufficient to be representatives of the whole country. To this end the Australian farmer can plan the year and plant crops accordingly, knowing that in their state various weather situations are to be avoided or capitalised upon.

There is an overview for each year including extreme events by month, an estimated rainfall anomalies stable and graphs for each year, the seasonal rainfall summary overview for each part of the state for each year, an estimated maximum temperatures anomalies table and maximum temperature trend graph or each state and for each year, and a maximum estimated temperature seasonal overview for each year and state. Every state in the Commonwealth is described in detail, and all questions may be answered.

For instance, almost every 10 years there have been floods in NSW, for instance 1990, 2000, 2010 and 2021, so what year are we expecting to get back to drier years or drought? Where will the Murray Darling be flooding in 2026? What states will be affected by it? When does La Nina finish? How will the southern states react this year to the relentless Southern Antarctic weather that cuts through the Australian Bight, and sometimes feeds Sydney and coastal NSW towns with cool winds?

General trends for each state are discussed per month along with graphs and tables, including rainfall stats, and, at the end of the book are one-month-per-page rainfall figures. A weather diary runs through each state's summary, so that extreme weather events can be averted.

At the end of the book there is a chapter of what Australian weather actually is, the history of cycles, what the Sun, Moon and planets contribute to present and past weather, and how this will affect farming in the future. The El Nino/La Nina situation is discussed, and the reader is shown how to work out future years.

This new look almanac has come about from the questions and enquiries over the years that we have received in the process of preparing and producing these books for the Australian farmer. It is to be hoped that near the end of 2026, the 2027-2028 book will become available, followed by the 2029-2030 and almanac equivalents. We sincerely hope this book will prove to be a useful aid to your farming operation.

2025 GROWING SEASON (APRIL-JUNE)

(Based on estimated rainfalls)

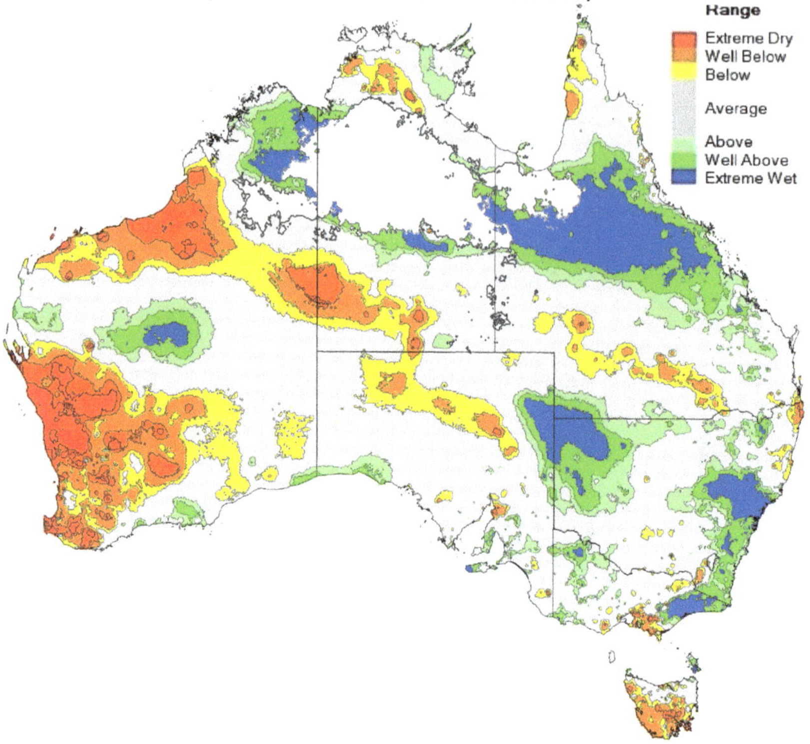

2025 HARVESTING SEASON (OCT-DEC)

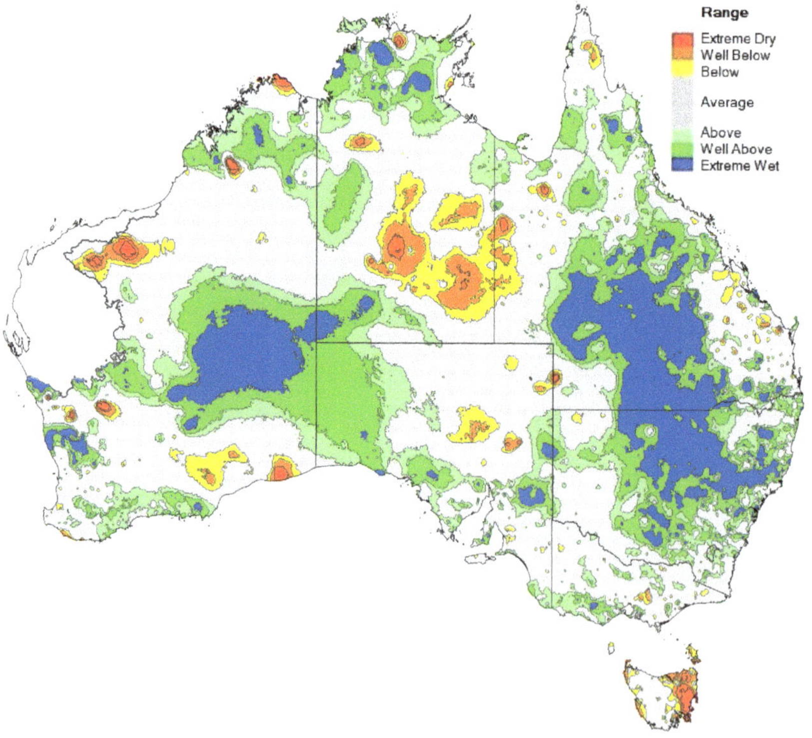

2026 GROWING SEASON (APRIL-JUNE)

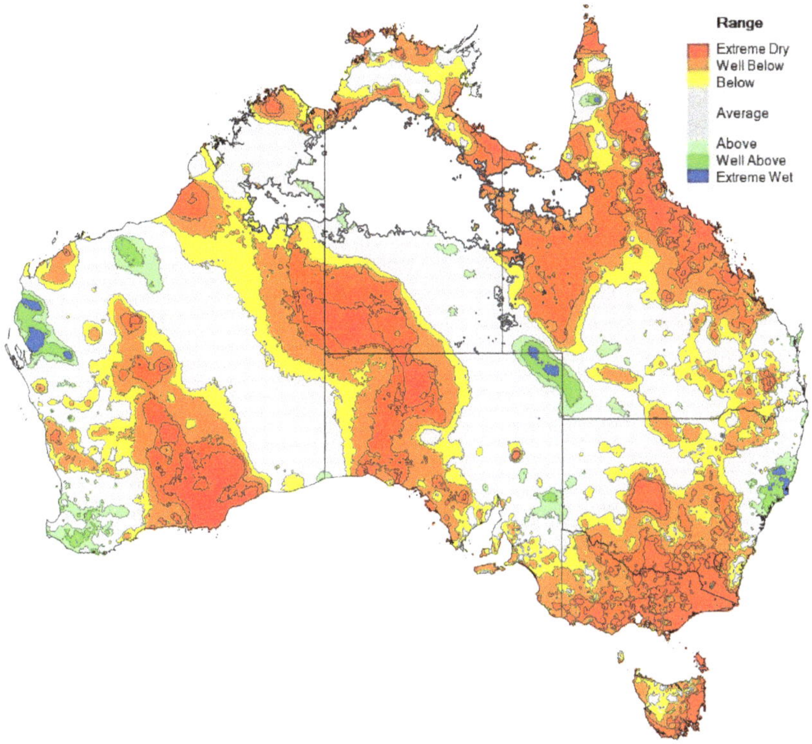

2026 HARVESTING SEASON (OCT-DEC)

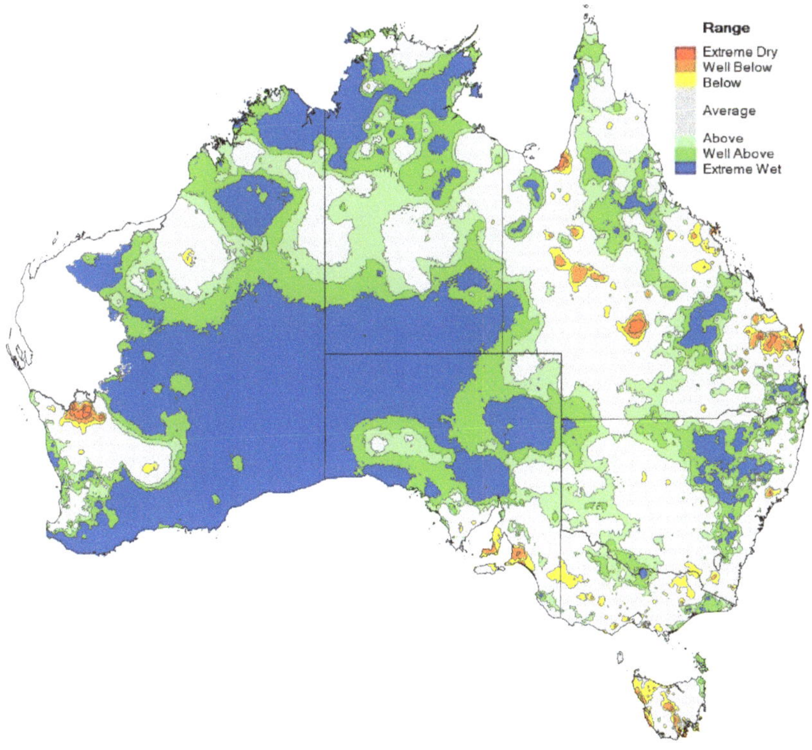

AUSTRALIAN WEATHER 2025

GENERAL OVERVIEW 2025

Of particular significance this year, the development of the monsoonal trough may be weaker than normal over Western Australia and parts of the Northern Territory, resulting in a deficit in rains across large parts of the country, particularly during February in Western Australia. Therefore, despite the promise of monsoonal rains over the first half of this year, widespread rains may not arrive in any significance until November, with even the winter/spring months of July to October being usually drier than normal.

Nevertheless, rainfall amounts by the end of the year may be average to above average over northern and central parts of the country, average to below average in the southwest of the continent, and a mix of wetter and drier locations in the southeast.

The year ahead may be generally warmer than usual, with particularly hotter and drier than normal conditions throughout most of the Murray Darling Basin, South Australia, New South Wales, Victoria and northern districts of Tasmania with many parts of these States enduring drought conditions. After an unusually hot first half of the year, with the chance of record high temperatures possibly developing in the west towards the end of summer and in the east midway through autumn, a cooler change may arrive from around July onwards. This cooler change may be mostly due to a series of low-pressure systems around this time.

MAJOR WEATHER EVENTS

General 2025

- ***Heatwaves*** - This year may see regular high temperature spells rather than extreme heatwaves, particularly over most of eastern and central regions of the country. The last week of January may be unusually hot in the south and east of the country with extremely hot temperatures affecting many parts of Queensland, Victoria, New South Wales, South Australia, and Tasmania, with a subsequent enhanced risk for bush fires. Extreme heat spells are likely to extend into February across most of southern and western Australia, as well as in the Northern Territory and Tasmania. Unseasonably high temperatures may be particularly significant at the end of summer in the west of the country, in the east nearer the end of autumn, and in the south at the end of spring.

  A major heatwave may affect the tropics in mid-May, extending over much of the tropical Australian regions of Western Australia, the Northern Territory and Queensland, as well as into parts of southern Queensland and northern New South Wales. Mid-March may see a more localised heatwave event along the western coast of Western Australia with the possibility of some districts seeing maximum temperatures rise into the upper 40's. Regions of significance for enduring unusually high temperatures may be the Murray-Darling Basin, and most of South Australia, Victoria, northern Tasmania, and New South Wales.

- ***Drought*** - This year's prolonged spells of high temperatures, heatwaves, and a weak monsoon season may be a strong contributing factor to the development and persistence of drought conditions over many parts of the country. At first, an unusually drier than normal summer may see the development of drought conditions becoming established in many parts of the country by the end of February, particularly in Western Australia. By the end of April and into early May, drought conditions may have spread into many parts of the country, particularly in the far southwest and southeast of the continent, including in the Murray-Darling Basin, through all of Victoria and even in the northern half of Tasmania. Despite some promising wet spells around mid-year, including in the normally dry season areas of the tropics, the period from mid-July to the end of October may be particularly dry across the southern half of the country, and it may not be until November that relief may arrive.

- ***Cyclones*** - Overall, this summer season (2024-2025), may see few cyclones directly impacting Australia directly, despite the potential for up to 10 cyclones developing within the monsoonal trough. Only 3 or 4 cyclones are anticipated to make actual landfall in Australia this year, with the remainder mostly expected to meander around the surrounding oceans, or make landfall in other nations. One cyclone may even arrive around the late-May-early-June period, reflecting an unusually cooler and wet spell over the tropics during their normal dry season. Overall, one cyclone may cross the Northern Territory and Queensland around mid-February, another two cyclonic systems may affect northern Western Australia, both around the mid-to-end-of March.

January 2025

Date	Day	Australia Overall - January 2025	Moon
1st	Wed	Very dry soil and extremely low rainfall over the past 12 months in Tasmania sees a very high bushfire risk over the State at this time.	
2nd	Thur		
3rd	Fri		
4th	Sat	Cold outbreak in the southeast of the continent, with chance of snow in the higher reaches of Tasmania, Victoria and in the New South Wales Snowy Mountains.	
5th	Sun		
6th	Mon		
7th	Tue	Cooler over the northeast of the country with heavy rain over the Northern Territory Top End regions over the next few days.	**1Q XhN**
8th	Wed	Lightning strikes around the South Coast of Western Australia possible around this time, possibly sparking some local fires.	**P13**
9th	Thu	Storms in South Australia around this time may cause fires and blackouts. A tropical low may form south of the Indonesian archipelago around this time and meander around the northwest coast of Western Australia over the next few days, and may become a named cyclonic system in a few days' time.	
10th	Fri	Severe storm activity with thunderstorms and heavy rain in the southeast of the country affecting much of southern Queensland to New South Wales-Canberra. Lightning strikes may start bush fires further south in the Victorian and South Australian states.	
11th	Sat	A trough crossing the southeast of the country produces widespread storms and heavy rain at times over much of Queensland, New South Wales and Victoria over the next couple of days.	
12th	Sun	A tropical cyclone forming near the north coast of Western Australia around this time and crossing the coast over the next couple of days. This system may result in unusually cooler than normal temperatures for this time of year over the next few days. This system may extend tropical moisture and heavy rains across the Western Australian Goldfields, Eucla, and South East Coastal districts including in the Esperance area.	
13th	Mon	Chance of bush fires in Victoria as a result of lightning strikes from passing storms.	**^**
14th	Tue	Unusually hot spell in the south and east of the country affecting many parts of Queensland, Victoria, South Australia and Tasmania.	**F**
15th	Wed	Extreme heat in South Australia, Victoria, and Tasmania around this time may exacerbate any fires currently occurring.	
16th	Thur		

Date	Day	Australia Overall - January 2025	Moon
17th	Fri	Chance that drought conditions may be widespread over New South Wales, with very high overnight temperatures in the southeast of the country adding to the discomfort.	
18th	Sat	Monsoonal trough brings widespread storms over the tropical north of the country including northern Western Australia, Queensland, and the Northern Territory throughout much of the coming week. This may cause widespread flooding in some of the upper river catchments in the southwest of Queensland and into the river basins in the far west.	
19th	Sun		
20th	Mon	Chance of bush fires in the central parts of South Australia continuing around this time, particularly in the outskirts near Adelaide as well as a high risk of fires in Victoria.	**XhS**
21st	Tue		**A**
22nd	Wed	A high fire risk continues in Victoria around this time, while New South Wales may be very hot, with a mix of storms and bushfires, and Queensland having a mix of strong storms, but also dust storms.	**3Q**
23rd	Thu		
24th	Fri	Monsoonal rains over the Northern Territory and Western Australia intensifies around this time, with severe storms drifting into the western half of Queensland. This may bring further flooding risks into the southwest of Queensland over the coming days, as well as some parts in the North Coast and Central Coast regions over the coming month ahead.	
25th	Sat		
26th	Sun	High bush fire risk in Victoria and South Australia around this time.	
27th	Mon		**V**
28th	Tue		
29th	Wed		
30th	Thur	Outbreaks of widespread and often heavy rains in parts of the Northern Territory, South Australia, Victoria, and New South Wales may produce flooding in a number of regions, particularly along river catchments.	**N**
31st	Fri		

February 2025

Date	Day	Australia Overall - February 2025	Moon
1st	Sat		
2nd	Sun		**P10**
3rd	Mon		**XhN**
4th	Tue		
5th	Wed		**1Q**
6th	Thu		
7th	Fri		
8th	Sat		
9th	Sun		**^**
10th	Mon		
11th	Tue	Heat wave temperatures and a high bush fire risks is likely to prevail over the southern half of Western Australia around this time, with the next several weeks possibly being unusually hotter than is typical for this time of year, not only in Western Australia, but also in South Australia and Tasmania.	
12th	Wed		
13th	Thur	Heavy monsoonal rains persist over the northern half of the country, with flooding still likely along the Queensland northern and central coasts over the next several weeks.	**F**
14th	Fri		
15th	Sat	Another tropical low may form northwest of the Northern Territory Arnhem Lands around this time and while threaten to turn cyclonic over the next few days as it traverses through the Gulf of Carpentaria, it is not expected to fully form into a structured cyclone.	
16th	Sun		**XhS**
17th	Mon		
18th	Tue	Moderate to major flooding may continue to be a concern in numerous river systems in Queensland and persist a risk for a few more weeks.	**A**
19th	Wed		
20th	Thu		
21st	Fri		**3Q**

Date	Day	Australia Overall - February 2025	Moon
22nd	Sat		
23rd	Sun		V
24th	Mon		
25th	Tue		
26th	Wed		
27th	Thur		
28th	Fri		N

March 2025

Date	Day	Australia Overall - March 2025	Moon
1st	Sat		
2nd	Sun		P7 XhN
3rd	Mon		
4th	Tue		
5th	Wed		
6th	Thu	An area of low pressure in the Arafura Sea may begin to form into a tropical low around this time as it builds near the Top End of the Northern Territory. This may turn fully cyclonic in a couple of days' time.	
7th	Fri		1Q
8th	Sat	An area of low pressure in the Arafura Sea a couple of days ago, may form into a tropical cyclone around this time near the Top End of the Northern Territory and intensify over the next couple of days as it traverses the Joseph Bonaparte Gulf before crossing the Western Australian Kimberley coast in a few days' time. By around the 3rd-4th day of this cyclone, it may reach its peak as it batters the Port Hedland area as it moves deeper into the Pilbara.	^
9th	Sun	Supercell storms may batter New South Wales in districts west of the Tablelands over the next couple of days. Chance of severe hailstorms also in ACT around this time.	
Date	Day	Australia Overall - March 2025	Moon
10th	Mon	The following 3-4 weeks may be unusually drier than normal in Queensland while the remainder of the country, particularly in west and central Australia, will be wetter than normal. Temperatures may be milder than normal overnight, while daytime maximum temperatures may be below normal in many of the wetter areas of the country, due to heavy cloud cover.	
11th	Tue	Storms in the Kakadu National Park Northern Territory may produce tornado formation winds and bring some damage.	
12th	Wed		
13th	Thur	An active monsoonal trough may continue to prevail over the north of the country around this time, with subsequent flood warnings along Queensland river systems continuing over the coming couple of weeks.	
14th	Fri		F
15th	Sat		XhS
16th	Sun	Another low-pressure system with cyclonic potential may form in the Christmas Island vicinity around this time but after briefly intensifying as it drifts off the Western Australian coasts, weaken in a few days' time as it eventually crosses the Pilbara coast near Port Hedland.	
17th	Mon		
18th	Tue		A
19th	Wed	A tropical cyclone that may have battered the northern regions of Western Australia and the Northern Territories over the previous couple of weeks may finally weaken around this time, and may be downgraded before dissipating in a few more days.	
20th	Thu		
21st	Fri		
22nd	Sat		
23rd	Sun		3Q V
24th	Mon		
25th	Tue		
26th	Wed		

Date	Day	Australia Overall - March 2025	Moon
27th	Thur		
28th	Fri		
29th	Sat		N XhN
30th	Sun		P4
31st	Mon		

April 2025

Date	Day	Australia Overall - April 2025	Moon
1st	Tue		
2nd	Wed	Another low-pressure system with cyclonic potential may once more form to the west of the Western Australian Kimberley and Pilbara and drift around the northwest coastlines of the State over the coming week. However, this system may only briefly turn cyclonic before downgrading swiftly without making landfall.	
3rd	Thu		
4th	Fri		^
5th	Sat		1Q
6th	Sun		
7th	Mon		
8th	Tue		
9th	Wed		
10th	Thur	The next 3-4 weeks may be generally warm and dry over most of the country, although a major rain event may affect southern parts of the continent in about 4 weeks' time, particularly in South Australia. This unusual arrival of drier conditions may be because the tropical wet season may come to an early end over the next couple of weeks, due to persistent prevailing high pressure prevailing over the southeast of the country.	
11th	Fri	Chance of brief flooding along the Paroo River system in Queensland around this time.	XhS
12th	Sat	A very strong northerly wind flow associated with an intense low-pressure system crosses the southeast of the country around this time with storms, heavy rain and blustery and damaging wind gusts likely through the southeast of South Australia, western parts of Victoria and later over Tasmania.	
13th	Sun		F
14th	Mon	Severe storms likely in the Metropolitan region of New South Wales.	A
15th	Tue		
16th	Wed		
17th	Thu		
18th	Fri		
19th	Sat		V
20th	Sun		
21st	Mon		3Q
22nd	Tue		
23rd	Wed		
24th	Thur		
25th	Fri	Western Australia may see the start of the cooler season around this time, along with the first significant rains in the southern half of the State.	
26th	Sat		XhN
27th	Sun		
28th	Mon		P3 N
29th	Tue		
30th	Wed		

May 2025

Date	Day	Australia Overall - May 2025	Moon
1st	Thu		
2nd	Fri	Storms around the southeast of Queensland to the coastal regions of New South Wales may produce some flash flooding as well as the chance of waterspouts off shore.	^

Date	Day	Australia Overall - May 2025	Moon
3rd	Sat		
4th	Sun	A significant break in the dry spell around this time may see widespread rainfall in central parts of Western Australia over the next couple of days. A band of tropical moisture may interact with a developing low-pressure system crossing through the Great Australian Bight, resulting in some significant rain events over most of New South Wales, Victoria, and South Australia during the next 3-4 days.	
5th	Mon		1Q
6th	Tue		
7th	Wed		
8th	Thur		
9th	Fri		XhS
10th	Sat	Many parts of Australia, particularly in the southwest and southeast of the continent, but also into parts of Queensland, may be under drought conditions by around this time and with this situation continuing despite the return of beneficial rain systems over the coming month.	
11th	Sun	The next 3-4 weeks may be unusually milder than is normal for this time of year, particularly in the eastern half of the country where both maximum and minimum temperatures are likely to be well above normal. Only some southern parts of Western Australia may be generally slightly cooler than what is normal for this time of year. It may be drier than normal in some parts of Queensland and Western Australia, while much wetter than average elsewhere. In particular, Tasmania may be up to 50% wetter than what is typical around this time of the year. The good rains expected in New South Wales and western Victoria over the coming few weeks, may indicate the early recovery from the recent drought situations. Unusually, the next 3-4 weeks may also be wetter than normal in large areas of the northern tropics.	A
12th	Mon		
13th	Tue		F
14th	Wed		
15th	Thur		
16th	Fri		V
17th	Sat		
18th	Sun		
19th	Mon		
20th	Tue		3Q
21st	Wed		
22nd	Thu		
23rd	Fri		XhN
24th	Sat	Unstable winds may develop over the southeast of the continent with the chance to form into tornadoes in some places, particularly at risk in the Geelong district of Victoria.	
25th	Sun		
26th	Mon	A late season tropical low with cyclonic potential may form in the Solomon Sea around this time, but swiftly weaken as it moves westward and brushes the far north of Cape York Peninsula in a couple of days' time.	P5
27th	Tue		N
28th	Wed		
29th	Thur		^
30th	Fri		
31st	Sat	Unsettled winds in the Adelaide region may briefly produce funnel formation events.	

June 2025

Date	Day	Australia Overall - June 2025	Moon
1st	Sun		
2nd	Mon		
3rd	Tue		1Q
4th	Wed		
5th	Thu		XhS
6th	Fri		
7th	Sat		A
8th	Sun		

Date	Day	Australia Overall - June 2025	Moon
9th	Mon		
10th	Tue		
11th	Wed	The next 3-4 weeks may see up to 5 intense depressions traverse the eastern coastal regions of New South Wales, with a number of storms associated with these depressions possibly causing damage to infrastructure and property, including in the Hunter and Central Coast of the Sydney districts. As such, these next few weeks may be unusually colder than is normal for this time of year. It may also be significantly wetter than normal over much of northern Australia as well as along the eastern coast, while in the southwest of Western Australia, across Tasmania, over most of South Australia, and western parts of Victoria and New South Wales, the coming few weeks may be generally dry.	F
12th	Thur		V
13th	Fri		
14th	Sat		
15th	Sun		
16th	Mon		
17th	Tue		
18th	Wed	An intense low-pressure system off the eastern coast of New South Wales directs severe storms with strong winds and gales and torrential rain over the eastern half of the State, with the chance of extensive flooding and damage to infrastructure and property likely.	
19th	Thu		3Q XhN
20th	Fri		
21st	Sat		
22nd	Sun		
23rd	Mon		P8
24th	Tue		
25th	Wed		N
26th	Thur	An east coast low-pressure system brings storms with gales and localised flooding onto the Central and South Coasts of New South Wales. In contrast, widespread fogs may affect many parts of Victoria and Tasmania, while frosts continue in South Australia.	^
27th	Fri	It may be unseasonably cold and wet in the Northern Territory and Queensland states around this time, with the colder change moving southwards into New South Wales, Victoria, and Tasmania over the next several days.	
28th	Sat		
29th	Sun	Outbreaks of heavy rain in New South Wales and Victoria may produce some localised flooding in the Gippsland and Newcastle districts.	
30th	Mon	An intense depression off the eastern coast of the country brings a wintry blast with storms, widespread snow, rain, and high seas over the next few days into the south-eastern quarter of the country from Tasmania to southern Queensland.	

July 2025

Date	Day	Australia Overall - July 2025	Moon
1st	Tue		
2nd	Wed	A strong cold front brings unstable winds, rain, and raises dust in Western Australia over the next few days.	XhS
3rd	Thu		1Q
4th	Fri		
5th	Sat		A
6th	Sun	Another east coast low-pressure system brings further snow into the Snowy Mountains over the next few days.	
7th	Mon	A depression off the southeast coast may extend heavy rain over Victoria, possibly producing further localised flooding in the Gippsland region over the next few days.	
8th	Tue	Another depression to the southwest of the country extends colder temperatures some heavy rain for a time into the southern Western Australia districts over the next several days.	
9th	Wed		
10th	Thur	An outbreak of heavy rain in the Urandangie area of Queensland around this time may cause some brief minor flooding along the Georgina River system.	V

Date	Day	Australia Overall - July 2025	Moon
11th	Fri	A warm front crosses Victoria and may produce some significant rain along the southern coasts. Further west, another front crosses the South West and Central Coastal regions of Western Australia producing some severe storm activity, including in the Perth region over the next couple of days.	F
12th	Sat	Chance of unseasonal storms crossing the northwest of Western Australia.	
13th	Sun	Winds may be very fresh and blustery at times over the southern parts of the country from Western Australia to Victoria through the Bight, and over Tasmania.	
14th	Mon	Unusually warm spell in New South Wales, as well as in South Australia and the Northern Territory.	
15th	Tue	The next 3-4 weeks may see temperatures return to more seasonal levels for this time of year, while unseasonably drier than normal conditions may prevail through most of the northern two-thirds of the continent, with the lower third receiving most of the precipitation. Most of the Northern Territory, the Western Australian Kimberley and inland desert regions may be almost completely dry as well as northern parts of South Australia, and the west of New South Wales. Even much of eastern Queensland will be drier than normal, and parts of the Murray-Darling Basin may only receive around half of their normal rainfall compared to what is typical for this time of year.	
16th	Wed		
17th	Thu	The coming 3-4 months may be very dry in many parts of the country, particularly in the south-eastern parts of the continent, with only the occasional patchy rain spell now and then, which may not be enough to ease drought conditions.	XhN
18th	Fri		3Q
19th	Sat		
20th	Sun		
21st	Mon		P11
22nd	Tue		
23rd	Wed		^
24th	Thur		
25th	Fri		N
26th	Sat		
27th	Sun		
28th	Mon		
29th	Tue		
30th	Wed		XhS
31st	Thu		

August 2025

Date	Day	Australia Overall - August 2025	Moon
1st	Fri		
2nd	Sat		1Q A
3rd	Sun		
4th	Mon		
5th	Tue		
6th	Wed		V
7th	Thur		
8th	Fri		
9th	Sat		F
10th	Sun		
11th	Mon	The next 3-4 weeks may be particularly wet along the eastern coasts, while most other parts of the continent will be predominately dry. It may also be an unusually warm period in most southern parts of the country, but unusually cooler in the tropics.	
12th	Tue	Unstable winds over Tasmania around this time may produce a tornado formation in the south of the State.	
13th	Wed		XhN
14th	Thu		
15th	Fri		P12
16th	Sat		3Q
17th	Sun		

Date	Day	Australia Overall - August 2025	Moon
18th	Mon		
19th	Tue		^
20th	Wed		
21st	Thur		
22nd	Fri		
23rd	Sat		N
24th	Sun		
25th	Mon		
26th	Tue		XhS
27th	Wed		
28th	Thu		
29th	Fri		
30th	Sat		A
31st	Sun		1Q

September 2025

Date	Day	Australia Overall - September 2025	Moon
1st	Mon		
2nd	Tue	A low-pressure system around the Sunshine Coast around this time may produce intense rainfall into the region, leading to some flash flooding in local streams in the southeast of the State over the next several days, including along the Noosa, Mary, Maroochy, and Mooloolah Rivers.	V
3rd	Wed		
4th	Thur		
5th	Fri		
6th	Sat		
7th	Sun		
8th	Mon		F
9th	Tue		XhN
10th	Wed		
11th	Thu	Exceptionally drier than normal conditions may continue to prevail over large parts of the country in the coming 3-4 weeks, with a number of areas only likely to receive around half their normal rainfall amounts. The central northern parts of New South Wales may be significantly drier than usual, including in the Murray-Darling Basin, and similarly in South Australia, Western Australia and the Northern Territory, although the latter two states may be closer to their dry season patterns. Only Queensland and Tasmania can expect close to normal rainfall over the next 3-4 weeks, and small areas, mostly in the southeast of Queensland and around the North Coast and in the Central Coast region of Western Australia may be wetter than normal. Accompanying the drier than normal conditions will be slightly warmer than average temperatures over the next 3-4 weeks, particularly in the southern half of the country.	P9
12th	Fri		
13th	Sat		
14th	Sun		3Q
15th	Mon	Heavy rain around the Sunshine Coast in Queensland may cause some flooding along the Maroochy River system and nearby creeks over the next couple of days.	^
16th	Tue		
17th	Wed		
18th	Thur		
19th	Fri		
20th	Sat		
21st	Sun		
22nd	Mon		N XhS
23rd	Tue		
24th	Wed		
25th	Thu		
26th	Fri		A

Date	Day	Australia Overall - September 2025	Moon
27th	Sat		
28th	Sun		
29th	Mon		
30th	Tue		1Q V

October 2025

Date	Day	Australia Overall - October 2025	Moon
1st	Wed		
2nd	Thur		
3rd	Fri		
4th	Sat		
5th	Sun		
6th	Mon		
7th	Tue		F XhN
8th	Wed		
9th	Thu		P6
10th	Fri		
11th	Sat	The next 3-4 weeks may continue to be warmer and drier than normal over most of the country, with only around a third to half the normal rainfall expected compared to what is normal for this time of year, and temperatures up to 1-2°C warmer than average. Overall, both maximum and minimum temperatures will be above normal almost everywhere, with only some parts around the south coast of Western Australia, western districts of Tasmania and a few districts on the tropical coasts slightly cooler than average. The next few weeks may be notably drier than average in Victoria in particular, where only around 50-60% of their normal rainfall is anticipated. Only some parts of the southwest of Western Australia, western districts in Tasmania, and south-eastern Queensland districts may see above normal rainfall for the second month in a row.	
12th	Sun		
13th	Mon		^
14th	Tue		3Q
15th	Wed		
16th	Thur		
17th	Fri		
18th	Sat	Chance of severe thunderstorms around the Upper North Coast of New South Wales around this time that may see large hail and damaging wind gusts.	
19th	Sun		
20th	Mon		XhS
21st	Tue		
22nd	Wed		N
23rd	Thu		
24th	Fri		A
25th	Sat		
26th	Sun		
27th	Mon		V
28th	Tue		
29th	Wed		
30th	Thur		1Q
31st	Fri		

November 2025

Date	Day	Australia Overall - November 2025	Moon
1st	Sat		
2nd	Sun		
3rd	Mon		XhN

Date	Day	Australia Overall - November 2025	Moon
4th	Tue		
5th	Wed	Storms in the Northern Rivers region may produce tornado activity in some areas, including in the Lismore district.	
6th	Thur		F P1
7th	Fri		
8th	Sat		
9th	Sun		^
10th	Mon		
11th	Tue	The next 3-4 weeks may finally bring above average rainfall over much of the country, except in Tasmania and the southwest of Western Australia, seeing an almost reversal in the pattern to the previous two months. These rains may finally bring some relief to the rain deficient areas in the eastern half of the country. Generally, the next few weeks will be cooler than normal over the northern half of the continent, while the southern half may be warmer.	
12th	Wed		3Q
13th	Thu		
14th	Fri		
15th	Sat		
16th	Sun		XhS
17th	Mon		
18th	Tue	Severe thunderstorm activity crossing the Murray Valley region may produce tornadoes in its path.	
19th	Wed		
20th	Thur		N A
21st	Fri	As the monsoonal trough strengthens, a weak low may form within it just north of Cocos Island with the potential to turn cyclonic over the coming days. However, it may just drift around the Indian Ocean and Timor Sea without making landfall in Western Australia.	
22nd	Sat		
23rd	Sun		V
24th	Mon		
25th	Tue	Another potential cyclonic system may form to the east of Australia near Papua New Guinea around this time, and track in a westerly course for a time, towards the far north of Queensland. However, again, this cyclonic system is not expected to make landfall before weakening and dissipating back into the monsoonal trough over the next few days.	
26th	Wed		
27th	Thu		
28th	Fri	Heavy rainfall possible around this time in the interior of Queensland and bring outbreaks of flooding around the Diamantina Lakes and nearby river systems over the next several days.	1Q
29th	Sat		
30th	Sun	A heightened bush fire risk may have developed by around this time in the Victorian Gippsland region.	XhN

December 2025

Date	Day	Australia Overall - December 2025	Moon
1st	Mon		
2nd	Tue		
3rd	Wed		
4th	Thur		P2
5th	Fri		F
6th	Sat		^
7th	Sun		
8th	Mon	Thunderstorms in Queensland around this time may cause flooding along the Warrego and Paroo river systems over the next few days.	
9th	Tue		
10th	Wed		

Date	Day	Australia Overall - December 2025	Moon
11th	Thu	The next 3-4 weeks may be see around average to above average rainfall over most of the country, reducing the rainfall deficiencies particularly in the east of the continent. Many regions may see up to 40-50% more rainfall than is typical for this time of year, and some parts of Western Australia and New South Wales up to 75-80% more rainfall. Overall, temperatures may be near normal for this time of year almost everywhere, with overnight minimum temperatures generally above average, while daytime maximum temperatures slightly cooler than normal.	
12th	Fri		**3Q**
13th	Sat		**XhS**
14th	Sun		
15th	Mon	A low-pressure system in the Coral Sea may extend torrential rain, gale force winds and subsequent flooding and damage to property and infrastructure in parts of the southeast of Queensland and into the northeast of New South Wales over the next few days.	
16th	Tue	Storms with lightning strikes may cause bushfires in the southeast of South Australia around this time and over the coming couple of weeks, including on Kangaroo Island.	
17th	Wed		**A**
18th	Thur		
19th	Fri	Storms in the Sydney area of New South Wales around this time may produce large hail.	
20th	Sat		**N V**
21st	Sun	Chance of funnel formation clouds forming in the Hunter Valley region around this time.	
22nd	Mon		
23rd	Tue		
24th	Wed		
25th	Thu	A band of rain may cross the west of Queensland around this time producing heavy falls and possible renewed flooding along the Barcoo River and some local river systems on the Darling Downs.	
26th	Fri		
27th	Sat		**XhN**
28th	Sun		**1Q**
29th	Mon		
30th	Tue	A storm complex in the Central Highlands of Victoria may produce brief tornado formations, particularly in the Ballarat area.	
31st	Wed		

NORTHERN TERRITORY 2025

GENERAL TRENDS

Year Overall:

Slightly drier by around 5-10% less than average
Slightly cooler but within 0.5°C of the average

Summer:

Drier by around 30% below the average
Wettest – Second half of January
Driest – Last week of February
Warmer by around 0.5°C above the average
Hottest – Second week of December and mid-February
Coolest – First week of January

Autumn:

Wetter by up to 35-40% more than the average
Wettest – Second half of March
Driest – First half of May
Warmer by up to 1°C above the average
Hottest – First half of March and second week of May
Coolest – Second week of April and third week of May

Winter:

Considerably wetter by up to 200% more than the average
Wettest – Last week of June
Driest – Second half of July to first half of August
Cooler by around 1-1.5°C below the average
Hottest – First week of June and fourth week of August
Coolest – Second and third weeks of July

Spring

Drier by around 30-35% below the average
Wettest –Mid-November to end of the month
Driest – Throughout September
Slightly warmer but within 0.5°C of the average
Hottest – Mid-October and second week of November
Coolest – First half of September

Major Weather Events

January 2025

Date	Day	Northern Territory - January 2025	Darwin	Moon
1st	Wed	Heavy rain in the Victoria River region, particularly in the Keep River Ranges.	Showers	
2nd	Thur	Very heavy rain likely in some parts of the Darwin-Daly region over the next couple of days.	Showers	
3rd	Fri	Heavy rain throughout the Alice Springs region while mostly dry over the Top End.		
4th	Sat	Very heavy rain in the Darwin-Daly region, particularly in the Larrimah and Jabiru areas.	Showers	
5th	Sun	Very wet and hot over the Top End, including in the Darwin-Daly region with some very heavy rain likely in the Katherine area to Douglas River areas over the next couple of days.	Showers	
6th	Mon		Rain, Th	
7th	Tue	Heavy rain in the Top End districts over the next couple of days, particularly around Jabiru and Gove.	Rain	1Q XhN
8th	Wed	Cool conditions continue. Heavy rain at times in the Gulf Country and along the Central Coast.	Rain	P13
9th	Thu	High bushfire risk in the Barkly district around this time. In contrast, storm activity across the Top End brings rain and some heavy downpours at times in the Darwin-Daly region, particularly in the Jabiru area as well as in the Gulf of Carpentaria and on Centre Island.	Rain, Th	
10th	Fri	Widespread heavy rain over the Top End over the next couple of days.	Rain, Th	
11th	Sat	Very hot in the Alice Springs region.	Rain	
12th	Sun	Spells of rain in the Arnhem Land and Kakadu regions.	Showers	
13th	Mon	A band of tropical moisture from a monsoonal system that crossed the north of WA brings heavy rain for a time into the Alice Springs district over the next couple of days.	Rain	^
14th	Tue		Showers	F
15th	Wed	Very hot overnight in the Alice Springs region, and in the Jervois area.	Rain	
16th	Thur	Very hot over the Top End around this time.	Rain, Th	
17th	Fri	The monsoonal trough stabilises over the Top End around this time and maintains it position for much of the upcoming month ahead.	Rain, Th	
18th	Sat			
19th	Sun	Daily outbreaks of monsoonal rains over the northern half of the State over the remainder of the month.	Showers	
20th	Mon		Showers, Th	XhS
21st	Tue		Showers, Th	A
22nd	Wed	Monsoonal rain activity south of the Barkly Tablelands over the next 7-9 days.	Rain, Th	3Q
23rd	Thu		Showers	
24th	Fri			
25th	Sat	Milder spell in the Alice Springs region.	Showers, Th	
26th	Sun		Showers, Th	
27th	Mon		Rain, Th	V
28th	Tue	Very heavy rain in the Alice Spring region over the next few days, particularly in the Numery and Arltunga districts.	Showers	
29th	Wed	Milder temperatures in the Alice Springs region.	Showers	
30th	Thur	Very strong wind gusts around Centre Island and the Roper-McArthur region. Heavy rain in the Atula district.		N
31st	Fri	Very strong and blustery winds around the Gulf of Carpentaria, particularly across Centre Island.		

February 2025

Date	Day	Northern Territory - February 2025	Darwin	Moon
1st	Sat	Unusually cool overnight in the Alice Springs region. Rain over the Top End over the next couple of days.	Rain	
2nd	Sun		Rain	P10
3rd	Mon		Showers, Th	XhN
4th	Tue		Showers	
5th	Wed	Thunderstorms and very heavy rainfall in the Arnhem region over the next few days.	Showers, Th	1Q
6th	Thu			
7th	Fri	Heavy rain across the Top End and through the Roper-McArthur regions.	Rain, Th	
8th	Sat	Monsoonal rains persist in most regions north of the Barkly Tablelands over the next week ahead.	Showers	

Date	Day	Northern Territory - February 2025	Darwin	Moon
9th	Sun	A tropical low may form near the Arnhem Lands and drift through the Gulf of Carpentaria over the next few days, possibly reaching cyclone strength within a week, before making landfall over Cape York Peninsula in QLD.	Showers	^
10th	Mon			
11th	Tue	Very hot in the Roper-McArthur-Douglas River region, Victoria-Daly, and Barkly Tablelands to through the Alice Springs region with maximum temperatures into the low 40's in numerous places.	Showers, Th	
12th	Wed	Strong winds around the Top End northern coasts, particularly across McCluer Island.	Showers, Th	
13th	Thur		Rain, Th	F
14th	Fri		Rain, Th	
15th	Sat	A tropical cyclone that formed in the Gulf of Carpentaria a few days earlier, may make landfall over Cape York Peninsula in QLD around this time before swifty moving off into the Coral Sea. Heavy rain in the Roper-McArthur to the Top End over the next few days.	Rain, Th	
16th	Sun		Rain, Th	XhS
17th	Mon		Showers, Th	
18th	Tue	Brief respite in monsoonal rains over the next couple of days, except around the far northern coastal fringes of the Top End.	Rain, Th	A
19th	Wed		Th, Hail	
20th	Thu	Chance of storms with hail and damaging winds in the Kakadu to Darwin-Daly region, particularly around the Adelaide River area. Strong gusty winds also in the Roper-McArthur and Centre Island regions. Very hot in the Alice Springs region.	Th	
21st	Fri		Showers	3Q
22nd	Sat	Unusually cool overnight in the Alice Springs region.	Rain	
23rd	Sun	Monsoonal rain respite over much of the Top End regions over the remainder of the month, except for some rain activity mostly around the far north coastal fringes.	Rain, Th	V
24th	Mon		Rain, Th	
25th	Tue			
26th	Wed			
27th	Thur	A tropical storm near the Darwin-Daly region may bring gale force winds, outbursts of torrential downpours, and possible brief funnel formation winds over the next couple of days.	Gales	
28th	Fri	Very hot overnight in the Alice Springs region.	Th	N

March 2025

Date	Day	Northern Territory - March 2025	Darwin	Moon
1st	Sat	Monsoonal rains continue over the Top End regions daily over the coming month.	Rain	
2nd	Sun	Squally winds across the Kakadu to Darwin-Daly regions may cause localised wind damage, with particular risks around the Crab Claw Island area.	Rain, Th	P7 XhN
3rd	Mon	Near gale force winds likely over the Darwin-Daly region may bring further damage and cooler temperatures.	Showers, Th, Gales	
4th	Tue		Showers	
5th	Wed		Rain	
6th	Thu		Rain, Th	
7th	Fri		Rain, Th	1Q
8th	Sat	A convergence of the monsoonal trough crosses the Arnhem, bringing severe storms and heavy rain with chance of localised flash flooding in some parts of the region.	Rain	^
9th	Sun	A severe tropical cyclone may form over the Top End around this time and slowly drift westward through the Josephine Bonaparte Gulf and into the north of WA in a few days.	Rain, Th	
10th	Mon	Rain over the Top End, with heavy falls and milder temperatures in the Kakadu - Arnhem region, particularly around Jabiru and Warruwi.	Rain	
11th	Tue	Unsettled winds in the Kakadu to north Katherine region may develop into a tornado crossing the region, bringing localised damage in its parth, with a heightened risk in the Mary River and Wildman River areas. Heavy rain over the Top End over the next few days. Downpours in the Darwin-Daly region.	Showers	
12th	Wed	Squalls over the Arnhem, Darwin-Daly, and northern Roper-McArthur districts with flash flooding likely. Heavy rain in the Jabiru district with chance of flash-flooding. Very hot in the Alice Springs region over the next few days.	Rain, Th	
13th	Thur	Very warm overnight in the Victoria River region.	Rain, Th	
14th	Fri		Rain	F
15th	Sat	Very strong and blustery winds in the Rabbit Flat district.	Rain	XhS

Date	Day	Northern Territory - March 2025	Darwin	Moon
16th	Sun	Some rain across the Barkly Tablelands.	Showers	
17th	Mon		Showers	
18th	Tue		Showers	A
19th	Wed	Showers over the Barkly and spread by tomorrow into the northern half of the Alice Springs region.	Showers, Th	
20th	Thu		Rain, Th	
21st	Fri		Showers, Th	
22nd	Sat	Showers into the west of the Alice springs region over the next couple of days.	Rain, Th, Fg	
23rd	Sun			3Q V
24th	Mon		Showers	
25th	Tue		Rain	
26th	Wed	Outbreaks of heavy rain spreads throughout the State over the next few days. Heavy rain at times in the Alice Springs region.	Showers	
27th	Thur	Very heavy rain in the Roper-McArthur region.	Showers	
28th	Fri	Widespread rain continues through the Alice Springs region, particularly in the western half, over the next few days until the end of the month.		
29th	Sat		Rain	N XhN
30th	Sun		Showers	P4
31st	Mon	Chance of thunderstorms, heavy rain and flash flooding in the Katherine region over the next couple of days.	Rain, Th	

April 2025

Date	Day	Northern Territory - April 2025	Darwin	Moon
1st	Tue	The monsoonal trough continues to bring heavy rain activity over the Top End for the next week before swiftly dissipating early in the second week.	Rain	
2nd	Wed		Rain	
3rd	Thu	A storm crossing through the Tennant Creek region may bring heavy downpours at times. Widespread rain activity across the State over the next couple of days.		
4th	Fri		Rain	^
5th	Sat			1Q
6th	Sun		Showers	
7th	Mon	Unusually cool in the Kulgera area. Widespread rain throughout the western half of the State at first, spreading eastward during tomorrow, before contracting back up over the Top End and clearing by the end of the week.		
8th	Tue		Showers	
9th	Wed		Showers	
10th	Thur	Unusually cold overnight in the Alice Springs region.	Rain, Th	
11th	Fri	Overnight minimum temperatures may drop below 10°C in the Alice Springs region over the next couple of days.	Showers	XhS
12th	Sat			
13th	Sun	Very warm on Centre Island overnight. Strong winds through the Gulf of Carpentaria, with gusts across Centre Island.	Showers	F
14th	Mon	Some rain around the West Arnhem Lands.		A
15th	Tue	Spotty outbreaks of rain linger around the Top End over the next 2-3 days.		
16th	Wed			
17th	Thu	Very hot in the Arnhem Lands.		
18th	Fri			
19th	Sat		Showers	V
20th	Sun	Very heavy rain in the Arnhem Lands particularly in the Maningrida area.	Rain	
21st	Mon			3Q
22nd	Tue	Chance of showers in the Barkly Tablelands over the next 3-4 days.		
23rd	Wed			
24th	Thur			
25th	Fri		Showers	
26th	Sat	Unusually warm overnight in the West Arnhem.		XhN
27th	Sun			
28th	Mon			P3 N
29th	Tue			
30th	Wed	Unusually cool overnight in the Katherine, Arnhem and Roper-McArthur regions over the next few days.		

May 2025

Date	Day	Northern Territory - May 2025	Darwin	Moon
1st	Thu	Mostly dry over the Top End throughout this month.		
2nd	Fri			^
3rd	Sat			
4th	Sun			
5th	Mon			1Q
6th	Tue			
7th	Wed			
8th	Thur	Unusually cool in the Alice Springs region, including in the Kulgera and Yulara districts.		
9th	Fri	Very hot in the Timber Creek region, with maximum temperatures in the upper 30's.		XhS
10th	Sat			
11th	Sun	Unusually hotter than normal over the next few weeks.		A
12th	Mon	Very hot in the Darwin-Daly region with maximum temperatures in the upper 30's in many districts. Chance of a cloud band crossing the central part of the continent bringing some rain along the southern border districts of the Alice Springs region.		
13th	Tue			F
14th	Wed	Very hot in the Timber Creek and Daly Waters regions over the next couple of days.	Fg	
15th	Thur	Mostly dry throughout the week in all regions.	Fg	
16th	Fri	Very warm overnight over the northern coasts of the Top End.		V
17th	Sat			
18th	Sun			
19th	Mon		Fg	
20th	Tue			3Q
21st	Wed			
22nd	Thu	Showers in the Tanami Desert and Alice Springs region.	Showers	
23rd	Fri	Chance of isolated showery spells in the Arnhem Lands over the next couple of days, with possible heavy falls.	Rain	XhN
24th	Sat			
25th	Sun	Some showers through the Alice Springs region over the next couple of days.		
26th	Mon			P5
27th	Tue	Pockets of heavy rain in the Victoria River region, particularly along the Wickham River catchment and with rain into the Alice Springs region, easing to showers by tomorrow.		N
28th	Wed			
29th	Thur	Cold overnight at Yulara.		^
30th	Fri			
31st	Sat			

June 2025

Date	Day	Northern Territory - June 2025	Darwin	Moon
1st	Sun	The dry season is well established by this time and dominates most of the State throughout the month.		
2nd	Mon			
3rd	Tue	Very low minimum temperatures over the next few days throughout the Alice Springs region with many districts not reaching into double digits overnight.		1Q
4th	Wed	Chance of isolated showers in the Barkly over the next couple of days.		
5th	Thu			XhS
6th	Fri			
7th	Sat			A
8th	Sun	Isolated showers possible in the eastern border fringes of the Barkly and Alice Springs regions.		
9th	Mon			
10th	Tue	Unusually cold in the Alice Springs region, particularly in the Arltunga to Alice Springs area. Strong winds with gusts around the Darwin-Daly region, including over McCluer Island. Chance of some showers around the West Arnhem coastal fringes.		
11th	Wed	Unusually cooler than average and slightly wetter than normal over the next 3-4 weeks compared to what is typical for this time of year.		F
12th	Thur			V
13th	Fri			
14th	Sat			

Date	Day	Northern Territory - June 2025	Darwin	Moon
15th	Sun			
16th	Mon			
17th	Tue	Very warm overnight over the Top End region, particularly in the Arnhem Lands.		
18th	Wed			
19th	Thu			3Q XhN
20th	Fri			
21st	Sat			
22nd	Sun			
23rd	Mon			P8
24th	Tue			
25th	Wed	Chance of showers in the Barkly Tablelands over the next 3-4 days.		N
26th	Thur	Very cold overnight in the Alice Springs region with chance of frosts in Alice Springs town itself.		^
27th	Fri			
28th	Sat	Band of widespread rain activity across the eastern half of the State over the next few days, more widespread through the Alice Springs region. Heavy rain in the Arnhem Lands.	Showers	
29th	Sun	Very strong gusty winds through the Rabbit Flat region.		
30th	Mon	Unusually cool spell across the State, particularly in the Roper-McArthur and Tennant Creek regions, including around the Larrimah district. Heavy rain at times in the Katherine, Timber Creek, and Victoria River regions.	Showers	

July 2025

Date	Day	Northern Territory - July 2025	Darwin	Moon
1st	Tue	A band of rain crosses the State from the Victoria River region to the Roper-McArthur region with showers extending into the south of the State.		
2nd	Wed			XhS
3rd	Thu	Chance of frosts in the Jervois area.		1Q
4th	Fri	Frost likely overnight at Yulara.		
5th	Sat	Another cloud band may bring rain from the Victoria River to the Roper-McArthur across Barkly Tablelands and with some isolated showers into the Alice Springs region over the next couple of days.	Showers	A
6th	Sun			
7th	Mon	Some pockets of frost in the Alice Springs region overnight, including in the Curtin Springs district. Dry throughout the State for the remainder of this month.		
8th	Tue			
9th	Wed	Very cool overnight around the western Top End to the Roper-McArthur regions over the next couple of days.		
10th	Thur		Fg	V
11th	Fri	Mostly cooler and dry over the next 3-4 weeks. Very cool overnight in the Darwin-Daly region, particularly around the Channel Point area.		F
12th	Sat			
13th	Sun		Fg	
14th	Mon	Unusually warm overnight in the West Arnhem.	Fg	
15th	Tue	Chance of frosts in the Alice Springs region.	Fg	
16th	Wed		Fg	
17th	Thu			XhN
18th	Fri			3Q
19th	Sat			
20th	Sun			
21st	Mon	Very low minimum temperatures in the Darwin-Daly region, particularly in the Mango Farm area.		P11
22nd	Tue			
23rd	Wed			^
24th	Thur			
25th	Fri	Cold day in the Alice Springs region, particularly in the Kulgera district.		N
26th	Sat			
27th	Sun			
28th	Mon			
29th	Tue			
30th	Wed	Cold with overnight frosts in the Alice Springs region over the next few nights.		XhS
31st	Thu	Frosts in Alice Springs area.		

August 2025

Date	Day	Northern Territory - August 2025	Darwin	Moon
1st	Fri	Dry throughout the State over most of this coming month with just a couple of interjections of wet spells.		
2nd	Sat			1Q A
3rd	Sun			
4th	Mon	Some brief showers in the West Arnhem Lands likely, particularly around Alyangula.	Fg	
5th	Tue		Fg	
6th	Wed			V
7th	Thur		Fg	
8th	Fri		Fg	
9th	Sat	Very warm day in the Darwin-Daly region, particularly around the Douglas River area.	Fg	F
10th	Sun	Unusually warm overnight in the West Arnhem. Unusually warm with unsettled and blustery winds for a time at Yulara.	Fg	
11th	Mon			
12th	Tue	Chance of frost pockets in Alice Springs.	Fg	
13th	Wed	Cold day in the Alice Springs region, particularly in the Arltunga district.	Fg	XhN
14th	Thu		Fg	
15th	Fri			P12
16th	Sat			3Q
17th	Sun	Chance of some frost pockets in the Alice Springs region, including in sheltered parts of Rabbit Flats to Curtin Springs districts.		
18th	Mon			
19th	Tue			^
20th	Wed	Very cold overnight in the Alice Springs region with chance of frosts in some districts, including around the Arltunga districts.		
21st	Thur			
22nd	Fri			
23rd	Sat			N
24th	Sun	Hot spell over the Top End over the next few days.	Fg	
25th	Mon		Fg	
26th	Tue			XhS
27th	Wed	An unseasonal cloud band may develop over the Top End to the Roper-McArthur and Barkly bringing colder temperatures and scattered outbreaks of rain through the regions. Chance of some heavy falls for a brief time in the Elsey district, as well as in pockets in the Mataranka and Katherine districts and along the Hodgson and Douglas River catchments. Strong blustery winds at times over McCluer Island.	Showers	
28th	Thu	Shower activity may continue in the Douglas River district and over the Top end..		
29th	Fri	Chance of frosts in the Alice Springs region, including in the Kulgera district.		
30th	Sat			A
31st	Sun			1Q

September 2025

Date	Day	Northern Territory - September 2025	Darwin	Moon
1st	Mon	The dry season persists throughout this month, with all regions mostly dry each day.		
2nd	Tue	Warmer temperatures return over the northern half of the State once more.		V
3rd	Wed			
4th	Thur			
5th	Fri			
6th	Sat		Fg	
7th	Sun			
8th	Mon	Chance of some isolated showers on the Barkly Tablelands.		F
9th	Tue		Fg	XhN
10th	Wed			
11th	Thu		Fg	P9
12th	Fri	Strong gusty winds in the south of the Alice Springs region.		
13th	Sat			
14th	Sun		Fg	3Q

Date	Day	Northern Territory - September 2025	Darwin	Moon
15th	Mon			^
16th	Tue	Unusually cold overnight in the Alice Springs region, with chance of some patches of frost, particularly in the Arltunga area.	Fg	
17th	Wed	Unusually cold day in the Timber Creek district. Chance of isolated showers around the Barkly Tablelands.		
18th	Thur			
19th	Fri			
20th	Sat	Cooler in the Alice Springs region, particularly around the Kulgera area.	Fg	
21st	Sun			
22nd	Mon	Strong winds around Yulara.		N XhS
23rd	Tue			
24th	Wed			
25th	Thu			
26th	Fri		Fg	A
27th	Sat		Fg	
28th	Sun		Fg	
29th	Mon		Fg	
30th	Tue		Fg	1Q V

October 2025

Date	Day	Northern Territory - October 2025	Darwin	Moon
1st	Wed	The dry season persists throughout this month, with all regions mostly dry each day.		
2nd	Thur		Fg	
3rd	Fri	Unusually hot in the Victoria River - Darwin-Daly regions of the State with maximum temperatures possible into the upper 30's to early 40's, particularly in the Bradshaw area over the next couple of days.	Fg	
4th	Sat		Fg	
5th	Sun	Very hot in some parts of the Alice Spring region, with some districts having maximum temperatures into the high 30's.		
6th	Mon	Unusually warm overnight in the northern half of the State, particularly in the Victoria-Daly region. Strong gusty winds at times over the Top End, particularly in the Kaduku area to the northern Arnhem.		
7th	Tue			F XhN
8th	Wed	Heavy rain in Darwin area over the next couple of days.	Showers	
9th	Thu		Rain	P6
10th	Fri			
11th	Sat	Unusually cold overnight in the Darwin-Daly region, particularly in the Tindal area.	Fg	
12th	Sun	Warmer temperatures return across the State over the next few weeks with daily maximas regularly reaching into the upper 30's to low 40's.	Showers	
13th	Mon			^
14th	Tue			3Q
15th	Wed	Chance of brief rain in the Darwin-Daly region.		
16th	Thur			
17th	Fri	Showers possible in the Victoria-Daly region and in the southeast of the Alice Springs region over the next couple of days.		
18th	Sat			
19th	Sun	Very warm overnight in the Daly Waters region.		
20th	Mon	Isolated showers likely in the Roper-McArthur and over the eastern Barkly Tablelands.		XhS
21st	Tue			
22nd	Wed	Squally storms over the Top End districts over the next couple of days may bring some localised damage due to wind gusts.		N
23rd	Thu			
24th	Fri		Showers, Th	A
25th	Sat	Unusually cold overnight in some parts of the Alice Springs region, particularly in the Arltunga area. Very high maximum temperature in the Black Point area.		
26th	Sun			
27th	Mon		Showers, Th	V
28th	Tue		Showers	
29th	Wed			
30th	Thur			1Q
31st	Fri			

November 2025

Date	Day	Northern Territory - November 2025	Darwin	Moon
1st	Sat	Dry throughout the State over the next few days.	Show ers	
2nd	Sun		Show ers	
3rd	Mon			XhN
4th	Tue	A band of rain crosses the Victoria River to Darwin-Daly regions over the next couple of days.		
5th	Wed	Heavy rain in the Timber Creek district.		
6th	Thur	Very strong winds with gusts in the Alice Springs region.		F P1
7th	Fri		Show ers, Th	
8th	Sat	Very warm overnight in the Daly Waters region.	Show ers, Th	
9th	Sun			^
10th	Mon	The passage of a storm with damaging wind gusts and large hail may cross through the Alice Springs / Simpsons Desert region over the next couple of days. Cold day in the Alice Springs region, particularly in the Arltunga district.		
11th	Tue	Regular afternoon storms return over the next 3-4 weeks, particularly over the northern half of the State and daily maximum temperatures into their early 40's become established almost everywhere. Rain in the west of the Alice Springs region over the next few days, including heavy rain in the Watarrka area.	Rain	
12th	Wed		Rain, Th	3Q
13th	Thu	Heavy rain in the west of the Alice Springs region from the Tamani Desert to the Curtin Springs areas with showers persisting over the next couple of days.	Rain, Th	
14th	Fri	Cool spell in the Alice Springs region over the nex few days. Unusually cool day at Curtin Springs to the Watarrka areas.	Show ers	
15th	Sat		Th	
16th	Sun	Cold overnight in the Alice Springs region, particularly in the Arltunga district.	Show ers	XhS
17th	Mon		Rain, Th	
18th	Tue	Mostly dry throughout the State over the coming week, except for some outbreaks of rain at times mosty in the Darwin-Daly region.	Th	
19th	Wed		Rain, Th	
20th	Thur			N A
21st	Fri		Show ers, Th	
22nd	Sat		Show ers, Th	
23rd	Sun		Show ers, Th	V
24th	Mon	Blustery winds in the Roper-McArthur.	Rain, Th	
25th	Tue			
26th	Wed	Rain in the Top End over the next few days. Heavy rain in the Darwin-Daly region likely, particularly along the Adelaide River catchment.	Show ers, Th	
27th	Thu		Th	
28th	Fri		Show ers, Th	1Q
29th	Sat	Showers in the Arnhem Lands to the Darwin-Daly spreading into the Roper-McArthur by tomorrow.	Show ers, Th	
30th	Sun			XhN

December 2025

Date	Day	Northern Territory - December 2025	Darwin	Moon
1st	Mon	Dust storms may develop in the Simpson Desert. Very warm in Ali Curung overnight. Shower activity over the Top End and pockets of rain in the central Alice Springs region.	Show ers, Th	
2nd	Tue	Chance of storms crossing the Victoria River region bringing localised damage in some places.	Rain, Th	
3rd	Wed	Very heavy rain in Darwin with chance of localised flooding.		
4th	Thur	Showers along the northern coastal fringes of the Top End.		P2
5th	Fri		Show ers	F
6th	Sat		Th	^
7th	Sun	Blustery winds in the Darwin-Daly region over the next couple of days possibly causing some localised wind damage in some places.		
8th	Mon	Very strong winds in the Victoria-Daly region, particularly in the Bradshaw area.		
9th	Tue		Rain	
10th	Wed	Very hot in the Alice Springs region, with maximum temperatures around the mid-40's in numerous districts, including at Curtin Springs.		
11th	Thu	Very hot over the next couple of weeks before a cooler change brings daily temperatures back to close to normal by New Year.		

Date	Day	Northern Territory - December 2025	Darwin	Moon
12th	Fri	A deep heat-low may develop over the Top End around this time bringing a very hot spell across the State over the next couple of weeks. Very hot in the Barkly to the Darwin-Daly regions, including the Elliot and Brunette Downs districts, with maximum temperatures in the upper 40's likely. Very hot in the far southeast of the Alice Springs region, including around Ali Curung.		**3Q**
13th	Sat	Hot overnight in many districts, including in the Brunette Downs area and in Darwin city.		**XhS**
14th	Sun	The monsoonal trough begins to activate around this time with storm events steadily becoming more frequent between now and the end of the year, particularly over the top End to the Roper-McArthur regions and along the Barkly Tablelands.	Rain, Th	
15th	Mon			
16th	Tue	A severe squall line crosses the Darwin-Daly region, including in Darwin itself.	Rain, Th, Gusts	
17th	Wed	Very strong winds, possibly to gale force at times, in the Alice Springs region, particularly around Yulara.	Th	**A**
18th	Thur	Outbreaks of showers in most parts of the State.	Showers, Th	
19th	Fri		Rain	
20th	Sat	Heavy rain over the Top End and down through the Roper-McArthur region.	Showers, Th	**N V**
21st	Sun	Brief cool spell in the far south of the State, particularly milder around Yulara to Curtin Springs areas in the next couple of days.	Rain, Th	
22nd	Mon			
23rd	Tue			
24th	Wed			
25th	Thu	Monsoonal rains over the northern half of the State over the next couple of days, as well as extending over the Barkly Tablelands, and directing isolated shower activity into the Alice Springs region.	Showers, Th	
26th	Fri			
27th	Sat	Spotty outbreaks of monsoonal rains daily throughout the State until the end of the year.		**XhN**
28th	Sun			**1Q**
29th	Mon		Showers, Th	
30th	Tue		Th	
31st	Wed	Widespread shower activity through most regions of the State.	Rain, Th	

NORTHERN TERRITORY GRAPHS AND TABLES 2025

Seasonal Breakdown:

State Overall Rainfall Timing and Estimated mm Trend Graphs
State Overall Maximum Temperature Trend Graphs
Regional Rainfall Expectations Tables - mm and Trends
Regional Maximum and Minimum Temperature Trends Tables

Monthly Breakdown:

Rainfall:
Regional Expectations by mm
Regional Expectations by mm Variance to Average
Regional Variance Trend by mm
Regional Estimated Expectations by % of Variance to Average
Regional Estimatedl Variance Trend by %

Maximum & Minimum Temperature Trends:
Regional Estimated Average Temperature
Regional Estimated Degree of Variation from the Average
Regional Estimated Temperature Trend

Seasonal Breakdowns

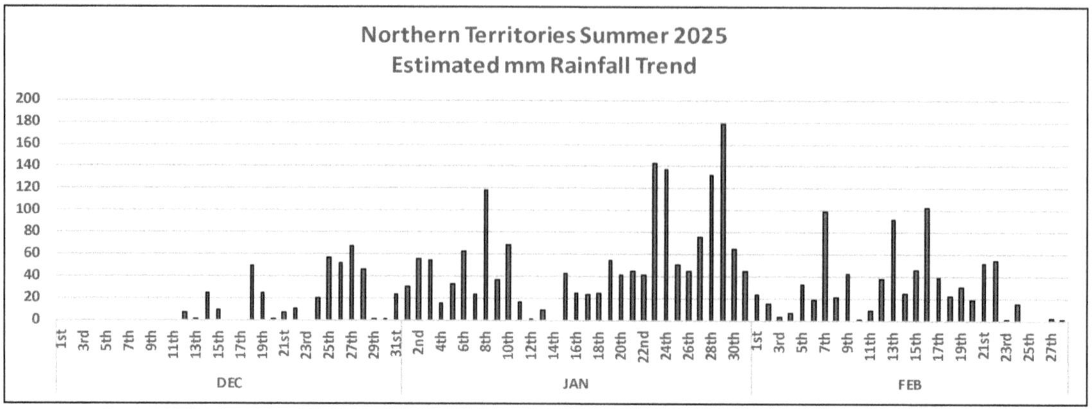

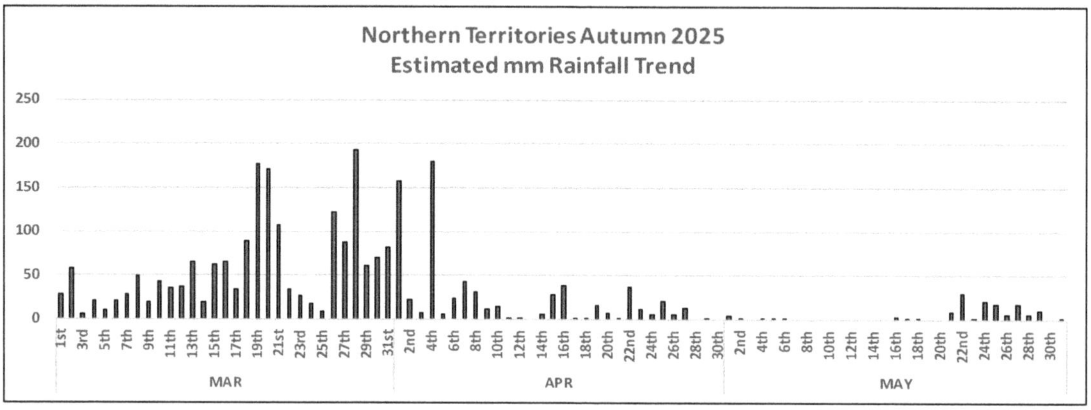

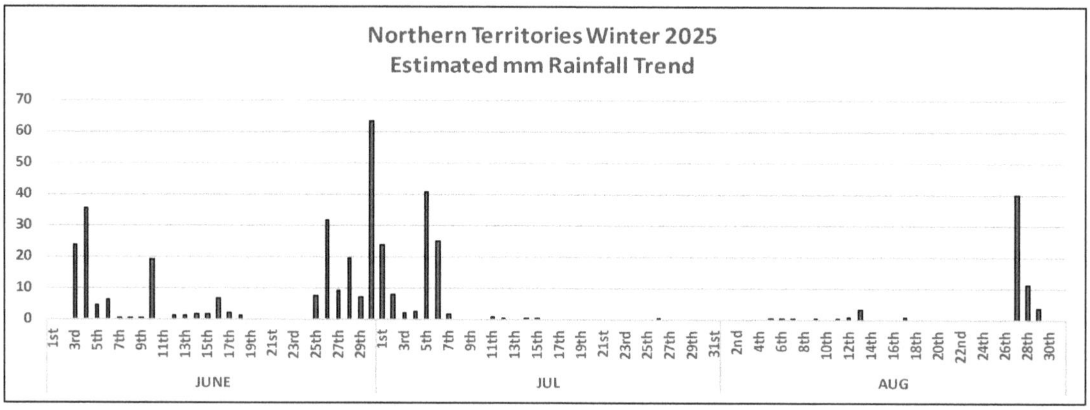

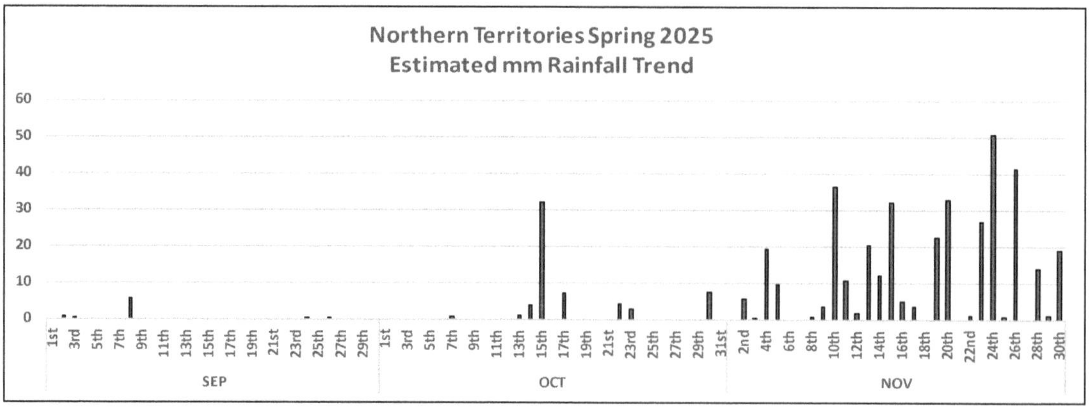

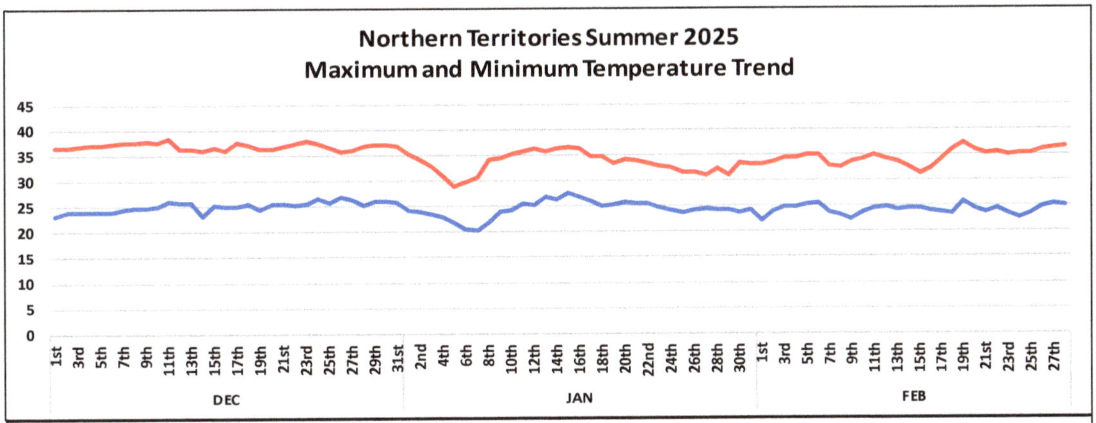

Northern Territories Summer 2025
Maximum and Minimum Temperature Trend
DEC
JAN
FEB

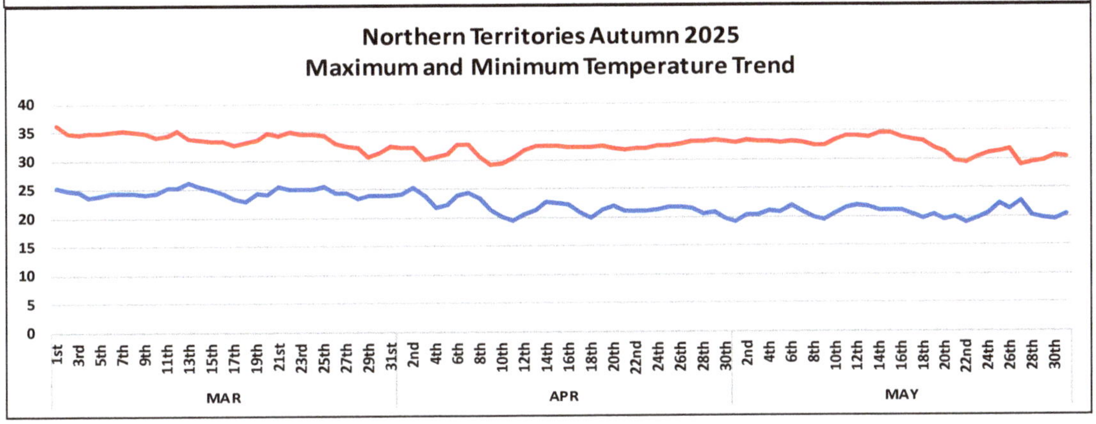

Northern Territories Autumn 2025
Maximum and Minimum Temperature Trend
MAR
APR
MAY

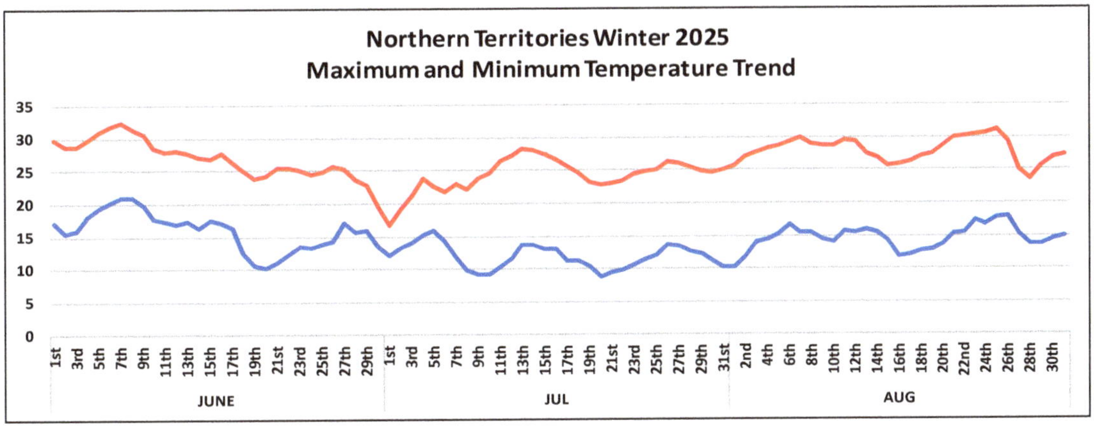

Northern Territories Winter 2025
Maximum and Minimum Temperature Trend
JUNE
JUL
AUG

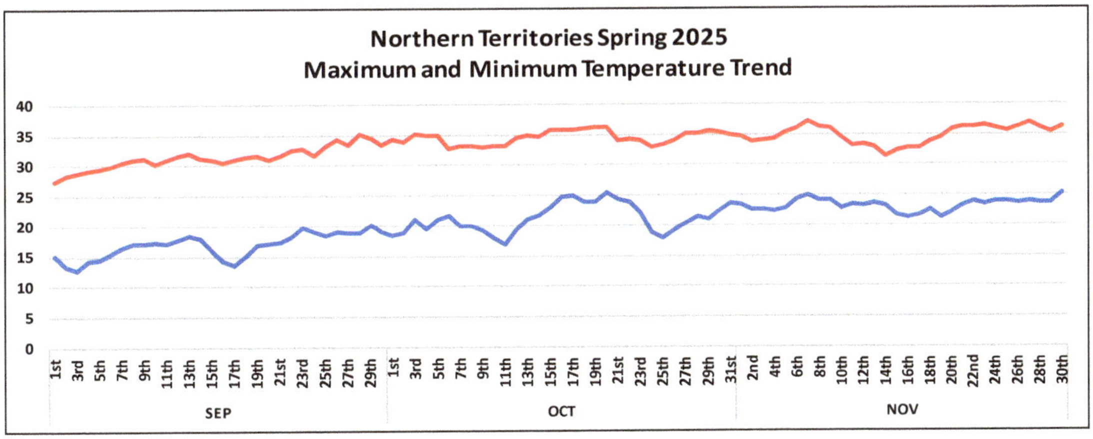

Northern Territories Spring 2025
Maximum and Minimum Temperature Trend
SEP
OCT
NOV

2025 NT - Estimated Seasonal Rainfall Expectations (mm)

REGION	Arnhem Lands	Darwin Daly	Katherine	Roper-McArthur	Barkly	Alice Springs	Victoria River	NT
Town Sample	Gove	Darwin	Katherine	Centre Island	Tennant Ck	Alice Springs	Victoria River	Overall
Summer	463	795	507	435	292	130	283	2905
Autumn	594	879	490	281	116	72	242	2674
Winter	215	4	40	21	33	10	86	410
Spring	42	166	123	6	26	34	40	438
Overall	1246	1753	1166	747	479	255	683	6328

2025 NT - Estimated Seasonal Rainfall Variation from the Average (mm)

REGION	Arnhem Lands	Darwin Daly	Katherine	Roper-McArthur	Barkly	Alice Springs	Victoria River	NT
Town Sample	Gove	Darwin	Katherine	Centre Island	Tennant Ck	Alice Springs	Victoria River	Overall
Summer	-306	-320	-296	-185	-22	9	-154	-1275
Autumn	-1	412	263	-57	36	-2	106	756
Winter	158	-5	38	11	24	-25	80	281
Spring	-17	-54	-13	-38	-38	-27	-43	-229
Overall	-235	-59	-3	-265	13	-36	20	-566

2025 NT - Estimated Seasonal Rainfall Trend Variation by 30mm

REGION	Arnhem Lands	Darwin Daly	Katherine	Roper-McArthur	Barkly	Alice Springs	Victoria River	NT
Town Sample	Gove	Darwin	Katherine	Centre Island	Tennant Ck	Alice Springs	Victoria River	Overall
Summer	drr	drr	drr	drr	av	av	drr	drr
Autumn	av	wtr	wtr	av	wtr	av	wtr	wtr
Winter	wtr	av	wtr	av	av	av	wtr	wtr
Spring	av	drr	av	drr	drr	av	drr	drr
Overall	drr	av	av	drr	av	av	av	drr

2025 NT - Estimated Seasonal Rainfall Expectations by %

REGION	Arnhem Lands	Darwin Daly	Katherine	Roper-McArthur	Barkly	Alice Springs	Victoria River	NT
Town Sample	Gove	Darwin	Katherine	Centre Island	Tennant Ck	Alice Springs	Victoria River	Overall
Summer	-40%	-29%	-37%	-30%	-7%	7%	-35%	-31%
Autumn	0%	88%	116%	-17%	45%	-3%	78%	39%
Winter	277%	-53%	1516%	108%	284%	-72%	1287%	218%
Spring	-29%	-24%	-9%	-85%	-59%	-44%	-52%	-34%
Overall	-16%	-3%	0%	-26%	3%	-13%	3%	-8%

2025 NT - Estimated Seasonal Rainfall Trend Variation by 25%

REGION	Arnhem Lands	Darwin Daly	Katherine	Roper-McArthur	Barkly	Alice Springs	Victoria River	NT
Town Sample	Gove	Darwin	Katherine	Centre Island	Tennant Ck	Alice Springs	Victoria River	Overall
Summer	drr	drr	drr	drr	av	av	drr	drr
Autumn	av	wtr	wtr	av	wtr	av	wtr	wtr
Winter	wtr	drr	wtr	wtr	wtr	drr	wtr	wtr
Spring	drr	av	av	drr	drr	drr	drr	drr
Overall	av	av	av	drr	av	av	av	av

2025	NT - Estimated Seasonal Maximum Temperature Average and Trend							
Region Town Sample	Arnhem Lands Gove	Darwin Daly Darwin	Katherine Katherine	Roper-McArthu Centre Island	Barkly Tennant Ck	Alice Springs Alice Springs	Victoria River Victoria River	NT Overall
Summer Est	32.0	33.0	36.2	34.3	36.4	35.9	37.4	35.0
Trend	av	a-w	wmr	av	av	av	a-w	av
Autumn Est	30.3	32.3	33.7	32.3	33.5	31.9	34.7	32.7
Trend	av	av	av	a-w	wmr	wmr	a-w	a-w
Winter Est	27.2	30.5	29.2	26.1	23.6	20.4	27.9	26.4
Trend	clr	av	clr	a-c	clr	av	clr	clr
Spring Est	31.4	33.3	36.3	32.6	34.2	30.9	36.6	33.6
Trend	av	av	a-c	av	av	av	a-c	av

2025	NT - Estimated Seasonal Minimum Temperature Average and Trend							
Region Town Sample	Arnhem Lands Gove	Darwin Daly Darwin	Katherine Katherine	Roper-McArthu Centre Island	Barkly Tennant Ck	Alice Springs Alice Springs	Victoria River Victoria River	NT Overall
Summer Est	24.4	25.2	24.8	27.1	24.5	21.5	24.8	24.6
Trend	av	av	av	av	av	a-w	av	av
Autumn Est	23.6	24.2	21.8	25.8	22.0	17.0	21.8	22.3
Trend	av	a-w	wmr	a-w	wmr	wmr	wmr	wmr
Winter Est	20.6	19.5	13.9	18.5	12.1	3.3	11.5	14.2
Trend	av	av	av	a-c	clr	clr	av	a-c
Spring Est	20.6	23.4	21.1	24.1	20.9	13.5	19.7	20.5
Trend	a-c	a-c	clr	av	av	a-c	clr	clr

Monthly Breakdowns

2025	NORTHERN TERRITORY - Estimated Rainfall Expectations (mm)							
REGION	Arnhem Lands	Darwin Daly	Katherine	Roper-McArthur	Barkly	Alice Springs	Victoria River	NT
Town Sample	Gove	Darwin	Katherine	Centre Island	Tennant Ck	Alice Springs	Victoria River	Overall
January	274	369	208	201	265	121	255	1693
February	119	249	191	227	18	0	7	812
March	350	727	311	217	14	37	191	1847
April	186	134	178	32	99	18	47	694
May	59	19	0	32	3	17	4	133
June	166	0	2	9	10	7	51	244
July	31	4	6	5	23	0	35	106
August	19	0	32	6	0	3	0	60
September	6	1	0	0	0	0	0	7
October	2	47	3	0	1	0	7	60
November	35	117	120	6	25	34	33	371
December	1	85	114	11	20	18	52	301
Overall	1246	1753	1166	747	479	255	683	6328

2025	NORTHERN TERRITORY - Estimated Rainfall Variation from the Average (mm)							
REGION	Arnhem Lands	Darwin Daly	Katherine	Roper-McArthur	Barkly	Alice Springs	Victoria River	NT
Town Sample	Gove	Darwin	Katherine	Centre Island	Tennant Ck	Alice Springs	Victoria River	Overall
January	-32	-96	-105	-35	142	81	98	53
February	-161	-124	-39	-25	-104	-40	-154	-647
March	75	391	131	-30	-40	1	83	612
April	-36	26	138	-50	81	-2	25	182
May	-41	-6	-6	23	-5	-2	-2	-37
June	135	-2	1	4	7	-7	48	185
July	11	3	6	3	19	-17	32	59
August	12	-6	31	4	-1	-2	-1	37
September	0	-16	-4	-1	-7	-7	-4	-38
October	-12	-18	-35	-6	-17	-21	-12	-122
November	-5	-19	25	-31	-14	1	-27	-69
December	-183	-192	-147	-121	-48	-23	-67	-780
Overall	-235	-59	-3	-265	13	-36	20	-566

2025	NORTHERN TERRITORY - Estimated Rainfall Trend Variation by 10mm							
REGION	Arnhem Lands	Darwin Daly	Katherine	Roper-McArthur	Barkly	Alice Springs	Victoria River	NT
Town Sample	Gove	Darwin	Katherine	Centre Island	Tennant Ck	Alice Springs	Victoria River	Overall
January	drr	drr	drr	drr	wtr	wtr	wtr	wtr
February	drr	drr	drr	drr	drr	drr	drr	drr
March	wtr	wtr	wtr	drr	drr	av	wtr	wtr
April	drr	wtr	wtr	drr	wtr	av	wtr	wtr
May	drr	av	av	wtr	av	av	av	drr
June	wtr	av	av	av	av	av	wtr	wtr
July	wtr	av	wtr	av	wtr	drr	wtr	wtr
August	wtr	av	wtr	av	av	av	av	wtr
September	av	drr	av	av	av	av	av	drr
October	drr	drr	drr	av	drr	drr	drr	drr
November	av	drr	wtr	drr	drr	av	drr	drr
December	drr	drr	drr	drr	drr	drr	drr	drr
Overall	drr	av	av	drr	av	av	av	drr

2025	NORTHERN TERRITORY - Estimated Rainfall Expectations by %							
REGION	Arnhem Lands	Darwin Daly	Katherine	Roper-McArthur	Barkly	Alice Springs	Victoria River	NT
Town Sample	Gove	Darwin	Katherine	Centre Island	Tennant Ck	Alice Springs	Victoria River	Overall
January	-10%	-21%	-34%	-15%	114%	204%	62%	3%
February	-57%	-33%	-17%	-10%	-85%	-100%	-96%	-44%
March	27%	117%	72%	-12%	-74%	4%	77%	50%
April	-16%	24%	344%	-61%	435%	-9%	112%	35%
May	-41%	-23%	-100%	256%	-58%	-9%	-31%	-22%
June	437%	-100%	129%	64%	206%	-49%	1932%	316%
July	58%	267%	504%	135%	427%	-100%	1157%	126%
August	178%	-100%	1700%	191%	-100%	-41%	-100%	157%
September	4%	-94%	-100%	-64%	-100%	-100%	-100%	-84%
October	-89%	-28%	-92%	-100%	-96%	-100%	-64%	-67%
November	-12%	-14%	26%	-84%	-36%	4%	-44%	-16%
December	-100%	-69%	-56%	-92%	-70%	-56%	-56%	-72%
Overall	-16%	-3%	0%	-26%	3%	-13%	3%	-8%

2025	NORTHERN TERRITORY - Estimated Rainfall Trend Variation by 25%							
REGION	Arnhem Lands	Darwin Daly	Katherine	Roper-McArthur	Barkly	Alice Springs	Victoria River	NT
Town Sample	Gove	Darwin	Katherine	Centre Island	Tennant Ck	Alice Springs	Victoria River	Overall
January	av	av	drr	av	wtr	wtr	wtr	av
February	drr	drr	av	av	drr	drr	drr	drr
March	wtr	wtr	wtr	av	drr	av	wtr	wtr
April	av	av	wtr	drr	wtr	av	wtr	wtr
May	av	av	av	wtr	av	av	av	av
June	wtr	drr	wtr	wtr	wtr	drr	wtr	wtr
July	wtr	wtr	wtr	wtr	wtr	drr	wtr	wtr
August	wtr	drr	wtr	wtr	drr	drr	drr	wtr
September	av	drr	drr	drr	drr	drr	drr	drr
October	drr	drr	drr	drr	drr	drr	drr	drr
November	av	av	wtr	drr	drr	av	drr	av
December	drr	drr	drr	drr	drr	drr	drr	drr
Overall	av	av	av	drr	av	av	av	av

2025	NORTHERN TERRITORIES - Estimated Average Maximum Temperatures							
Region	Arnhem Lands	Darwin Daly	Katherine	Roper-McArthu	Barkly	Alice Springs	Victoria River	NT Estimates
Town Sample	Gove	Darwin	Katherine	Centre Island	Tennant Ck	Alice Springs	Victoria River	Overall
January	32	33	35	34	34	34	34	**34**
February	32	32	35	34	36	36	38	**35**
March	30	31	34	33	37	37	36	**34**
April	31	32	33	32	32	30	34	**32**
May	30	33	34	32	32	29	35	**32**
June	28	32	30	28	24	18	28	**27**
July	26	29	27	24	21	19	25	**24**
August	28	31	31	27	26	24	30	**28**
September	30	33	34	30	31	28	34	**31**
October	31	34	38	33	36	33	37	**35**
November	33	34	37	35	36	31	38	**35**
December	34	34	37	36	39	36	40	**37**
Overall	**30**	**32**	**34**	**31**	**32**	**30**	**34**	**32**

2025	NORTHERN TERRITORY - Estimated Average Minimum Temperatures							
Region	Arnhem Lands	Darwin Daly	Katherine	Roper-McArthu	Barkly	Alice Springs	Victoria River	NT Estimates
Town Sample	Gove	Darwin	Katherine	Centre Island	Tennant Ck	Alice Springs	Victoria River	Overall
January	25	25	24	27	23	22	25	**24**
February	25	25	25	27	24	20	24	**24**
March	24	24	24	26	25	23	25	**25**
April	24	25	22	26	21	14	21	**22**
May	22	23	19	25	20	14	20	**21**
June	23	22	17	21	13	3	14	**16**
July	19	17	11	16	10	2	9	**12**
August	20	19	15	19	14	5	12	**15**
September	18	21	16	21	17	10	15	**17**
October	20	24	22	24	22	14	21	**21**
November	23	25	25	27	23	16	24	**23**
December	25	26	25	28	26	21	25	**25**
Overall	22	23	20	24	20	14	19	**20**

2025	NORTHERN TERRITORY - Estimated Degree of Variation from the Maximum Average							
Region	Arnhem Lands	Darwin Daly	Katherine	Roper-McArthu	Barkly	Alice Springs	Victoria River	NT
Town Sample	Gove	Darwin	Katherine	Centre Island	Tennant Ck	Alice Springs	Victoria River	Overall
January	-0.4	0.9	0.4	0.5	-2.8	-3.0	-2.3	-1.0
February	0.1	0.8	0.9	0.8	-0.1	0.8	2.0	0.8
March	-1.3	-0.8	-0.7	-0.2	2.2	4.5	0.5	0.6
April	-0.5	-0.3	-0.8	0.1	0.0	1.2	-1.1	-0.2
May	0.5	1.1	1.9	2.4	4.1	5.5	2.9	2.6
June	-0.1	1.3	0.1	0.8	-0.6	-1.4	-0.9	-0.1
July	-2.1	-1.7	-3.6	-2.6	-3.9	-0.8	-4.5	-2.8
August	-1.1	-0.8	-1.6	-0.8	-1.3	1.0	-1.8	-0.9
September	-0.8	0.0	-2.1	-0.5	-0.7	1.0	-1.6	-0.7
October	-0.7	0.4	-0.1	-0.2	0.7	2.2	-0.7	0.2
November	0.1	0.2	-0.2	0.2	-0.7	-2.9	-0.2	-0.5
December	1.1	1.5	1.3	1.1	2.1	0.4	2.0	1.4
Overall	-0.4	0.2	-0.4	0.1	-0.1	0.7	-0.5	-0.1

2025	NORTHERN TERRITORY - Estimated Degree of Variation from the Minimum Average							
Region	Arnhem Lands	Darwin Daly	Katherine	Roper-McArthu	Barkly	Alice Springs	Victoria River	NT
Town Sample	Gove	Darwin	Katherine	Centre Island	Tennant Ck	Alice Springs	Victoria River	Overall
January	-0.2	0.3	0.2	0.4	-1.7	0.2	-0.2	-0.1
February	0.5	0.5	0.4	0.4	-0.4	-1.0	-0.5	0.0
March	0.4	-0.3	1.0	0.1	1.6	5.0	1.5	1.3
April	0.6	0.8	1.4	0.5	0.6	1.3	1.1	0.9
May	-0.3	1.2	1.8	2.3	3.7	6.2	3.8	2.7
June	1.8	1.9	2.5	1.5	-0.4	-1.7	1.8	1.1
July	-1.1	-1.9	-3.1	-2.6	-2.5	-2.3	-2.2	-2.2
August	0.5	-1.0	-0.4	-0.9	-0.4	-0.9	-1.2	-0.6
September	-1.2	-2.4	-3.6	-1.1	-0.9	-0.6	-3.5	-1.9
October	-1.3	-0.4	-1.6	-0.4	0.2	-0.5	-1.7	-0.8
November	-0.6	-0.2	-0.3	0.0	-0.8	-1.5	-0.7	-0.6
December	0.2	0.4	-0.3	1.1	1.3	0.9	-0.5	0.4
Overall	-0.1	-0.1	-0.2	0.1	0.0	0.4	-0.2	0.0

2025	NORTHERN TERRITORIES - Estimated Maximum Temperature Trend							
Region	Arnhem Lands	Darwin Daly	Katherine	Roper-McArthu	Barkly	Alice Springs	Victoria River	NT
Town Sample	Gove	Darwin	Katherine	Centre Island	Tennant Ck	Alice Springs	Victoria River	Overall
January	av	a-w	av	av	clr	clr	clr	a-c
February	av	a-w	a-w	a-w	av	a-w	wmr	a-w
March	clr	a-c	a-c	av	wmr	wmr	a-w	a-w
April	av	av	a-c	av	av	wmr	clr	av
May	av	wmr	wmr	wmr	wmr	wmr	wmr	wmr
June	av	wmr	av	a-w	a-c	clr	a-c	av
July	clr	clr	clr	clr	clr	a-c	clr	clr
August	clr	a-c	clr	a-c	clr	a-w	clr	a-c
September	a-c	av	clr	av	a-c	a-w	clr	a-c
October	a-c	av	av	av	a-w	wmr	a-c	av
November	av	av	av	av	a-c	clr	av	av
December	wmr	wmr	wmr	wmr	wmr	av	wmr	wmr
Overall	av	av	av	av	av	a-w	av	av

2025	NORTHERN TERRITORY - Estimated Minimum Temperature Trend							
Region	Arnhem Lands	Darwin Daly	Katherine	Roper-McArthu	Barkly	Alice Springs	Victoria River	NT
Town Sample	Gove	Darwin	Katherine	Centre Island	Tennant Ck	Alice Springs	Victoria River	Overall
January	av	av	av	av	clr	av	av	av
February	av	av	av	av	av	a-c	av	av
March	av	av	a-w	av	wmr	wmr	wmr	wmr
April	a-w	a-w	wmr	av	a-w	wmr	wmr	a-w
May	av	wmr	wmr	wmr	wmr	wmr	wmr	wmr
June	wmr	wmr	wmr	wmr	av	clr	wmr	wmr
July	clr	clr	clr	clr	clr	clr	clr	clr
August	av	a-c	av	a-c	av	a-c	clr	a-c
September	clr	clr	clr	clr	a-c	a-c	clr	clr
October	clr	av	clr	av	av	av	clr	a-c
November	av	av	av	av	a-c	clr	a-c	a-c
December	av	av	av	wmr	wmr	a-w	av	av
Overall	av	av	av	av	av	av	av	av

QUEENSLAND
2025

GENERAL TRENDS

Year Overall:	Drier by around 10-15% less than average Slightly cooler but within 0.5°C of the average
Summer:	Slightly drier by around 10-15% below the average Wettest – First half of February Driest – Second half of January Slightly cooler but within 0.5°C of the average Hottest – First half of December and third week of January Coolest – Second week of January
Autumn:	Drier by around 30-35% below the average Wettest – First week and last weeks of March Driest – Last week of April through to end of first week of May Warmer by up to 1.5-2°C above the average Hottest – Second half of March Coolest – Last week of April to mid-May
Winter:	Wetter by up to 60-65% more than the average Wettest – Second half of June to end of first week of July Driest – Second half of July and second and third weeks of August Cooler by around 1°C below the average Hottest –First half of June and second and fourth weeks of August Coolest – End of June to first week of July
Spring:	Slightly wetter but within 5% of the average Wettest –Mid September and second half of November Driest – Last ten days of September to first half of October Slightly cooler by around 0.5-1°C below the average Hottest – Third week of October and second week of November Coolest – First half of September

Major Weather Events
January 2025

Date	Day	Queensland - January 2025	Brisbane	Moon
1st	Wed	Heavy rain in the North Coast-Barron region and in some parts of the Northwest, particularly around Camooweal.		
2nd	Thur	Occasional strong and blustery winds in the Darling Downs.		
3rd	Fri			
4th	Sat	Chance of dust storms in the Central West. Heavy rain at times in the Warrego and in some parts of the Far Southwest.	Showers	
5th	Sun	A passing trough directs a cool southerly flow through the southeast with unusually low temperatures likely to prevail over the next few days. Chance of downpours for a time in the Wide Bay-Burnett and Brisbane-Southeast Coast regions.	Rain	
6th	Mon	Heavy rain at times along the Central Coast and into the Central Highlands. Blustery at times in the Wide Bay-Burnett region.	Rain	
7th	Tue	Unusually cool due to widespread cloud over the next couple of days in the Central parts of the State to the Southwest. Heavy rain in the Upper Carpentaria, North Coast Barron and Herbert regions, and in the Wide Bay-Burnett areas.		1Q XhN
8th	Wed	Cool conditions continue.		P13
9th	Thu	A small depression crosses the Gulf of Carpentaria bringing storms, unstable winds, and some heavy rain with downpours into the Gulf Country, the Upper Carpentaria, Far North and North Coast-Barron regions of the State.		
10th	Fri	Further storms and rain bring heavy falls in the far north again, as well as in the Gulf Country and along the Central Coast.		
11th	Sat	The next 3-4 weeks may see an active monsoonal system bring good rainfalls into the Channel Country, Warrego, and Central West, including the chance of flooding along the inland river catchments. Overall, the next 3-4 weeks may see near average rainfall in the tropics, but drier conditions elsewhere. Maximum temperatures may be overall cooler than is normal for this time of year, while overnight minimums may be generally warmer. Widespread storms and heavy rain over the next couple of days, particularly in the Central West.		
12th	Sun	Severe thunderstorms with strong gusty winds in the Central West. Very hot in the State over the coming week with many districts seeing maximum temperatures into the upper 30's to early 40's, dropping into the late teens to low 20's overnight. Very hot in the Lower Western, particularly in the Boulia area, with maximum temperatures into the mid-40's likely.	Rain	
13th	Mon	Unusually mild day on the Darling Downs.	Rain	^
14th	Tue		Rain	F
15th	Wed	Unusually cool overnight in the North Burnett - Port Curtis region. Showers in the Gulf country turn to rain spreading into the Far North over the coming week and down into the Carpentaria by the end of the week.	Show ers	
16th	Thur	Very warm overnight in the Lower Western, particularly in the Birdsville district.	Show ers	
17th	Fri			
18th	Sat			
19th	Sun	Warm spell throughout the State over the coming week. Unusually mild day on the Darling Downs.		
20th	Mon	Unstable winds in the Western Downs may form into tornado activity for a time, possibly bringing damage to local property.		XhS
21st	Tue	Very hot in the Far Southwest - Lower Western over the next few days with daily maximas into the mid-40's in a number of districts. Regular outbreaks of heavy monsoonal rains over the northern and western halves of the State over the coming week.		A
22nd	Wed	Very warm over the next couple of nights in the Lower Western, particularly in the Birdsville and Bedourie districts. Very heavy rain likely in the Lower Carpentaria region, particularly in the Toorak area, which may cause some flash flooding.		3Q
23rd	Thu	Widespread severe thunderstorm activity and strong gusty winds in the Central West and Southeast of the State over the next couple of days with chance of large hail.		
24th	Fri	Severe thunderstorms and gusty winds likely in the Central West.		
25th	Sat	Mostly dry spell over the coming few days in the east of the State from the Central Coast to the South Coast and into the adjacent inland regions.		
26th	Sun			
27th	Mon	Outbreaks of heavy monsoonal rains persist in the north and west of the State and continue for the remainder of the month, becoming more intense and heavier, particularly in the Channel Country, at the end of the month.	Show ers	V
28th	Tue		Show ers	
29th	Wed		Show ers	
30th	Thur	A monsoonal low tracks from the Northern Territory into the southwest of the State over the next couple of days bringing some significant rainfall through the Channel Country, Warrego, and Central west, with the chance of flooding in the local river catchments.		N
31st	Fri	Very heavy rain likely in the Glengyle district with chance of localised flooding.		

February 2025

Date	Day	Queensland - February 2025	Brisbane	Moon
1st	Sat	The monsoonal trough invigorates throughout the State over the coming week, bringing scattered rain into most districts over the next 4-5 days, before easing and contracting back to the Gulf Country and Cape York Peninsula. Very heavy rain in the Lower Western, particularly in the Bedourie district, with chance of flash flooding.		
2nd	Sun			P10
3rd	Mon		Show ers	XhN
4th	Tue		Rain, Th	
5th	Wed	Thunderstorm activity with large hail, strong winds gusts, and possible tornado formations in the Moreton.		1Q
6th	Thu	High maximum temperatures return over the coming week throughout most of the State.		
7th	Fri	Mostly dry throughout the southern half of the State over the coming few days.		
8th	Sat	Unusually cool overnight throughout much of the south and east of the State, with some parts of the Darling Downs particularly cooler than normal, with overnight minimums possibly dropping into the single digits, particularly in the Stanthorpe and Applethorpe areas.	Rain, Th	
9th	Sun			^
10th	Mon	Very heavy rain in the Herbert-North Coast region, particularly in the Ingham district, possibly causing flooding along the Hawkins Creek.		
11th	Tue	The next 3-4 weeks may be wetter than normal over the Cape York Peninsula to the central tropical coastal districts and adjacent inland parts, while drier than normal in the west and southeast. Maximum temperatures may be slightly cooler than average in the central and northern parts of the State, while slightly warmer elsewhere.	Rain	
12th	Wed	A low pressure system within the monsoonal trough crosses Cape York Peninsula over the next couple of days, bringing heavy falls to the region and some flooding along the regional river systems. Rainfall may be particularly heavy in the Roma and Hecate districts as well as heavy rain for a time in the Mackay to the Northern regions and up into the adjacent inland parts.	Show ers	
13th	Thur	Cool overnight in the Darling Downs over the next few days, particularly in the Applethorpe and Dalby districts.		F
14th	Fri	Mostly dry throughout the southern half of the State over the coming week.		
15th	Sat	A tropical low may form near the Arnhem Lands and drift through the Gulf of Carpentaria over the next few days, possibly reaching cyclone strength within a week, before making landfall over Cape York Peninsula in QLD.		
16th	Sun	Very heavy rain in the Atherton Tablelands around this time, with chance of localised flooding, particularly in the Topaz district.		XhS
17th	Mon	Very hot in the Warrego to the Lower Western, with maximum temperatures into the low 40's and minimum temperatures around the late 20's in a number of areas, including in the Cunnamulla and Birdsville districts.		
18th	Tue	Strong gusty winds at times in the Gulf Country - Northwest regions.		A
19th	Wed	Further thunderstorm activity and strong gusty winds in the Gulf Country to the Northwest, as well as thunderstorms ahead of an advancing trough line crossing the Sunshine Coast region with strong gusty winds at times. Very heavy raln likely in the northern Cape York Peninsula region. Very hot spell throughout most of the State over the next 3-4 days.		
20th	Thu	Showers develop over the southeast and turn to rain and storms over the next week in the coastal parts from the Fraser Coast to the Gold Coast.	Rain, Th	
21st	Fri	Chance of severe thunderstorms with strong winds, very heavy rain, and large hail at times in the Southwest Downs region.	Show ers	3Q
22nd	Sat	Supercell thunderstorm activity likely in the Capricornia region, with very large hail possible. Thunderstorms with hail and heavy rain also on the Darling Downs to the Maranoa. Chance of severe storm activity in the Gulf Country - Northwest, which may bring localised damage in the Breakaway Creek region.	Rain	
23rd	Sun	Cool spell over the next few days in the Darling Downs to the South Coast-Moreton regions, particularly in the Toowoomba and Maleny districts, with daily maximas only into the low 20's likely. Heavy rain spreads into the Port Curtis region around this time while continuing in the South Coast-Moreton regions and adjacent inland.	Show ers	V
24th	Mon	Thunderstorms and outbursts of very heavy rain across the Tropical North.	Show ers	
25th	Tue	Shower activity persists in the eastern half of the State until the end of the month, slowly easing.		
26th	Wed		Show ers	
27th	Thur	Thunderstorms and heavy rain at times over the Tropical North.	Show ers	
28th	Fri	Mostly dry in the Lower West, Channel Country and Darling Downs-Warrego regions, with showery outbreaks elsewhere.	Show ers	N

March 2025

Date	Day	Queensland - March 2025	Brisbane	Moon
1st	Sat	A relatively dry spell throughout much of the coming week throughout the State except for monsoonal showers continuing over Cape York Peninsula to the Gulf Country and down along the northeastern coastal fringes as far south as the Fraser Coast briefly.		
2nd	Sun			P7 XhN
3rd	Mon		Showers	
4th	Tue	Showers briefly spread to the Fraser Coast and coastal Wide Bay Burnett districts around this time.		
5th	Wed		Showers	
6th	Thu		Showers	
7th	Fri	Very hot in the Lower Western - Far Southwest, with maximum temperatures into the low 40's and minimum temperatures nearing the 30's in a number of areas, particularly in the Birdsville and Thargomindah districts.		1Q
8th	Sat	Hot spell spreads throughout most of the State over the next few days. Some isolated showers likely in the Channel Country.		^
9th	Sun	Rain in the eastern half of New South Wales briefly spread showers into the Darling Downs and Warrego for a couple of days.		
10th	Mon		Showers	
11th	Tue	The next 3-4 weeks may be generally drier than average for this time of year in the southeast of the State, while average to slightly above average rainfall elsewhere. Both maximum and minimum temperatures may be overall warmer than normal during this next few weeks in most districts, but nearer to the norm over Cape York.	Showers	
12th	Wed	Very hot over the coming week with maximum temperatures into the low 40's in many districts. Chance of patches of heavy rain around the Gold Coast about this time.		
13th	Thur	Very heavy rain over the next 4-5 days in the Barron-North Coast region. Dry over the next couple of days in the south of the State.		
14th	Fri	Very hot in the Lower Western and Warrego around this time, with maximum temperatures into the mid-40's, particularly in the Birdsville and Cunnamulla districts.		F
15th	Sat	Very warm overnight in the Far Southwest, particularly in the Quilpie district, with minimum temperatures remaining in the low 30's likely.	Showers	XhS
16th	Sun	Showers develop in the southeast around this time from the Fraser Coast to the Gold Coast at first, and spread westward over the Darling Downs, Warrego, and into the Gulf Country over the next few days.	Rain	
17th	Mon		Showers	
18th	Tue	Scattered thunderstorms and pockets of heavy rain in the southeast of the State from the Darling Downs to Brisbane.	Showers, Th	A
19th	Wed			
20th	Thu	Some lingering showers around the Fraser Coast, while clearing to mostly dry through southern regions for the next few days.		
21st	Fri	Very hot spell in the eastern coastal regions and into the adjacent inland regions over the next couple of days, with maximum temperatures into the low 40's and overnight minimums only briefly dropping below the 30's in many districts, including in parts of the Port Curtis, South Coast, Moreton, Mackay, Central Highlands, Maranoa, and Darling Downs regions.	Fg	
22nd	Sat	Mostly dry throughout the State, including monsoonal rains clearing to isolated showers in the Far North and Gulf Country.		
23rd	Sun	Severe thunderstorm activity in the Northwest and Gulf Country reinvigorates as well as thunderstorms and bursts of heavy rain on the North Tropical Coast and persist for much of the coming week. Particularly strong wind gusts in the Julia Creek region. Cool day in the Darling Downs.		3Q V
24th	Mon			
25th	Tue	Some shower activity in the Central West over the next couple of days.		
26th	Wed	Mostly dry in the east of the State from the Wide Bay Burnett to the Southeast-Moreton and across the Darling Downs and Warrego for the remainder of the month. Some shower activity briefly in the Channel Country.		
27th	Thur			
28th	Fri	Dry conditions spread throughout the inland regions of the Central West and persist until the end of the month. In contrast, showers persist through the coastal Gulf Country to the Far North.		
29th	Sat	Very heavy rain over the next 4-5 days in the Barron-North Coast region.		N XhN
30th	Sun			P4
31st	Mon	Showers spread briefly down from Cape York Peninsula into the Central Coast and Mackay regions.		

April 2025

Date	Day	Queensland - April 2025	Brisbane	Moon
1st	Tue	Mostly dry in the southeast of the State and the Central Highlands and Lowlands over the next couple of days.		
2nd	Wed			
3rd	Thu	Scattered showers and outbreaks of rain move into the southeast of the State over the next couple of days, with some heavier falls in the Warrego and Daring Downs for a time, with showers briefly moving up along the eastern coastal fringes to the Mackay in a couple of days time.		
4th	Fri		Show ers	^
5th	Sat	Mostly dry over the next couple of days throughout the State, except for some isolated showers lingering around the Gulf Country.		1Q
6th	Sun			
7th	Mon	Showers turn to heavy rain in the coastal parts of the Gulf Country around this time and showers move up over Cape York Peninsula over the next couple of days. Mostly dry elsewhere.		
8th	Tue	Unusually cool overnight in some parts of the Darling Downs, with minimum temperatures dropping into the single digits in some districts, particularly around the Warwick area.		
9th	Wed			
10th	Thur	Unusually cold overnight in the Darling Downs, Warrego, and Maranoa regions, with minimum temperatures in the low single digits, particularly in the Dalby, Cunnamulla, and Injune areas. Mostly dry throughout the State over the next few days except for some outbreaks of showers along the coastal fringes from the Far North to the Port Curtis regions.		
11th	Fri	The next 3-4 weeks may continue to be very dry in the southeast and central parts of the State, while average to slightly above average rainfall is likely elsewhere, with the tropical season coming to an end. Maximum temperatures may be above average over most of the interior, cooler over the Cape York Peninsula, and around average elsewhere, while minimum temperatures may be above normal in the southwest and Central Highlands, below average in the Goldfields and Warrego, and within a degree of what is typical for this time of year elsewhere.		XhS
12th	Sat			
13th	Sun	Unusually warm overnight in the Lower Carpentaria.		F
14th	Mon	Very hot day in the Lower Carpentaria region.		A
15th	Tue	Mostly dry spell throughout the State over the remainder of the month, with just the occasional brief showery interruption along the coastal fringes. Isolated showers possible in the coastal fringes around the Central Coast.		
16th	Wed			
17th	Thu	Chance of some isolated showers around the Fraser Coast to the Moreton region, and in the coastal fringes of the North Coast.	Show ers	
18th	Fri	Somer showers possible along the North Coast to the Central Coasts over the next few days.	Show ers	
19th	Sat			V
20th	Sun			
21st	Mon			3Q
22nd	Tue	Chance of heavy rain cross the tip of the northern Cape York Peninsula, with chance of localised flooding in the Bamaga district.		
23rd	Wed	Some isolated shower activity along the coastal fringes from the Far North to the Central Coast over the next few days.		
24th	Thur			
25th	Fri	Cooler overnight in some parts of the Moreton-South Coast and Port Curtis regions, with some districts minimum temperatures dropping into the single digits around this time.		
26th	Sat	Dry throughout the State over the remainder of the month.		XhN
27th	Sun	Very warm spell throughout most of the State over the next 3-4 days with daily maximas into the mid-to-upper 30's in many districts.		
28th	Mon			P3 N
29th	Tue			
30th	Wed			

May 2025

Date	Day	Queensland - May 2025	Brisbane	Moon
1st	Thu	Dry throughout the State over most of the coming week.		
2nd	Fri	Chance of thunderstorms and brief downpours in the Brisbane regon.	Showers, Th	^
3rd	Sat			
4th	Sun			
5th	Mon			1Q
6th	Tue	Very warm spell throughout most of the State over the next 4-5 days with daily maximas into the mid-to-upper 30's in many districts.		
7th	Wed	Cool over the next couple of nights in the North Coast regions from the Barron to the Herbert. A storm system in the southeast of the country spreads northwards towards Queensland, bringing some showers into the Channel Country around this time.		
8th	Thur	Unusually cool day in the Darling Downs, particularly in the Applethorpe district. Widespread rain throughout the southeast of the State moves up into the southern regions of Queensland over the next couple of days. Chance of heavy falls for a time in the Warrego, Darling Downs, and South Coast-Maranoa regions.	Showers	
9th	Fri	Rain activity in the southeast of the State eases to passing showers from the Port Curtis region to the South Coast-Maranoa, while clearing to mostly dry in the Warrego and Darling Downs. Dry elsewhere.		XhS
10th	Sat	Cold overnight in the Central Highlands, Maranoa, Warrego, and Darling Downs, with minimum temperatures in the single digits, particularly in the Tambo district. Mostly dry throughout the State over the coming week.		
11th	Sun	It may continue to be drier than average over the coming 3-4 weeks once more, with most other districts seeing closer to average rainfall, while both maximum and minimum temperatures may be warmer almost everywhere except on Cape York Peninsula, where temperatures may be slightly cooler than what is typical for this time of year.		A
12th	Mon	Cold overnight in some parts of the Moreton-South Coast with minimum temperatures in the single digits, particularly in the Kingaroy area.		
13th	Tue			F
14th	Wed	Unusually hot spell over much of the State during the next few days, including in the Lower Carpentaria with maximum temperatures into the upper 30's in some districts and overnight minimums in the late teens, particularly in the Julia Creek area.	Fg	
15th	Thur		Fg	
16th	Fri			V
17th	Sat	Chance of showers along the coastal regions from the North Coast to the Mackay over the coming week.		
18th	Sun	Some isolated showers possible around the Gold Coast to Brisbane districts over the next few days. Showers may be heavier for a time around the Mackay to Port Curtis coastal fringes.	Showers	
19th	Mon			
20th	Tue	Isolated showers likely from the Fraser Coast to the Gold Coast over the next few days.	Showers	3Q
21st	Wed	Cold day in the Darling Downs, particularly in the Applethorpe area.		
22nd	Thu	Some brief heavy falls likely around the Mackay coasal fringes over the next couple of days. Isolated coastal showers likely around the Gulf of Carpentaria.	Showers	
23rd	Fri			XhN
24th	Sat	Chance of some showers briefly in the Central Highlands.		
25th	Sun	Brief spell of rain in the Channel Country and the Central West to the Warrego over the next couple of days.		
26th	Mon	A late season tropical low that may form in the Solomon Sea around this time, may briefly brush the far north of the Cape York Peninsula over the next couple of days as it passes through the Torres Strait.		P5
27th	Tue	Showers turn to heavy rain for a time in the Channel Country over the next couple of days. Mostly dry elsewhere.		N
28th	Wed	Some scattered rain activity in the southeast of the State over the next couple of days. Mostly dry elsewhere.	Showers	
29th	Thur			^
30th	Fri	Brief showers possible around the Fraser Coast to the South Coast-Moreton while mostly dry elsewhere over the next couple of days.		
31st	Sat	Chance of a brief spell of heavy rain around the far northern tip of Cape York Peninsula around this time.	Fg	

June 2025

Date	Day	Queensland - June 2025	Brisbane	Moon
1st	Sun	Cold overnight temperatures during the next few nights in many parts of the southeast of the State, from the Port Curtis region and the Upper Western southwards, with many districts seeing overnight minimum temperatures in the single digits. Some outbreaks of rain in the Port Curtis to the South Coast-Moreton regions around this time, mostly dry elsewhere.		
2nd	Mon	Cold overnight in the Darling Downs, with chance of scattered frosts in some districts, including in the Stanthorpe and Applethorpe areas. Some showers around the Far North over the next few days.		
3rd	Tue			1Q
4th	Wed	Chance of some shower activity in the Central West and Gulf Country coastal fringes to the Cape York Peninsula over the next few days.	Show ers	
5th	Thu	Isolated showers in the southeast of the State over the next few days.	Show ers	XhS
6th	Fri	Very heavy rain in the Barron-North Coast region, particularly in the Cairns area.	Rain, Th	
7th	Sat		Show ers	A
8th	Sun	Showers in the Central West and Channel Country as well as in the North Peninsula to the Central Coast regions.	Fg	
9th	Mon	Mostly dry except for the chance of brief lingering showers around the North and Central Coasts and scattered rain activity associated with storm activity throughout New South Wales around this time spreads into the southern border districts of Queensland. Outbreaks of rain may briefly spread northwards into the Central Highlands for a time.	Show ers	
10th	Tue	Cold overnight in the Lower West, Warrego, and Far Southwest, with minimum temperatures in the low single digits in a number of districts. Chance of patches of frost overnight in some Darling Downs districts, including in the Applethorpe area. Mostly dry throughout the State over the next few days.		
11th	Wed	Unusually rain activity in the coming 3-4 weeks may see many parts of the State wetter than is typical for this time of year, and as a consequence of the increased cloud during this time, daily maximum temperatures may be colder than average almost everywhere, while minimum temperatures may also be below normal from the Northwest Gulf country to the Warrego and the remaining regions of the State minimums may be average to slightly above average.		F
12th	Thur	Mild spell throughout most of the State over the next 3-4 days, with daily maximas in the 20's to early 30's in many districts.		V
13th	Fri	Warm in the Upper Carpentaria region, with maximum temperatures in the early 30's in many districts. Spotty showers return over the eastern coastal regions from the North Coast to the South Coast and across the Darling Downs around this time, with rain activity becoming more intense over the coming days throughout the eastern half of the State.		
14th	Sat	Very warm in the southern parts of Cape York Peninsula over the next few days with maximum temperatures reaching into the mid-30's in some districts.		
15th	Sun	A wet front moves into the east of the State over the next 3-4 days bringing outbreaks of storms and heavy rain from the North Coast to the South Coast and into the adjacent inland regions.	Rain	
16th	Mon		Rain	
17th	Tue			
18th	Wed	Cold spell over the northern half of the State over the next 2-3 days, with a number of districts in the Carpentaria, Central Highlands and Lowlands, Upper Wester, and Southern Peninsula likely to see overnight minimas in the single digits and some areas near to frost levels.		
19th	Thu	Rain activity clears throughout the State bringing a mostly dry period over the next 5-6 days.		3Q XhN
20th	Fri			
21st	Sat			
22nd	Sun			
23rd	Mon			P8
24th	Tue	Frost patches likely over the next few nights in the Central Highlands, Darling Downs, Maranoa and Warrego regions. Chance of some brief showers developing around the North Coast.		
25th	Wed	Rain activity builds over the northern half of the State around this time and spreads throughout most regions with only the Warrego to the South Coast-Moreton and Gold Coast remaining mostly dry.		N
26th	Thur	Overnight frosts become more widespread in the Darling Downs, Maranoa, and Warrego around this time, with severe frosts possible in the Stanthorpe, Mitchell, and Warwick districts.		^
27th	Fri	Heavy rain in the Carpentaria to the Gulf Country over the next couple of days.		
28th	Sat			
29th	Sun	Rain activity eases to mostly dry in the Central West and Carpentaria, but showers continue elsewhere as well as briefly crossing the Darling Downs and Warrego.		
30th	Mon	Rain intensifies over the northern half of the State once again, with particularly heavy falls in the Central Lowlands to the Carpentaria.		

July 2025

Date	Day	Queensland - July 2025	Brisbane	Moon
1st	Tue	Cold over the next few nights in the southeast quarter of the State, with a number of districts nearing frost level minimas. Heavy rain and storms continue over the northern half of the State over the coming week.		
2nd	Wed			XhS
3rd	Thu		Showers	1Q
4th	Fri		Showers	
5th	Sat	Heavy rain over the northern half of the State continues, and spreads into the southeast quarter over the next couple of days.	Rain	A
6th	Sun	Heavy rain throughout most of the State around this time, except in the Channel Country and only some light showers lingering over Cape York Peninsula.	Rain	
7th	Mon	Rain activity clears throughout the north and central parts of the State around this time, but continues in the east from the Mackay to the South Coast-Moreton regions, with showers into the Darling Downs.		
8th	Tue	Single digit minimum temperatures likely over the next 2-3 nights in many districts from the North Coast to the Mackay and Port Curtis regions, Carpentaria, Central Highlands, and in the Moreton-South Coast regions.		
9th	Wed	Mostly dry throughout the State over the remainder of the month. Chance of some isolated showers lingering around the South Coast- Moreton.		
10th	Thur			V
11th	Fri	The next 3-4 weeks may continue to bring drier than normal conditions throughout the southern and central regions of the State, with northern districts wetter than average, and all other districts more within the norm for this time of year, while both maximum and minimum temperatures may be average to cooler than normal.		F
12th	Sat			
13th	Sun	Very warm spell throughout the southern half of the State over the next few days.		
14th	Mon			
15th	Tue	Dry conditions continue throughout the State.		
16th	Wed			
17th	Thu			XhN
18th	Fri			3Q
19th	Sat	Very mild overnight around the North Peninsula. In contrast, very cold day in the Darling Downs.		
20th	Sun	Cold overnight during the coming week to ten days in from the Carpentaria and the South Peninsula to the South Coast-Moreton and adjacent inland regions with many districts seeing minimum temperatures in the single digits and close to frost levels.		
21st	Mon			P11
22nd	Tue	Chance of frost patches in the Moreton - South Coast region.		
23rd	Wed			^
24th	Thur	Chance of patches of frosts in the Central Highlands, particularly in the Tambo area.		
25th	Fri			N
26th	Sat	Scattered frosts over the next couple of nights from the South Coast-Moreton to the Central Highlands, including in parts of the Maranoa, Port Curtis, and Darling Downs.		
27th	Sun			
28th	Mon			
29th	Tue	Cold with frost patches overnight in the Darling Downs, particularly in the Stanthorpe area.		
30th	Wed	Very mild overnight around the North Peninsula. Chance of some patches of frost in the South Coast-Moreton region, particularly in the Gympie district, and scattered through the Central Highlands, Maranoa, and Warrego over the next couple of days.		XhS
31st	Thu			

August 2025

Date	Day	Queensland - August 2025	Brisbane	Moon
1st	Fri	Mostly dry continues continue throughout the State over the next fortnight.		
2nd	Sat	Chance of some isolated showers around the North Coast and South Coast-Moreton over the next couple of days.	Showers	1Q A
3rd	Sun		Showers	
4th	Mon	A moist onshore flow brings showers over the North Tropical coast over the next couple of days, with some heavy rain at times in the Atherton Tablelands to the Cassowary Coast districts.		
5th	Tue	Warm days and mild nights around the North Peninsula over the next few days.		
6th	Wed	Warm spell from the Upper Carpentaria to the South Coast-Moreton regions over the next few days.		V
7th	Thur			
8th	Fri	Unusually warm in the Lower Carpentaria.		
9th	Sat	Warm spell in the Carpentaria, North Coast, Mackay, Central Highlands and Lowlands and the Upper Western over the next couple of days with maximum temperatures in the late 20's to the early 30's likely. Chance of some isolated showers in the Darling Downs.		F
10th	Sun	Unusually warm ovenright in the North Peninsula, particularly in the Coconut Island area.		
11th	Mon	With the exception of some much needed rain in he southeast of the State over the coming 3-4 weeks, almost all other regions may continue to see below average rainfall, while temperatures, both maximum and minimums may be fairly typical for this time of year.		
12th	Tue	Warm spell over the next few days in the southern half of the State, as well as into the Mackay to the Herbert-North Coast regions, with maximum temperatures into the low 30's at times in some districts, particularly in the Lower Western.		
13th	Wed	Chance of some isolated showers in the Channel Country.		XhN
14th	Thu			
15th	Fri	Cold overnight temperatures in the following 3-4 days throughout much of the State with few districts seeing minimum temperatures in the double digits. Some patches of frosts likely in the Darling Downs and South Coast-Moreton.		P12
16th	Sat			3Q
17th	Sun	Patches of frosts in the Darling Downs, Maranoa, and South Coast-Moreton.		
18th	Mon			
19th	Tue			^
20th	Wed	Warm spell throughout most of the State over the coming week, with many districts seeing daily maximum temperatures in the high 20's to the early 30's. Chance of some isolated showers on the North Coast around this time.		
21st	Thur			
22nd	Fri			
23rd	Sat			N
24th	Sun	Chance of some isolated showers in the Southern Peninsula to the North Coast over the next couple of days.		
25th	Mon		Fg	
26th	Tue	Very mild overnight in the North Peninsula region, particularly around Coconut and Horn Islands with minimum temperatures likely to sit around the mid-20's. A wet front crosses the State over the next few days bringing some isolated showers at first through the Central Lowlands and Central Highlands, becoming more intense and spreading over the next couple of days.	Showers	XhS
27th	Wed	A wet front brings a line of rain from the Gulf Country and Carpentaria, down across the Central Lowlands and Highlands and into the southeast of the State, including showers into the South Coast-Moreton over the next couple of days, easing to isolated by tomorrow.		
28th	Thu	Some showers over Cape York Peninsula over the next few days.		
29th	Fri	Cold day in the Darling Downs, particularly in the Applethorpe area. Storms and heavy rain return throughout the southeast of the State from the Mackay to the Darling Downs and adjacent coastal districts, including in the South Coast-Moreton over the next couple of days.	Rain	
30th	Sat		Showers	A
31st	Sun	Rain activity ease and contracts to the southeast districts from the Fraser Coast to the Moreton and Gold Coast. Mosty dry elsewhere.	Rain	1Q

September 2025

Date	Day	Queensland - September 2025	Brisbane	Moon
1st	Mon	A complex low pressure system near Fraser Island directs strong to gale force winds onto the Southeast Coast and Wide Bay-Burnett coasts, with heavy rain and scattered thunderstorms at times on the Sunshine Coast to the South Coast and Gold Coast over the next several days.	Showers	
2nd	Tue	Very hot spell on the North Peninsula around this time, with daily maximum temperatures in the upper 30's possible. Rain and thunderstorms spread deeper into the southeast quarter of the State over the next couple of days.	Rain, Th	V
3rd	Wed	Very heavy rain in the Sunshine Coast region, with chance of flash flooding in some districts.	Showers	
4th	Thur	Warmer spell over the State during the coming week, with a number of districts recording maximum temperatures into the low 30's, particularly on the Cape York Peninsula to the Western districts. Rain activity in the southeast quarter of the State contracts back to along the coastal fringes once more to the Fraser Coast with lingering showers in the South Coast-Moreton, Gold Coast over the next couple of days.	Showers	
5th	Fri	Some showers briefly spread northwards along the eastern coasts into the Port Curtis and Mackay regions.		
6th	Sat	Mostly dry throughout the State over the coming week.		
7th	Sun			
8th	Mon	Chance of isolated showers brushing the North Coast region over the next couple of days.		F
9th	Tue			XhN
10th	Wed			
11th	Thu	The next 3-4 weeks may see wetter than normal conditions in some parts of the Central Coast, Capricornia, Wide Bay, and Gulf Country regions, with most other regions continue to be drier than what is typical for this time of year. Temperatures however, over this coming 3-4 weeks, may be fairly typical for this time of year.		P9
12th	Fri			
13th	Sat	A band of showers from the Southern Peninsula to the Central Lowlands at first, spreading to scattered rain activity in the Cape York Peninsula to the Central Highlands and adjacent coastal districts over the next couple of days.		
14th	Sun		Rain	3Q
15th	Mon	Very heavy rain in the South Burnett region, particularly in the Melrose area with rain spreading into the southeast quarter of the State over the next few days. Cold and wet day on the Darling Downs, Port Curtis, Maranoa, and Central Highland regions.	Rain	^
16th	Tue	Very heavy rain in the Capricornia region.	Showers	
17th	Wed		Showers	
18th	Thur	Rain activity in the southeast of the State clears to coastal showers from the Mackay to the Moreton around this time.		
19th	Fri	Mostly dry throughout the State over the remainder of the month.		
20th	Sat			
21st	Sun	Chance of isolated showers in the Central Lowlands and Highlands.		
22nd	Mon			N XhS
23rd	Tue			
24th	Wed			
25th	Thu		Th	
26th	Fri	Chance of some isolated showers around the South Coast-Moreton.	Fg	A
27th	Sat	Very warm spell throughout most of the State over the next few days with daily maximas into the mid to upper 30's in many districts.		
28th	Sun			
29th	Mon			
30th	Tue	Very warm in the North Peninsula overnight.		1Q V

October 2025

Date	Day	Queensland - October 2025	Brisbane	Moon
1st	Wed	Very cold overnight in some parts of the Darling Downs, particularly in the Stanthorpe district which may see some frost patches in sheltered areas. Very warm in the next couple of days in the Carpentaria to the North Coast, with daily maximas into the upper 30's in a number of districts. Dry throughout most of the State over the coming week.		
2nd	Thur	Chance of brief outbursts of heavy rain in the Brisbane region, and hail showers in the Darling Downs over the next couple of days.	Showers	
3rd	Fri	Chance of brief showers around the North Coast-Barron.		
4th	Sat			
5th	Sun	Very hot spell across the southern half of the State over the next 4-5 days, particularly in the Lower Western around this time, with maximum temperatures possibly rising into the low 40's in some districts, including in the Birdsville area. Chance of some spotty showers briefly in the Central Lowlands.		
6th	Mon			
7th	Tue	Brief shower activity likely in a line from the Gulf Country to the Southeast.	Showers	F XhN
8th	Wed	Dry throughout most of the State over the coming week. Chance of some isolated drizzle patches lingering in the South Coast-Moreton to the Fraser Coast over the next couple of days.	Fg	
9th	Thu			P6
10th	Fri			
11th	Sat	The next 3-4 weeks may be warmer than is typical for this time of year and see average to above average rainfall in most regions except in the far southwest. Cold overnight in the South Coast-Moreton to the Darling Downs with overnight minimums in the single digits in many districts. Particularly cold overnight in the Stanthorpe district.		
12th	Sun	Hot spell throughout the State over the next couple of weeks. Some light showers around the North and Central Coasts over the next couple of days.	Fg	
13th	Mon			^
14th	Tue	Very hot spell in the Moreton-South Coast to the Port Curtis regions, with daily maximas into the mid to upper 30's in many districts. Chance of some spotty shower activity over the northern half of Cape York Peninsula.		3Q
15th	Wed	Very hot in the Upper Carpentaria with maximum temperatures into the upper 30's to early 40's possible, particularly in the Georgetown area. Isolated showers over the next couple of days in the Port Curtis region, while mostly dry elsewhere.		
16th	Thur	Strong gusty winds in the Wide Bay-Burnett region may cause localised damage.		
17th	Fri	A showery front crosses the State over the coming week, becoming more intense in a few days, before easing. Strong gusty winds at times in the Southeast regions from the Darling Downs to Brisbane regions, possibly accompanying thunderstorm activity. Some showers likely in the Central West.	Showers, Th	
18th	Sat	Widespread rain, thunderstorms and hail showers over the next few days in the Southeast of the State, including in the Darling Downs, Gold Coast, Brisbane, and Moreton areas. Gusty winds in the Maranoa.	Rain, Th	
19th	Sun	Rain, thunderstorms and hail showers the Southeast quarter including Brisbane to the Moreton regions and on the Gold Coast. Chance of very large hail pellets that may cause damage to property. Strong gusty winds over the Darling Downs and Maranoa which may cause some localised damage to property.	Rain, Th	
20th	Mon	Thunderstorms in the southeast of the State with rain in the Moreton to Brisbane areas and the chance of hail in the Darling Downs. Strong blustery winds in the Central West.	Rain, Th	XhS
21st	Tue	Hail showers likely in the Southeast from the Gold Coast to the Wide Bay-Burnett regions, including in Brisbane to South Burnett. Winds likely to be very gusty in the Southeast, from the Moreton to the Gold Coast regions, including over Fraser Bay and up into the Wide Bay-Burnett region. Gusty winds may bring some local damage in the Central Highlands.		
22nd	Wed	Chance of heavy rain on the Darling Downs and thunderstorms and hail showers in the Gold Coast to Brisbane region. Strong and blustery winds in the coastal districts from the Gold Coast to the Fraser Coast, including in the Moreton.	Th	N
23rd	Thu	Maximum temperatures may reach into the low 40's in some parts of the Carpentaria around this time. Heavy rain likely in the northern half of Cape York Peninsula, and along the coastal districts from the North Coast to the Fraser Coast. Mostly dry elsewhere.		
24th	Fri	Cold overnight in the Central Highland and Lowlands, Port Curtis, Darling Downs, Maranoa, and Warrego around this time with overnight minimums only in the single digits in many districts. Dry spell throughout the State over the next 4-5 days.		A
25th	Sat			
26th	Sun			
27th	Mon			V
28th	Tue			
29th	Wed	Isolated showers along the coastal fringes from the North Coast to the Fraser Coast over the next few days, while mostly dry elsewhere.		
30th	Thur			1Q
31st	Fri			

November 2025

Date	Day	Queensland - November 2025	Brisbane	Moon
1st	Sat	Dry throughout the State over the next few days.		
2nd	Sun			
3rd	Mon	Chance of isolated showers around the Central Highlands. Dry elsewhere, but with rain threatening throughout the southeast quarter of the State.		XhN
4th	Tue	Storm activity develops in the South East of the state, including in the Blackall-Tambo area and the Brisbane to Moreton regions, with possible accompanying thunderstorms, hail, and heavy rain at times over the next couple of days.	Rain, Th	
5th	Wed		Show ers, Th	
6th	Thur	Very hot over the next 3-4 days throughout the State with a number of districts seeing maximum temperatures into the upper 30's to low 40's. Rain activity in the southeast clears to mostly dry except for some lingering showers aorund the coastal fringes from the Gold Coast to the Fraser Coast over the next couple of days.		F P1
7th	Fri	Storm activity with strong and blustery winds over the southeast of the State over the next few days, with sudden outbursts of heavy rain in some parts of the Gold Coast to Brisbane and the Fraser Coast regions with the chance of localised flash flooding and wind damage particularly around the Gold Coast. Very hot day in the Lower Western with daily maximas into the mid-40's, particularly in the Boulia district.	Rain, Th	
8th	Sat	Severe thunderstorm activity in the southeast of the State, with heavy rain in the Wide Bay-Burnett region, and strong winds gusts, gales and hail in the Central Highlands. Gusty winds may cause some property and infrastructure damage in the Emerald region. Chance of storm activity in the Brisbane area that may cause some localised damage. Gale wind gusts also likely around the Port Curtis to South Coast regions. Mild overnight temperatures in the South Coast-Moreton to the Port Curtis regions.		
9th	Sun			^
10th	Mon	Rain activity clears in the eastern half of the State, while showers prevail through the Channel Country to the Central West.	Show ers	
11th	Tue	The next 3-4 weeks may be warmer and wetter than normal in most of the State, except for in the southwest once more, and with some parts of the Warrego and Central Highlands and Lowlands receiving well above their average rainfall as well as being much cooler than what is typical for this time of year. Some outbreaks of rain return in the southeast quarter over the next couple of days.		
12th	Wed	Warm spell over the coming week over much of the State, particularly in the eastern regions, with daily maximas into the mid to upper 30's.		3Q
13th	Thu	Showers in the Northwest to the Gulf Country and along the QLD-NSW border districts of the Warrego over the next couple of days.		
14th	Fri			
15th	Sat	Heavy rain likely in the Central Coast over the next couple of days and with heavy rain activity spreading into the southeast quarter of the State over the next 3-4 days. Cooler over the next few days in many parts of the State with minimum temperatures dropping into the low to mid-teens in many districts.	Rain	
16th	Sun	A band of thunderstorms crossing through the Port Curtis-South Coast region during the next few days bringing some heavy rain into the region, particularly in the Rockhampton area.		XhS
17th	Mon		Rain	
18th	Tue	Outbreaks of rain prevail throughout the eastern regions of the State over the next few days, including spreading westward into the Southwest, Darling Downs, Warrego, and Central Highlands and Lowlands to the Central Coast.	Show ers	
19th	Wed	Heavy rain possible in the Blackall-Tambo area.		
20th	Thur	Outbreaks of rain spread into the North Coast-Herbert around this time and into the southern districts of Cape York Peninsula.		N A
21st	Fri	Thunderstorms in the Southwest region.	Show ers	
22nd	Sat	Outbreaks of rain throughout many parts of the State over the next few days.	Show ers	
23rd	Sun	Cooler over the next couple of nights in many parts of the State with minimum temperatures dropping into the low to mid-teens in many districts.		V
24th	Mon	Storm activity with heavy downpours likely in the Gulf Country - Northwest over the next few days. Strong gusty winds accompanying the passage of thunderstorms may cause some damage in the Mt Isa and Julia Creek districts of the Carpentaria. Cold overnight on the Darling Downs, particularly in the Applethorpe district, where minimum temperatures may drop into the single digits.		
25th	Tue			
26th	Wed	Rain activity clears in the western regions, with a line of rain covering the remainder of the State from Cape York Peninsula to the Central Highlands over the next couple of days.	Rain	
27th	Thu	A tropical low that had formed near Papua New Guinea a few days ago, may begin to wander around the Coral Sea around this time, but is not expected to make any landfall in Queensland.	Show ers	

Date	Day	Queensland - November 2025	Brisbane	Moon
28th	Fri	Shower activity lingers around the eastern coasts of the State over the next couple of days from Cape York Peninsula to the South Coast-Moreton, while clearing to mostly dry elsewhere.		1Q
29th	Sat	Shower activity contracts to the northern coasts from the North Coast-Herbert to the Far North Peninsula, with mostly dry conditions elsewhere throughout the State.		
30th	Sun	Very hot in some parts of the Lower Western and Far Southwest regions, particularly in the Birdsville, Thargomindah, Windorah, and Bedourie districts, with maximum temperatures into the low 40's. Isolated showers around the North Coast and Far Northern Peninsula, mostly dry elsewhere.		XhN

December 2025

Date	Day	Queensland - December 2025	Brisbane	Moon
1st	Mon	Mild overnight temperatures during the coming week in many parts of the State with temperatures dropping into the mid-to-upper teens. Some showers around Cape York Peninsula and Mackay.	Showers	
2nd	Tue	Scattered rain activity in the Southeast quarter over the coming week. Heavy falls likely on the Granite Belt.	Showers	
3rd	Wed	Scattered rain with some heavy falls possible at times in the Moreton area.	Showers	
4th	Thur	A low pressure system crosses the southern half of the State around this time, bringing thunderstorms and heavy rain in many districts over the coming week, particularly in the Central Highlands and Maranoa, where there is the chance of downpours causing some localised flash-flooding.	Showers	P2
5th	Fri		Showers	F
6th	Sat		Showers	^
7th	Sun		Showers	
8th	Mon	Very hot spell over the next few days in the Upper and Lower Western and Carpentaria regions, with daily maximums into the low 40's in numerous districts.		
9th	Tue	Widespread rain activity persists over the eastern half of the State throughout the next 4-5 days.	Showers	
10th	Wed	Hail showers possible in the Tropical Far North. Strong blustery winds in the Mackay-Whitsunday to Brisbane regions. Very hot around this time in the Lower Carpentaria, with maximum temperatures into the mid-40's possible, particularly in the Century Mine district.		
11th	Thu	Strong and unsettled winds in the Mackay-Whitsunday area including over Hamilton Island. Very hot in the Carpentaria over the coming week with maximum temperatures into the lower 40's in many districts.	Showers	
12th	Fri	The next 3-4 weeks may be milder overnight than is normal for this time of year, but cooler during the day mostly as a result of significant rainfall events during this lookahead time, with wetter than average conditions in most areas except in some parts of the eastern coastal districts and adjacent inland parts from about Rockhampton southwards, which may receive closer to their average rainfall for this time of year. Very hot day in the Upper Western with maximum temperatures into the mid-to-upper 40's in a number of districts, including in the Camooweal area. Mostly dry in the west of the State.	Showers	3Q
13th	Sat	Heavy rain on the North Coast. Strong winds with gusts in the Northwest-Outback, as well as in the Gulf Country and Mackay-Whitsunday regions. Storms with possible hail showers and possible damaging winds in the Central Highlands over the next couple of days.		XhS
14th	Sun	Widespread outbreaks of rain over the northern half of the State, from the Carpentaria and Gulf Country to the Mackay and Cape York Peninsula. Scattered showers in the southeast districts.	Showers	
15th	Mon	Heavy rain possible on the Sunshine Coast with outbreaks of rain and thunderstorms throughout the southeast quarter of the State over the coming week, and showers along the coastal fringes from the Mackay to Cape York Peninsula.	Showers, Th	
16th	Tue	Very hot in the Central Highlands and Lowlands and Western regions over the next couple of days with maximum temperatures into the low 40's in many districts. Heavy rain briefly over the southern half of Cape York Peninsula.		
17th	Wed	Brief respite from rain over the Central Coastal regions, while showers persist over Cape York Peninsula and storm acitivty in the southeastern regions from the Fraser Coast to the South Coast-Moreton regions.	Rain, Th	A
18th	Thur	Storm activity including gusty winds, in the Gulf Country to the Northwest with thunderstorms and heavy rain at times, possibly bringing some localised flooding, particularly through local river systems including in the Leichhardt River catchment. Winds particularly strong at Mt Isa.		
19th	Fri	Mostly dry ver the northern half of the State with shower activity in the southern half.		

Date	Day	Queensland - December 2025	Brisbane	Moon
20th	Sat	Heavy rain likey in the Wide Bay-Burnett region, particularly in South Burnett.		N V
21st	Sun	Strong and unsettled winds in the Gulf Country - Northwest, particularly in Mt Isa. Rain and thunderstorms in the southeast quarter of the State over the next 3-4 days.	Rain, Th	
22nd	Mon	Heavy rain in the Brisbane region around this time.	Rain	
23rd	Tue	Strong winds with gusts in the Northwest - Outback, including in Cloncurry. A front brings a band of rain down through the State from the Gulf Country to the Southeast over the over the next few days.	Rain	
24th	Wed	Mostly dry along the eastern coastal regions over the next couple of days with rain activity through the inland regions.	Rain	
25th	Thu	Very hot in the Western districts around this time, with daily maximas into the mid-40's in some districts. Scattered rain activity throughout the western half of the State over the next couple of days.		
26th	Fri			
27th	Sat	Heavy rain likely in the Gulf Country, the far north of Cape York Peninsula and in the Darling Downs to the Moreton over the next couple of days.	Show ers	XhN
28th	Sun	Very strong and blustery winds at times in the Northwest - Outback region, including in Julia Creek and Camooweal areas. Very warm overnight in the Lower Western region with overnight minimums into the low 30's in some areas.		1Q
29th	Mon	Severe thunderstorm and strong gusty winds that may cause localised damage in the Gulf Country - Northwest regions, including in Mt Isa. Outbreaks of rain in most districts in the western half of the State for the remainder of the year, while coastal regions from the Central Coast to the South Coast remain mostly dry.		
30th	Tue	Heavy rain likely in the Southwest and Channel Country.		
31st	Wed			

QUEENSLAND GRAPHS AND TABLES 2025

Seasonal Breakdown:	State Overall Rainfall Timing and Estimated mm Trend Graphs State Overall Maximum Temperature Trend Graphs Regional Rainfall Expectations Tables - mm and Trends Regional Maximum and Minimum Temperature Trends Tables
Monthly Breakdown:	**Rainfall:** Regional Expectations by mm Regional Expectations by mm Variance to Average Regional Variance Trend by mm Regional Estimated Expectations by % of Variance to Average Regional Estimatedl Variance Trend by % **Maximum & Minimum Temperature Trends:** Regional Estimated Average Temperature Regional Estimated Degree of Variation from the Average Regional Estimated Temperature Trend

Seasonal Breakdowns

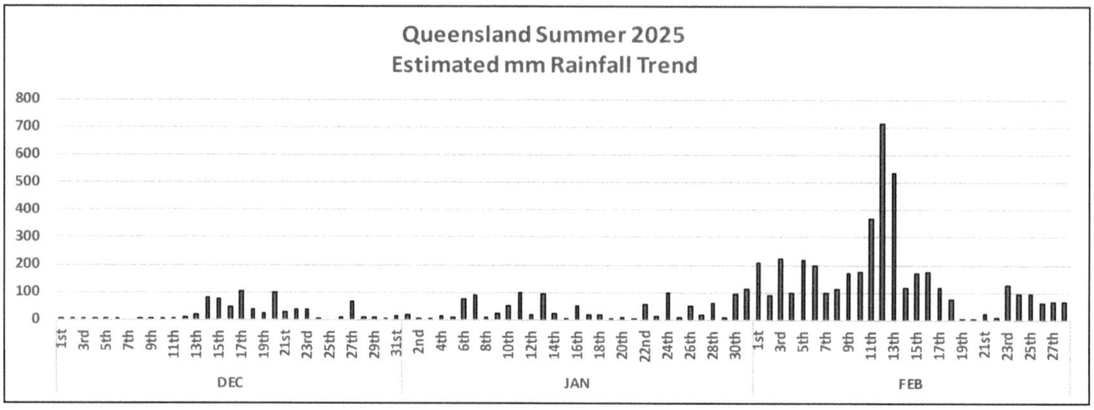

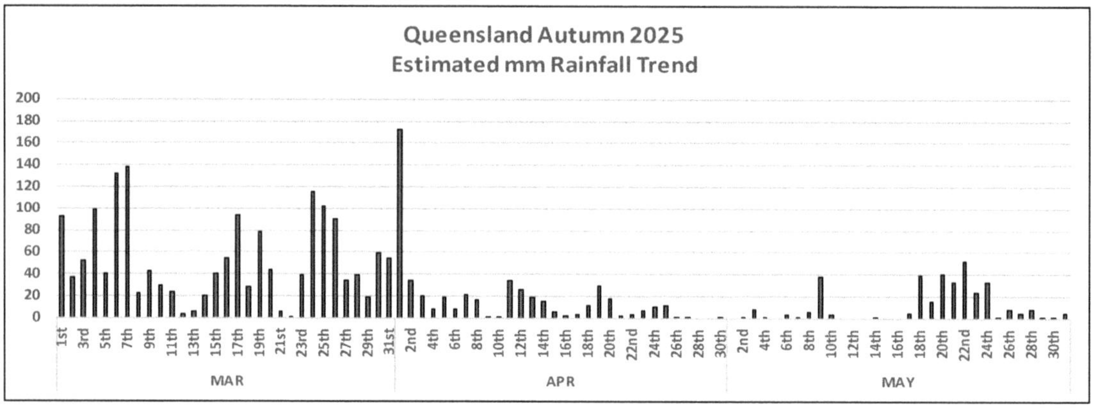

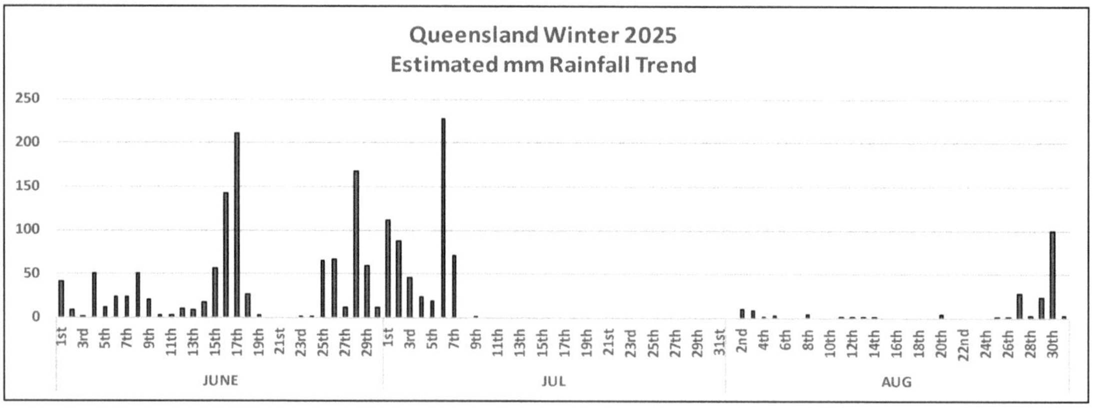

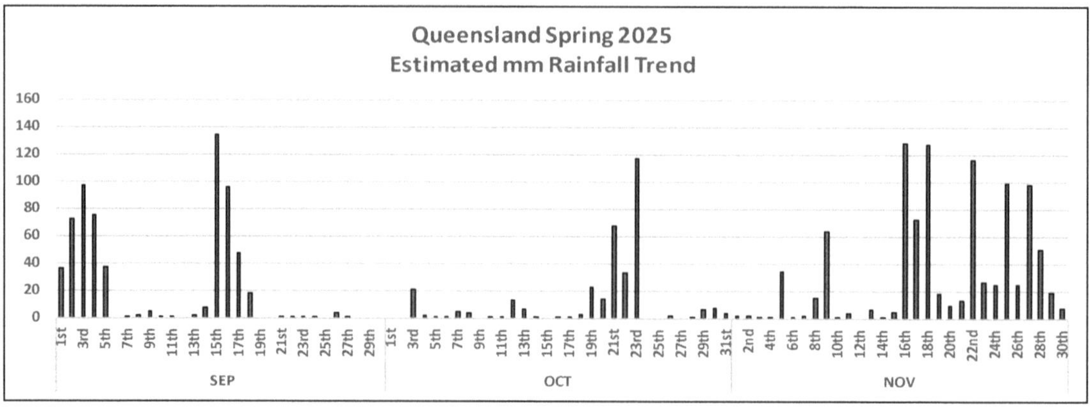

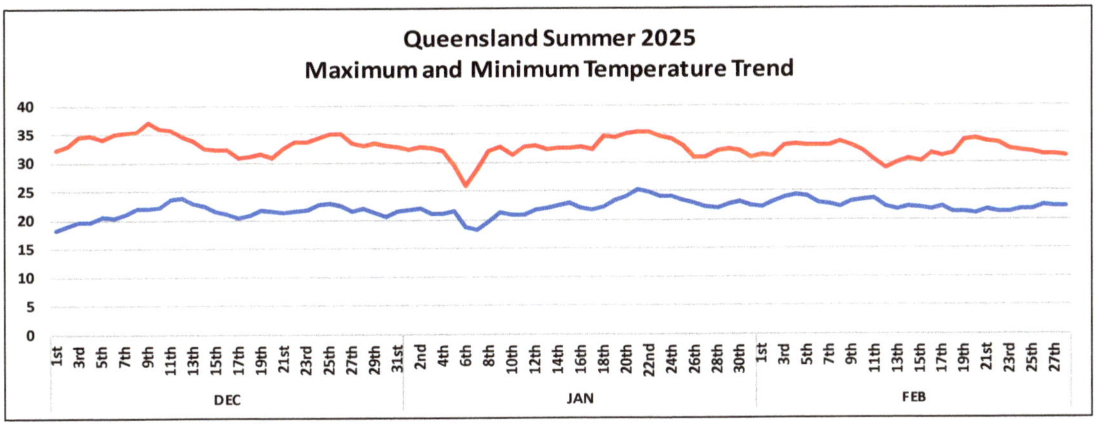
Queensland Summer 2025
Maximum and Minimum Temperature Trend
DEC
JAN
FEB

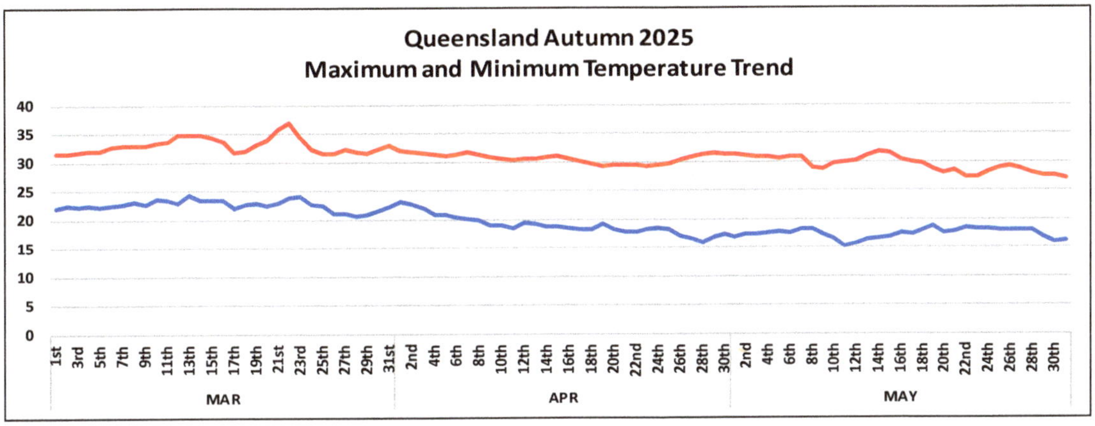
Queensland Autumn 2025
Maximum and Minimum Temperature Trend
MAR
APR
MAY

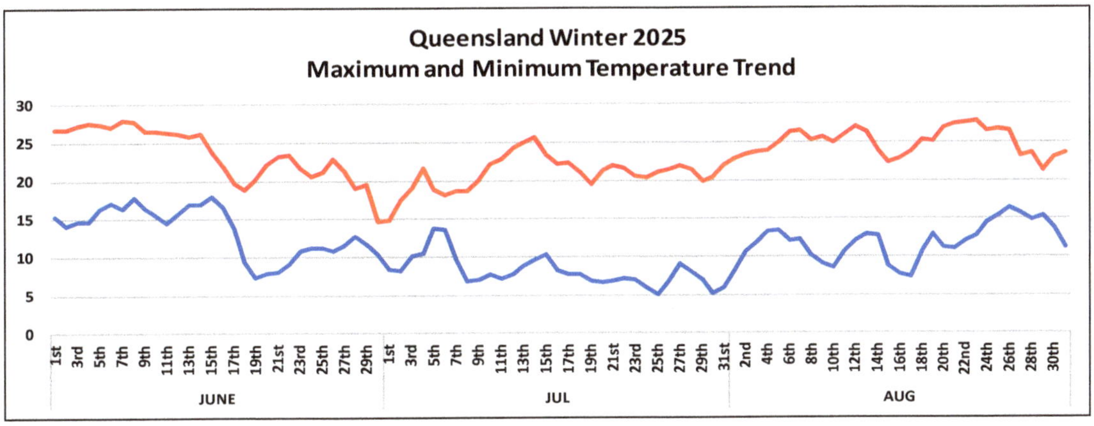
Queensland Winter 2025
Maximum and Minimum Temperature Trend
JUNE
JUL
AUG

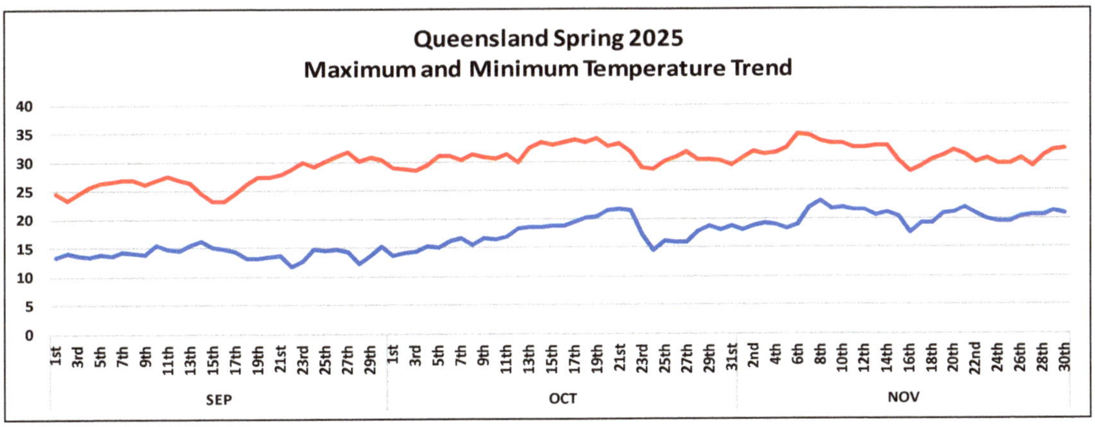
Queensland Spring 2025
Maximum and Minimum Temperature Trend
SEP
OCT
NOV

2025 — QLD - Estimated Seasonal Rainfall Expectations (mm)

REGION	Gulf Country	Northwest	Central West	Channel Country	Peninsula	Tropical Coast & Tablelnds	Goldfields / Upper Flinders	Herbet / Lower Burdekin	Central Coast-Whitsundays	Central Highlands	Capricornia	Maranoa / Warrego	Darling Dwns / Granite Belt	Wide Bay / Burnett	Southeast	QLD
Town Sample	Normanton	Mt Isa	Longreach	Birdsville	Weipa	Cairns	Charters Twrs	Townsville	Mackay	Emerald	Rockhampton	Charleville	Toowoomba	Maryborough	Brisbane	Overall
Summer	515	174	272	86	854	769	515	859	913	338	197	152	153	192	296	6285
Autumn	176	76	40	29	548	531	85	99	439	12	77	86	32	184	61	2474
Winter	61	117	80	0	40	190	211	116	268	162	181	47	93	138	177	1880
Spring	33	30	54	28	241	114	39	26	162	110	201	137	152	421	193	1942
Overall	853	470	459	133	1781	1581	857	1173	1777	691	644	508	454	925	748	13053

2025 — QLD - Estimated Seasonal Rainfall Variation from the Average (mm)

REGION	Gulf Country	Northwest	Central West	Channel Country	Peninsula	Tropical Coast & Tablelnds	Goldfields / Upper Flinders	Herbet / Lower Burdekin	Central Coast-Whitsundays	Central Highlands	Capricornia	Maranoa / Warrego	Darling Dwns / Granite Belt	Wide Bay / Burnett	Southeast	QLD
Town Sample	Normanton	Mt Isa	Longreach	Birdsville	Weipa	Cairns	Charters Twrs	Townsville	Mackay	Emerald	Rockhampton	Charleville	Toowoomba	Maryborough	Brisbane	Overall
Summer	-138	-126	71	18	-416	-236	144	175	225	85	-111	-42	-184	-204	-85	-825
Autumn	-20	-2	-61	-9	14	-162	-49	-120	80	-178	-109	-21	-182	-102	-316	-1235
Winter	47	101	33	-27	30	89	153	62	144	78	82	-27	-25	-4	12	748
Spring	-26	-41	-21	-4	124	-81	-50	-90	12	-26	52	24	-49	227	-17	34
Overall	-70	5	35	-32	-150	-412	204	101	455	29	-97	19	-416	-92	-383	-805

2025 — QLD - Estimated Seasonal Rainfall Trend Variation by 10mm

REGION	Gulf Country	Northwest	Central West	Channel Country	Peninsula	Tropical Coast & Tablelnds	Goldfields / Upper Flinders	Herbet / Lower Burdekin	Central Coast-Whitsundays	Central Highlands	Capricornia	Maranoa / Warrego	Darling Dwns / Granite Belt	Wide Bay / Burnett	Southeast	QLD
Town Sample	Normanton	Mt Isa	Longreach	Birdsville	Weipa	Cairns	Charters Twrs	Townsville	Mackay	Emerald	Rockhampton	Charleville	Toowoomba	Maryborough	Brisbane	Overall
Summer	drr	drr	wtr	av	drr	drr	wtr	wtr	wtr	wtr	drr	drr	drr	drr	drr	drr
Autumn	av	av	drr	av	av	drr	drr	drr	wtr	drr	drr	av	drr	drr	drr	drr
Winter	wtr	wtr	wtr	av	wtr	wtr	wtr	wtr	wtr	wtr	wtr	av	av	av	av	wtr
Spring	av	drr	av	av	wtr	drr	drr	drr	av	av	wtr	av	drr	wtr	av	av
Overall	av	av	av	av	drr	drr	wtr	wtr	wtr	av	av	av	drr	av	drr	drr

2025 — QLD - Estimated Seasonal Rainfall Expectations by %

REGION	Gulf Country	Northwest	Central West	Channel Country	Peninsula	Tropical Coast & Tablelnds	Goldfields / Upper Flinders	Herbet / Lower Burdekin	Central Coast-Whitsundays	Central Highlands	Capricornia	Maranoa / Warrego	Darling Dwns / Granite Belt	Wide Bay / Burnett	Southeast	QLD
Town Sample	Normanton	Mt Isa	Longreach	Birdsville	Weipa	Cairns	Charters Twrs	Townsville	Mackay	Emerald	Rockhampton	Charleville	Toowoomba	Maryborough	Brisbane	Overall
Summer	-21%	-42%	35%	26%	-33%	-23%	39%	26%	33%	33%	-36%	-22%	-55%	-51%	-22%	-12%
Autumn	-10%	-2%	-60%	-23%	3%	-23%	-36%	-55%	22%	-94%	-59%	-20%	-85%	-36%	-84%	-33%
Winter	333%	610%	68%	-99%	308%	89%	266%	117%	116%	94%	83%	-36%	-21%	-3%	7%	66%
Spring	-44%	-57%	-27%	-13%	105%	-41%	-56%	-78%	8%	-19%	35%	21%	-24%	117%	-8%	2%
Overall	-8%	1%	8%	-20%	-8%	-21%	31%	9%	34%	4%	-13%	4%	-48%	-9%	-34%	-6%

2025 QLD - Estimated Seasonal Rainfall Trend Variation by 25%

REGION	Gulf Country	Northwest	Central West	Channel Country	Peninsula	Tropical Coast & Tablelnds	Goldfields / Upper Flinders	Herbet / Lower Burdekin	Central Coast-Whitsundays	Central Highlands	Capricornia	Maranoa / Warrego	Darling Dwns / Granite Belt	Wide Bay / Burnett	Southeast	QLD
Town Sample	Normanton	Mt Isa	Longreach	Birdsville	Weipa	Cairns	Charters Twrs	Townsville	Mackay	Emerald	Rockhampton	Charleville	Toowoomba	Maryborough	Brisbane	Overall
Summer	av	drr	wtr	wtr	drr	av	wtr	wtr	wtr	wtr	drr	av	drr	drr	av	av
Autumn	av	av	drr	av	av	av	drr	drr	av	drr	drr	av	drr	drr	drr	drr
Winter	wtr	wtr	wtr	drr	wtr	wtr	wtr	wtr	wtr	wtr	wtr	drr	av	av	av	wtr
Spring	drr	drr	drr	av	wtr	drr	drr	drr	av	av	wtr	av	av	wtr	av	av
Overall	av	av	av	av	av	av	wtr	av	wtr	av	av	av	drr	av	drr	av

2025 QLD - Estimated Seasonal Maximum Temperature Average and Trend

Region	Gulf Country	Northwest	Central West	Channel Country	Peninsula	Tropical Coast & Tablelnds	Goldfields / Upper Flinders	Herbet / Lower Burdekin	Central Coast-Whitsundays	Central Highlands	Capricornia	Maranoa / Warrego	Darling Dwns / Granite Belt	Wide Bay / Burnett	Southeast	QLD
Town Sample	Normanton	Mt Isa	Longreach	Birdsville	Weipa	Cairns	Charters Twrs	Townsville	Mackay	Emerald	Rockhampton	Charleville	Toowoomba	Maryborough	Brisbane	Overall
Summer Est	34.2	35.1	33.2	31.2	30.8	29.5	31.6	32.7	36.0	36.1	38.5	35.0	27.8	30.2	28.0	32.7
Trend	wmr	av	a-c	av	a-c	av	av	clr	a-c	av	a-c	av	av	av	a-c	av
Autumn Est	32.3	33.8	30.9	30.0	30.6	27.6	30.7	32.4	34.2	34.5	34.1	31.9	26.2	28.7	27.4	31.0
Trend	av	av	a-w	a-w	wmr	wmr	wmr	wmr	wmr	wmr	wmr	wmr	wmr	wmr	wmr	wmr
Winter Est	30.7	28.6	23.3	25.7	24.8	21.1	23.0	22.2	23.1	23.8	21.8	20.2	16.8	22.3	21.4	23.2
Trend	a-c	clr	clr	av	a-c	a-c	a-c	clr	clr	clr	a-c	av	a-c	av	av	a-c
Spring Est	34.8	35.8	31.2	29.6	29.3	26.7	28.8	29.9	33.2	33.9	32.4	29.5	23.5	26.5	24.9	30.0
Trend	av	av	clr	av	av	a-c	a-c	clr	av	av	a-c	av	a-c	clr	a-c	a-c

2025 QLD - Estimated Seasonal Minimum Temperature Average and Trend

Region	Gulf Country	Northwest	Central West	Channel Country	Peninsula	Tropical Coast & Tablelnds	Goldfields / Upper Flinders	Herbet / Lower Burdekin	Central Coast-Whitsundays	Central Highlands	Capricornia	Maranoa / Warrego	Darling Dwns / Granite Belt	Wide Bay / Burnett	Southeast	QLD
Town Sample	Normanton	Mt Isa	Longreach	Birdsville	Weipa	Cairns	Charters Twrs	Townsville	Mackay	Emerald	Rockhampton	Charleville	Toowoomba	Maryborough	Brisbane	Overall
Summer Est	25.4	22.3	22.2	24.4	23.8	23.4	21.6	23.8	22.8	21.0	21.3	22.0	16.5	19.5	20.0	22.0
Trend	av	clr	av	a-c	av	av	a-c	av	av	a-c	a-c	a-w	a-c	a-c	a-c	av
Autumn Est	22.8	20.3	18.9	19.7	22.7	22.2	19.1	21.4	21.4	18.2	19.7	17.5	14.9	18.1	17.9	19.6
Trend	av	wmr	wmr	wmr	av	a-w	a-w	a-w	wmr	wmr	wmr	wmr	wmr	wmr	wmr	wmr
Winter Est	16.0	9.2	8.3	6.8	19.6	17.5	11.6	14.3	13.5	9.3	10.7	5.3	6.8	9.2	9.5	11.2
Trend	a-c	av	av	av	av	av	a-c	av	av	av	av	av	av	av	av	av
Spring Est	22.2	17.1	16.3	16.3	21.4	19.9	17.2	19.6	19.0	16.0	16.9	14.7	12.5	16.1	15.3	17.4
Trend	a-c	a-c	av	a-c	av	a-c	a-c	a-c	av	a-c	av	a-w	av	a-w	av	av

Monthly Breakdowns

2025 — QUEENSLAND - Estimated Rainfall Expectations (mm)

REGION	Gulf Country	Northwest	Central West	Channel Country	Peninsula	Tropical Coast & Tablelnds	Goldfields / Upper Flinders	Herbet / Lower Burdekin	Central Coast-Whitsundays	Central Highlands	Capricornia	Maranoa / Warrego	Darling Dwns / Granite Belt	Wide Bay / Burnett	Southeast	QLD
Town Sample	Normanton	Mt Isa	Longreach	Birdsville	Weipa	Cairns	Charters Twrs	Townsville	Mackay	Emerald	Rockhampton	Charleville	Toowoomba	Maryborough	Brisbane	Overall
January	244	122	74	53	115	83	29	53	68	77	34	58	33	22	83	1148
February	244	51	153	21	628	612	461	761	788	177	126	46	66	111	173	4417
March	130	59	39	19	450	339	78	85	164	1	56	55	4	115	45	1639
April	40	13	0	1	98	149	4	11	142	5	4	0	3	28	8	507
May	6	4	0	9	1	42	3	3	133	6	17	31	25	41	8	329
June	40	82	20	0	33	156	115	79	181	75	114	8	32	69	95	1099
July	20	31	42	0	3	7	93	32	83	71	57	25	29	41	53	588
August	1	4	19	0	3	27	2	4	3	16	10	14	33	27	30	192
September	6	0	10	0	2	11	7	2	51	35	57	20	51	279	107	639
October	0	0	6	0	47	30	15	14	46	22	25	22	44	45	17	332
November	27	30	39	28	193	73	16	10	65	53	119	95	57	97	69	971
December	95	73	58	2	208	51	32	118	53	153	25	133	79	49	63	1192
Overall	853	470	459	133	1781	1581	857	1173	1777	691	644	508	454	925	748	13053

2025 — QUEENSLAND - Estimated Rainfall Variation from the Average (mm)

REGION	Gulf Country	Northwest	Central West	Channel Country	Peninsula	Tropical Coast & Tablelnds	Goldfields / Upper Flinders	Herbet / Lower Burdekin	Central Coast-Whitsundays	Central Highlands	Capricornia	Maranoa / Warrego	Darling Dwns / Granite Belt	Wide Bay / Burnett	Southeast	QLD
Town Sample	Normanton	Mt Isa	Longreach	Birdsville	Weipa	Cairns	Charters Twrs	Townsville	Mackay	Emerald	Rockhampton	Charleville	Toowoomba	Maryborough	Brisbane	Overall
January	-16	5	1	28	-371	-296	-124	-197	-149	7	-44	-14	-67	-92	-31	-1360
February	-6	-55	81	-7	127	164	321	469	459	97	16	-22	-50	-46	28	1577
March	-28	4	6	2	43	-50	4	-25	24	-56	-23	15	-63	1	-76	-223
April	9	0	-43	-8	-9	-66	-26	-64	-4	-48	-51	-33	-67	-49	-149	-607
May	-2	-6	-24	-2	-20	-47	-27	-30	61	-73	-36	-3	-52	-54	-91	-405
June	31	74	0	-10	30	115	96	60	117	48	72	-23	-10	12	32	642
July	17	26	25	-11	2	-24	70	19	59	30	32	-1	-17	-2	-15	210
August	-1	1	8	-6	-1	-2	-13	-16	-32	1	-22	-2	2	-15	-5	-103
September	3	-5	-2	-6	-0	-24	-2	-9	15	19	30	-4	14	241	78	348
October	-11	-20	-16	-12	26	-18	-3	-12	18	-33	-27	-10	-23	-28	-55	-223
November	-18	-16	-3	14	98	-39	-45	-69	-22	-12	49	37	-39	14	-40	-91
December	-49	-4	2	-14	-75	-127	-47	-23	-90	51	-95	79	-43	-75	-60	-570
Overall	-70	5	35	-32	-150	-412	204	101	455	29	-97	19	-416	-92	-383	-805

2025	QUEENSLAND - Estimated Rainfall Trend Variation by 10mm															
REGION	Gulf Country	Northwest	Central West	Channel Country	Peninsula	Tropical Coast & Tablelnds	Goldfields / Upper Flinders	Herbet / Lower Burdekin	Central Coast-Whitsundays	Central Highlands	Capricornia	Maranoa / Warrego	Darling Dwns / Granite Belt	Wide Bay / Burnett	Southeast	QLD
Town Sample	Normanton	Mt Isa	Longreach	Birdsville	Weipa	Cairns	Charters Twrs	Townsville	Mackay	Emerald	Rockhampton	Charleville	Toowoomba	Maryborough	Brisbane	Overall
January	drr	av	av	wtr	drr	drr	drr	drr	drr	av	drr	drr	drr	drr	drr	drr
February	av	drr	wtr	av	wtr	wtr	wtr	wtr	wtr	wtr	wtr	drr	drr	drr	wtr	wtr
March	drr	av	av	av	wtr	drr	av	drr	wtr	drr	drr	wtr	drr	av	av	drr
April	av	av	drr	av	av	drr	drr	drr	av	drr	drr	drr	drr	drr	drr	drr
May	av	av	drr	av	drr	drr	drr	drr	wtr	drr	drr	av	drr	drr	drr	drr
June	wtr	wtr	av	av	wtr	wtr	wtr	wtr	wtr	wtr	wtr	drr	av	wtr	wtr	wtr
July	wtr	wtr	wtr	drr	av	drr	wtr	wtr	wtr	wtr	wtr	av	drr	av	drr	wtr
August	av	av	av	av	av	av	drr	drr	drr	av	drr	av	av	drr	av	drr
September	av	av	av	av	av	drr	av	av	wtr	wtr	wtr	av	wtr	wtr	wtr	wtr
October	drr	drr	drr	drr	wtr	drr	av	drr	wtr	drr	drr	av	drr	drr	drr	drr
November	drr	drr	av	wtr	wtr	drr	drr	drr	drr	drr	wtr	wtr	drr	wtr	drr	drr
December	drr	av	av	drr	drr	drr	drr	drr	drr	wtr	drr	wtr	drr	drr	drr	drr
Overall	av	av	av	av	drr	drr	wtr	wtr	wtr	av	av	av	drr	av	drr	drr

2025	QUEENSLAND - Estimated Rainfall Expectations by %															
REGION	Gulf Country	Northwest	Central West	Channel Country	Peninsula	Tropical Coast & Tablelnds	Goldfields / Upper Flinders	Herbet / Lower Burdekin	Central Coast-Whitsundays	Central Highlands	Capricornia	Maranoa / Warrego	Darling Dwns / Granite Belt	Wide Bay / Burnett	Southeast	QLD
Town Sample	Normanton	Mt Isa	Longreach	Birdsville	Weipa	Cairns	Charters Twrs	Townsville	Mackay	Emerald	Rockhampton	Charleville	Toowoomba	Maryborough	Brisbane	Overall
January	-6%	4%	2%	115%	-76%	-78%	-81%	-79%	-69%	10%	-57%	-19%	-67%	-80%	-27%	-54%
February	-2%	-52%	112%	-24%	25%	37%	231%	161%	140%	120%	15%	-32%	-43%	-29%	19%	56%
March	-18%	7%	16%	9%	11%	-13%	5%	-23%	17%	-98%	-29%	37%	-94%	1%	-63%	-12%
April	29%	0%	-99%	-87%	-8%	-31%	-85%	-86%	-3%	-91%	-93%	-100%	-96%	-63%	-95%	-55%
May	-20%	-58%	-98%	-19%	-96%	-52%	-90%	-91%	83%	-92%	-68%	-9%	-68%	-57%	-92%	-55%
June	333%	893%	1%	-96%	854%	281%	497%	314%	180%	175%	171%	-75%	-24%	21%	50%	141%
July	525%	457%	141%	-100%	100%	-76%	292%	140%	239%	72%	130%	-4%	-37%	-4%	-22%	55%
August	-29%	38%	72%	-100%	-28%	-7%	-87%	-79%	-92%	7%	-70%	-14%	6%	-36%	-13%	-35%
September	93%	-100%	-15%	-100%	-5%	-68%	-20%	-81%	43%	116%	113%	-16%	36%	627%	268%	119%
October	-100%	-98%	-75%	-100%	126%	-37%	-18%	-46%	67%	-60%	-52%	-31%	-34%	-38%	-77%	-40%
November	-40%	-34%	-6%	100%	103%	-35%	-73%	-87%	-25%	-18%	69%	65%	-41%	17%	-37%	-9%
December	-34%	-5%	3%	-88%	-26%	-71%	-59%	-16%	-63%	49%	-79%	146%	-35%	-60%	-49%	-32%
Overall	-8%	1%	8%	-20%	-8%	-21%	31%	9%	34%	4%	-13%	4%	-48%	-9%	-34%	-6%

2025	QUEENSLAND - Estimated Rainfall Trend Variation by 25%															
REGION	Gulf Country	Northwest	Central West	Channel Country	Peninsula	Tropical Coast & Tablelnds	Goldfields / Upper Flinders	Herbet / Lower Burdekin	Central Coast-Whitsundays	Central Highlands	Capricornia	Maranoa / Warrego	Darling Dwns / Granite Belt	Wide Bay / Burnett	Southeast	QLD
Town Sample	Normanton	Mt Isa	Longreach	Birdsville	Weipa	Cairns	Charters Twrs	Townsville	Mackay	Emerald	Rockhampton	Charleville	Toowoomba	Maryborough	Brisbane	Overall
January	av	av	av	wtr	drr	drr	drr	drr	drr	av	drr	av	drr	drr	drr	drr
February	av	drr	wtr	av	av	wtr	wtr	wtr	wtr	wtr	av	drr	drr	drr	av	wtr
March	av	av	av	av	av	av	av	av	av	drr	drr	wtr	drr	av	drr	av
April	wtr	av	drr	drr	av	drr	drr	drr	av	drr	drr	drr	drr	drr	drr	drr
May	av	drr	drr	av	drr	drr	drr	drr	wtr	drr	drr	av	drr	drr	drr	drr
June	wtr	wtr	av	drr	wtr	wtr	wtr	wtr	wtr	wtr	wtr	drr	av	av	wtr	wtr
July	wtr	wtr	wtr	drr	wtr	drr	wtr	wtr	wtr	wtr	wtr	av	drr	av	av	wtr
August	drr	wtr	wtr	drr	drr	av	drr	drr	drr	av	drr	av	av	drr	av	drr
September	wtr	drr	av	drr	av	drr	av	drr	wtr	wtr	wtr	av	wtr	wtr	wtr	wtr
October	drr	drr	drr	drr	wtr	drr	av	drr	wtr	drr	drr	drr	drr	drr	drr	drr
November	drr	drr	av	wtr	wtr	drr	drr	drr	av	av	wtr	wtr	drr	av	drr	av
December	drr	av	av	drr	drr	drr	drr	av	drr	wtr	drr	wtr	drr	drr	drr	drr
Overall	av	av	av	av	av	av	wtr	av	wtr	av	av	av	drr	av	drr	av

2025 QUEENSLAND - Estimated Average Maximum Temperatures

Region	Gulf Country	Northwest	Central West	Channel Country	Peninsula	Tropical Coast & Tablelnds	Goldfields / Upper Flinders	Herbet / Lower Burdekin	Central Coast-Whitsundays	Central Highlands	Capricornia	Maranoa / Warrego	Darling Dwns / Granite Belt	Wide Bay / Burnett	Southeast	QLD Estimates
Town Sample	Normanton	Mt Isa	Longreach	Birdsville	Weipa	Cairns	Charters Twrs	Townsville	Mackay	Emerald	Rockhampton	Charleville	Toowoomba	Maryborough	Brisbane	Overall
January	34	34	35	37	35	32	34	32	30	33	32	34	26	30	27	**32**
February	33	34	34	37	32	31	31	30	29	31	31	35	29	32	29	**32**
March	34	37	37	39	32	31	32	32	30	35	32	36	29	31	29	**33**
April	33	33	34	33	32	30	30	31	27	32	30	32	26	28	27	**31**
May	34	33	32	30	33	29	30	30	26	30	30	28	23	27	26	**29**
June	30	24	23	21	32	26	24	26	22	23	24	20	18	22	22	**24**
July	25	21	21	20	29	24	20	23	19	20	21	18	14	20	20	**21**
August	31	26	26	25	31	26	25	26	22	24	25	23	19	24	22	**25**
September	34	31	29	29	34	28	28	27	24	26	26	26	20	24	23	**27**
October	37	35	35	35	35	30	32	29	27	32	30	32	26	27	26	**31**
November	37	36	35	33	36	31	34	31	29	32	30	31	25	28	26	**32**
December	37	39	38	39	35	32	35	31	30	33	31	32	26	29	28	**33**
Overall	**33**	**32**	**32**	**32**	**33**	**29**	**30**	**29**	**26**	**29**	**28**	**29**	**23**	**27**	**25**	**29**

2025 QUEENSLAND - Estimated Average Minimum Temperatures

Region	Gulf Country	Northwest	Central West	Channel Country	Peninsula	Tropical Coast & Tablelnds	Goldfields / Upper Flinders	Herbet / Lower Burdekin	Central Coast-Whitsundays	Central Highlands	Capricornia	Maranoa / Warrego	Darling Dwns / Granite Belt	Wide Bay / Burnett	Southeast	QLD Estimates
Town Sample	Normanton	Mt Isa	Longreach	Birdsville	Weipa	Cairns	Charters Twrs	Townsville	Mackay	Emerald	Rockhampton	Charleville	Toowoomba	Maryborough	Brisbane	Overall
January	26	23	22	25	25	24	22	25	23	21	21	22	16	19	20	**22**
February	25	21	21	24	24	24	22	24	23	22	23	21	18	21	22	**22**
March	25	24	22	25	24	24	22	24	24	21	23	22	18	21	21	**23**
April	23	20	19	18	23	22	18	21	21	17	19	17	14	17	17	**19**
May	21	17	16	16	21	21	17	19	19	16	18	14	13	16	16	**17**
June	19	11	10	7	22	20	14	17	16	11	14	6	8	12	12	**13**
July	12	7	5	5	17	14	8	10	10	7	7	3	4	6	7	**8**
August	17	10	9	8	19	18	13	15	14	10	11	7	8	9	9	**12**
September	19	12	12	13	19	17	14	16	16	13	14	11	10	14	13	**14**
October	23	18	17	16	21	20	17	20	19	16	17	15	13	16	15	**18**
November	25	21	20	20	23	22	20	23	22	19	20	19	15	18	18	**20**
December	26	22	23	24	24	23	22	24	23	21	21	21	16	19	20	**22**
Overall	22	17	16	17	22	21	18	20	19	16	17	15	13	16	16	**18**

2025 QUEENSLAND - Estimated Degree of Variation from the Maximum Average

Region	Gulf Country	Northwest	Central West	Channel Country	Peninsula	Tropical Coast & Tablelnds	Goldfields / Upper Flinders	Herbet / Lower Burdekin	Central Coast-Whitsundays	Central Highlands	Capricornia	Maranoa / Warrego	Darling Dwns / Granite Belt	Wide Bay / Burnett	Southeast	QLD
Town Sample	Normanton	Mt Isa	Longreach	Birdsville	Weipa	Cairns	Charters Twrs	Townsville	Mackay	Emerald	Rockhampton	Charleville	Toowoomba	Maryborough	Brisbane	Overall
January	0.3	-2.8	-2.0	-3.4	3.3	0.5	0.2	0.1	0.0	-2.0	-0.4	-1.5	-2.3	-0.8	-1.8	-0.8
February	-0.5	-1.1	-1.5	-1.6	0.3	-0.2	-1.7	-1.3	-0.5	-2.6	0.0	0.6	1.9	1.3	0.1	-0.4
March	-0.3	2.8	2.3	3.6	0.1	0.4	0.3	0.9	0.9	1.9	1.3	3.7	3.0	1.2	0.9	1.5
April	-1.0	1.5	2.3	1.3	-0.4	0.3	0.0	0.8	0.2	2.3	1.6	3.7	2.8	0.8	1.4	1.2
May	2.3	4.7	4.7	3.8	1.1	1.7	2.8	2.0	2.2	3.9	3.6	4.6	3.4	2.7	2.5	3.1
June	0.2	-0.9	-0.6	-0.9	0.6	0.3	-0.7	0.2	0.1	-0.9	0.0	-0.3	0.5	0.1	0.9	-0.1
July	-4.5	-3.6	-3.2	-2.2	-1.9	-1.6	-4.6	-2.5	-2.0	-3.8	-2.8	-1.6	-2.7	-1.8	-1.2	-2.7
August	-1.0	-1.3	-0.4	0.5	-0.8	-0.2	-1.6	-0.2	-1.0	-1.1	0.0	0.9	-0.2	0.7	0.3	-0.4
September	-1.0	-0.9	-0.9	-0.1	-0.7	0.3	-2.4	-0.4	-1.3	-3.0	-1.5	-0.6	-2.4	-1.7	-1.3	-1.2
October	-0.2	0.1	1.0	1.0	-0.5	0.0	-0.7	0.0	-0.5	0.0	0.6	1.7	1.1	-0.2	0.3	0.2
November	-0.1	-0.5	-1.1	-3.2	-0.3	0.2	-0.5	0.1	-0.6	-1.6	-1.1	-1.6	-1.6	-1.4	-1.0	-0.9
December	0.1	1.6	0.3	-0.2	0.6	0.6	0.3	-0.4	-0.2	-1.5	-1.1	-3.0	-1.9	-1.2	-0.5	-0.4
Overall	-0.5	0.0	0.1	-0.2	0.1	0.1	-0.7	-0.1	-0.2	-0.7	0.0	0.6	0.1	0.0	0.1	-0.1

2025 QUEENSLAND - Estimated Degree of Variation from the Minimum Average

Region	Gulf Country	Northwest	Central West	Channel Country	Peninsula	Tropical Coast & Tablelnds	Goldfields / Upper Flinders	Herbet / Lower Burdekin	Central Coast-Whitsundays	Central Highlands	Capricornia	Maranoa / Warrego	Darling Dwns / Granite Belt	Wide Bay / Burnett	Southeast	QLD
Town Sample	Normanton	Mt Isa	Longreach	Birdsville	Weipa	Cairns	Charters Twrs	Townsville	Mackay	Emerald	Rockhampton	Charleville	Toowoomba	Maryborough	Brisbane	Overall
January	0.3	-1.1	-0.9	-2.2	0.2	0.2	-0.1	0.3	-0.4	-1.6	-1.1	0.3	-1.7	-1.5	-1.7	-0.7
February	-0.1	-1.9	-1.2	-1.2	0.1	0.1	-0.1	0.1	-0.2	-0.3	0.3	-0.2	0.4	0.8	0.4	-0.2
March	0.0	1.8	1.6	2.9	0.3	0.9	1.0	1.2	1.5	0.8	1.6	3.2	1.0	1.4	1.0	1.4
April	0.0	1.3	2.2	0.5	0.0	0.1	-0.2	0.5	0.9	0.2	1.0	2.8	0.3	0.6	0.4	0.7
May	1.6	3.6	3.8	4.8	-0.4	0.7	1.9	1.1	1.9	3.0	3.4	4.8	3.5	3.0	3.0	2.6
June	2.0	1.0	2.1	-1.0	2.3	2.4	1.8	2.6	2.2	1.1	2.9	0.3	0.7	2.1	1.3	1.6
July	-4.0	-1.8	-2.2	-1.6	-1.7	-3.1	-3.9	-3.4	-3.2	-2.4	-2.9	-1.4	-2.5	-3.0	-2.2	-2.6
August	0.2	-0.5	0.9	-0.5	0.7	0.8	0.3	0.8	0.1	-0.1	0.5	1.0	0.6	0.1	-0.6	0.3
September	-1.1	-1.9	-0.3	-0.2	-0.4	-1.2	-1.4	-1.6	-0.7	-0.9	0.5	0.7	-0.8	1.4	-0.3	-0.5
October	-0.4	-0.3	-0.6	-1.2	-0.4	-0.7	-0.8	-1.1	-0.6	-0.8	-0.3	0.2	0.3	0.9	-0.7	-0.4
November	-0.6	-0.5	0.0	-1.5	0.0	0.2	0.0	0.5	0.1	-0.4	0.1	0.8	-0.2	0.6	-0.2	-0.1
December	-0.2	-0.7	0.4	0.0	-0.1	-0.1	0.1	0.4	-0.2	-0.8	-0.4	0.0	-0.4	-0.3	-0.2	-0.2
Overall	-0.2	-0.1	0.5	-0.1	0.1	0.0	-0.1	0.1	0.1	-0.2	0.5	1.0	0.1	0.5	0.0	0.2

2025	QUEENSLAND - Estimated Maximum Temperature Trend															
Region	Gulf Country	Northwest	Central West	Channel Country	Peninsula	Tropical Coast & Tablelnds	Goldfields / Upper Flinders	Herbet / Lower Burdekin	Central Coast-Whitsundays	Central Highlands	Capricornia	Maranoa / Warrego	Darling Dwns / Granite Belt	Wide Bay / Burnett	Southeast	QLD
Town Sample	Normanton	Mt Isa	Longreach	Birdsville	Weipa	Cairns	Charters Twrs	Townsville	Mackay	Emerald	Rockhampton	Charleville	Toowoomba	Maryborough	Brisbane	Overall
January	av	clr	clr	clr	wmr	av	av	av	av	clr	av	clr	clr	a-c	clr	a-c
February	av	clr	clr	clr	av	av	clr	clr	av	clr	av	a-w	wmr	wmr	av	av
March	av	wmr	wmr	wmr	av	av	av	a-w	a-w	wmr	wmr	wmr	wmr	wmr	a-w	wmr
April	a-c	wmr	wmr	wmr	av	av	av	a-w	av	wmr	wmr	wmr	wmr	a-w	wmr	wmr
May	wmr	wmr	wmr	wmr	wmr	wmr	wmr	wmr	wmr	wmr	wmr	wmr	wmr	wmr	wmr	wmr
June	av	a-c	a-c	a-c	a-w	av	a-c	av	av	a-c	av	av	av	av	a-w	av
July	clr	clr	clr	clr	clr	clr	clr	clr	clr	clr	clr	clr	clr	clr	clr	clr
August	a-c	clr	av	av	a-c	av	clr	av	a-c	clr	av	a-w	av	a-w	av	av
September	a-c	a-c	a-c	av	a-c	av	clr	av	clr	clr	clr	a-c	clr	clr	clr	clr
October	av	av	a-w	a-w	av	av	a-c	av	av	av	a-w	wmr	wmr	av	av	av
November	av	av	clr	clr	av	av	av	av	a-c	clr	clr	clr	clr	clr	a-c	a-c
December	av	wmr	av	av	a-w	a-w	av	av	av	clr	clr	clr	clr	clr	av	av
Overall	av	av	av	av	av	av	a-c	av	av	a-c	av	a-w	av	av	av	av

2025	QUEENSLAND - Estimated Minimum Temperature Trend															
Region	Gulf Country	Northwest	Central West	Channel Country	Peninsula	Tropical Coast & Tablelnds	Goldfields / Upper Flinders	Herbet / Lower Burdekin	Central Coast-Whitsundays	Central Highlands	Capricornia	Maranoa / Warrego	Darling Dwns / Granite Belt	Wide Bay / Burnett	Southeast	QLD
Town Sample	Normanton	Mt Isa	Longreach	Birdsville	Weipa	Cairns	Charters Twrs	Townsville	Mackay	Emerald	Rockhampton	Charleville	Toowoomba	Maryborough	Brisbane	Overall
January	av	clr	a-c	clr	av	av	av	av	av	clr	clr	av	clr	clr	clr	a-c
February	av	clr	clr	clr	av	av	av	av	av	av	av	av	av	a-w	av	av
March	av	wmr	wmr	wmr	av	a-w	a-w	wmr	wmr	a-w	wmr	wmr	a-w	wmr	a-w	wmr
April	av	wmr	wmr	av	av	av	av	av	a-w	av	a-w	wmr	av	a-w	av	a-w
May	wmr	wmr	wmr	wmr	av	a-w	wmr	wmr	wmr	wmr	wmr	wmr	wmr	wmr	wmr	wmr
June	wmr	wmr	wmr	a-c	wmr	wmr	wmr	wmr	wmr	wmr	wmr	av	a-w	wmr	wmr	wmr
July	clr	clr	clr	clr	clr	clr	clr	clr	clr	clr	clr	clr	clr	clr	clr	clr
August	av	av	a-w	av	a-w	a-w	av	a-w	av	av	av	a-w	a-w	av	a-c	av
September	clr	clr	av	av	av	clr	clr	clr	a-c	a-c	av	a-w	a-c	wmr	av	av
October	av	av	a-c	clr	av	a-c	a-c	clr	a-c	a-c	av	av	av	a-w	a-c	av
November	a-c	av	av	clr	av	av	av	av	av	av	av	a-w	av	a-w	av	av
December	av	a-c	av	av	av	av	av	av	av	a-c	av	av	av	av	av	av
Overall	av	av	av	av	av	av	av	av	av	av	av	a-w	av	av	av	av

NEW SOUTH WALES / ACT 2025

GENERAL TRENDS

Year Overall:	Slightly wetter but within 5-10% of the average Slightly warmer but within 0.5°C of the average
Summer:	Drier by around 30-35% below the average Wettest – Second half of February Driest – Second half of January Warmer by around 0.5-1°C above the average Hottest – First week of December and second half of January Coolest – First week of January
Autumn:	Slightly wetter but within 5% of the average Wettest –Second week of March and first half of May Driest – Second half of April Warmer by up to 2-2.5°C above the average Hottest – Mid-March Coolest – Second week of May and last week of May
Winter:	Wetter by up to 50% more than the average Wettest – Second half of June Driest – Third week of August Slightly cooler but within 0.5°C of the average Hottest – First week of June and mid-August Coolest – Second half of July
Spring:	Drier by around 20-25% below the average Wettest –Second and third weeks of November Driest – Second week of September and first half of October Slightly warmer but within 0.5°C of the average Hottest – Second half of November Coolest – Mid-September

Major Weather Events

January 2025

Date	Day	New South Wales / ACT - January 2025	Sydney	Canberra	Moon
1st	Wed	Outbreaks of rain in the North Coast the adjacent Tableland and along the South Coast to the Illawarra. A wet front moves across the western half of the State over the next couple of days.			
2nd	Thur		Showers	Showers, Th	
3rd	Fri	Rain clears in the west while moves over the Tablelands and into the regions to the east and up along the coasts over the next couple of days, with heavy rain at times in the Metropolitan to the Hunter and Northern Tablelands.	Rain, Th		
4th	Sat	Strong winds to gale force at times in the Snowy Mountains over the next couple of days, as well as briefly for a time around Broken Hill.			
5th	Sun	Very heavy rain in the Far Northwest and in some parts of the Lower Darling. Strong and unsettled winds in the Snowy Mountains. Wet in the Illawarra to the South Coast with occasional downpours.			
6th	Mon	Heavy rain in the Illawarra. Occasional downpours in the Hunter Valley and up into the Central Tablelands and in the Sydney regions.			
7th	Tue	Severe cold spell for the next few days over the State with the chance of frost and snow in southern parts of the State to the Snowy Mountains possible. Heavy rain at times in the Hunter Valley and the Mid North Coast to the Central Tablelands. Strong winds around the South Coast over the next couple of days.	Showers, Th		**1Q XhN**
8th	Wed		Showers, Th	Rain, Th	**P13**
9th	Thu	Rain may be heavy for a time in some parts of the Illawarra.		Showers, Th	
10th	Fri		Showers	Th	
11th	Sat	Thunderstorms with heavy rain, strong wind gusts, and hail over the next few days in the southeast of the State, including in the Sydney to the Illawarra and South Coast areas, which may bring some damage to property. Chance of flash flooding and very large hail in the Canberra area and in the Southern and Central Tablelands. Possible frost in the Charlotte Pass area of the Snowy Mountains.	Rain		
12th	Sun	Patchy outbreaks of heavy rain and hail with the passage of storms through the central east of the State. Chance of large hail in the Southern Tablelands.			
13th	Mon	A trough over the west extends instability over much of the State with widespread scattered thunderstorms prevailing. Some outbursts of heavy rain in the Central Tablelands and possible storms in the Broken Hill area may bring cause localised flash flooding in some parts of the Lower Western. Storms may bring hail and flash flooding to the Sydney region, as well as into ACT, and the Illawarra to the South Coast, with distruptions to infrastructure and damage to local property possible. Heavy rain with downpours also in the Mid North Coast to the Northern Tablelands.	Showers		**^**
14th	Tue	The next 3-4 weeks may be hotter than normal for this time of year, with the western half of the State much wetter than average, while the eastern half may be drier than normal. Widespread storms and heavy rain, particularly in the Central Tablelands, where showers of large hail may occur.			**F**
15th	Wed	Storms on the Central Tablelands may produce large size hail.			
16th	Thur	Unusually cool day in the Northern Tablelands to the North Coast regions, with maximum temperatures in the mid to upper teens likely in a number of districts.		Showers	
17th	Fri	Showers through the south and west of the State.	Showers	Showers	
18th	Sat	Showers clear in the west but move over the Tablelands to the coast.			
19th	Sun	Mostly dry in the State over the coming week with only a few exceptions. Chance of brief rain along the North Coast over the next couple of days.			
20th	Mon				**XhS**
21st	Tue	Cool respite over the next couple of nights in the Far Northwest.			**A**
22nd	Wed	Chance of frosts overnight in the Snowy Mountains, particularly in the Charlotte Pass district, and an unusually cool day in the Upper North Coast, where maximum temperatures may only reach into the low 20's in a number of areas.			**3Q**
23rd	Thu	Very hot spell in the western half of the State over the next few days with maximum temperatures reaching into the mid-40's in many districts.			
24th	Fri	Very hot day in the State, particularly in the Far Northwest region, where maximum temperatures in the mid to upper 40's are possible, including in the Wilcannia district and in the low 40's in many districts in the Riverina to the Southwest Slopes.			
25th	Sat	Storm with strong wind gusts in the Central Wesst Slopes including the Coonabarabran area. Very hot in the Hunter, Southwest Slopes, Southern and Central Tableland regions around this time, where maximum temperatures in the low to mid 40's are likely in a number of districts.			
26th	Sun	Pockets of rain in the Northern Tablelands to the Northwest over the next couple of days.			
27th	Mon			Th	**V**
28th	Tue	Storm with hail and damaging winds on the Central West Plains. Some patches of heavy rain in the Upper and Lower Western.		Showers, Th	
29th	Wed	Very strong winds with gusts in the Southwest and Central West Slopes, which may cause power outages and damage to property, particularly in the Wagga Wagga and Trangie areas. Outbreaks of rain through the Lower Western, across the Riverina and into the Southern Tablelands to the South Coast.			
30th	Thur	Pockets of heavy rain on the Central Tablelands. Very warm overnight on the Southwest Plains and Slopes.			**N**
31st	Fri	Strong blustery winds for a time over the Central Tablelands.			

February 2025

Date	Day	New South Wales / ACT - February 2025	Sydney	Canberra	Moon
1st	Sat	Heavy rain and possible pockets of localised flash flooding in the Lower Western, particularly in the Broken Hill area around this time.			
2nd	Sun		Showers	Showers	**P10**
3rd	Mon	Showers in the Central West with heavier rain into the Central Tablelands and into the Hunter and North Coast regions over the next couple of days. Very heavy rain in the Far Northwest around this time, particularly in the Tibooburra area. Very hot in the Metropolitan region around this time, with maximum temperatures into the low 40's likely in many areas.	Showers		**XhN**
4th	Tue		Showers		
5th	Wed				**1Q**
6th	Thu	Mostly dry throughout the State over the next 5-6 days.			
7th	Fri				
8th	Sat	Very hot over the next couple of days in the North Coast region with maximum temperatures into the upper 30's to early 40's likely in many districts.			
9th	Sun	Windy with possible thunderstorms for a time on the Mid-North Coast region, particularly in the Taree area and very strong and blustery winds around the South Coast. Unusually cold day in the Snowy Mountains, particularly in the Thredbo district, where maximum temperatures may not reach into double digits.			**^**
10th	Mon	Cooler minimum temperatures overnight in the Tablelands, with a number of districts likely to see overnight temperatures dropping into the low single digits. Chance of patches of frost in the Southern Tablelands.	Showers		
11th	Tue	Very cold with possible patches of frost in the Snowy Mountains, particularly in the Perisher Valley area.	Showers		
12th	Wed	Chance of rain in the Northwest to the North Coast regions, as well as showers in the Greater Sydney region to the Illawarra.	Th	Showers, Th	
13th	Thur	The next 3-4 weeks may see fairly average rainfall for this time of year despite active inland troughs bringing above normal amounts of thunderstorms, and with both maximum and minimum temperatures warmer than average.		Showers, Th	**F**
14th	Fri	The passage of a very unstable storm system crosses the Central West Plains to the Central Tablelands, which has the potential to turn tornadic, brings severe wind gusts and hail, with chance of very large hail stones in the Orange district. Heavy rain and localised flooding in the Northwest Slopes, particularly in the Tamworth area.			
15th	Sat	Mostly dry over the next 3-4 days.			
16th	Sun	A very hot spell settles over the State over the coming week, with many districts likely to reach maximum temperatures into the upper 20's to 30's in eastern areas, and from the upper 30's to early 40's in western districts.			**XhS**
17th	Mon	Very warm night in the Upper Darlings, particularly in the Cobar district, where minimum temperatures may remain near the late 20's to early 30's.	Showers, Th		
18th	Tue	Very hot on the Southwest Plains, with many districts possibly seeing maximum temperatures in the mid 40's, including in the Ivanhoe area.	Showers		**A**
19th	Wed	Pockets of rain in the northeast quarter of the State over the next couple of days.		Showers, Th	
20th	Thu	Heavy rain for a time in the Hunter, with accompanying strong winds and gusts which may cause some damage to property, possibly in the Scone area. Blustery in the Central West Plains and chance of hail showers.		Showers, Th	
21st	Fri	Stormy in the Central Tablelands with hail, heavy rain, and possible flash flooding. Heavy rain in the South Coast districts.	Rain	Showers	**3Q**
22nd	Sat	Storm brings heavy rain and some hail sowers around the Mid-North Coast to Northwest Slopes areas.	Rain		
23rd	Sun	Stormy air with heavy rain and strong wind gusts in the Central and Northern Tablelands which may cause pockets of flooding and power outages. Chance of large size hail showers with potential to cause damage to property.	Showers		**V**
24th	Mon	Patches of showers in the northeastern quarter of the State. Unusually cold day in the Snowy Mountains, particularly in the Thredbo district, where maximum temperatures may not reach into double digits. Heavy rain likely over the next couple of days from the Metropolitan to the South Coast regions.		Th	
25th	Tue	Chance of hail in the Central Tablelands and rain in the Lower Western.			
26th	Wed	Isolated showers in the Lower Western, Riverina and South West Slopes. Very cold over the next couple of nights in the Snowy Mountains, with some areas, including at Perisher, possibly being cold enough to form frost patches. It may also be unusually cool overnight on the Upper North Coast.		Th	
27th	Thur	Showers along the Tablelands and on the North Coast.			
28th	Fri	Heavy rain with flash flooding in some parts of the Mid-North Coast to the Northern Tablelands regions.		Showers	**N**

March 2025

Date	Day	New South Wales / ACT - March 2025	Sydney	Canberra	Moon
1st	Sat	Some shower activity on most days in the coming week in the eastern half of the State.		Th	
2nd	Sun			Th	**P7 XhN**
3rd	Mon			Rain, Fg	
4th	Tue	Patches of heavy rain likely in the South Coast and along the Tablelands.		Th	
5th	Wed	Isolated showers east of the Tablelands.		Rain, Th	
6th	Thu		Showers, Th	Th	
7th	Fri	Another warmer spell across the State over the next couple of days, particularly in the eastern districts, with maximum temperatures settling in the low to mid 30's in many places.	Rain		**1Q**
8th	Sat	Bursts of heavy rain in the Southern Tablelands likey that may cause some localised flooding. Chance of hail in the Central Tablelands.	Rain	Showers	**^**
9th	Sun	Stormy with heavy rain throughout the eastern half of the State. Heavy rain in the Illawarra and North Coast regions over the next couple of days and storms with severe wind gusts in the Metropolitan, particularly in the Penrith area.		Rain, Fg	
10th	Mon	Storms crossing the Central Tablelands to the Mid-North Coast brings heavy falls at times with chance of hail showers in the Mid-North Coast.	Rain, Th	Th	
11th	Tue	Storms with strong winds and heavy rain in the Mid-North Coast to the Northern Tablelands and over the Northwest Plains.	Showers	Showers	
12th	Wed	A supercell storm crosses the Southern Tablelands to ACT region and may bring heavy rain, localised flooding and large hail.		Fg	
13th	Thur	Storms with outbreaks of heavy rain, hail and strong gusty winds cross the Northwest Slopes, Northern Rivers, Hunter, Metropolitan, Illawarra, and Central Tablelands including in ACT, bringing some flash flooding in in the regions, including in parts of Canberra, the Metropolitan, and Grafton and chance of hail with damage potential on the Northwest Slopes.			
14th	Fri	Storms with heavy rain, hail, and strong blustery winds in the Metropolitan region. Unsettled winds in the Northwest Slopes may generate tornadic activity. Strong gusty winds in the Central West Slopes region. Very heavy rain likely in the Northern Tablelands, particularly in the Inverell district.	Rain, Th		**F**
15th	Sat	Hail showers in the Northern Rivers region, with very large hail possible in the Grafton area. A very hot spell settles over the State during the next 4-5 days, with many districts seeing daily maximum temperatures into the mid to late 30's and some, particularly in the Riverina region and the inland western parts, into the low to mid 40's.	Rain, Th		**XhS**
16th	Sun	The next 3-4 weeks may see fairly average rainfall across most of the State, except in some parts of the Murray Darling Basin, which may be unusually wetter than normal, and once again, the following few weeks may be warmer than is typical for this time of year, both for maximum and minimum temperatures. It may be exceptionally hot around this time in the Far Northwest to the Lower Darling regions, particularly in the Wilcannia and Menindee areas, with maximum temperatures into the mid-40's.	Showers		
17th	Mon	Storms with very strong wind gusts in the Illawarra. Very warm overnight in the Far Northwest, particularly in the Tibooburra district, where the minimum temperatures may near 30°C.	Showers		
18th	Tue	Storm with strong wind gusts and some heavy rain in the Mid-North Coast to the Illawarra regions which may see some damage to property in the Illawarra, Metropolitan and up to the Hunter.	Showers, Th	Showers	**A**
19th	Wed				
20th	Thu				
21st	Fri	Storms with outbreaks of heavy rain in the Northern Tablelands, and storms with heavy rain, hail, and gusty winds at times in the Illawarra to the Metropolitan areas which may bring damage to property.			
22nd	Sat	Clearer spell over the next 3-4 days. Very cold overnight in the Snowy Mountains, with frosts in a number of districts, including in the Charlottes Pass area.	Showers		
23rd	Sun		Showers	Fg	**3Q V**
24th	Mon			Fg	
25th	Tue			Showers	
26th	Wed	Showers in the southern half of the State.		Showers	
27th	Thur	Outbreaks of rain spread in a line from the Upper Western to the South Coast.	Showers	Showers	
28th	Fri	Heavy rain around the Mid-North Coast to the Illawarra at times.	Showers	Showers, Th	
29th	Sat	Outbreaks of rain over the southern half of the State over the next couple of days.	Showers, Fg	Rain, Fg	**N XhN**
30th	Sun				**P4**
31st	Mon				

April 2025

Date	Day	New South Wales / ACT - April 2025	Sydney	Canberra	Moon
1st	Tue				
2nd	Wed				
3rd	Thu	A wet front crosses the State over the next couple of days, with widespread precipitation over the western half of the State at first, spreading into the eastern half by tomorrow.	Rain, Th	Showers	
4th	Fri	Storms with very heavy downpours, and hail, in the Metropolitan and Hunter which may cause localised flash flooding in some districts. Lightning may cause some power outages, particularly in the Hunter.	Showers		^
5th	Sat	Rain clears to coastal showers around this time, with some lingering heavy falls at times around the North Coast.	Showers		1Q
6th	Sun			Fg	
7th	Mon	Very strong and blustery winds for a time around the Hunter coastline.			
8th	Tue	Widespread rain through the western half of the State once more, with some heavy falls at times likely in the Far Northwest.	Showers		
9th	Wed	Showers along the Tablelands.			
10th	Thur	Mostly dry throughout the State over the next week.			
11th	Fri				XhS
12th	Sat	Very cold in the Snowy Mountains around this time, with maximum temperatures barely reaching above zero, particularly in the Thredbo district.	Showers		
13th	Sun		Rain, Th		F
14th	Mon	Chance of some isolated showers around the Metropolitan region over the next couple of days. Cold in the Riverina over the next few nights, with overnight minimums in the low single digits in many districts.	Showers, Th	Fg	A
15th	Tue	The next 3 weeks may be unusually drier than is normal for this time of year, before some good rainfall can be expected early in May with the overall expectation of the next 3-4 weeks tallying up to around average rainfall. It may also continue to be warmer than normal throughout most of the State over the coming 3-4 weeks.	Showers		
16th	Wed	A storm over the next couple of days in the Metropolitan brings heavy rain and chance of flash flooding in some areas.	Rain		
17th	Thu	Mild overnight in the Southern Tablelands.	Rain		
18th	Fri	Showers prevail along the eastern coastal districts over the next couple of days.	Showers	Fg	
19th	Sat			Fg	V
20th	Sun	Mostly dry throughout the State for the remainder of this month. Very strong and gusty winds around this time in the coastal areas of the North Coast region. Frosts likely in the Southern Tablelands-Snowy Mountains region.			
21st	Mon			Fg	3Q
22nd	Tue	Chance of isolated showers in some coastal parts mostly in the Mid-North Coast to the Hunter regions.		Fg	
23rd	Wed				
24th	Thur	Cold overnight in the Metropolitan region to the Central Tablelands, with many areas likely to drop into single digits for a time.			
25th	Fri		Showers, Th		
26th	Sat	Cold overnight in the Hunter around this time with minimum temperatures dropping into the single digits in many districts. Chance of frost patches in some sheltered parts of Murrurundi.	Showers	Fg	XhN
27th	Sun	Chance of some shower activity in the North Coast and Central Tablelands. Very warm spell throughout most of the State over the next few days, with many districts seeing maximum temperatures into the late 20's to the early 30's.			
28th	Mon	Unusually hot day in the Far Northwest, particularly in the Tibooburra area with maximum temperatures into the mid-30's likely.			P3 N
29th	Tue		Fg		
30th	Wed			Fg	

May 2025

Date	Day	New South Wales / ACT - May 2025	Sydney	Canberra	Moon
1st	Thu	Mostly dry over the next couple of days.			
2nd	Fri		Rain, Th		^
3rd	Sat	Patches of rain down through the eastern coastal districts over the next 3-4 days. Strong gusty winds at times in the Northern Rivers region, particularly around Grafton. Mild overnight temperatures and warm maximum temperatures in the Lower North Coast region, while in contrast, overnight temperatures may be unusually low in the Southwest Slopes over the next couple of days.	Rain		
4th	Sun	Chance of heavy rain and flash flooding risks in the Metropolitan area over the next couple of days, particularly in the Narrabeen district. Patches of heavy rain also likely in the Illawarra.	Rain	Showers	

Date	Day	New South Wales / ACT - May 2025	Sydney	Canberra	Moon
5th	Mon		Showers		1Q
6th	Tue	An intense depression crosses the State over the next few days bringing widespread and frequently heavy rain throughout many regions. Chance of flooding along many river catchments, as well as possible further flooding outbreaks in the Metropolitan region.			
7th	Wed	Cold in the Snowy Mountains over the next few days, with maximum temperatures in the single digits in many areas, including at Threadbo and Charlottes Pass. Scattered outbreaks of rain in the central and southern coastal regions over the next couple of days. Rain may be very heavy for a time in the Illawarra, with chance of brief flash flooding, particularly in the Bellambi area.	Showers	Rain	
8th	Thur		Showers	Showers	
9th	Fri	A band of rain moves into the west and crosses the State over the next few days, bringing some welcome rain into many districts, but still wetter in the west, while less rainfall in the east. Storm and heavy rain activity clear out of northern parts of the State around this time, but continue to prevail in the southern regions for a couple of more days including in the Lower Western, Riverina, Southwest Slopes, Southern Tablelands and to the South Coasts. Overnight severe frosts likely in the Snowy Mountains, particularly in the Charlottes Pass area, and cold in the South Coast region with overnight minimums dropping to near the frost point in some districts.		Showers	XhS
10th	Sat			Fg	
11th	Sun	Mostly dry over the next week with just a few exceptions. Chance of isolated showers in the coastal districts.			A
12th	Mon	Cold spell in the Far Northwest to the Upper North Coast region over the next couple of nights with overnight minimum temperatures hovering around the 10°C level.			
13th	Tue				F
14th	Wed	Chance of isolated showers in the far southeast of the State.			
15th	Thur	Overall, the next 3-4 weeks may be unusually hotter than what is typical for this time of year, and while rainfall amounts may be closer to average overall, much of this expected rainfall can be expected over one brief but significant rain event towards the end of May.			
16th	Fri	Unsually warm spell across the State over the next few days with daily maximums into the upper 20's to the early 30's in many districts.			V
17th	Sat			Fg	
18th	Sun	Unusually warm day in the North Coast, particularly in the Casino area.			
19th	Mon	A band of rain crosses the State bringing a line of showers from the Far Northwest to the Central Tablelands to the Hunter-Mid-North Coast regions over the next couple of days.	Showers		
20th	Tue			Fg	3Q
21st	Wed	Showers in the northeast quarter of the State and in the Central West, while mosty dry elsewhere. Very mild overnight temperatures in the Central West Slopes and Plains.			
22nd	Thu	Chance of very heavy rain for a time in the Mid-North Coast region around this time.	Showers	Fg	
23rd	Fri	Some showers in the Upper and Lower Western regions. Severe frosts in the Snowy Mountains, particularly in the Charlottes Pass area.	Fg	Fg	XhN
24th	Sat				
25th	Sun	A band of rain and storms cross the State over the next 4-5 days, with showers at first over the western half of the State before tracking eastward over the next couple of days.			
26th	Mon		Fg	Fg	P5
27th	Tue	Rain activity intensifies throughout the State with heavy falls almost everywhere around the next couple of days. Chance of flash flooding in susceptible districts.	Showers	Rain	N
28th	Wed		Rain	Showers	
29th	Thur	Hail showers in the Lower Western. Storm and rain activity contract to the east of the Tablelands around this time, while clearing in the Plains west of the Tablelands.			^
30th	Fri	Mostly dry throughout the State over the next couple of days.			
31st	Sat	Unusually mild overnight in the Lower North Coast region.		Rain	

June 2025

Date	Day	New South Wales / ACT - June 2025	Sydney	Canberra	Moon
1st	Sun	Some shower activity spreads up from Victoria into the Riverina, Southern Tablelands and South Coast districts for a time.			
2nd	Mon	Mostly dry throughout the State over the next 5-6 days.		Fg	
3rd	Tue	Snow in the Snowy Mountains over the next few days.			1Q
4th	Wed			Fg	
5th	Thu			Fg	XhS
6th	Fri	Chance of heavy snow in the Charlottes Pass to Spencers Creek areas.		Fg	
7th	Sat	Cold overnight with scattered frosts in many districts in the north of the State including in the Northern Tablelands and Northwest Slopes.		Fg	A
8th	Sun	Rain spreads across the State over the next couple of days, with showers at first in the Lower and Upper Western regions, before becoming widespread over the next day.	Showers	Rain	
9th	Mon	Widespread rain and storms throughout the State.			
10th	Tue	Mostly dry over the next 2-3 days.			

Date	Day	New South Wales / ACT - June 2025	Sydney	Canberra	Moon
11th	Wed	Snowflurries in the Snowy Mountains.			F
12th	Thur	Snow in Bimberi, Cabramurra, Perisher, Charlottes Pass and Thredbo. Winds likely to be very strong and blustery in the Snowy Mountains around this time.			V
13th	Fri	Isolated showers in the Lower Western to the Riverina and up into the Southern Tablelands and Illawarra. Light snow in the NSW-ACT ski fields over the next couple of days. Scattered frosts overnight in many districts through the Central and Southern Slopes and Tablelands to the South Coast regions.		Fg	
14th	Sat	The next 3-4 weeks may see a series of up to 4-5 low pressure systems pass by the NSW coast resutling in considerably higher than normal rainfall than what is normal for this time of year over the eastern half of the State, as well as colder than normal temperatures. In the west however, while temperatures may also be much colder than is typical for this time of year, rainfall amounts may be below normal, and very much drier than average in the Far Northwest. The last week of June and into the first week of July, may also bring significant snowfalls into the Southern Tablelands, including on the Snowy Mountains.		Fg	
15th	Sun	Mild overnight temperatures in the Far Northwest region.			
16th	Mon	Chance of heavy rain in the far northeast of the State over the next few days, associated with an intense storm system coming down through the east of Queensland.	Rain, Th		
17th	Tue	Outbreaks of heavy rain spreads down through the North Coast to the Hunter over the next couple of days, with showers spreading over the Northern and Central Tablelands to the adjacent Plains. Unusually mild day in the Upper North Coast region.	Showers		
18th	Wed	A low pressure system brings further storms and heavy rain into the Hunter over the next couple of days.	Showers		
19th	Thu				3Q XhN
20th	Fri	Brief showers likely along the eastern coasts from the Hunter to the Illawarra.		Fg	
21st	Sat	Very strong and gusty winds likely in the Sydney region.			
22nd	Sun	Chance of isolated showers in the Lower Western to the South Coast along the Victorian border fringes. Very heavy rain in the Hunter, with chance of flash flooding, particularly along the Wyong River catchment.		Fg	
23rd	Mon			Showers, Fg	P8
24th	Tue	Snowflurries in the Snowy Mountains. Chance of some rain along the South Coast and spreading up through the Illawarra to the Hunter over the next couple of days.	Rain	Showers	
25th	Wed	The next couple of weeks may see good snowfalls in the Southern Tablelands, particularly in the Snowy Mountains.	Rain	Rain	N
26th	Thur	Light snow in the Snowy Mountains over the next couple of days. Heavy rain over the next couple of days in the Illawarra to the Hunter regions at times.	Rain	Showers	^
27th	Fri	Light snow spreads into the ACT ski regions. Heavy snow likely in the Queanbeyan area.	Showers		
28th	Sat		Rain	Showers	
29th	Sun	Outbreaks of rain in most districts from the Tablelands to the east coasts, particularly in the North Coast region at first, and by tomorrow contracting into the Metropolitan to the South Coast regions.	Rain	Rain	
30th	Mon	Rain may be very heavy at times in the Illawarra to the South Coast.			

July 2025

Date	Day	New South Wales / ACT - July 2025	Sydney	Canberra	Moon
1st	Tue	Snow in the NSW-ACT ski fields over the next couple of days.			
2nd	Wed	Severe frosts in the Snowy Mountains over the next week. Unusually low maximum temperatures over the next few days in the Northern Tablelands.			XhS
3rd	Thu	Brief snow flurries in the NSW-ACT ski fields. Cooler spell spreads over the State around this time and persists for several days with scattered pockets of frost in numerous districts.	Rain		1Q
4th	Fri	Snow in the NSW-ACT ski fields.	Rain, Fg		
5th	Sat	Severe frosts overnight in the Perisher Valley area.	Rain	Rain	A
6th	Sun	Heavy rain at times in most districts from the Tablelands to the eastern coasts over the next few days.	Rain	Rain	
7th	Mon			Showers	
8th	Tue	Scattered rain activity briefly spreads into the Plains adjacent to the Tablelands.	Showers		
9th	Wed	Heavy snow in the NSW-ACT ski fields over the next couple of days. Unusually cold spell in the South Coast region.			
10th	Thur	Light shower activity in the east and south of the State over the coming week.			V
11th	Fri	Light snow in the NSW-ACT ski fields over the next couple of days.	Showers	Showers	F
12th	Sat				
13th	Sun	Snow in the NSW-ACT ski fields.		Showers	
14th	Mon	Wet spell across the southern half of the State over the next few days with heavy rain at times likely in the Southern Tablelands. Chance of heavy snow in the NSW-ACT ski fields. Overall, the next 3-4 weeks may be drier than average for this time of year, with temperatures nearer to normal.		Showers	
15th	Tue	Light snow in the NSW-ACT ski fields.		Showers	
16th	Wed	Snowflurries in the Snowy Mountains. Mild temperatures in the mid-20's over the next few days throughout most districts east of the Tablelands and along the coasts.		Showers	
17th	Thu	Snow in the NSW-ACT ski fields. In contrast, it may be an unusually warm day and mild night in some parts of the North Coast, particularly in the Lismore district. Heavy rain at times in the Riverina.		Rain	XhN

Date	Day	New South Wales / ACT - July 2025	Sydney	Canberra	Moon
18th	Fri	Heavy snow in the NSW-ACT ski fields. Rain and storms spread throughout the southern half of the State and up along the Tablelands to the North Coast over the next couple of days.	Rain, Th		3Q
19th	Sat	Snow in the NSW-ACT ski fields.	Rain		
20th	Sun	Heavy snow in the NSW-ACT ski fields over the next couple of days.	Showers		
21st	Mon	Chance of thunderstorms, hail showers and very strong winds with possible gale gusts at times in the Metropolitan region. Very heavy rain at times in the South Coast region.	Showers	Fg	P11
22nd	Tue	Mostly dry across the State over the remainder of the month.		Fg	
23rd	Wed	Chance of isolated showers in the Southern Tablelands.			^
24th	Thur			Fg	
25th	Fri	Light snow in the NSW-ACT ski fields over the next couple of days.		Fg	N
26th	Sat				
27th	Sun	Snowflurries in the Snowy Mountains.		Showers	
28th	Mon	Brief showery spell through the Riverina to the Southwest Slopes and adjacent Tablelands.			
29th	Tue	Very cold over the next few nights in the Northern Tablelands to the Northwest Slopes regions with scattered frosts in many districts.			
30th	Wed	Snow in the NSW-ACT ski fields over the next couple of days as well as in parts of the Central Tablelands, possibly including in the Orange and Oberon districts.	Showers		XhS
31st	Thu	Very cold in the Snowy Mountains with daily maximum temperatures settling in the frost levels in many districts, particularly in the Thredbo area.	Showers	Fg	

August 2025

Date	Day	New South Wales / ACT - August 2025	Sydney	Canberra	Moon
1st	Fri	Snowflurries in the Snowy Mountains. Mostly dry elsewhere throughout the State over most of the coming week ahead.		Fg	
2nd	Sat	Scattered frost patches spread into many districts throughout the State over the next few days.	Fg	Fg	1Q A
3rd	Sun			Fg	
4th	Mon	Severe frosts overnight in the Northern Tablelands, particularly in the Glen Innes district.		Fg	
5th	Tue	Very cold with severe frosts overnight in the Snowy Mountains, particularly in the Charlotte Pass region, with double-digit level frosts likely.		Showers	
6th	Wed	Showers move into the southern half of the State around this time, and briefly spread up to the North Coast over the following day.	Th	Th	V
7th	Thur	Thunderstorms with lightning, bursts of heavy rain, and hail in the Southwest Slopes. Milder spell over the northern and western regions of the State over the next few days with maximum temperatures into the mid-20's in many districts.			
8th	Fri	Some showers in the Lower Western, mostly dry elsewhere. Very wearm spell in the Riverina.		Fg	
9th	Sat	Light snow in the NSW-ACT ski fields over the next couple of days. Showers briefly spread up through the western Plains to the Northern Tablelands.		Fg	F
10th	Sun			Fg	
11th	Mon	Some showers in the southern half of the State over the next couple of days.			
12th	Tue		Showers	Showers	
13th	Wed	A band of rain crosses the State bringing a line of showers from the Far Northwest to the Central Tablelands and travels down into the Southern Tablelands to the South Coast briefly.	Showers		XhN
14th	Thu	Showers linger in the Central West to the adjacent Tablelands and over to the coasts near the Hunter to the Metropolitan. The next 3-4 weeks may be warmer than average for this time of year, and while the northeast quarter of the State may be wetter than average, most other parts of the State may see significantly less rainfall than normal.		Showers	
15th	Fri	Mostly dry spell throughout the State over the next 7-10 days. Snowflurries in the Snowy Mountains. Mild temperatures over the next couple of days in the northeast of the State.	Showers		P12
16th	Sat	Frosts and light snow in the NSW-ACT ski fields over the next couple of days.		Fg	3Q
17th	Sun	Scattered frosts through the Inland Plains and western parts of the State up into the Tablelands over the coming week.			
18th	Mon	Snow in the NSW-ACT ski fields.			
19th	Tue	Snowflurries in the NSW-ACT ski fields.			^
20th	Wed				
21st	Thur				
22nd	Fri				
23rd	Sat				N
24th	Sun	Snowflurries in the NSW-ACT ski fields. Isolated showers likely in the eastern coastal districts from the South Coast to the North Coast over the next couple of days.	Showers		
25th	Mon		Showers	Showers, Fg	
26th	Tue	Some showers in the North Coast to the Northern Tablelands.	Showers, Fg	Rain, Fg	XhS
27th	Wed	Scattered rain activity in the east of the State from the Tablelands to the Coast, with some heavy falls likely in the Mid-North Coast and South Coast districts. Overnight frosts in the next couple of days in the western half of the State.			
28th	Thu	Snowflurries in the NSW-ACT ski fields.			
29th	Fri	Chance of storms with hail showers in the Riverina and developing heavy rain in the northeast of the North Coast region.	Rain	Showers	
30th	Sat	Rain activity intensifies over the next couple of days throughout the northeast of the State with heavy falls from the North Coast to the Illawarra and into the adjacent Tablelands.	Showers		A
31st	Sun		Showers		1Q

September 2025

Date	Day	New South Wales / ACT - September 2025	Sydney	Canberra	Moon
1st	Mon	Occasional shower activity prevails along the eastern coasts over the next 3-4 days, while remaining mostly dry elsewhere.	Showers		
2nd	Tue		Showers	Fg	V
3rd	Wed	Very heavy rain with chance of flash flooding in the Lower North Coast region, particularly in the Coffs Harbour districts.	Rain, Fg	Showers	
4th	Thur		Showers, Fg		
5th	Fri	Mostly dry spell throughout the State over the next 7-9 days. Severe frosts likely around this time in the Snowy Mountains region, particularly at Charlotte Pass.			
6th	Sat				
7th	Sun				
8th	Mon				F
9th	Tue	Mild spell over the next couple of days in the eastern coastal regions including in the Hunter to the Central Tablelands.			XhN
10th	Wed			Showers	
11th	Thu				P9
12th	Fri	Unusually hot spell over the next couple of days in the west of the State, including in the Southwest Plains and Riverina regions, and particularly in the Ivanhoe, Hillston, Wyalong, Narrandera and Griffith districts.			
13th	Sat	Very strong and unsettled winds through the South Tablelands with chance of gale gusts at times, particularly in the Snowy Mountains districts.			
14th	Sun	Some showers around the eastern coastal regions from the Hunter to the Illawarra. Overall, the next 3-4 weeks may be drier than average for this time of year, particularly in the Murray Darling Basin, with maximum temperatures above average, and minimum temperatures near normal.	Showers		3Q
15th	Mon	Shower activity spreads up along the coasts to the North Coast to the South Coast over the next 4-5 days. Some light frosts likely in the Riverina region over the next couple of nights.	Showers	Showers	^
16th	Tue		Rain		
17th	Wed	Very cold day with severe overnight frosts in the Snow Mountains, particularly in the Thredbo area as well as in some parts of the Southwest Slopes over the next couple of days.	Rain	Showers	
18th	Thur	Unusually cold day in the Northern Tablelands.	Rain	Showers	
19th	Fri			Fg	
20th	Sat	Mostly dry throughout the State over the remainder of the month, with just a couple of exceptions.		Showers	
21st	Sun	Some showers along the southern border areas, into the Riverina and the Southern Tablelands and briefly to the South Coast over the next couple of days.			
22nd	Mon			Fg	N XhS
23rd	Tue				
24th	Wed	Light snow in the NSW-ACT ski fields.	Showers, Th, Sn	Showers	
25th	Thu	Isolated showers possible along the South Coast and North Coast regions.			
26th	Fri			Showers	A
27th	Sat	Strong gusty winds in the Hunter with chance of causing localised damage. Snowflurries in the Snowy Mountains and light snow in the ACT ski fields. Wintry showers in the South Coast to the adjacent Tablelands.			
28th	Sun			Fg	
29th	Mon		Showers	Showers	
30th	Tue	Chance of light showers briefly in the south of the State through the Riverina to the Southern Tablelands. Chance of light snow in the NSW-ACT ski fields.	Showers	Fg	1Q V

October 2025

Date	Day	New South Wales / ACT - October 2025	Sydney	Canberra	Moon
1st	Wed	Mostly dry over the next couple of weeks. Some brief showers possible in the South Coast to the Metropolitan around this time. Chance of frosts over the next few nights in the Northern Tablelands and in some parts of the Riverina.			
2nd	Thur				
3rd	Fri	Chance of some showers on the North Coast, with isolated showers reaching into the Central West for a brief time.			
4th	Sat				
5th	Sun	Thunderstorms in the Mid-North Coast, which may bring heavy rain and some flash flooding in the region, possibly in the Coffs Harbour area.		Fg	
6th	Mon	Heavy rain around this time in the Northern Rivers region. Chance of frosts over the next few nights in the Tablelands, including in some parts of the Northern Tablelands.	Showers	Showers, Fg	
7th	Tue				F XhN
8th	Wed	Very warm spell throughout most of the State over the coming week with many districts seeing daily maximas into the upper 20's to early 30's. However, chance of some pockets of frost in the eastern districts of the Riverina.			

Date	Day	New South Wales / ACT - October 2025	Sydney	Canberra	Moon
9th	Thu	Chance of isolated showers in the South Coast to the Southern Tablelands. Very hot spell in the Upper Darling/Far Northwest region, with some districts possible seeing maximum temperatures into the mid-30's.			P6
10th	Fri				
11th	Sat	Light snow in the NSW-ACT ski fields over the next couple of days.			
12th	Sun				
13th	Mon	Snowflurries in the NSW-ACT ski fields. Very strong and blustery winds in the Snowy Mountains.			^
14th	Tue	The next 3-4 weeks may continue to be warmer than what is normal for this time of year, and with below average rainfall in most parts of the State.			3Q
15th	Wed	Mostly dry conditions continue throughout most of the State over the next 3-4 days accompanied by very warm temperatures in most districts, with daily maximas reaching into the upper 20's to mid-30's being common.			
16th	Thur		Showers, Th		
17th	Fri		Showers		
18th	Sat	Brief showery spell over the next couple of days around the North Coast.	Showers, Th		
19th	Sun				
20th	Mon	Very cold spell over the next couple of days in the Snowy Mountains, particularly in the Thredbo district where maximum temperatures may barely rise above the frost levels.			XhS
21st	Tue	A supercell of thunderstorms with heavy rain, large hail, flash flooding, and damaging winds over the next couple of days in the Mid-North Coast and Northern Rivers regions and up into the adjacent Northern Tablelands. Hail stones may be unusually large at times. Severe frosts likely in the Snowy Mountains, particularly at Perisher and very cold overnight with chance of scattered frosts in some parts of the Riverina over the next couple of days.	Rain, Th	Showers	
22nd	Wed	Storms with heavy rain and hail in the Mid-North Coast and Northern Rivers regions may continue to bring disruption and damage across many areas in these regions. Storm activity may be particularly damaging in Grafton and Lismore. Chance of frosts overnight in sheltered parts of the Central West Slopes.	Showers, Th	Showers	N
23rd	Thu	Shower activity continues along the eastern coastal fringes.			
24th	Fri	Severe thunderstorms and strong blustery winds in the Northwest Plains, with chance of damaging wind gusts.			A
25th	Sat	Thunderstorms, heavy rain, and hail from the Hunter through the Metropolitan and into the Illawarra. Light snow in the NSW-ACT ski regions.			
26th	Sun	A settled period arrives around this time bringing mostly dry conditions throughout the State over the remainder of this month. Very cold over the next couple of nights in the southern half of the State including in the Riverina to the South Coast regions, including the chance of patches of frost.			
27th	Mon	Unusually cold overnight overnight temperatures spread into the northern half of the State, including the chance of patches of frost in the Hunter to the Northwest Slopes.			V
28th	Tue				
29th	Wed				
30th	Thur	Scattered frosts over the next couple of nights in the Southern Tablelands and including the Snowy Mountains.			1Q
31st	Fri				

November 2025

Date	Day	New South Wales / ACT - November 2025	Sydney	Canberra	Moon
1st	Sat	Dry conditions continue for the next couple of days.			
2nd	Sun	Very warm spell over the northern half of the State over the next couple of days with a number of districts seeing maximum temperatures into the upper 30's.			
3rd	Mon	A wet spell moves across the State over the next few days, starting in the Upper Western at first, before spreading throughout most of the State, later contracting to the coasts in a few days time. Unusually mild overnight temperatures in the Lower Darling region and very warm day in the Riverina to the Southwest Slopes regions with maximum temperatures into the upper 30's in some districts.	Rain	Showers	XhN
4th	Tue	Very hot in the Southwest Plains with some districts likely to see maximum temperatures into the upper 30's.	Showers, Th	Rain	
5th	Wed	Heavy rain along the Tablelands to the east coast.	Showers, Th	Rain, Th, Fg	
6th	Thur		Fg	Fg	F P1
7th	Fri	Thunderstorms with large hailstones and heavy rain likely on the Northern Tablelands.	Showers, Th	Showers	
8th	Sat	Scattered thunderstorms and squalls across the Metropolitan, Northern Rivers and Hunter regions, with chance of localised flash flooding. Unsettled winds may generate tornado activity within the thunderstorms supercell, which may bring some damage in the Northern Rivers region. Heavy rain also in the Northern Tablelands to the Northwest Slopes.			
9th	Sun				^
10th	Mon	Storms and hail showers in the Riverina and in the Central West Plains, affecting Wyalong. Very strong and blustery winds in the Snowy Mountains, possibly to gale force for a time.			

Date	Day	New South Wales / ACT - November 2025	Sydney	Canberra	Moon
11th	Tue	Thunderstorms with bursts of intense rain, possible flash flooding, and large hail likely in the Northern Rivers region, including in the Lismore district.	Showers, Th	Showers	
12th	Wed	Thunderstorms and heavy rain on the Northern Tablelands, with chance of pockets of flash flooding, with particular risk in the Elsmore area.	Rain, Th	Showers	**3Q**
13th	Thu	Widespread storm activity spreads throughout the State over the next couple of days with heavy rain in many districts and the chance of flash flooding particularly at risk along river catchments.	Rain, Th	Rain	
14th	Fri	The next 3-4 weeks may be warmer than normal for this time of year, particularly in the south and west, as well as wetter overall, with much higher than average rainfall in some parts of the Murray Darling Basin.	Showers	Showers	
15th	Sat	Storm activity eases over the northern half of the State, but persist in the Riverina to the South Coast regions.			
16th	Sun	Very warm spell over the next couple of days in the northeast of the State.	Rain		**XhS**
17th	Mon	Storms and heavy rain rebuilds over the next few days along the east of the State from the coast to the adjacent Tablelands. Very strong and blustery winds for a time along the South Coast.	Rain		
18th	Tue	Very cold in the Snowy Mountains over the next couple of days with overnight frosts and maximum temperatures barely rising above zero, including in the Thredbo area.	Rain		
19th	Wed	Cool over the next couple of nights in the southern half of the State, with minimum temperatures in the single digits.	Showers	Rain	
20th	Thur				**N A**
21st	Fri	Very wet in the Mid-North Coast around this time, with chance of flash flooding in some areas, including in the Port Macquarie district, lingering showers elsewhere along the coasts.		Showers	
22nd	Sat	Mostly dry throughout the State over the next 4-5 days.			
23rd	Sun			Fg	**V**
24th	Mon				
25th	Tue	Cool spell over the next few nights in many parts of the State.			
26th	Wed				
27th	Thu	Chance of some showers briefly in the Central to Northern Tablelands.		Th	
28th	Fri	Showers in the Southern Tablelands to the adjacent coastal districts over the next couple of days.		Showers, Th	**1Q**
29th	Sat	Showers may briefly spread up along the Central to the Northern Tablelands over the next couple of days. Dry elsewhere.	Showers, Th		
30th	Sun	Very warm day in the Illawarra.		Showers	**XhN**

December 2025

Date	Day	New South Wales / ACT - December 2025	Sydney	Canberra	Moon
1st	Mon	Storm activity spreads rain across the State into most districts over the next few days.	Showers, Th	Showers, Th	
2nd	Tue	Thunderstorms and chance of flash flooding and hail in the Southern Tablelands over the next couple of days. Windy with strong gusts in the Riverina, particularly around the Griffith area. Very hot spell over the next few days in the northern half of the State.		Rain, Th	
3rd	Wed	Thunderstorms and outbreaks of heavy rain in the Southern Tablelands, as well as in the Hunter. Chance of flash flooding in some parts of the Hunter. Very hot west of the Tablelands over the next couple of days. Extremely hot in the Southwest Plains with some districts possibly having maximum temperatures reach up into the mid-40's including in the Ivanhoe district.	Showers	Showers	
4th	Thur	Thunderstorms and heavy rain with flash flooding risks in the Mid-North Coast to the Hunter regions.	Showers		**P2**
5th	Fri	The passage of slow-moving thunderstorms over the next couple of days cross the Mid-North Coast region, bringing heavy rain in many parts, as well as some flash flooding in localised areas.	Showers	Th	**F**
6th	Sat	Pockets of heavy rain likely in inland districts west of the Tablelands over the next couple of days.	Rain		**^**
7th	Sun	Heavy rain and scattered thunderstorms in the Southwest Slopes with the chance of localised flash flooding in some places.	Showers		
8th	Mon	Scattered thunderstorms and outbreaks of heavy rain cross the Upper Western to the Riverina over the next couple of days, with the chance of flash flooding in some places, including in the Cobar and Yanco districts.	Showers		
9th	Tue				
10th	Wed	Outbreaks of heavy rain and thunderstorms move into the Lower Western region, with possible flash flooding in the Broken Hill area.		Rain, Th	
11th	Thu	A slow moving band of thunderstorms through the southeast of the State may bring heavy rain and chance of flash flooding in some parts of the Southern Tablelands to the South Coast.	Showers	Rain	
12th	Fri	Thunderstorms and heavy rain with chance of flash flooding in the Northwest and Southwest Slopes, and into the adjacent Hunter and Illawarra regions. Winds may be very strong and cause localised damage in the Southwest Slopes.		Th	**3Q**
13th	Sat	Scattered thunderstorms cross the Metropolitan to the Southern Tablelands including in ACT, with outbreaks of associated heavy rain with the storms. Chance of flash flooding in ACT and into the Southern Tablelands.	Rain, Th	Rain, Th	**XhS**

Date	Day	New South Wales / ACT - December 2025	Sydney	Canberra	Moon
14th	Sun	The next 3-4 weeks ahead may be much wetter and cooler than is normal for this time of year. Storm activity becomes widespread across most of the State around this time. Thunderstorms and heavy rain with chance of flash flooding in some places on the Central West Slopes including in Dubbo.	Showers, Th		
15th	Mon	Thunderstorm activity with heavy rain and possible flash flooding outbreaks move into the Northern Tablelands.	Showers, Th		
16th	Tue	A band of thunderstorms with heavy rain and some pockets of flash flooding in the Central Tablelands, Riverina, Hunter, Mid-North Coast, to the Illawarra. A squally line of thunderstorms may cause some damage in the Berrigan region of the Riverina. Chance of damaging winds and possible large hail showers at times in the Hunter and Mid-North Coast regions.	Showers		
17th	Wed	A cell of thunderstorms cross the Metropolitan region, bursts of heavy rain and large hail and possible pockets of flash flooding in northern parts of Sydney to the Central Coast. Further thunderstorms with damaging lightning, heavy rain, flash flooding, and large hail also into the Hunter to the Mid-North Coast.	Rain, Th	Showers, Th	**A**
18th	Thur	An unstable band of scattered thunderstorms cross the Central West Slopes, Central Tablelands, Metropolitan and Illawarra regions bringing widespread storms, gusty winds, heavy rain with localised flooding and showers of hail at times. Chance of damage to property and infrastructure with the passage of the storms.		Fg	
19th	Fri	Strong winds with gusts and scattered thunderstorms in the Northwest Plains, particularly in the Moree area.	Showers, Th		
20th	Sat	A line of thunderstorms cross the Illawarra to the Metropolitan and the adjacent Southern Highlands, with heavy rain, flash flooding, lightning strikes and hail likely, with chance of damage to some property and infrastructure. Storms and damaging wind gusts likely in the Northwest and Central West Slopes, including affecting Tamworth. Unusually warm overnight in the Far Northwest to the Northern Tablelands with minimum temperatures into the upper 20's in many districts.	Showers		**N V**
21st	Sun	A sequence of further thunderstorms with heavy rain cross the Mid-North Coast to the Hunter and into the Metropolitan region over the next couple of days, with chance of flash flooding once more, and major hail storms and gusty winds which may bring damage to property and infrastructure.	Rain		
22nd	Mon				
23rd	Tue	Thunderstorms with hail showers, heavy rain and flash flooding potential in the Northern Tablelands.			
24th	Wed	Rain activity in the east of the State eases to isolated bringing a brief respite of mostly dry conditions over the next couple of days. Cold day in the Snowy Mountains.			
25th	Thu			Showers	
26th	Fri	Widespread storms and rain returns throughout most of the State over the next couple of days, although remaining mostly dry along the western borders.	Rain, Th	Showers	
27th	Sat				**XhN**
28th	Sun	Isolated showers in the coastal districts, while a brief drier spell elsewhere.			**1Q**
29th	Mon	Another wet spell crosses the State over the next few days, bringing widespread outbreaks of rain. Heavy rain and scattered thunderstorms in the Mid-North Coast region with chance of localised flooding and some hail showers. Chance of thunderstorms damaging wind gusts in the Central West Plains, particularly in the Coonamble area.		Showers, Fg	
30th	Tue			Rain	
31st	Wed			Rain	

NEW SOUTH WALES / ACT GRAPHS AND TABLES 2025

Seasonal Breakdown – State Overall Rainfall Timing and Estimated mm Trend Graphs
State Overall Maximum Temperature Trend Graphs
Regional Rainfall Expectations Tables - mm and Trends
Regional Maximum and Minimum Temperature Trends Tables

Monthly Breakdown - **Rainfall:**
Regional Expectations by mm
Regional Expectations by mm Variance to Average
Regional Variance Trend by mm
Regional Estimated Expectations by % of Variance to Average
Regional Estimatedl Variance Trend by %

Maximum & Minimum Temperature Trends:
Regional Estimated Average Temperature
Regional Estimated Degree of Variation from the Average
Regional Estimated Temperature Trend

Seasonal Breakdowns

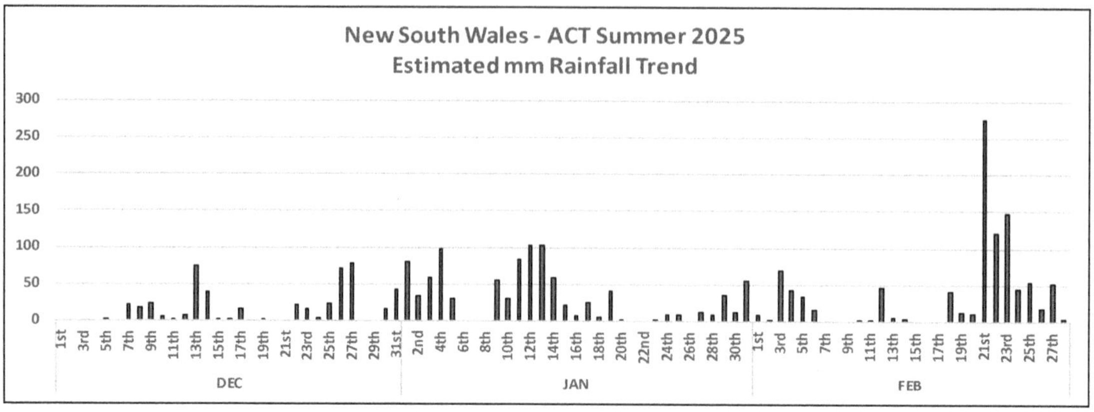

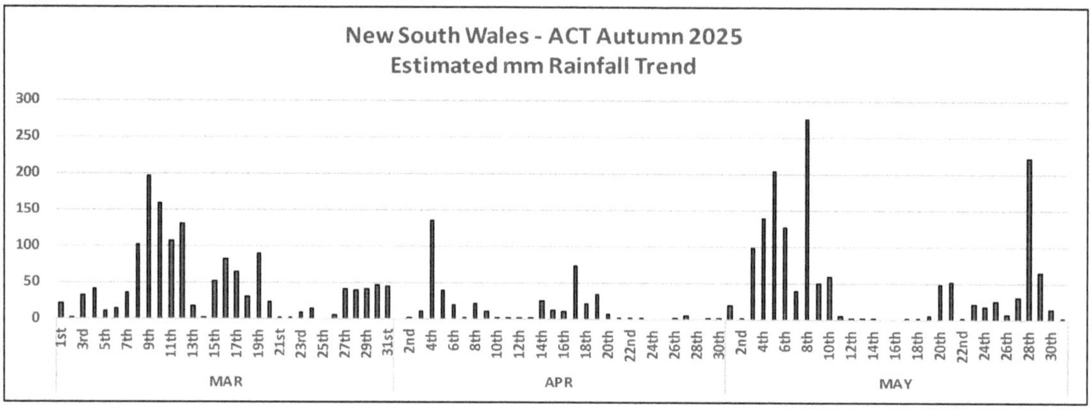

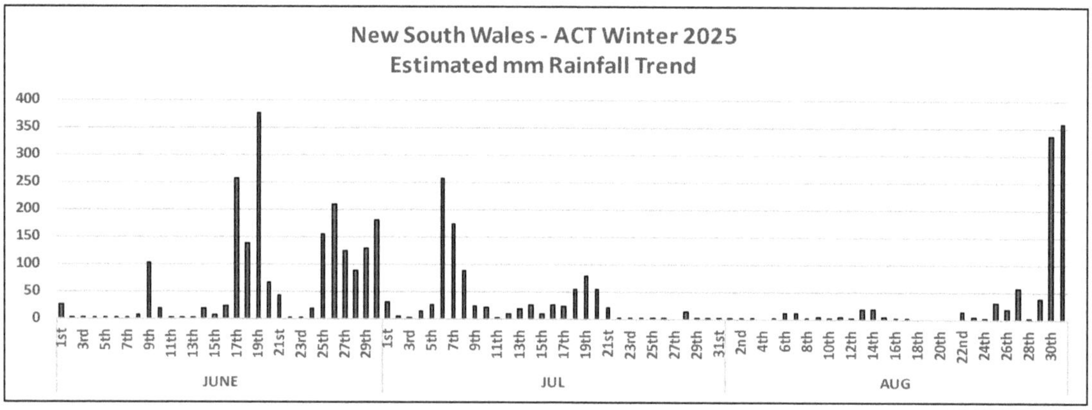

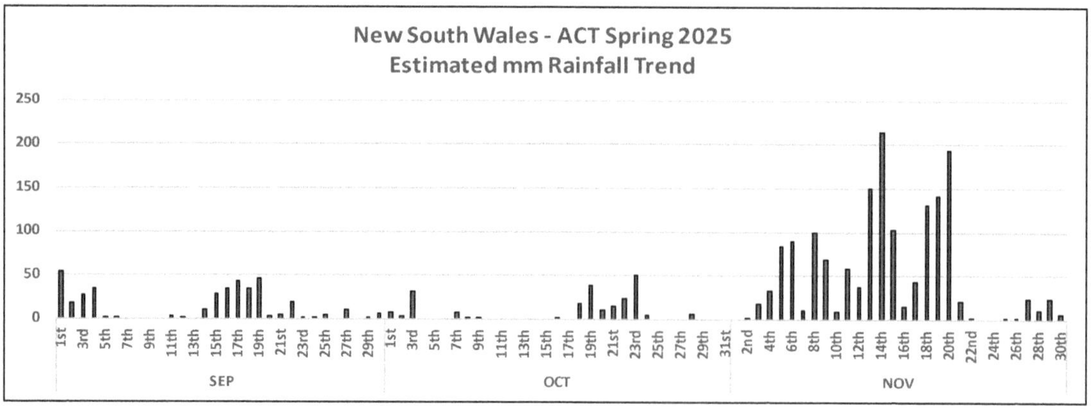

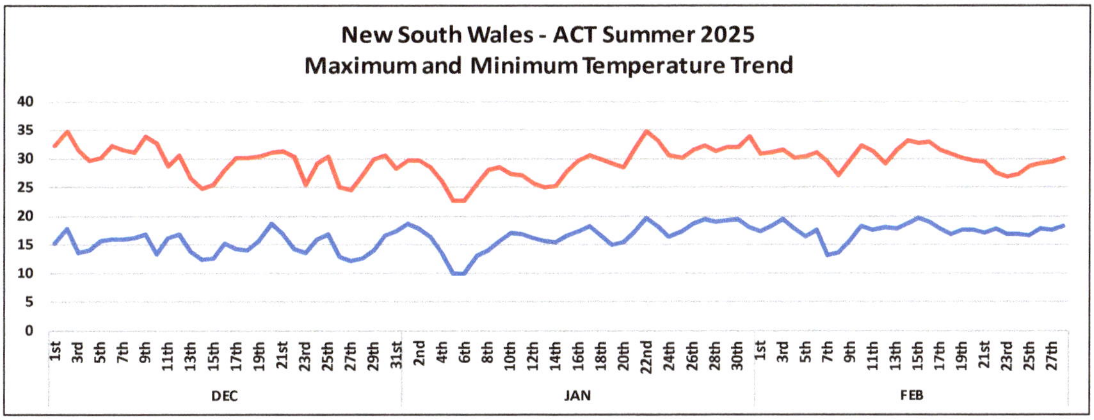
New South Wales - ACT Summer 2025
Maximum and Minimum Temperature Trend
DEC
JAN
FEB

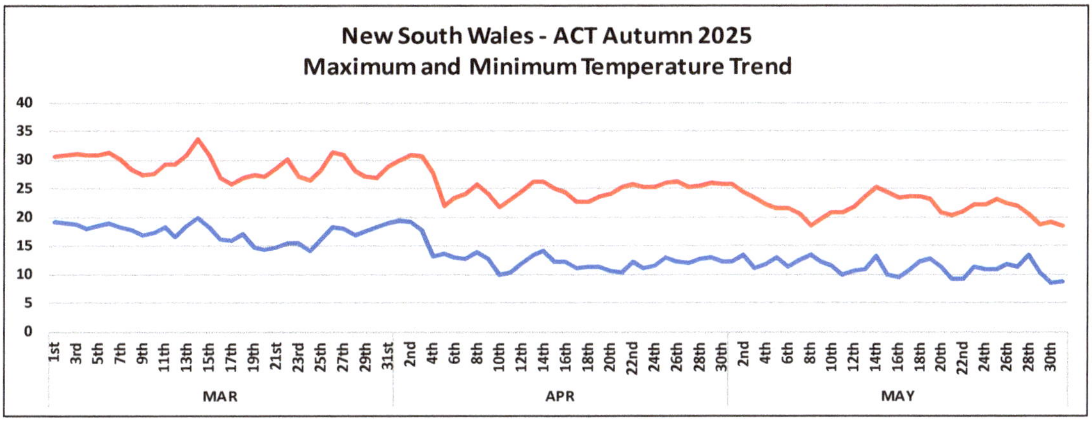
New South Wales - ACT Autumn 2025
Maximum and Minimum Temperature Trend
MAR
APR
MAY

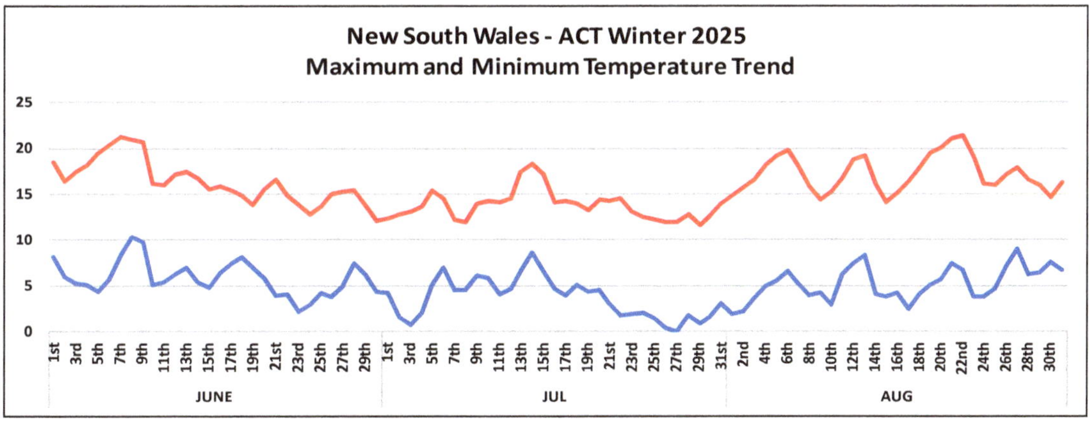
New South Wales - ACT Winter 2025
Maximum and Minimum Temperature Trend
JUNE
JUL
AUG

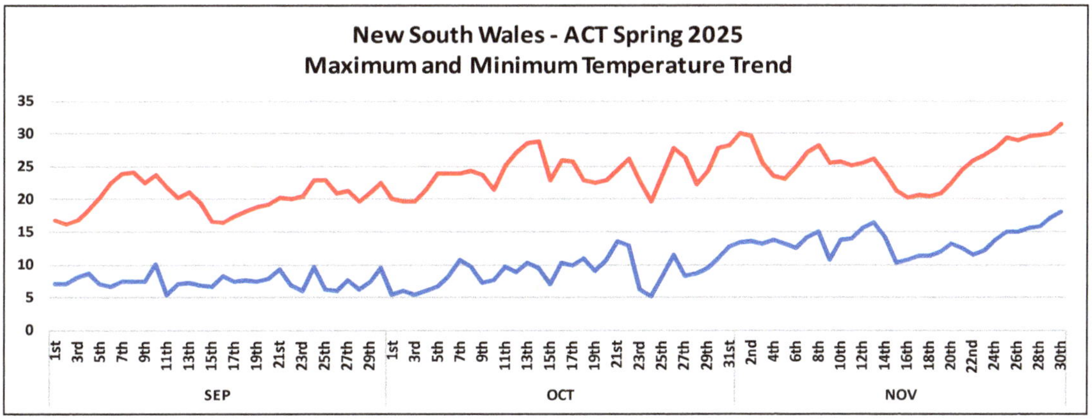
New South Wales - ACT Spring 2025
Maximum and Minimum Temperature Trend
SEP
OCT
NOV

2025 NSW / ACT - Estimated Seasonal Rainfall Expectations (mm)

REGION	Upper West	Lower West	Riverina	NW Slopes & Plains	Northern Tablelds	CW Slopes & Plains	Central Tablelds	SW Slopes & Plains	Southern Tablelds	Northern Rivers	Mid-North Coast	Hunter	Metropolitan	Illawarra	South Coast	ACT	NSW / ACT
Town Sample	Wilcannia	Broken Hill	Hay	Moree	Armidale	Dubbo	Orange	Wagga Wagga	Goulburn	Grafton	Coffs Harbour	Newcastle	Sydney	Wollongong	Bega	Canberra	Overall
Summer	17	140	49	66	261	87	156	57	93	314	266	208	179	174	297	92	2456
Autumn	96	92	76	127	197	159	209	152	166	257	247	480	303	471	188	188	3406
Winter	33	32	66	103	142	170	180	97	231	243	424	603	709	446	278	158	3916
Spring	28	34	24	60	146	48	115	49	127	267	485	170	186	201	144	96	2179
Overall	220	322	321	503	854	555	731	439	722	1112	1449	1470	1500	1481	980	642	13299

2025 NSW / ACT - Estimated Seasonal Rainfall Variation from the Average (mm)

REGION	Upper West	Lower West	Riverina	NW Slopes & Plains	Northern Tablelds	CW Slopes & Plains	Central Tablelds	SW Slopes & Plains	Southern Tablelds	Northern Rivers	Mid-North Coast	Hunter	Metropolitan	Illawarra	South Coast	ACT	NSW / ACT
Town Sample	Wilcannia	Broken Hill	Hay	Moree	Armidale	Dubbo	Orange	Wagga Wagga	Goulburn	Grafton	Coffs Harbour	Newcastle	Sydney	Wollongong	Bega	Canberra	Overall
Summer	-59	81	-35	-148	-32	-78	-82	-74	-95	-116	-256	-58	-92	-179	76	-72	-1218
Autumn	22	34	-17	13	55	31	34	26	29	-61	-322	119	-11	101	-18	64	99
Winter	-27	-22	-33	6	14	29	-98	-65	86	82	181	355	463	188	130	20	1308
Spring	-48	-27	-67	-101	-39	-98	-119	-98	-42	96	159	-47	-25	-95	-49	-76	-676
Overall	-66	90	-45	-83	105	-25	-194	-127	83	32	-212	378	458	204	213	44	855

2025 NSW / ACT - Estimated Seasonal Rainfall Trend Variation by 10mm

REGION	Upper West	Lower West	Riverina	NW Slopes & Plains	Northern Tablelds	CW Slopes & Plains	Central Tablelds	SW Slopes & Plains	Southern Tablelds	Northern Rivers	Mid-North Coast	Hunter	Metropolitan	Illawarra	South Coast	ACT	NSW / ACT
Town Sample	Wilcannia	Broken Hill	Hay	Moree	Armidale	Dubbo	Orange	Wagga Wagga	Goulburn	Grafton	Coffs Harbour	Newcastle	Sydney	Wollongong	Bega	Canberra	Overall
Summer	drr	wtr	drr	drr	drr	drr	drr	drr	drr	drr	drr	drr	drr	drr	wtr	drr	drr
Autumn	av	wtr	av	av	wtr	wtr	wtr	av	av	drr	drr	wtr	av	wtr	av	wtr	wtr
Winter	av	av	drr	av	av	av	drr	drr	wtr	wtr	wtr	wtr	wtr	wtr	wtr	av	wtr
Spring	drr	av	drr	drr	drr	drr	drr	drr	drr	wtr	wtr	drr	av	drr	drr	drr	drr
Overall	av	av	av	av	wtr	av	drr	drr	av	av	drr	wtr	wtr	wtr	wtr	av	wtr

2025 NSW / ACT - Estimated Seasonal Rainfall Expectation by %

REGION	Upper West	Lower West	Riverina	NW Slopes & Plains	Northern Tablelds	CW Slopes & Plains	Central Tablelds	SW Slopes & Plains	Southern Tablelds	Northern Rivers	Mid-North Coast	Hunter	Metropolitan	Illawarra	South Coast	ACT	NSW / ACT
Town Sample	Wilcannia	Broken Hill	Hay	Moree	Armidale	Dubbo	Orange	Wagga Wagga	Goulburn	Grafton	Coffs Harbour	Newcastle	Sydney	Wollongong	Bega	Canberra	Overall
Summer	-77%	138%	-42%	-69%	-11%	-47%	-34%	-56%	-51%	-27%	-49%	-22%	-34%	-51%	35%	-44%	-33%
Autumn	30%	59%	-18%	11%	38%	24%	20%	20%	21%	-19%	-57%	33%	-3%	27%	-9%	51%	3%
Winter	-46%	-41%	-34%	6%	11%	21%	-35%	-40%	59%	51%	74%	143%	188%	73%	88%	14%	50%
Spring	-63%	-44%	-74%	-63%	-21%	-67%	-51%	-67%	-25%	56%	49%	-22%	-12%	-32%	-25%	-44%	-24%
Overall	-23%	39%	-12%	-14%	14%	-4%	-21%	-22%	13%	3%	-13%	35%	44%	16%	28%	7%	7%

2025 NSW / ACT - Estimated Seasonal Rainfall Trend Variation by 25%

REGION	Upper West	Lower West	Riverina	NW Slopes & Plains	Northern Tablelds	CW Slopes & Plains	Central Tablelds	SW Slopes & Plains	Southern Tablelds	Northern Rivers	Mid-North Coast	Hunter	Metropolitan	Illawarra	South Coast	ACT	NSW / ACT
Town Sample	Wilcannia	Broken Hill	Hay	Moree	Armidale	Dubbo	Orange	Wagga Wagga	Goulburn	Grafton	Coffs Harbour	Newcastle	Sydney	Wollongong	Bega	Canberra	Overall
Summer	drr	wtr	drr	drr	av	drr	drr	drr	drr	drr	drr	av	drr	drr	wtr	wtr	drr
Autumn	wtr	wtr	av	av	wtr	av	av	av	av	av	drr	wtr	av	wtr	av	wtr	av
Winter	drr	drr	drr	av	av	av	drr	drr	wtr	wtr	wtr	wtr	wtr	wtr	wtr	av	wtr
Spring	drr	drr	drr	drr	av	drr	drr	drr	av	wtr	wtr	av	av	drr	av	drr	av
Overall	av	wtr	av	av	av	av	av	av	av	av	av	wtr	wtr	av	wtr	av	av

2025 NSW/ACT - Estimated Seasonal Maximum Temperature Average and Trend

Region	Upper West	Lower West	Riverina	NW Slopes & Plains	Northern Tablelds	CW Slopes & Plains	Central Tablelds	SW Slopes & Plains	Southern Tablelds	Northern Rivers	Mid-North Coast	Hunter	Metropolitan	Illawarra	South Coast	ACT	NSW / ACT
Town Sample	Wilcannia	Broken Hill	Hay	Moree	Armidale	Dubbo	Orange	Wagga Wagga	Goulburn	Grafton	Coffs Harbour	Newcastle	Sydney	Wollongong	Bega	Canberra	Overall
Summer Est	35.6	33.2	33.1	34.2	26.4	34.1	27.4	32.8	28.7	29.2	26.8	28.1	25.6	25.3	23.9	29.9	29.6
Trend	av	av	av	a-w	a-w	wmr	wmr	wmr	wmr	a-c	av	wmr	av	a-w	clr	a-w	a-w
Autumn Est	29.9	27.1	27.5	29.9	22.0	27.6	20.8	26.4	22.4	27.2	25.2	25.4	24.5	23.8	22.9	23.5	25.4
Trend	wmr	wmr	wmr	wmr	wmr	wmr	wmr	wmr	wmr	wmr	wmr	wmr	wmr	wmr	a-w	wmr	wmr
Winter Est	17.6	15.7	15.8	17.7	13.0	15.1	9.6	13.9	11.4	21.4	19.7	18.0	17.8	17.3	16.7	12.4	15.8
Trend	a-c	a-c	a-c	clr	av	clr	clr	av	clr	av	av	av	a-w	av	av	clr	av
Spring Est	27.8	25.7	25.3	27.8	20.3	25.3	18.6	23.6	19.9	26.5	23.8	24.0	22.9	22.3	20.5	20.8	23.4
Trend	av	av	a-c	av	av	av	av	wmr	av	av	av	wmr	a-w	a-w	clr	a-c	av

2025 NSW/ACT - Estimated Seasonal Minimum Temperature Average and Trend

Region	Upper West	Lower West	Riverina	NW Slopes & Plains	Northern Tablelds	CW Slopes & Plains	Central Tablelds	SW Slopes & Plains	Southern Tablelds	Northern Rivers	Mid-North Coast	Hunter	Metropolitan	Illawarra	South Coast	ACT	NSW / ACT
Town Sample	Wilcannia	Broken Hill	Hay	Moree	Armidale	Dubbo	Orange	Wagga Wagga	Goulburn	Grafton	Coffs Harbour	Newcastle	Sydney	Wollongong	Bega	Canberra	Overall
Summer Est	19.0	19.7	16.9	19.8	11.6	18.3	12.8	16.6	9.3	18.5	18.2	18.6	19.1	16.5	13.6	13.9	16.4
Trend	av	av	av	av	clr	a-w	a-w	a-w	clr	av	a-c	av	a-w	clr	a-c	av	av
Autumn Est	15.5	15.5	14.0	15.9	8.3	14.5	9.5	12.9	11.1	16.2	16.3	17.0	17.2	15.2	11.2	10.2	13.8
Trend	wmr	wmr	wmr	wmr	av	wmr	wmr	wmr	wmr	wmr	wmr	wmr	wmr	av	wmr	wmr	wmr
Winter Est	6.0	4.2	3.9	4.4	0.0	3.6	1.9	2.9	3.5	7.4	7.9	9.7	9.6	7.9	3.8	2.2	4.9
Trend	a-w	av	av	a-c	clr	av	wmr	av	wmr	wmr	av	av	a-w	clr	a-w	wmr	av
Spring Est	12.9	11.4	10.5	12.2	4.9	8.6	6.1	7.4	3.2	13.4	13.6	14.8	14.7	12.2	7.8	6.6	10.0
Trend	wmr	a-c	a-w	a-c	clr	a-c	av	av	clr	a-w	av	a-w	wmr	clr	a-c	av	av

Monthly Breakdowns

2025 — NEW SOUTH WALES / ACT - Estimated Rainfall Expectations (mm)

REGION	Upper West	Lower West	Riverina	NW Slopes & Plains	Northern Tablelds	CW Slopes & Plains	Central Tablelds	SW Slopes & Plains	Southern Tablelds	Northern Rivers	Mid-North Coast	Hunter	Metropolitan	Illawarra	South Coast	ACT	NSW / ACT
Town Sample	Wilcannia	Broken Hill	Hay	Moree	Armidale	Dubbo	Orange	Wagga Wagga	Goulburn	Grafton	Coffs Harbour	Newcastle	Sydney	Wollongong	Bega	Canberra	Overall
January	14	123	47	40	56	73	68	36	52	156	94	10	60	58	51	34	971
February	2	16	0	20	145	12	66	19	29	57	104	104	68	66	243	56	1005
March	14	13	17	67	113	70	112	60	100	132	110	116	91	189	126	118	1450
April	12	9	11	1	23	17	10	11	5	24	25	119	74	68	21	4	433
May	70	70	47	59	61	72	87	81	60	102	112	244	138	214	40	66	1523
June	29	29	28	49	32	107	41	26	124	50	102	474	466	261	112	62	1992
July	1	3	27	16	23	41	108	45	89	36	116	39	110	114	137	77	981
August	3	1	11	38	87	21	31	26	18	158	207	90	133	71	29	19	943
September	0	0	2	0	10	0	5	7	21	56	98	39	57	54	19	14	383
October	0	5	0	4	18	0	3	1	4	40	66	30	15	19	8	3	217
November	28	29	22	55	118	48	108	41	102	171	320	101	114	127	117	79	1579
December	48	25	109	153	167	93	93	87	118	132	94	104	174	239	76	111	1822
Overall	220	322	321	503	854	555	731	439	722	1112	1449	1470	1500	1481	980	642	13299

2025 — NEW SOUTH WALES / ACT - Estimated Rainfall Variation from the Average (mm)

REGION	Upper West	Lower West	Riverina	NW Slopes & Plains	Northern Tablelds	CW Slopes & Plains	Central Tablelds	SW Slopes & Plains	Southern Tablelds	Northern Rivers	Mid-North Coast	Hunter	Metropolitan	Illawarra	South Coast	ACT	NSW / ACT
Town Sample	Wilcannia	Broken Hill	Hay	Moree	Armidale	Dubbo	Orange	Wagga Wagga	Goulburn	Grafton	Coffs Harbour	Newcastle	Sydney	Wollongong	Bega	Canberra	Overall
January	-14	104	19	-39	-55	10	-8	-2	-11	9	-66	-62	-20	-47	-9	-21	-212
February	-28	-3	-30	-48	62	-33	-8	-22	-29	-107	-117	-22	-51	-84	150	3	-365
March	-18	-12	-13	4	72	32	54	19	48	-7	-117	3	2	69	46	73	256
April	-9	-4	-17	-21	-29	-26	-44	-27	-38	-74	-155	-1	-45	-70	-54	-37	-650
May	49	51	13	29	12	25	25	34	18	20	-50	118	32	102	-10	28	494
June	2	12	-8	12	-4	64	-34	-27	73	4	-4	357	366	158	58	18	1049
July	-17	-19	-4	-17	-32	-10	7	-14	44	-26	33	-35	37	47	78	29	101
August	-12	-15	-21	12	50	-25	-71	-24	-32	104	151	33	60	-17	-6	-27	158
September	-21	-17	-29	-36	-38	-40	-72	-42	-30	11	38	-19	-5	-22	-34	-38	-394
October	-30	-18	-35	-48	-47	-49	-75	-52	-50	-19	-37	-40	-49	-80	-51	-48	-728
November	3	8	-3	-17	46	-9	28	-4	37	104	158	12	29	7	36	10	445
December	29	4	83	86	69	37	4	35	51	12	-47	35	101	141	8	55	701
Overall	-66	90	-45	-83	105	-25	-194	-127	83	32	-212	378	458	204	213	44	855

2025 NEW SOUTH WALES / ACT - Estimated Rainfall Trend Variation by 10mm

REGION	Upper West	Lower West	Riverina	NW Slopes & Plains	Northern Tablelds	CW Slopes & Plains	Central Tablelds	SW Slopes & Plains	Southern Tablelds	Northern Rivers	Mid-North Coast	Hunter	Metropolitan	Illawarra	South Coast	ACT	NSW / ACT
Town Sample	Wilcannia	Broken Hill	Hay	Moree	Armidale	Dubbo	Orange	Wagga Wagga	Goulburn	Grafton	Coffs Harbour	Newcastle	Sydney	Wollongong	Bega	Canberra	Overall
January	drr	wtr	wtr	drr	drr	av	av	av	drr	av	drr	drr	drr	drr	av	drr	drr
February	drr	av	drr	drr	wtr	drr	av	drr	drr	drr	drr	drr	drr	drr	wtr	av	drr
March	drr	drr	drr	av	wtr	wtr	wtr	wtr	wtr	av	drr	av	av	wtr	wtr	wtr	wtr
April	av	av	drr	drr	drr	drr	drr	drr	drr	drr	drr	av	drr	drr	drr	drr	drr
May	wtr	wtr	wtr	wtr	wtr	wtr	wtr	wtr	wtr	wtr	drr	wtr	wtr	wtr	av	wtr	wtr
June	av	wtr	av	wtr	av	wtr	drr	drr	wtr	av	av	wtr	wtr	wtr	wtr	wtr	wtr
July	drr	drr	av	drr	drr	av	av	drr	wtr	drr	wtr	drr	wtr	wtr	wtr	wtr	wtr
August	drr	drr	drr	wtr	wtr	drr	drr	drr	drr	wtr	wtr	wtr	wtr	drr	av	drr	wtr
September	drr	drr	drr	drr	drr	drr	drr	drr	drr	wtr	wtr	drr	av	drr	drr	drr	drr
October	drr	drr	drr	drr	drr	drr	drr	drr	drr	drr	drr	drr	drr	drr	drr	drr	drr
November	av	av	av	drr	wtr	av	wtr	av	wtr	wtr	wtr	wtr	wtr	av	wtr	av	wtr
December	wtr	av	wtr	wtr	wtr	wtr	av	wtr	wtr	wtr	drr	wtr	wtr	wtr	av	wtr	wtr
Overall	av	av	av	av	wtr	av	drr	drr	av	av	drr	wtr	wtr	wtr	wtr	av	wtr

2025 NEW SOUTH WALES / ACT - Estimated Rainfall Expectation by %

REGION	Upper West	Lower West	Riverina	NW Slopes & Plains	Northern Tablelds	CW Slopes & Plains	Central Tablelds	SW Slopes & Plains	Southern Tablelds	Northern Rivers	Mid-North Coast	Hunter	Metropolitan	Illawarra	South Coast	ACT	NSW / ACT
Town Sample	Wilcannia	Broken Hill	Hay	Moree	Armidale	Dubbo	Orange	Wagga Wagga	Goulburn	Grafton	Coffs Harbour	Newcastle	Sydney	Wollongong	Bega	Canberra	Overall
January	-51%	540%	70%	-49%	-49%	16%	-11%	-6%	-17%	6%	-41%	-86%	-25%	-45%	-15%	-38%	-18%
February	-93%	-16%	-100%	-71%	75%	-73%	-10%	-54%	-50%	-65%	-53%	-18%	-43%	-56%	163%	6%	-27%
March	-57%	-49%	-42%	7%	177%	81%	93%	45%	93%	-5%	-51%	2%	3%	58%	58%	163%	21%
April	-41%	-31%	-60%	-95%	-56%	-61%	-82%	-71%	-88%	-76%	-86%	-1%	-38%	-51%	-72%	-90%	-60%
May	240%	260%	36%	99%	24%	54%	40%	71%	43%	24%	-31%	93%	30%	91%	-20%	72%	48%
June	7%	77%	-21%	31%	-11%	146%	-45%	-51%	146%	10%	-3%	304%	366%	154%	108%	40%	111%
July	-97%	-88%	-14%	-51%	-58%	-19%	7%	-24%	98%	-43%	40%	-47%	52%	69%	131%	61%	11%
August	-79%	-93%	-67%	44%	131%	-54%	-70%	-48%	-64%	194%	271%	57%	81%	-19%	-17%	-59%	20%
September	-100%	-99%	-94%	-99%	-78%	-100%	-94%	-85%	-59%	25%	63%	-33%	-7%	-29%	-64%	-74%	-51%
October	-99%	-77%	-100%	-92%	-72%	-100%	-96%	-99%	-92%	-32%	-36%	-57%	-77%	-81%	-86%	-95%	-77%
November	11%	40%	-13%	-24%	63%	-16%	35%	-9%	58%	156%	97%	14%	34%	6%	45%	15%	39%
December	152%	21%	308%	130%	70%	65%	5%	68%	75%	10%	-33%	51%	137%	143%	11%	97%	63%
Overall	-23%	39%	-12%	-14%	14%	-4%	-21%	-22%	13%	3%	-13%	35%	44%	16%	28%	7%	7%

2025 NEW SOUTH WALES / ACT - Estimated Rainfall Trend Variation by 25%

REGION	Upper West	Lower West	Riverina	NW Slopes & Plains	Northern Tablelds	CW Slopes & Plains	Central Tablelds	SW Slopes & Plains	Southern Tablelds	Northern Rivers	Mid-North Coast	Hunter	Metropolitan	Illawarra	South Coast	ACT	NSW / ACT
Town Sample	Wilcannia	Broken Hill	Hay	Moree	Armidale	Dubbo	Orange	Wagga Wagga	Goulburn	Grafton	Coffs Harbour	Newcastle	Sydney	Wollongong	Bega	Canberra	Overall
January	drr	wtr	wtr	drr	drr	av	av	av	av	av	drr	drr	av	drr	av	drr	av
February	drr	av	drr	drr	wtr	drr	av	drr	drr	drr	drr	av	drr	drr	wtr	av	drr
March	drr	drr	drr	av	wtr	wtr	wtr	wtr	wtr	av	drr	av	av	wtr	wtr	wtr	av
April	drr	drr	drr	drr	drr	drr	drr	drr	drr	drr	drr	av	drr	drr	drr	drr	drr
May	wtr	wtr	wtr	wtr	av	wtr	wtr	wtr	wtr	av	drr	wtr	wtr	wtr	av	wtr	wtr
June	av	wtr	av	wtr	av	wtr	drr	drr	wtr	av	av	wtr	wtr	wtr	wtr	wtr	wtr
July	drr	drr	av	drr	drr	av	av	av	wtr	drr	wtr	drr	wtr	wtr	wtr	wtr	av
August	drr	drr	drr	wtr	wtr	drr	drr	drr	drr	wtr	wtr	wtr	wtr	av	av	drr	av
September	drr	drr	drr	drr	drr	drr	drr	drr	drr	av	wtr	drr	av	drr	drr	drr	drr
October	drr	drr	drr	drr	drr	drr	drr	drr	drr	drr	drr	drr	drr	drr	drr	drr	drr
November	av	wtr	av	av	wtr	av	wtr	av	wtr	wtr	wtr	av	wtr	av	wtr	av	wtr
December	wtr	av	wtr	wtr	wtr	wtr	av	wtr	wtr	av	drr	wtr	wtr	wtr	av	wtr	wtr
Overall	av	wtr	av	av	av	av	av	av	av	av	av	wtr	wtr	av	wtr	av	av

2025 NEW SOUTH WALES / ACT - Estimated Average Maximum Temperature

Region	Upper West	Lower West	Riverina	NW Slopes & Plains	Northern Tablelds	CW Slopes & Plains	Central Tablelds	SW Slopes & Plains	Southern Tablelds	Northern Rivers	Mid-North Coast	Hunter	Metropolitan	Illawarra	South Coast	ACT	NSW / ACT
Town Sample	Wilcannia	Broken Hill	Hay	Moree	Armidale	Dubbo	Orange	Wagga Wagga	Goulburn	Grafton	Coffs Harbour	Newcastle	Sydney	Wollongong	Bega	Canberra	Overall
January	34	33	33	32	26	33	27	33	29	28	26	28	26	25	24	30	**29**
February	36	33	33	36	27	35	27	33	29	31	28	29	26	26	25	30	**30**
March	36	33	32	34	25	32	25	31	26	29	27	28	26	26	25	28	**29**
April	30	27	27	30	22	28	21	27	23	27	25	25	24	23	23	23	**25**
May	25	22	23	26	19	23	17	21	18	25	24	23	23	22	21	19	**22**
June	17	15	16	18	15	16	11	15	12	22	20	18	18	18	18	13	**16**
July	16	14	14	15	10	13	7	12	9	20	19	16	16	16	15	11	**14**
August	20	18	17	20	15	16	11	15	12	23	20	19	19	18	18	13	**17**
September	25	23	22	24	17	21	15	19	16	24	22	21	21	20	19	17	**20**
October	28	26	25	29	22	26	19	23	20	28	25	26	24	24	20	21	**24**
November	31	28	29	30	22	29	22	28	23	28	24	25	24	23	22	24	**26**
December	34	32	30	31	23	30	23	29	23	28	26	26	25	23	22	25	**27**
Overall	**27**	**25**	**25**	**27**	**20**	**25**	**19**	**24**	**20**	**26**	**24**	**24**	**23**	**22**	**21**	**21**	23

2025 NEW SOUTH WALES / ACT - Estimated Average Minimum Temperature

Region	Upper West	Lower West	Riverina	NW Slopes & Plains	Northern Tablelds	CW Slopes & Plains	Central Tablelds	SW Slopes & Plains	Southern Tablelds	Northern Rivers	Mid-North Coast	Hunter	Metropolitan	Illawarra	South Coast	ACT	NSW / ACT
Town Sample	Wilcannia	Broken Hill	Hay	Moree	Armidale	Dubbo	Orange	Wagga Wagga	Goulburn	Grafton	Coffs Harbour	Newcastle	Sydney	Wollongong	Bega	Canberra	Overall
January	19	20	19	20	11	18	13	17	11	18	18	18	19	16	14	14	**17**
February	19	20	17	20	13	19	13	18	13	20	19	20	21	18	15	15	**17**
March	19	20	18	20	12	19	13	18	14	19	19	19	20	18	14	14	**17**
April	14	13	12	15	7	13	8	11	13	16	16	16	16	15	11	9	**13**
May	13	13	12	13	6	12	7	10	7	14	14	15	15	13	8	8	**11**
June	6	5	5	6	0	5	2	4	7	10	9	10	10	9	5	2	**6**
July	5	3	2	2	-1	3	1	2	2	5	6	9	8	6	3	1	**4**
August	7	5	5	5	0	4	3	3	1	7	8	11	10	8	3	3	**5**
September	10	8	8	9	3	6	3	4	3	11	12	13	13	11	5	3	**7**
October	12	10	9	11	3	6	5	5	3	13	13	15	14	11	6	6	**9**
November	16	16	15	17	9	14	10	13	4	17	16	17	17	14	12	10	**14**
December	18	19	17	19	12	17	12	16	9	19	18	19	19	17	14	14	**16**
Overall	13	13	12	13	6	11	8	10	7	14	14	15	15	13	9	8	**11**

2025 NEW SOUTH WALES / ACT - Estimated Degree of Variation from the Maximum Average

Region	Upper West	Lower West	Riverina	NW Slopes & Plains	Northern Tablelds	CW Slopes & Plains	Central Tablelds	SW Slopes & Plains	Southern Tablelds	Northern Rivers	Mid-North Coast	Hunter	Metropolitan	Illawarra	South Coast	ACT	NSW / ACT
Town Sample	Wilcannia	Broken Hill	Hay	Moree	Armidale	Dubbo	Orange	Wagga Wagga	Goulburn	Grafton	Coffs Harbour	Newcastle	Sydney	Wollongong	Bega	Canberra	Overall
January	-2.9	-0.9	-2.1	-1.8	-0.8	-0.9	-0.8	0.7	0.3	-2.5	-1.0	2.5	-0.4	0.0	-3.6	-0.9	-0.9
February	0.2	0.4	-0.2	2.2	2.0	2.4	1.0	2.0	2.2	0.7	1.3	3.7	0.5	1.4	-1.9	1.8	1.2
March	3.8	3.5	2.5	2.7	1.9	2.9	1.7	3.5	2.5	1.0	0.9	3.7	1.7	1.9	0.6	2.3	2.3
April	2.3	2.4	2.3	2.6	1.9	2.5	2.3	4.1	2.4	1.2	0.8	2.3	1.9	1.2	0.3	2.3	2.1
May	3.1	2.5	3.0	3.1	3.3	3.2	2.6	3.8	2.4	2.2	2.4	2.6	3.2	2.2	1.2	2.7	2.7
June	-0.9	-1.3	0.2	-0.9	1.8	-0.2	-0.3	1.1	0.0	1.4	1.0	0.9	1.3	0.4	0.8	-0.1	0.3
July	-1.9	-1.5	-2.1	-3.6	-2.7	-3.0	-2.7	-0.9	-2.4	-1.1	-0.3	-0.6	-0.1	-1.3	-1.7	-2.2	-1.8
August	0.2	0.3	-0.1	-0.3	0.9	-1.1	-0.4	0.4	-0.9	0.1	0.5	1.2	1.0	0.2	-0.4	-0.8	0.1
September	0.4	0.9	0.1	-0.6	-1.2	-0.2	-0.1	1.6	-0.4	-1.2	-0.5	0.7	0.5	-0.2	-1.2	-1.4	-0.2
October	-0.6	0.4	-1.8	1.3	1.4	0.8	0.8	1.3	0.3	1.5	1.8	3.4	1.9	2.0	-1.7	-0.4	0.8
November	-1.5	-0.3	-0.4	-1.1	-0.5	-0.1	-0.2	2.5	0.2	-1.0	-0.7	1.9	0.1	0.8	-1.4	-0.9	-0.2
December	-1.2	0.0	-1.8	-2.4	-1.7	-2.0	-2.4	-0.4	-3.1	-1.3	-0.7	1.1	-0.6	-0.7	-2.7	-3.2	-1.4
Overall	0.1	0.5	0.0	0.1	0.6	0.4	0.1	1.7	0.3	0.1	0.4	2.0	0.9	0.7	-1.0	-0.1	0.4

2025 NEW SOUTH WALES / ACT - Estimated Degree of Variation from the Minimum Average

Region	Upper West	Lower West	Riverina	NW Slopes & Plains	Northern Tablelds	CW Slopes & Plains	Central Tablelds	SW Slopes & Plains	Southern Tablelds	Northern Rivers	Mid-North Coast	Hunter	Metropolitan	Illawarra	South Coast	ACT	NSW / ACT
Town Sample	Wilcannia	Broken Hill	Hay	Moree	Armidale	Dubbo	Orange	Wagga Wagga	Goulburn	Grafton	Coffs Harbour	Newcastle	Sydney	Wollongong	Bega	Canberra	Overall
January	-0.7	-1.1	0.4	-0.7	-2.5	0.0	-0.1	0.7	-2.2	-0.9	-1.6	-1.5	0.2	-2.8	-0.7	0.0	-0.9
February	0.1	-0.3	-0.3	0.7	0.0	1.1	0.8	1.1	0.0	0.9	-0.1	0.7	1.6	-1.2	-0.1	1.4	0.4
March	3.7	4.0	3.1	2.8	0.9	3.8	3.5	4.0	3.7	1.1	0.7	1.1	2.5	-0.4	1.3	2.8	2.4
April	2.9	1.5	2.1	1.8	-1.4	2.8	1.9	1.7	7.1	1.9	0.9	0.9	1.7	-0.9	1.7	1.6	1.8
May	5.0	5.2	5.1	4.5	1.8	5.2	4.9	4.3	4.1	5.0	2.4	3.2	3.5	-0.1	2.6	5.6	3.9
June	0.3	-0.2	0.6	-0.4	-2.1	0.1	0.8	0.0	5.8	2.3	0.4	0.0	1.0	-2.0	1.1	1.4	0.6
July	0.2	-0.9	-1.2	-2.1	-2.1	-0.4	0.5	-0.9	2.0	0.0	-1.4	0.3	0.2	-3.7	0.9	1.3	-0.5
August	1.5	0.0	0.9	0.1	-1.3	0.6	1.9	-0.3	0.7	1.5	0.0	1.2	1.2	-2.3	0.4	2.1	0.5
September	1.5	-0.6	1.3	-0.3	-2.0	-0.4	0.1	-1.1	0.2	1.0	0.6	1.3	1.7	-1.7	-0.6	0.1	0.1
October	0.7	-1.6	-1.0	-1.6	-4.2	-3.4	-0.8	-2.8	-2.6	0.0	-0.7	0.6	0.6	-2.6	-2.1	0.2	-1.3
November	1.7	0.4	1.5	0.1	-1.3	0.7	1.7	2.3	-4.5	0.9	0.0	0.9	1.3	-1.4	0.3	0.7	0.3
December	0.3	0.5	1.3	-0.2	0.3	1.3	1.6	2.4	-1.5	1.3	0.2	0.8	1.0	-0.8	0.7	1.2	0.7
Overall	1.4	0.6	1.1	0.4	-1.2	0.9	1.4	1.0	1.1	1.2	0.1	0.8	1.4	-1.7	0.5	1.5	0.7

2025 NEW SOUTH WALES / ACT - Estimated Maximum Temperature Trend

Region	Upper West	Lower West	Riverina	NW Slopes & Plains	Northern Tablelds	CW Slopes & Plains	Central Tablelds	SW Slopes & Plains	Southern Tablelds	Northern Rivers	Mid-North Coast	Hunter	Metropolitan	Illawarra	South Coast	ACT	NSW / ACT
Town Sample	Wilcannia	Broken Hill	Hay	Moree	Armidale	Dubbo	Orange	Wagga Wagga	Goulburn	Grafton	Coffs Harbour	Newcastle	Sydney	Wollongong	Bega	Canberra	Overall
January	clr	a-c	clr	clr	a-c	a-c	a-c	a-w	av	clr	a-c	wmr	av	av	clr	a-c	a-c
February	av	av	av	wmr	wmr	wmr	a-w	wmr	wmr	a-w	wmr	wmr	av	wmr	clr	wmr	wmr
March	wmr	wmr	wmr	wmr	wmr	wmr	wmr	wmr	wmr	a-w	a-w	wmr	wmr	wmr	a-w	wmr	wmr
April	wmr	wmr	wmr	wmr	wmr	wmr	wmr	wmr	wmr	wmr	a-w	wmr	wmr	wmr	av	wmr	wmr
May	wmr	wmr	wmr	wmr	wmr	wmr	wmr	wmr	wmr	wmr	wmr	wmr	wmr	wmr	wmr	wmr	wmr
June	a-c	clr	av	a-c	wmr	av	av	wmr	av	wmr	a-w	a-w	wmr	av	a-w	av	av
July	clr	clr	clr	clr	clr	clr	clr	a-c	clr	clr	av	a-c	av	clr	clr	clr	clr
August	av	av	av	av	a-w	clr	av	av	a-c	av	av	wmr	a-w	av	av	a-c	av
September	av	a-w	av	a-c	clr	av	av	wmr	av	clr	av	a-w	av	av	clr	clr	av
October	a-c	av	clr	wmr	wmr	a-w	a-w	wmr	av	wmr	wmr	wmr	wmr	wmr	clr	av	a-w
November	clr	av	av	clr	av	av	av	wmr	av	a-c	a-c	wmr	av	a-w	clr	a-c	av
December	clr	av	clr	clr	clr	clr	clr	av	clr	clr	a-c	wmr	a-c	a-c	clr	clr	clr
Overall	av	av	av	av	a-w	av	av	wmr	av	av	av	wmr	a-w	a-w	a-c	av	av

2025 NEW SOUTH WALES / ACT - Estimated Minimum Temperature Trend

Region	Upper West	Lower West	Riverina	NW Slopes & Plains	Northern Tablelds	CW Slopes & Plains	Central Tablelds	SW Slopes & Plains	Southern Tablelds	Northern Rivers	Mid-North Coast	Hunter	Metropolitan	Illawarra	South Coast	ACT	NSW / ACT
Town Sample	Wilcannia	Broken Hill	Hay	Moree	Armidale	Dubbo	Orange	Wagga Wagga	Goulburn	Grafton	Coffs Harbour	Newcastle	Sydney	Wollongong	Bega	Canberra	Overall
January	a-c	clr	av	a-c	clr	av	av	a-w	clr	a-c	clr	clr	av	clr	a-c	av	a-c
February	av	av	av	a-w	av	wmr	a-w	wmr	av	a-w	av	a-w	wmr	clr	av	wmr	av
March	wmr	wmr	wmr	wmr	a-w	wmr	wmr	wmr	wmr	wmr	a-w	wmr	wmr	av	wmr	wmr	wmr
April	wmr	wmr	wmr	wmr	clr	wmr	wmr	wmr	wmr	wmr	a-w	a-w	wmr	a-c	wmr	wmr	wmr
May	wmr	wmr	wmr	wmr	wmr	wmr	wmr	wmr	wmr	wmr	wmr	wmr	wmr	av	wmr	wmr	wmr
June	av	av	a-w	av	clr	av	a-w	av	wmr	wmr	av	av	a-w	clr	wmr	wmr	a-w
July	av	a-c	clr	clr	clr	av	av	a-c	wmr	av	clr	av	av	clr	a-w	wmr	av
August	wmr	av	a-w	av	clr	a-w	wmr	av	a-w	wmr	av	wmr	wmr	clr	av	wmr	av
September	wmr	a-c	wmr	av	clr	av	av	clr	av	a-w	a-w	wmr	wmr	clr	a-c	av	av
October	a-w	clr	a-c	clr	clr	clr	a-c	clr	clr	av	a-c	a-w	a-w	clr	clr	av	clr
November	wmr	av	wmr	av	clr	a-w	wmr	wmr	clr	a-w	av	a-w	wmr	clr	av	a-w	av
December	av	av	wmr	av	av	wmr	wmr	wmr	clr	wmr	av	a-w	a-w	a-c	a-w	wmr	a-w
Overall	wmr	a-w	wmr	av	clr	a-w	wmr	a-w	wmr	wmr	av	a-w	wmr	clr	av	wmr	a-w

VICTORIA 2025

GENERAL TRENDS

Year Overall:

Slightly drier by around 5-10% below the average
Warmer by around 0.5-1°C above the average

Summer:

Drier by around 45-50% less than average
Wettest – End of January
Driest – Second and third weeks of February
Warmer by around 1°C above the average
Hottest – Third week of December, end of January and mid-February
Coolest – First week of January

Autumn:

Slightly wetter by around 5-10% more than the average
Wettest – Second week of May
Driest – Third week of March and second half of April
Warmer by up to 2.5-3°C above the average
Hottest – First half of March
Coolest – Last three weeks of May

Winter:

Slightly wetter by around 10-15% more than the average
Wettest – Mid June and mid-July
Driest – Last week of August
Slightly cooler but within 0.5°C of the average
Hottest – First week of June and first and third weeks of August
Coolest – First week of July and first week of August

Spring:

Drier by around 40-45% less than average
Wettest –Mid-November
Driest – First half of September
Slightly warmer but within 0.5°C of the average
Hottest – Second half of November
Coolest – Second half of September to mid-October

Major Weather Events

January 2025

Date	Day	Victoria - January 2025	Melbourne	Moon
1st	Wed	Strong winds in the southeast of the State from the Grampians across to the Alpine region and down to the southeast coast. Strong bush fire risk continues for at least the next ten days, especially in the northeast of the State. Heavy rain with chance of localised flooding in some Central districts as well as in West Gippsland and on the West Coast. Unusually cool in the Upper Northeast and Wimmera districts.	Showers	
2nd	Thur	Very low maximum temperatures and occasional strong wind gusts in the West Gippslands and in the higher reaches of the south Wimmera region over the next couple of days. Chance of downpours in the Mt Baw Baw area. Scattered rain activity spreads from the western borders to across the State over the next couple of days.		
3rd	Fri	Strong winds with occasional gusts at times in the West Gippsland, Western Plains and West Coast regions. Lingering showers in the Lakes region.	Showers	
4th	Sat	Strong winds and scattered showers in the southeast of the State over the next couple of days. Heavy rain and a high tornado risk with the passage of a thunderstorm in the Melbourne vicinity. Very low temperatures in the Alpine regions including in the Upper Northeast and West Gippsland may see some frosts over the next couple of nights and chance of snowflurries in the Victorian Alps.	Rain, Th, Tornado	
5th	Sun	Severe cold spell for the next few days over the State with the chance of snow in the Dandenong Ranges and frosts in the Upper Northeast higher reaches. Unsettled blustery winds in the Upper Northeast to West Gippsland over the next couple of days.	Showers	
6th	Mon	Clearing to a mostly dry spell over the next few days.	Showers	
7th	Tue			**1Q XhN**
8th	Wed	Chance of frosts in the upper reaches of the Upper Northeast, including on Mt Hotham.		**P13**
9th	Thu	Occasional strong wind gusts in West Gippsland.	Showers	
10th	Fri	Lightning strikes may start fires, possibly along the VIC-SA border areas. A thunderstorm in the Wimmera region may cause some brief localised flooding. Heavy rain also in the North Mallee and Central regions. Unusually low maximum temperatures continue to affect a number of areas in the South Wimmera area. Very blustery at times in the West Gippsland.	Showers	
11th	Sat	Widespread thunderstorms storms and heavy rain in the Grampians and Macedon Ranges, with chance of flash flooding in some areas. Scattered shower activity elsewhere over the next couple of days.	Showers	
12th	Sun	The next 3-4 weeks may be dominated by mostly high pressure systems, and overall be slightly wetter and warmer than is normal for this time of year, with most of the rainfall expected over the last week of this month and into the start of next month.		
13th	Mon	Dry over the next few days.		**^**
14th	Tue			**F**
15th	Wed	Very hot spell in the Barwon to the Gippslands over the next few days, with unusually high overnight minimum temperatures likely, particularly in the Bairnsdale and Portland districts.		
16th	Thur	Strong and blustery winds which may cause some property damage in the Swan Hill area of the Lodden Mallee region. Some shower activity in the Southwest, later spreading across the State by tomorrow.		
17th	Fri			
18th	Sat	Lingering showers in the Gippslands and Central Highlands over the next couple of days.	Showers	
19th	Sun			
20th	Mon	Mostly dry over the coming week.		**XhS**
21st	Tue			**A**
22nd	Wed		Showers, Th	**3Q**
23rd	Thu	Mostly dry across the State although chance of some isolated showers in the Central Highlands.		
24th	Fri			
25th	Sat			
26th	Sun	Thunderstorms with blustery winds and large hail in the Melbourne area and Gippsland, that may cause some damage to property. Very hot in the Mallee region, with maximum temperatures likely to reach into the lower 40's.		
27th	Mon	Very warm and humid night in the Melbourne area as well as in the Hume region, with chance of some isolated showers in the Central Highlands.		**V**
28th	Tue	A cold front crosses the State over the next few days bringing widespread rain and storms that may cause flash flooding for a time in the Northwest Mallee, particularly around Mildura.	Rain	
29th	Wed	Heavy rain at times in the Grampians.	Showers	
30th	Thur	Very heavy rain in the Casterton area of the Barwon region, with chance of flash flooding.	Rain	**N**
31st	Fri	Heavy rain persists throughout the State, with particular heavy falls along the southwest coasts.	Showers	

February 2025

Date	Day	Victoria - February 2025	Melbourne	Moon
1st	Sat	Rain eases to showers.		
2nd	Sun	Mostly dry over the nexxt 3-4 days.		**P10**
3rd	Mon			**XhN**
4th	Tue			
5th	Wed			**1Q**
6th	Thu	Unusually cool spell in the Alpine region over the next couple of days with some locations maximum temperatures possibly not reaching into the double digits and chance of sub-zero minimas overnight, particularly on Mt Hotham. Chance of isolated showers in the eastern half of the State over the next couple of days.		
7th	Fri	Very cold overnight in the Loddon-Mallee region, particularly in the Euhuca district.		
8th	Sat			
9th	Sun	Isolated showers in the Central Highlands over the next few days likely. Mostly dry elsewhere.		**^**
10th	Mon			
11th	Tue	The next 3-4 weeks may be hotter and drier than is typical for this time of year, with very long dry spells and only brief passing wet spells, the wettest being towards the middle of the second week of March. It may be particularly hot and humid towards the end of February and into early March as a thunderstorm embedded trough crosses the State.		
12th	Wed	Unusually cold overnight over much of the State, particularly in the Yarra Ranges. Dry spell throughout the State over the coming week.		
13th	Thur	Very high maximum temperatures throughout most of the State over the next several days.		**F**
14th	Fri			
15th	Sat	Very hot day across the State, particularly in the Loddon-Mallee and Wimmera regions, where daily maximas may reach into the low to mid-40's in some districts, particularly in the Mildura area.	Showers	
16th	Sun		Showers	**XhS**
17th	Mon	Brief cooler spell in the southwest of the State, particularly in the Weeaproinah area.		
18th	Tue	Cooler overnight in the west of the State over the next couple of days.		**A**
19th	Wed			
20th	Thu		Showers, Th	
21st	Fri	Stormy air in the Gippslands region with very strong and blustery winds and outbursts of heavy rain including the chance of brief localised flooding, particularly in the Genoa district over the next couple of days.		**3Q**
22nd	Sat	Showers spread into the Alpine and Loddon-Mallee regions for a time.		
23rd	Sun	Scattered storms across the Gippslands and Lodden Mallee regions with chance of flash flooding in some places.	Showers, Th	**V**
24th	Mon			
25th	Tue	Heavy rain in the Hume region, particularly in the Tarrawingee area. Very hot over the next 3-4 days in the Melbourne region, with daily maximas into the high 30's.	Showers, Th	
26th	Wed	Conditions in the Melbourne region over the next couple of weeks may be considerably humid with an almost tropical atmosphere prevailing. Maximum temperatures may also be considerably high in the Gippslands, Wimmera, and Western Plains and Coast regions over the next few days.	Showers, Th	
27th	Thur	Humid with isolated showers in the Southwest to southern Wimmera.		
28th	Fri	Very warm overnight in the Southwest, particularly around Cape Otway.	Showers	**N**

March 2025

Date	Day	Victoria - March 2025	Melbourne	Moon
1st	Sat	Heavy rain and hail for a time in the Northwest Mallee, including around Mildura. Showers elsewhere. Chance of strong gusty winds that may cause damage to property.		
2nd	Sun			**P7 XhN**
3rd	Mon	Showers in the west of the State and chance of hail showers in the Melbourne area that may cause some damage to property.		
4th	Tue	Showers clear in the west and spread into the east.	Showers, Th	
5th	Wed	Chance of large hail in the Gippsland region, particularly around the Maffra area.	Showers	
6th	Thu	Light showers activity throughout the State.		
7th	Fri	Very cool in the Alpine region overnight. Showers persist east of the Alpine region.		**1Q**
8th	Sat	A storm in the Barwon region may bring large hail and tornadic formation winds, with particular risk in the Mirranatwa area.		**^**
9th	Sun	Cool spell in the Alpine region, particularly at Mt Baw Baw and Mt Hotham.	Showers	

Date	Day	Victoria - March 2025	Melbourne	Moon
10th	Mon	Pockets of heavy rain with chance of flash flooding in the Gippslands, particularly in the Mt Nowa Nowa and Orbost districts.		
11th	Tue	Over the next few weeks until mid-April, temperatures, both day and overnight, may be warmer than is typical for this time of year, while rainfall may be around the average to slightly below average over most of the State, except in the Alpine districts, which can expect above average rainfall.		
12th	Wed	Lingering showers in the Gippsland.		
13th	Thur	Very hot spell across the State with maximum temperatures into the upper 30's common in many districts. It may be particularly hot in the Echuca area, with maximum temperatures once more up into the early 40's. Mostly dry over the coming week.	Showers, Th	
14th	Fri			**F**
15th	Sat			**XhS**
16th	Sun	Some frosts likely overnight in the Alpine region.		
17th	Mon			
18th	Tue		Showers	**A**
19th	Wed			
20th	Thu	Very warm around the West Coast about this time, with maximum temperatures in the mid-30's.	Fg	
21st	Fri			
22nd	Sat	Chance of isolated showers briefly in the Central Highlands over the next couple of days.		
23rd	Sun			**3Q V**
24th	Mon	Cool night in the Gippslands, particularly in the Bairnsdale district.		
25th	Tue		Showers	
26th	Wed	Unusually mild overnight in the Lodden-Mallee region, particularly in the Mildura district. A band of showers crosses the State over the next couple of days.	Showers	
27th	Thur			
28th	Fri	Cold overnight in the Upper North region, particularly in the Bendigo district.		
29th	Sat		Showers	**N XhN**
30th	Sun	A band of rain crosses over the northern border regions of the State bringing outbreaks of rain from the northern Mallee across the High Country, and into the Lakes region over the next couple of days.	Showers	**P4**
31st	Mon	Very heavy rain in the Alpine to the Lakes regions, particularly around Mt Buller with chance of localised flooding.	Fg	

April 2025

Date	Day	Victoria - April 2025	Melbourne	Moon
1st	Tue	Very hot across the State over the next couple of days with many districts seeing maximum temperatures into the upper 30's prevailing. Also, very warm over the next couple of nights in the Melbourne area, with minimum temperatures possibly sitting in the mid-to-upper 30's in some places.		
2nd	Wed		Rain, Th	
3rd	Thu	A wet front crosses the State bringing outbreaks of rain in most districts over the next couple of days.	Showers, Th	
4th	Fri	A cold spell moves over the State around this time, with overnight minimums unlikely to reach double digits in many parts of the State over the next couple of nights. Very cold in the Alpine region overnight, with sub-zero minimum temperatures prevailing, particularly on Mt Hotham and Mt Baw Baw.		**^**
5th	Sat			**1Q**
6th	Sun			
7th	Mon	Scattered showers develop in the west of the State spreading to most districts before becoming isolated in a couple of days time.	Rain	
8th	Tue	Cold overnight in the Grampians.	Showers	
9th	Wed	Very cold day in the Alpine region, particularly at Mt Hotham, where maximum temperatures may not be far off from frost level. Strong blustery winds at times in the Gippslands, particularly around the coasts.	Showers	
10th	Thur	Cold overnight in the Lower Northeast.	Showers	
11th	Fri	The next 3-4 weeks may bring above average temperatures throughout the State and above average rainfall in most districts, with only the Central and West Gippslands being closer to average. Generally, high pressure may prevail over the next couple of weeks, before a westerly change brings a series of fronts across the State during the half of this lookahead period.		**XhS**
12th	Sat	Mostly dry over the coming week except for some brief showery outbreaks mostly in the Central Highlands.		

Date	Day	Victoria - April 2025	Melbourne	Moon
13th	Sun			F
14th	Mon			A
15th	Tue			
16th	Wed	Cooler overnight temperatures settle in in many parts of the State over the next few nights with many districts dropping into the single digits overnight. Chance of isolated drizzly showers briefly along the coasts.		
17th	Thu	Cold overnight with frosts in the Alpine region, particularly at Mt Hotham.		
18th	Fri	Mostly dry spell over the coming week to 10 days.		
19th	Sat	Warm spell across the State over the next couple of days.		V
20th	Sun			
21st	Mon	Unusually warm night in the Mallee.		3Q
22nd	Tue		Fg	
23rd	Wed	Warm spell across the State over the next few days. Chance of isolated showers in the far northwest of the Mallee.		
24th	Thur	Particularly warm day in the Wimmera where maximum temperatures in the early 30's can be anticipated.	Showers, Th	
25th	Fri	Some showers briefly in the Gippsland to the Lakes region and in the Central Highlands.		
26th	Sat	Dry spell over the remainder of the month.		XhN
27th	Sun	Cold overnight in the Northeast to Central parts.		
28th	Mon			P3 N
29th	Tue			
30th	Wed			

May 2025

Date	Day	Victoria - May 2025	Melbourne	Moon
1st	Thu	A wet front crosses the State over the next few days bringing some shower activity in most districts, but particularly in the eastern half.	Showers	
2nd	Fri		Fg	^
3rd	Sat	Cold day in the Alpine region, with maximum temperatures possibly not reaching into the double digits.		
4th	Sun			
5th	Mon			1Q
6th	Tue	Cold overnight in the Gippslands to Central regions.		
7th	Wed	An intense front moves across the State over the next few days bringing stormy conditions with widespread rain throughout with heavy falls at times, particularly into central parts.	Rain	
8th	Thur	Very heavy rain for a time in the Grampians region. Strong and blustery winds in the Gippsland.	Rain	
9th	Fri		Showers	XhS
10th	Sat			
11th	Sun	The next upcoming 3-4 weeks may be wetter and warmer than is typical for this time of year, particularly in the Alpine regions, where some good snowfall volumes can also be expected. Overall, pre-frontal rain-bearing west to northwest cloud bands may prevail over the next few weeks.		A
12th	Mon	Another front moves over the State during the next few days bringing further widespread outbreaks of rain.	Showers, Th	
13th	Tue	Heavy rain for a time in the Melbourne and Hume regions with chance of pockets of flash flooding and hail. Chance of a line of microburst winds crossing the Southern Mallee, Woodend area which may cause damage in its path. Strong blustery winds also in the Central Highlands, particularly in the Eildon area, with chance of damage and accompanying outbursts of heavy rain followed by flash-flooding. Thunderstorms with gusty winds in the Hume region.	Showers	F
14th	Wed			
15th	Thur	Mild minimum temperatures overnight in the Gippslands. Storm activity begins to ease over the State around this time, with showers still lingering around the southern coasts for the next couple of days.		
16th	Fri			V
17th	Sat	Mild maximum temperatures in the Lodden-Mallee region.		
18th	Sun	A brief spell of showers brushes the southwest of the State.		
19th	Mon	Mostly dry over the next 4-5 days.		
20th	Tue	Cold over the next couple of nights in the Mallee.		3Q
21st	Wed	Milder maximum temperatures in the Alpine region over the next couple of days.	Fg	
22nd	Thu	Very cold overnight in the Wimmera and Gippslands.		
23rd	Fri	Milder maximum temperatures over the State in the next couple of days.		XhN

Date	Day	Victoria - May 2025	Melbourne	Moon
24th	Sat	Chance of tornado system forming attached to a convective cell as it crosses the Barwon region around the Lara area. Very hot day in the Mallee. Very warm day in the Lodden-Mallee particularly in the Mildura district.	Showers	
25th	Sun	Widespread rain across the State over the next couple of days. Heavy rain in the Grampians. Unusually warm in the far southwest of the State.	Showers	
26th	Mon			P5
27th	Tue	A band of widespread storms and heavy rain crosses the State over the next few days. Storms with heavy rain, large hail, and blustery winds in the Lodden Mallee region, which may cause damage to property and crops, particularly in the Ouyen area. Severe hail storm possible in the Heathmere area.	Rain	N
28th	Wed	Very heavy rain in the Castlemaine district.		
29th	Thur	Shower activity persists in most districts for the next few days until the end of the month and into the start of next month.		^
30th	Fri	Colder in the Alpine region over the coming days with regular frosts likely in many areas.	Showers	
31st	Sat			

June 2025

Date	Day	Victoria - June 2025	Melbourne	Moon
1st	Sun	Widespread precipitation persists across the State over the next couple of days. Chance of snowflurries on the Victoria Alps.		
2nd	Mon			
3rd	Tue	Showers persist along the coasts but clear in the north of the State.		1Q
4th	Wed			
5th	Thu			XhS
6th	Fri	Very cold overnight in the Hume to Mallee regions with some patches of frost likely in sheltered parts and very cold with widespread frosts in the Gippslands.		
7th	Sat	Some showers develop in the western half of the State around this time with the arrival of an intense cold front, which brings widespread and often heavy rain into most districts over the next few days.		A
8th	Sun	Strong winds with gusts possibly to gale force at times in the Victoria Alpine region. Mild overnight temperatures in the Gippslands.	Rain	
9th	Mon	Cold overnight with light snow and frosts in the Victoria Alpine region and snowfields over the next couple of days. Widespread heavy precipitation in most districts, with chance of localised flooding along river catchments.		
10th	Tue	Very cold with severe frosts likely in the Alpine region, particularly on Mt Hotham. Shower activity persists over much of the State over the coming week.	Showers	
11th	Wed	The next 3-4 weeks may be cold, and while unusually drier than normal in the north, the west, particularly in the Gippslands, may be wetter than normal for this time of year.		F
12th	Thur	Snowflurries in the Victorian Alps.	Fg	V
13th	Fri	Milder spell in the Yarra Ranges and the Melbourne area over the next couple of days.		
14th	Sat			
15th	Sun		Showers	
16th	Mon			
17th	Tue	Some spotty shower activity on most days over the remainder of the month, particularly in the east and along the southern coasts.		
18th	Wed			
19th	Thu	Milder spell in the north of the State over the next couple of days including in the Hume to the Alpine region.		3Q XhN
20th	Fri	Chance of frosts on the Western Plains.	Showers	
21st	Sat		Showers	
22nd	Sun		Showers	
23rd	Mon	Very cold in the Alpine region, with maximum temperatures unlike to rise above frost level throughout. Snowflurries in Falls Creek to Mt Hotham.	Showers	P8
24th	Tue	Very cold across the State with light snow in the Victorian Alps and daytime temperatures possibly not rising above frost level in the Hume region. Chance of frosts in the Melbourne suburbs.	Fg	
25th	Wed	Widespread fog across Melbourne, slow to clear and likely to return after dark. Mild daytime temperatures in the Hume region. Scattered frosts over the next couple of nights in the West Coast, Wimmera and Mallee regions.	Fg	N
26th	Thur	Mild overnight on Gabo Island.	Fg	^

Date	Day	Victoria - June 2025	Melbourne	Moon
27th	Fri		Showers	
28th	Sat	Light snow in the Victorian Alps over the next couple of days. Very cold day in the Barwon region.		
29th	Sun	Chance of some heavy rain for a time in the East Gippslands to the Lakes region.	Showers	
30th	Mon	Snowflurries in Falls Creek to Mt Hotham.	Showers	

July 2025

Date	Day	Victoria - July 2025	Melbourne	Moon
1st	Tue	Snow in the Victorian Alps and ski fields. Showers elsewhere over the next couple of days.		
2nd	Wed	Very cold overnight in the Alpine region, with severe frosts likely, particularly in the Mt Hotham area.		XhS
3rd	Thu	Cold with scattered frosts over the next couple of nights in the Central, Mallee and Wimmera regions.		1Q
4th	Fri			
5th	Sat	Some patches of frost in the Gippsland.		A
6th	Sun	Snowflurries in Falls Creek to Mt Hotham. Light shower activity in the Wimmera, spreading through the rest of the State over the next couple of days.	Rain	
7th	Mon	Snow in the Victorian Alps.	Rain	
8th	Tue	Milder maximum temperatures. Very heavy rain in the Gippslands with chance of pockets of flash flooding possible. Strong and blustery winds with gale gusts at times in the Alpine region, particularly at Mt Hotham.	Showers	
9th	Wed	Light snow in the Victoria Alpine region and snowfields over the next couple of days.	Showers	
10th	Thur	Scattered rain activity throughout most of the State daily over the next 8-10 days.	Showers	V
11th	Fri	The next 3-4 weeks may bring average to above average rainfall for this time of year, and near average temperatures. At times, snow may fall to low levels and some frosts may be severe in places, and fogs widespread. Snow in the Victorian Alps over the next couple of days. Very heavy rain likely around this time in the Southwest of the State.		F
12th	Sat		Showers	
13th	Sun	Snowflurries in Falls Creek to Mt Hotham. Very strong and unsettled winds across the Gippsland region.		
14th	Mon	Snow in the Victoria Alps over the next 3-4 days. Heavy rain in the southwest of the State over the next couple of days.	Showers	
15th	Tue	Heavy precipitation spreads from the Southwest, through the Central Highlands and eastward into the High Country and Northeast regions over the next few days with showers elsewhere.	Showers	
16th	Wed		Showers, Fg	
17th	Thu		Showers, Fg	XhN
18th	Fri	Snowflurries in Falls Creek to Mt Hotham.		3Q
19th	Sat	Mostly dry in the west of the State with showers persisting in the Gippsland to the Lakes regions over the next couple of days.	Fg	
20th	Sun		Fg	
21st	Mon	Showers clear in the east while return in the west, and spread throughout most of the State once more over the next few days.		P11
22nd	Tue	Light snow in the Victoria Alpine region and snowfields over the few of days.	Showers	
23rd	Wed		Showers	^
24th	Thur		Showers	
25th	Fri	Very cold with severe frosts in the Alpine region.		N
26th	Sat	Snowflurries in the Victorian Alps.		
27th	Sun	Light snow in the Victorian Alpine region and snowfields over the next few days. Heavy precipitation elsewhere, particularly in central districts.	Rain	
28th	Mon	Very cold in the Alpines around this time, with maximum temperatures unlikely to rise above frost levels on a number of Alps, including Mt Hotham.	Showers	
29th	Tue		Showers	
30th	Wed	Mostly dry over the next couple of days.	Fg	XhS
31st	Thu	Very cold spell throughout the State over the next few days, with widespread frosts overnight.	Fg	

August 2025

Date	Day	Victoria - August 2025	Melbourne	Moon
1st	Fri	Mostly dry over the next few days except for some isolated showery outbreaks mostly in the Central Highlands.		
2nd	Sat			1Q A
3rd	Sun			
4th	Mon	Warm spell across the State over the next couple of days with maximum temperatures in the late teens to early 20's in most districts. Particularly mild in the Mildura district.		
5th	Tue	Mild overnight in the Gippsland with light showers in central parts of the State.	Showers, Th	
6th	Wed	Light snow in the Victorian Alps, and some showers elsewhere over the next couple of days.	Showers	V
7th	Thur			
8th	Fri	Light showers in the west of the State over the next couple of days.	Showers	
9th	Sat			F
10th	Sun	Mostly dry day except for chance of isolated drizzly showers in the Gippsland.	Showers	
11th	Mon	The next 3-4 weeks may be much drier than is typical for this time of year, with maximum temperatures subsequently above normal, and minimum temperatures nearer to their seasonal average. Snowflurries in the Victorian Alps over the next couple of days and light showers elsewhere.		
12th	Tue			
13th	Wed	Light snow in the Victoria Alps over the next few days with light showers elsewhere, briefly contracting to the coastal fringes before returning to widespread in a couple of days.	Showers	XhN
14th	Thu	Very cold in inland parts with scattered frosts including in the Lodden-Mallee and Wimmera regions over the next couple of days.	Showers	
15th	Fri		Showers	P12
16th	Sat			3Q
17th	Sun			
18th	Mon	Light showers daily in most parts of the State over the next week.		
19th	Tue			^
20th	Wed	Mild night in the Lakes Entrance district. Very strong and blustery winds for a time across the Grampians, particularly around Mt mayiam.	Showers	
21st	Thur		Showers	
22nd	Fri	Heavy precipitation likely in the Alpine region around this time. Snowflurries in the Falls Creek to Mt Hotham region.	Showers	
23rd	Sat	Cold spell across the State over the next couple of days, with scattered frosts in many districts. Very cold day in the Alpine region, particularly on Mt Hotham, with both maximum and minimum temperatures are expected remain in the frost levels.	Showers, Fg	N
24th	Sun			
25th	Mon	Mostly dry over the next couple of days.		
26th	Tue			XhS
27th	Wed	Chance of heavy rain in the South West region with isolated precipitation elsewhere over the next couple of days.	Showers	
28th	Thu	Scattered frosts across the State over the next few days.	Showers	
29th	Fri	Mostly dry throughout the State until the end of the month.		
30th	Sat			A
31st	Sun			1Q

September 2025

Date	Day	Victoria - September 2025	Melbourne	Moon
1st	Mon	Isolated showers in the Gippsland, mostly dry elsewhere.		
2nd	Tue	Mostly dry throughout the State over the next week.		V
3rd	Wed			
4th	Thur			
5th	Fri	Chance of isolated showers along the southern coastal fringes.		
6th	Sat	Unusually warm spell across the State over the next few days, with a number of districts possibly seeing maximum temperatures into the mid to upper 20's.		
7th	Sun	Mild overnight temperatures in the Hume region.		
8th	Mon			F
9th	Tue	Very warm day in the Lodden-Mallee region, particularly in the Mildura district, where maximum temperatures into the upper 20's to early 30's is expected.		XhN
10th	Wed	The passage of a cold front may bring very strong winds in the Barwon district that may cause some localised damage. Mild overnight temperatures in the Melbourne area. Brief showery spells scattered over the State during the next couple of days.		

Date	Day	Victoria - September 2025	Melbourne	Moon
11th	Thu	Generally, the next 3-4 weeks may be drier than average, with a number of districts struggling to even reach half of their norm for this time of year. As a consequence of the drier conditions, maximum daily temperatures may be warmer than average, while overnight minimums may be slightly cooler than normal.		**P9**
12th	Fri	Overnight frosts likely in the Lodden-Mallee region, particularly in the Echuca district as well as in patches in the Gippsland.		
13th	Sat	Very cold spell in the Alpine region over the next couple of days. Maximum temperatures may not rise out of frost levels on a number of peaks. Showers over the southern half of the State.	Showers	
14th	Sun	Very cold at Mt Hotham, with severe overnight frosts. Showers linger in the Gippsland.		**3Q**
15th	Mon	Very cold in the Barwon and Gippslands around this time with widespread frosts likely. Mostly dry throughout the State over the next several days.		**^**
16th	Tue			
17th	Wed			
18th	Thur			
19th	Fri		Showers	
20th	Sat	A wet front crosses the State over the next couple of days bringing showers into most districts.	Showers	
21st	Sun		Showers	
22nd	Mon	Very wet day in the Alpine regon, particularly around Mt Baw Baw.		**N XhS**
23rd	Tue	Very warm day in the Lodden-Mallee region, particularly in the Mildura district, where maximum temperatures into the upper 20's to early 30's is expected.		
24th	Wed	Very strong winds in the Alpine region, particularly around Mt Hotham. Scattered showers over the southern half of the State over the next couple of days.	Showers	
25th	Thu			
26th	Fri	A wet front moves across the State over the next couple of days, with precipitation at first in the eastern higher reaches, spreading throughout the State for a time, and then contracting back to the Gippsland and Lakes region in a couple of days time.	Showers	**A**
27th	Sat	Snow in the Victorian Alpine regions and ski fields.		
28th	Sun			
29th	Mon		Showers	
30th	Tue	Widespread showers across the State.		**1Q V**

October 2025

Date	Day	Victoria - October 2025	Melbourne	Moon
1st	Wed	Mostly dry over the next 4-5 days except for some isolated showers briefly in central parts.	Showers	
2nd	Thur			
3rd	Fri	Milder change over the State during the next few days.		
4th	Sat			
5th	Sun	Very warm day in the Lodden-Mallee region, particularly in the Mildura district, where maximum temperatures into the upper 20's are expected.		
6th	Mon	Rain in the west of the State brings mild overnight temperatures in the Lodden-Mallee region.		
7th	Tue	Isolated showers across the State, turning to outbreaks of rain over the next few days in most districts.	Showers	**F XhN**
8th	Wed	Light snow in the Victorian Alpine region and snowfields over the next couple of days.	Showers	
9th	Thu	Very cold in the Gippsland and Wimmera regions, including in the Grampians, with overnight frosts and cold daytime temperatures.		**P6**
10th	Fri	Milder temperatures in the East Gippsland.		
11th	Sat	Another 3-4 weeks of below average rainfall can be expected ahead with some areas possibly not seeing any preciptation during the coming few weeks. Subsequently, accompanying the drier conditions will be above average maximum and minimum temperatures during this same period.		
12th	Sun	Heavy rain around this time in the Alpine region, particularly at Mt Buller.		
13th	Mon	Snowflurries in the Falls Creek to Mt Hotham region.	Showers	**^**
14th	Tue	Isolated patches of showers in many districts over the next 3-4 days.		**3Q**
15th	Wed		Showers, Th	
16th	Thur			
17th	Fri	Very cold over the next few nights in the Hume to Wimmera regions of the State, with scattered frosts in many districts.		
18th	Sat	Mostly dry over the next 3-4 days across the State.		
19th	Sun			
20th	Mon			**XhS**
21st	Tue		Showers	

Date	Day	Victoria - October 2025	Melbourne	Moon
22nd	Wed	A cold front crosses the State bringing some showers into most districts and snowflurries in the Falls Creek to Mt Hotham region and possibly Mt Baw Baw over the next couple of days.	Showers	N
23rd	Thu			
24th	Fri	Mostly dry throughout the State for the remainder of the month.		A
25th	Sat			
26th	Sun	Thunderstorms with severe winds gusts in the East Gippslands which may cause some damage to property. Very cold spell in the Alpine regon over the next few days. Very cold on Mt Baw Baw around this time, where maximum temperatures may not rise above frost levels.		
27th	Mon	Very cold in the West Coast and Alpine regions overnight, with severe frost level temperatures on Mt Hotham and possible patches of frost in the Wimmera region. Chance of isolated showers in the Gippsland.		V
28th	Tue	Very strong and blustery winds with gusts across the Gippslands. Cold overnight in the Barwon and Gippsland regions with chance of pockets of frost.		
29th	Wed			
30th	Thur			1Q
31st	Fri	Warm change across the State during the next few days, where some districts may see maximum temperatures into the upper 30's for a time, particularly in the Hume, Wimmera, Lodden-Mallee, and Melbourne regions.		

November 2025

Date	Day	Victoria - November 2025	Melbourne	Moon
1st	Sat	Unusually mild temperatures over the next couple of nights across much of the State, and particularly in the Mt Nowa Nowa district.	Showers	
2nd	Sun	The passage of a cold front brings scattered showers across the State, quickly clearing.		
3rd	Mon			XhN
4th	Tue	Showers in the Gippsland over the next few days.	Showers	
5th	Wed			
6th	Thur	Warm spell in the Gippsland over the next couple of days.		F P1
7th	Fri	Storm activity across the State. Heavy rain and scattered hail showers in the Hume region. Hail may be very large at times, particularly in the Myrtleford, Bright, and Beechworth areas. Strong to gale force winds may cause damage in a number of regions.		
8th	Sat	Isolated showers over the next few days.		
9th	Sun			^
10th	Mon			
11th	Tue	Rain may return in abundance over the next 3-4 weeks with most regions receiving above average amounts compared to what is average for this time of year. Maximum and minimum temperatures may also be above average during this period.	Showers	
12th	Wed	An intense wet front crosses the State over the next 3-4 days bringing widespread storms and heavy rain in many districts.		3Q
13th	Thu	Heavy rain moves into the west of the State.	Rain, Th	
14th	Fri	Very heavy rain in the State, with chance of some pockets of localised flooding, particularly in the Southwest region and along the southern coasts. Winds may be very strong and blustery around the coasts at this time, as well as into the Grampians.	Showers	
15th	Sat	Cold spell over the State around this time, including in the Alpine region over the next few days. Very cold at Mt Baw Baw and Mt Hotham where maximum temperatures may barely lift above frost levels and most alpine regions having overnight frosts in the next couple of days.		
16th	Sun	Clearer spell across the State over the next few days.		XhS
17th	Mon			
18th	Tue			
19th	Wed	Heavy rain in the Melbourne region with chance of flash flooding, particularly in the Little River area.	Rain, Th	
20th	Thur	Cold overnight in the Gippslands, Western Plains and West Coast regions over the next few days.	Fg	N A
21st	Fri			
22nd	Sat	Mostly dry over the next 5-6 days.		
23rd	Sun		Fg	V
24th	Mon			
25th	Tue			
26th	Wed	Hot spell in the central to western parts of the State, with daily maximums into the upper 30's common and into the early 40's in some areas.		

Date	Day	Victoria - November 2025	Melbourne	Moon
27th	Thu		Showers, Th	
28th	Fri	Showers in the eastern half of the State over the next couple of days.		**1Q**
29th	Sat	Very hot in the Loddon-Mallee, Central, and Hume regions over the next few days, with a number of districts likely to see maximum temperatures reach into the low 40's and overnight minimas into the mid-20's.		
30th	Sun	High temperatures spread to almost all regions throughout the State around this time.	Rain	**XhN**

December 2025

Date	Day	Victoria - December 2025	Melbourne	Moon
1st	Mon	A wet front crosses the State over the next 3 days bringing heavy rain with its passage, in the western half of the State at first, and crossing into the east later.	Rain	
2nd	Tue		Showers	
3rd	Wed			
4th	Thur	Mostly dry over the next 3-4 days across the State.		**P2**
5th	Fri			**F**
6th	Sat			**^**
7th	Sun			
8th	Mon	Heavy rain and chance of flash flooding and property damage in the Melbourne and Grampian regions, including in the Geelong and Mount Egerton areas.		
9th	Tue	Showers over the eastern half of the State during the next few days.		
10th	Wed			
11th	Thu	The next 3-4 weeks may continue to bring above normal rainfall over much of the State as well as warmer temperatures compared to what is typical for this time of year.		
12th	Fri		Showers	**3Q**
13th	Sat	Thunderstorms, hail, and heavy rain in the Hume to Melbourne regions, with pockets of flash flooding likely, and gusty winds that may cause damage to property. Chance of tornado formation winds developing with the storms, particularly at risk in the Loch area.	Showers, Th	**XhS**
14th	Sun	Rain activity contracts to the eastern half of the State around this time.		
15th	Mon	Mostly dry over the nex couple of days except for the chance of some showers in the Central Highlands.		
16th	Tue			
17th	Wed	Showers over the eastern half of the State during the next couple of days.		**A**
18th	Thur			
19th	Fri	Mostly dry over the next few days.		
20th	Sat	Chance of some showers around the southwest coasts.		**N V**
21st	Sun			
22nd	Mon	Chance of isolated showers in the Gippsland.		
23rd	Tue			
24th	Wed	A pocket of rain likely in the far northwest of the Mallee.	Showers	
25th	Thu	Showers over the southern half of the State.	Showers	
26th	Fri	Scattered rain activity over the eastern half of the State during the next couple of days.		
27th	Sat			**XhN**
28th	Sun			**1Q**
29th	Mon	Low pressure crosses the State bringing a cooler, blustery, and very wet spell over the next few days with widespread rain activity prevailing and many districts likely to see up to twice as much rainfall than normal for this time of year.	Showers, Th	
30th	Tue	Severe thunderstorms with flash flooding, gusty winds, heavy rain and large hail likelly in the Melbourne, Lodden Mallee, and Grampian regions, with chance of damage to property. Chance of tornado formations forming in the Barwon to Grampians regions.	Rain, Th, Fg	
31st	Wed	Severe thunderstorms in the Lodden Mallee, Grampians, Wimmera, and Hume regions, with outbreaks of heavy rain, pockets of flash-flooding and chance of large hail in some places. Chance of tornadic activity in the Lodden Mallee, particularly in the Elmore area.	Rain, Th	

VICTORIA GRAPHS AND TABLES 2025

Seasonal Breakdown –
State Overall Rainfall Timing and Estimated mm Trend Graphs
State Overall Maximum Temperature Trend Graphs
Regional Rainfall Expectations Tables - mm and Trends
Regional Maximum and Minimum Temperature Trends Tables

Monthly Breakdown -
Rainfall:
Regional Expectations by mm
Regional Expectations by mm Variance to Average
Regional Variance Trend by mm
Regional Estimated Expectations by % of Variance to Average
Regional Estimatedl Variance Trend by %

Maximum & Minimum Temperature Trends:
Regional Estimated Average Temperature
Regional Estimated Degree of Variation from the Average
Regional Estimated Temperature Trend

Seasonal Breakdowns

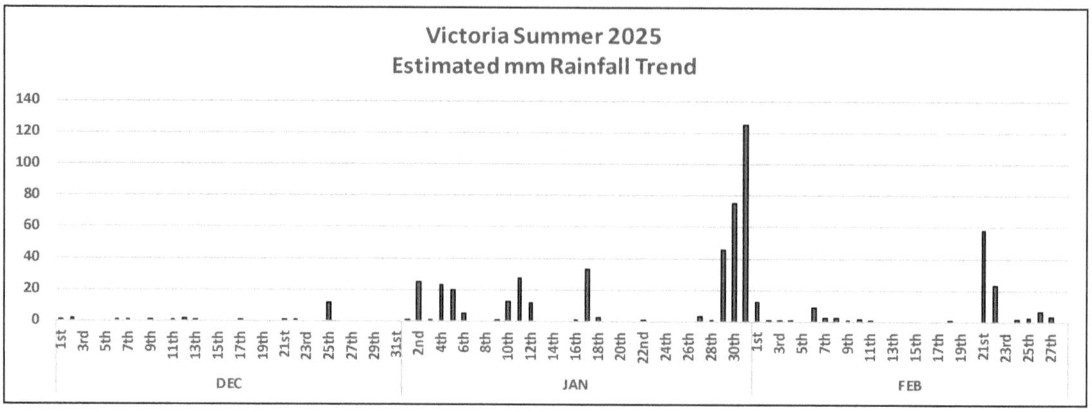

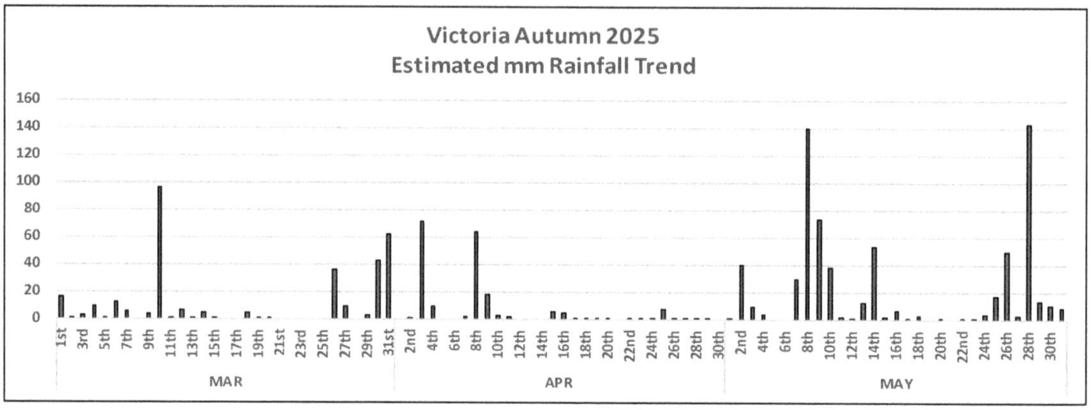

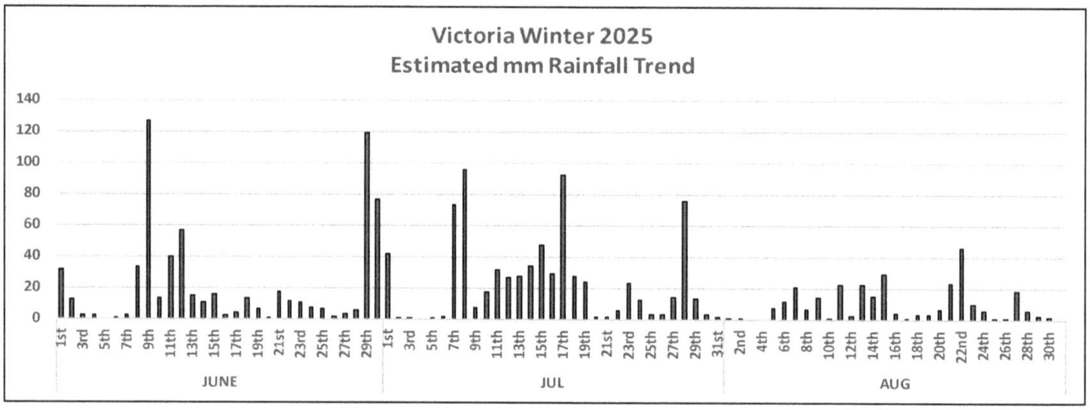

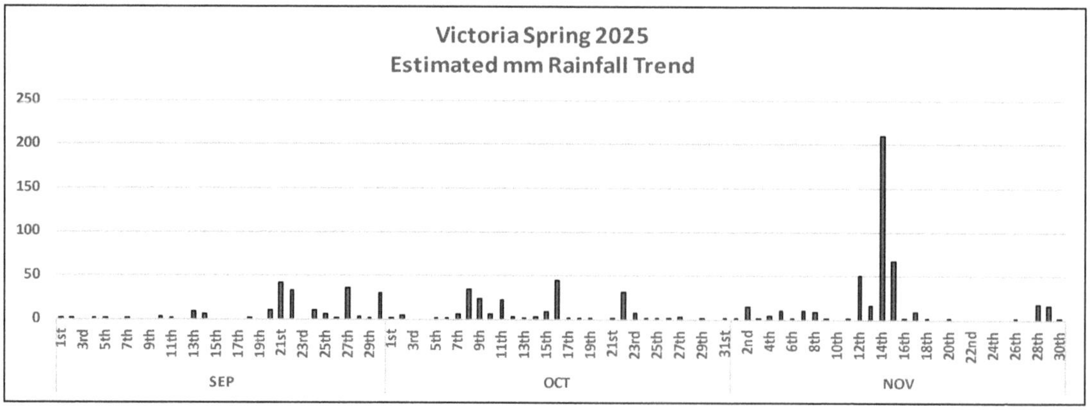

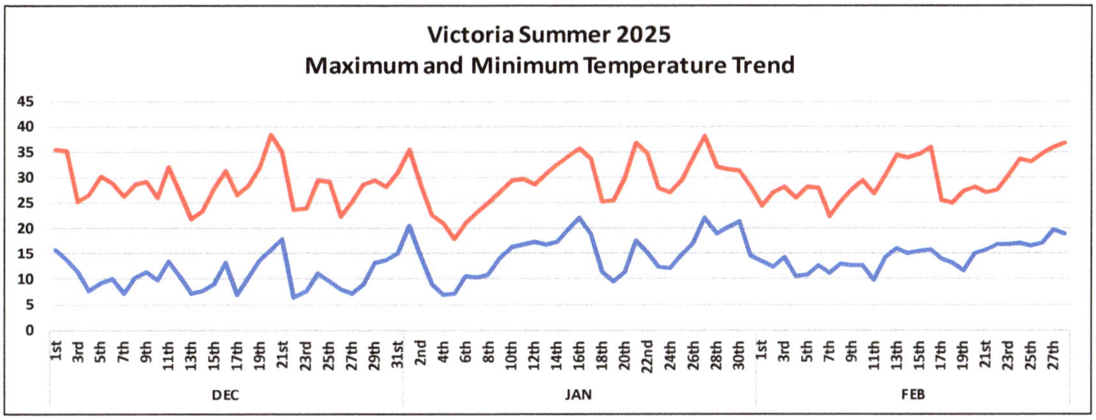
Victoria Summer 2025
Maximum and Minimum Temperature Trend
DEC
JAN
FEB

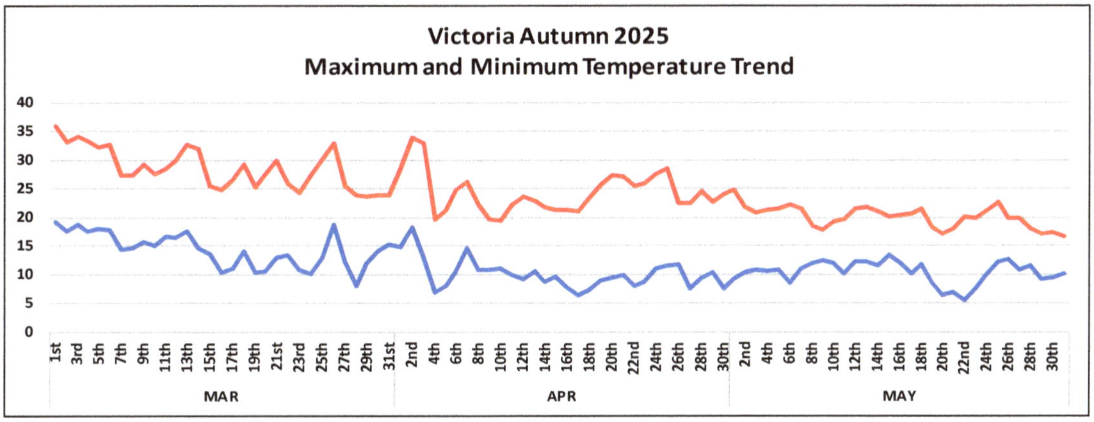
Victoria Autumn 2025
Maximum and Minimum Temperature Trend
MAR
APR
MAY

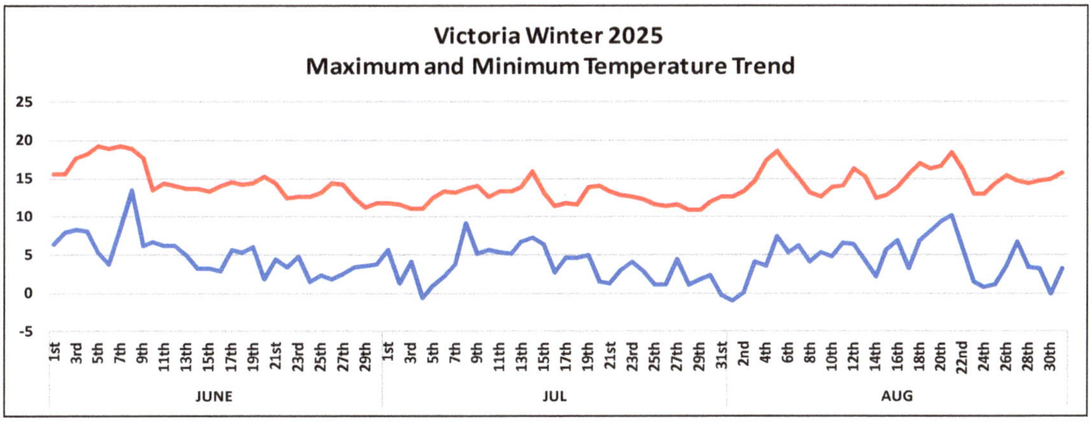
Victoria Winter 2025
Maximum and Minimum Temperature Trend
JUNE
JUL
AUG

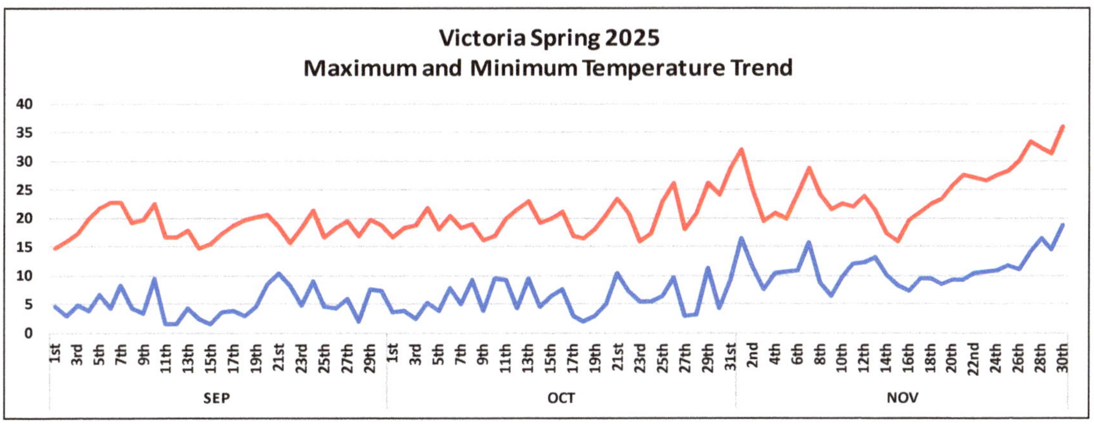
Victoria Spring 2025
Maximum and Minimum Temperature Trend
SEP
OCT
NOV

2025 VICTORIA - Estimated Seasonal Rainfall Expectations (mm)

REGION	Mallee	Wimmera	Western District	Northern Country	North Central	Central	Northeast	W & S Gippsland	East Gippsland	VIC
Town Sample	Mildura	Horsham	Warrnambool	Echuca	Bendigo	Melbourne	Benalla	Traralgon	Bairnsdale	Overall
Summer	62	78	124	24	43	54	28	68	75	555
Autumn	62	115	140	122	153	106	145	176	154	1173
Winter	42	113	363	99	123	134	179	210	405	1669
Spring	21	67	231	33	55	75	73	154	127	835
Overall	212	417	888	323	407	432	491	655	803	4629

2025 VIC - Estimated Seasonal Rainfall Variation from the Average (mm)

REGION	Mallee	Wimmera	Western District	Northern Country	North Central	Central	Northeast	W & S Gippsland	East Gippsland	VIC
Town Sample	Mildura	Horsham	Warrnambool	Echuca	Bendigo	Melbourne	Benalla	Traralgon	Bairnsdale	Overall
Summer	-5	2	7	-55	-53	-95	-92	-85	-90	-466
Autumn	13	30	-19	37	39	-31	4	13	16	102
Winter	-34	-40	101	-25	-56	-7	-40	30	253	181
Spring	-59	-54	33	-79	-85	-100	-103	-64	-68	-579
Overall	-60	-18	152	-77	-121	-171	-164	-58	152	-366

2025 VIC - Estimated Seasonal Rainfall Trend Variation by 10mm

REGION	Mallee	Wimmera	Western District	Northern Country	North Central	Central	Northeast	W & S Gippsland	East Gippsland	VIC
Town Sample	Mildura	Horsham	Warrnambool	Echuca	Bendigo	Melbourne	Benalla	Traralgon	Bairnsdale	Overall
Summer	av	av	av	drr	drr	drr	drr	drr	drr	drr
Autumn	av	wtr	av	wtr	wtr	drr	av	av	av	wtr
Winter	drr	drr	wtr	av	drr	av	drr	wtr	wtr	wtr
Spring	drr	drr	wtr	drr	drr	drr	drr	drr	drr	drr
Overall	av	av	wtr	av	drr	drr	drr	av	wtr	drr

2025 VIC - Estimated Seasonal Rainfall Expectation by %

REGION	Mallee	Wimmera	Western District	Northern Country	North Central	Central	Northeast	W & S Gippsland	East Gippsland	VIC
Town Sample	Mildura	Horsham	Warrnambool	Echuca	Bendigo	Melbourne	Benalla	Traralgon	Bairnsdale	Overall
Summer	-7%	3%	6%	-70%	-55%	-64%	-77%	-56%	-55%	-46%
Autumn	27%	36%	-12%	43%	34%	-23%	3%	8%	11%	9%
Winter	-45%	-26%	38%	-20%	-31%	-5%	-18%	17%	166%	12%
Spring	-74%	-45%	17%	-71%	-61%	-57%	-59%	-29%	-35%	-41%
Overall	-22%	-4%	21%	-19%	-23%	-28%	-25%	-8%	23%	-7%

2025 VIC - Estimated Seasonal Rainfall Trend Variation by 25%

REGION	Mallee	Wimmera	Western District	Northern Country	North Central	Central	Northeast	W & S Gippsland	East Gippsland	VIC
Town Sample	Mildura	Horsham	Warrnambool	Echuca	Bendigo	Melbourne	Benalla	Traralgon	Bairnsdale	Overall
Summer	av	av	av	drr	drr	drr	drr	drr	drr	drr
Autumn	wtr	wtr	av	wtr	wtr	av	av	av	av	av
Winter	drr	drr	wtr	av	drr	av	av	av	wtr	av
Spring	drr	drr	av	drr	drr	drr	drr	drr	drr	drr
Overall	av	av	av	av	av	drr	av	av	av	av

2025 VIC - Estimated Seasonal Maximum Temperature Average and Trend

Region Town Sample	Mallee Mildura	Wimmera Horsham	Western District Warrnambool	Northern Country Echuca	North Central Bendigo	Central Melbourne	Northeast Benalla	West and South Gippsland Traralgon	East Gippsland Bairnsdale	VIC Overall
Summer Est	32.6	30.5	24.7	32.2	29.8	27.4	31.9	28.0	25.9	29.2
Trend	a-w	av	a-w	wmr	a-w	wmr	a-w	wmr	a-w	wmr
Autumn Est	27.0	25.3	22.0	25.7	24.2	23.3	25.8	23.6	22.3	24.3
Trend	wmr	wmr	wmr	wmr	wmr	wmr	wmr	wmr	wmr	wmr
Winter Est	16.1	14.2	13.5	14.3	13.0	13.5	13.8	14.3	14.6	14.1
Trend	av	av	av	av	av	av	av	av	av	av
Spring Est	25.3	22.4	17.6	23.6	21.3	20.8	21.9	19.7	19.6	21.4
Trend	wmr	av	a-c	wmr	av	wmr	av	av	av	av

2025 VIC - Estimated Seasonal Minimum Temperature Average and Trend

Region Town Sample	Mallee Mildura	Wimmera Horsham	Western District Warrnambool	Northern Country Echuca	North Central Bendigo	Central Melbourne	Northeast Benalla	West and South Gippsland Traralgon	East Gippsland Bairnsdale	VIC Overall
Summer Est	16.8	12.3	10.9	14.2	13.3	15.6	14.5	11.9	12.0	13.2
Trend	a-w	av	a-c	av	av	wmr	wmr	av	av	av
Autumn Est	13.4	10.1	10.8	12.0	11.1	14.2	11.4	10.8	10.8	11.6
Trend	wmr	wmr	wmr	wmr	wmr	wmr	wmr	wmr	wmr	wmr
Winter Est	4.0	3.0	5.9	3.6	2.9	7.8	3.0	3.8	5.4	4.6
Trend	a-c	av	av	a-c	av	wmr	av	av	a-w	av
Spring Est	10.0	5.2	7.0	7.4	6.4	11.2	5.9	6.6	7.4	7.4
Trend	av	a-c	a-c	clr	a-c	wmr	a-c	a-c	av	av

Monthly Breakdowns

2025	VICTORIA - Estimated Rainfall Expectations (mm)									
REGION	Mallee	Wimmera	Western District	Northern Country	North Central	Central	Northeast	W & S Gippsland	East Gippsland	VIC
Town Sample	Mildura	Horsham	Warrnambool	Echuca	Bendigo	Melbourne	Benalla	Traralgon	Bairnsdale	Overall
January	59	71	100	22	29	41	22	51	16	412
February	2	6	20	1	13	9	6	14	53	124
March	11	12	12	29	27	20	37	86	86	319
April	6	4	44	9	9	38	27	28	25	191
May	46	99	84	84	116	48	81	62	43	663
June	17	44	174	44	41	29	75	33	192	649
July	14	43	98	37	65	88	81	125	184	736
August	11	26	91	18	17	17	23	52	29	284
September	3	13	75	7	19	12	10	42	12	192
October	5	12	67	4	4	22	8	57	25	203
November	13	42	90	22	32	41	55	55	91	440
December	26	44	34	46	34	67	66	50	48	415
Overall	212	417	888	323	407	432	491	655	803	4629

2025	VICTORIA - Estimated Rainfall Variation from the Average (mm)									
REGION	Mallee	Wimmera	Western District	Northern Country	North Central	Central	Northeast	W & S Gippsland	East Gippsland	VIC
Town Sample	Mildura	Horsham	Warrnambool	Echuca	Bendigo	Melbourne	Benalla	Traralgon	Bairnsdale	Overall
January	37	46	62	-6	-2	-4	-22	1	-34	78
February	-12	-13	-9	-19	-8	-31	-28	-26	-1	-146
March	-3	-7	-34	8	-1	-21	0	35	44	22
April	-7	-19	-6	-18	-24	-12	-16	-27	-31	-160
May	23	56	21	47	64	2	20	5	3	240
June	-6	-5	90	5	-18	-18	7	-24	131	163
July	-13	-10	12	-8	5	43	4	63	131	227
August	-15	-25	-2	-23	-43	-33	-50	-9	-8	-208
September	-25	-35	-7	-30	-31	-41	-56	-33	-45	-302
October	-21	-29	3	-34	-43	-37	-52	-14	-36	-264
November	-13	10	37	-15	-11	-22	5	-16	12	-14
December	-5	13	-16	14	-9	3	24	-12	-14	-2
Overall	-60	-18	152	-77	-121	-171	-164	-58	152	-366

2025	VICTORIA - Estimated Rainfall Trend Variation by 10mm									
REGION	Mallee	Wimmera	Western District	Northern Country	North Central	Central	Northeast	W & S Gippsland	East Gippsland	VIC
Town Sample	Mildura	Horsham	Warrnambool	Echuca	Bendigo	Melbourne	Benalla	Traralgon	Bairnsdale	Overall
January	wtr	wtr	wtr	av	av	av	drr	av	drr	wtr
February	drr	drr	av	drr	av	drr	drr	drr	av	drr
March	av	av	av	av	av	drr	av	wtr	wtr	av
April	av	drr	av	drr	drr	drr	drr	drr	drr	drr
May	wtr	wtr	wtr	wtr	wtr	av	wtr	av	av	wtr
June	av	av	wtr	av	drr	drr	av	drr	wtr	wtr
July	drr	av	wtr	av	av	wtr	av	wtr	wtr	wtr
August	drr	drr	av	drr	drr	drr	drr	av	av	drr
September	drr	drr	av	drr	drr	drr	drr	drr	drr	drr
October	drr	drr	av	drr	drr	drr	drr	drr	drr	drr
November	drr	av	wtr	drr	drr	drr	av	drr	wtr	av
December	av	wtr	drr	wtr	av	av	wtr	drr	drr	av
Overall	av	av	wtr	av	drr	drr	drr	av	wtr	drr

2025 VICTORIA - Estimated Rainfall Expectation by %

REGION	Mallee	Wimmera	Western District	Northern Country	North Central	Central	Northeast	W & S Gippsland	East Gippsland	VIC
Town Sample	Mildura	Horsham	Warrnambool	Echuca	Bendigo	Melbourne	Benalla	Traralgon	Bairnsdale	Overall
January	163%	184%	161%	-20%	-7%	-10%	-50%	1%	-67%	23%
February	-84%	-67%	-30%	-93%	-39%	-77%	-82%	-65%	-1%	-54%
March	-23%	-35%	-74%	37%	-4%	-51%	1%	70%	103%	7%
April	-53%	-82%	-12%	-68%	-72%	-24%	-38%	-49%	-55%	-46%
May	104%	129%	33%	125%	123%	3%	33%	8%	7%	57%
June	-25%	-10%	108%	13%	-30%	-38%	10%	-42%	213%	34%
July	-48%	-19%	14%	-17%	8%	97%	5%	102%	243%	45%
August	-58%	-49%	-2%	-55%	-72%	-66%	-69%	-15%	-22%	-42%
September	-90%	-74%	-8%	-80%	-62%	-78%	-85%	-44%	-79%	-61%
October	-82%	-70%	4%	-90%	-91%	-63%	-87%	-20%	-59%	-56%
November	-50%	29%	70%	-40%	-26%	-35%	10%	-23%	15%	-3%
December	-16%	40%	-32%	46%	-20%	5%	58%	-20%	-23%	0%
Overall	-22%	-4%	21%	-19%	-23%	-28%	-25%	-8%	23%	-7%

2025 VICTORIA - Estimated Rainfall Trend Variation by 25%

REGION	Mallee	Wimmera	Western District	Northern Country	North Central	Central	Northeast	W & S Gippsland	East Gippsland	VIC
Town Sample	Mildura	Horsham	Warrnambool	Echuca	Bendigo	Melbourne	Benalla	Traralgon	Bairnsdale	Overall
January	wtr	wtr	wtr	av	av	av	drr	av	drr	av
February	drr	drr	drr	drr	drr	drr	drr	drr	av	drr
March	av	drr	drr	wtr	av	drr	av	wtr	wtr	av
April	drr	drr	av	drr	drr	av	drr	drr	drr	drr
May	wtr	wtr	wtr	wtr	wtr	av	wtr	av	av	wtr
June	av	av	wtr	av	drr	drr	av	drr	wtr	wtr
July	drr	av	av	av	av	av	av	wtr	wtr	wtr
August	drr	drr	av	drr	drr	drr	drr	av	av	drr
September	drr	drr	av	drr	drr	drr	drr	drr	drr	drr
October	drr	drr	av	drr	drr	drr	drr	av	drr	drr
November	drr	wtr	wtr	drr	drr	drr	av	av	av	av
December	av	wtr	drr	wtr	av	av	wtr	av	av	av
Overall	av	av	av	av	av	drr	av	av	av	av

2025 VICTORIA - Estimated Average Maximum Temperatures

Region	Mallee	Wimmera	Western District	Northern Country	North Central	Central	Northeast	West and South Gippsland	East Gippsland	VIC
Town Sample	Mildura	Horsham	Warrnambool	Echuca	Bendigo	Melbourne	Benalla	Traralgon	Bairnsdale	Overall
January	32	30	26	31	29	28	32	28	27	**29**
February	33	30	24	33	30	28	32	29	26	**30**
March	32	30	24	31	29	27	31	28	25	**29**
April	27	26	22	26	24	23	26	23	22	**24**
May	22	20	19	20	19	20	21	19	20	**20**
June	16	15	14	15	14	14	15	15	16	**15**
July	15	13	12	13	11	12	12	13	13	**13**
August	18	15	14	15	14	14	14	15	15	**15**
September	22	19	17	20	18	18	18	18	18	**19**
October	24	21	17	22	20	20	21	18	19	**20**
November	29	27	20	28	26	24	27	23	22	**25**
December	31	29	22	29	28	25	29	24	23	**27**
Overall	**25**	**23**	**19**	**24**	**22**	**21**	**23**	**21**	**20**	**22**

2025 VICTORIA - Estimated Average Minimum Temperatures

Region	Mallee	Wimmera	Western District	Northern Country	North Central	Central	Northeast	West and South Gippsland	East Gippsland	VIC
Town Sample	Mildura	Horsham	Warrnambool	Echuca	Bendigo	Melbourne	Benalla	Traralgon	Bairnsdale	Overall
January	18	14	13	16	15	17	16	14	14	**15**
February	17	13	12	15	14	17	15	13	13	**14**
March	16	12	12	15	14	17	15	13	14	**14**
April	12	9	9	10	9	13	10	10	9	**10**
May	12	10	11	10	9	13	10	10	9	**10**
June	5	3	6	4	4	8	5	4	6	**5**
July	3	3	5	3	2	7	2	3	5	**4**
August	5	3	6	4	2	8	3	4	5	**4**
September	7	3	6	4	4	10	3	4	5	**5**
October	8	4	6	6	4	10	4	6	6	**6**
November	14	9	9	12	11	14	11	10	11	**11**
December	16	11	11	15	14	16	15	12	12	**14**
Overall	11	8	9	10	9	12	9	9	9	**9**

2025 VICTORIA - Estimated Degree of Variation from the Maximum Average

Region	Mallee	Wimmera	Western District	Northern Country	North Central	Central	Northeast	West and South Gippsland	East Gippsland	VIC
Town Sample	Mildura	Horsham	Warrnambool	Echuca	Bendigo	Melbourne	Benalla	Traralgon	Bairnsdale	Overall
January	-0.4	-1.3	1.8	0.4	-0.8	1.0	-0.4	1.6	1.2	0.3
February	1.0	0.0	-0.8	2.8	0.7	1.6	1.1	2.0	0.7	1.0
March	3.2	2.9	1.2	3.4	3.3	2.4	3.3	3.5	1.2	2.7
April	3.3	3.3	2.4	3.6	2.7	3.0	3.1	2.5	1.1	2.8
May	3.3	2.5	2.6	2.9	2.3	3.1	3.2	2.6	2.2	2.7
June	-0.2	0.3	0.0	0.7	0.7	0.7	0.8	1.2	0.7	0.5
July	-0.9	-0.9	-1.1	-0.9	-1.3	-1.5	-0.8	-1.0	-1.8	-1.1
August	0.6	-0.1	-0.3	0.0	-0.6	-0.1	0.0	0.0	-0.4	-0.1
September	1.8	1.3	0.7	1.5	0.8	1.7	0.3	0.7	0.0	1.0
October	0.2	-1.0	-1.5	0.1	-1.3	0.7	-0.7	-1.6	-0.8	-0.7
November	1.4	1.0	-0.9	2.4	1.6	1.7	0.9	1.2	0.0	1.0
December	0.6	0.5	-0.3	0.2	0.4	0.3	-0.7	-0.2	-0.9	0.0
Overall	1.2	0.7	0.4	1.4	0.7	1.2	0.9	1.1	0.3	0.9

2025 VICTORIA - Estimated Degree of Variation from the Minimum Average

Region	Mallee	Wimmera	Western District	Northern Country	North Central	Central	Northeast	West and South Gippsland	East Gippsland	VIC
Town Sample	Mildura	Horsham	Warrnambool	Echuca	Bendigo	Melbourne	Benalla	Traralgon	Bairnsdale	Overall
January	1.4	1.3	1.4	1.0	0.7	2.9	1.6	1.0	0.6	1.3
February	0.4	0.5	-0.1	-0.1	-0.2	2.6	1.3	0.5	0.5	0.6
March	2.5	1.6	1.5	2.4	2.5	4.0	4.2	1.9	2.3	2.5
April	1.6	1.4	0.3	0.7	1.4	2.8	2.6	0.8	0.7	1.4
May	4.7	4.2	3.2	3.8	4.2	4.4	5.2	3.0	2.7	3.9
June	-0.5	-0.3	0.6	-0.2	0.5	2.3	1.9	-0.3	1.0	0.5
July	-1.9	-0.6	-0.3	-1.2	-0.5	1.4	-1.1	-0.4	1.0	-0.4
August	-0.4	-0.2	0.3	-1.0	-0.3	2.1	-0.2	-0.2	0.9	0.1
September	-0.2	-1.1	-0.5	-2.1	-0.7	2.5	-1.6	-1.9	-0.6	-0.7
October	-1.3	-1.8	-1.1	-2.9	-2.4	1.6	-2.9	-1.5	-1.5	-1.5
November	1.7	0.1	-0.6	1.2	1.3	3.6	1.8	0.3	1.4	1.2
December	1.3	0.6	0.3	1.3	1.6	3.7	3.1	1.1	0.8	1.5
Overall	0.8	0.5	0.4	0.3	0.7	2.8	1.3	0.4	0.8	0.9

2025 VICTORIA - Estimated Maximum Temperature Trend

Region	Mallee	Wimmera	Western District	Northern Country	North Central	Central	Northeast	West and South Gippsland	East Gippsland	VIC
Town Sample	Mildura	Horsham	Warrnambool	Echuca	Bendigo	Melbourne	Benalla	Traralgon	Bairnsdale	Overall
January	av	clr	wmr	av	a-c	a-w	av	wmr	wmr	av
February	a-w	av	a-c	wmr	a-w	wmr	wmr	wmr	a-w	a-w
March	wmr	wmr	wmr	wmr	wmr	wmr	wmr	wmr	wmr	wmr
April	wmr	wmr	wmr	wmr	wmr	wmr	wmr	wmr	wmr	wmr
May	wmr	wmr	wmr	wmr	wmr	wmr	wmr	wmr	wmr	wmr
June	av	av	av	a-w	a-w	a-w	a-w	wmr	a-w	av
July	a-c	a-c	clr	a-c	clr	clr	a-c	a-c	clr	clr
August	a-w	av	av	av	a-c	av	av	av	av	av
September	wmr	wmr	a-w	wmr	a-w	wmr	av	a-w	av	a-w
October	av	a-c	clr	av	clr	a-w	a-c	clr	a-c	a-c
November	wmr	a-w	a-c	wmr	wmr	wmr	a-w	wmr	av	a-w
December	a-w	av	av	av	av	av	a-c	av	a-c	av
Overall	wmr	a-w	av	wmr	a-w	wmr	a-w	wmr	av	a-w

2025 VICTORIA - Estimated Minimum Temperature Trend

Region	Mallee	Wimmera	Western District	Northern Country	North Central	Central	Northeast	West and South Gippsland	East Gippsland	VIC
Town Sample	Mildura	Horsham	Warrnambool	Echuca	Bendigo	Melbourne	Benalla	Traralgon	Bairnsdale	Overall
January	wmr	wmr	wmr	a-w	a-w	wmr	wmr	a-w	a-w	wmr
February	av	av	av	av	av	wmr	wmr	av	av	a-w
March	wmr	wmr	wmr	wmr	wmr	wmr	wmr	wmr	wmr	wmr
April	wmr	wmr	av	a-w	wmr	wmr	wmr	a-w	a-w	wmr
May	wmr	wmr	wmr	wmr	wmr	wmr	wmr	wmr	wmr	wmr
June	av	av	a-w	av	av	wmr	wmr	av	a-w	av
July	clr	a-c	av	clr	av	wmr	clr	av	a-w	av
August	av	av	av	av	av	wmr	av	av	a-w	av
September	av	clr	av	clr	a-c	wmr	clr	clr	a-c	a-c
October	clr	clr	clr	clr	clr	wmr	clr	clr	clr	clr
November	wmr	av	av	wmr	wmr	wmr	wmr	av	wmr	wmr
December	wmr	a-w	av	wmr	wmr	wmr	wmr	wmr	a-w	wmr
Overall	a-w	av	av	av	a-w	wmr	wmr	av	a-w	a-w

TASMANIA 2025

GENERAL TRENDS

Year Overall:

Slightly drier by around 5-10% less than average
Warmer by around 0.5-1°C above the average

Summer:

Drier by around 15-20% below the average
Wettest – Mid-January
Driest – Mid-February
Slightly warmer but within 0.5°C of the average
Hottest – Last week of February
Coolest – First week of January and second week of February

Autumn:

Slightly drier but within 10% of the average
Wettest – Mid-May and last week of May
Driest – Second half of April
Warmer by up to 2-2.5°C above the average
Hottest – First half of March
Coolest – Second week of May to the end of the month

Winter:

Slightly wetter but within 5% of the average
Wettest –Second week of June and third week of August
Driest – First week of July
Slightly cooler but within 0.5°C of the average
Hottest – First week of June
Coolest – Last week of June

Spring:

Drier by around 20-25% below the average
Wettest –Throughout October
Driest – First week of September and Last half of November
Average temperatures overall
Hottest – Second half of November
Coolest – Mid-September and second week of October

Major Weather Events - TAS 2025

January 2025

Date	Day	Tasmania - January 2025	Hobart	Moon
1st	Wed	Strong winds around Tasman Island. Unusually cool in the West Coast districts over the next couple of days. Chance of heavy rain for a short time around the North and West Coasts.	Showers	
2nd	Thur	Unusually cool minimum temperatures and strong winds at times in the Southeast of the State	Showers	
3rd	Fri	Strong winds at times around the coastal fringes across the State. Unusually cold maximum temperature likely in the West Coast higher reaches.	Rain	
4th	Sat	Strong winds from the northwest of Tasmania down through the west coasts and into the southeast of the State over the next couple of days. Rain may be heavy for a time in the Southeast.	Showers	
5th	Sun	Severe cold spell for the next few days over the State with the chance of snow in the higher ranges of the State. Chance of outbreaks of heavy rain and wind gusts in the Southeast. Blustery winds also around the North and East Coasts.		
6th	Mon	Cold maximum temperatures persist around the West Coast to the Southeast over the next few days.	Showers	
7th	Tue			**1Q XhN**
8th	Wed			**P13**
9th	Thu	Occasional strong wind gusts along the North Coast.		
10th	Fri	Unusually cool in the Southeast.		
11th	Sat	Regular moderate to heavy rain on and off during the coming week.		
12th	Sun			
13th	Mon	Chance of snow flurries in the Central Highlands.		^
14th	Tue	Unsually warm in the State over the next couple of days. However, chance of light snow in the Central Highlands and flurries on Mt Mawson.		**F**
15th	Wed	Chance of snow flurries in the Central Highlands.	Rain	
16th	Thur		Rain	
17th	Fri		Showers	
18th	Sat			
19th	Sun			
20th	Mon			**XhS**
21st	Tue			**A**
22nd	Wed	Unusually warm in the State over he next couple of days.		**3Q**
23rd	Thu			
24th	Fri	Unusually warm spell over the State over the next few days.	Fg	
25th	Sat	Regular moderate to heavy rain on and off during the coming week.		
26th	Sun	Unsually high temperature in the Central Highlands to the Derwent Valley over the next couple of days.	Showers, Th	
27th	Mon		Showers, Th	**V**
28th	Tue			
29th	Wed	Heavy rain in the Derwent Valley to the Southern Midlands, with chance of localised flooding.	Showers	
30th	Thur		Showers	**N**
31st	Fri	Thunderstorms cross the southeast of the State. Heavy rain in the Hobart region, with chance of localised flooding causing damage to property and infrastructure. Warmer than average and humid almost everywhere.	Rain	

February 2025

Date	Day	Tasmania - February 2025	Hobart	Moon
1st	Sat	Heavy rain in the Southeast, including in Hobart.		
2nd	Sun			**P10**
3rd	Mon			**XhN**
4th	Tue			
5th	Wed		Showers	**1Q**
6th	Thu	Unusually cool spell, particularly in the Southeast, with chance of frosts on Mt Wellington.	Showers	
7th	Fri			
8th	Sat			
9th	Sun	Very strong winds to possible gale force in the Southeast, including over Mt Wellington.		^
10th	Mon			

Date	Day	Tasmania - February 2025	Hobart	Moon
11th	Tue	Chance of frosts in the Central Plateau region, including in Liawenee.		
12th	Wed			
13th	Thur			F
14th	Fri			
15th	Sat			
16th	Sun	Chance of snow flurries in the Central Highlands.		XhS
17th	Mon			
18th	Tue			A
19th	Wed			
20th	Thu	Thunderstorms and heavy rain in the Northeast. Gale force winds possible over Hogan Island.		
21st	Fri			3Q
22nd	Sat			
23rd	Sun			V
24th	Mon			
25th	Tue			
26th	Wed	Thunderstorms with heavy rain, hail, and strong gusty winds in the Derwent Valley to Hobart regions which may cause outbreaks of flash flooding and damage to property. Possible flash flooding in Howrah and Sandford.	Rain, Th	
27th	Thur	Scattered thunderstorms cross the State over the next couple of days, particularly in the southeast.		
28th	Fri	Unusually warm on Flinders Island and in the Southeast to the Derwent Valley regions.	Showers	N

March 2025

Date	Day	Tasmania - March 2025	Hobart	Moon
1st	Sat			
2nd	Sun			P7 XhN
3rd	Mon			
4th	Tue	Thunderstorms with heavy rain in the Northeast and Northern regions. Unusually warmer than normal in the Northwest region.	Showers	
5th	Wed		Showers, Th	
6th	Thu	Thunderstorms in the Southeast with heavy rain that may cause some local flash flooding.		
7th	Fri			1Q
8th	Sat		Showers	^
9th	Sun		Showers	
10th	Mon	Very heavy rain in the eastern coastal regions.	Showers	
11th	Tue	A low pressure system near Tasmania may bring scattered rain across the eastern half of the State over the nex couple of days.	Showers	
12th	Wed		Showers	
13th	Thur	Very hot across the State.	Showers, Th	
14th	Fri		Showers	F
15th	Sat		Showers	XhS
16th	Sun	Unusually cool overnight. Chance of frosts overnight in the Derwent Valley, Mt Wellington and Central Plateau.		
17th	Mon		Showers	
18th	Tue			A
19th	Wed			
20th	Thu			
21st	Fri	Gale force winds likely in the southeast of the State, with severe gusts at Mt Wellington.		
22nd	Sat			
23rd	Sun			3Q V
24th	Mon			
25th	Tue	Very hot across the State.	Showers	
26th	Wed			
27th	Thur			
28th	Fri	Unusually cool overnight, particularly in the Central Highlands.		
29th	Sat			N XhN
30th	Sun			P4
31st	Mon			

April 2025

Date	Day	Tasmania - April 2025	Hobart	Moon
1st	Tue	Very hot across the State over the next couple of days.		
2nd	Wed	Cold front crosses the State bringing high rainfall in the Western Highlands. In contrast, very hot on Flinders Island and in the Northwest of the State.	Show ers	
3rd	Thu			
4th	Fri	Unusually cool over the next couple of nights, particularly in central parts.		^
5th	Sat			1Q
6th	Sun			
7th	Mon		Show ers	
8th	Tue	Cold front crosses the State bringing high rainfall in the Western Highlands.	Show ers	
9th	Wed	Very heavy rain at Mt Read.	Show ers	
10th	Thur			
11th	Fri			XhS
12th	Sat			
13th	Sun			F
14th	Mon		Show ers	A
15th	Tue	Very cold with overnight frosts in the Central Plateau.		
16th	Wed			
17th	Thu	Pockets of frost across the State, particularly in the Central Plateau.		
18th	Fri			
19th	Sat	Unusually high maximas possible over the next couple of days. Very hot in the Midlands.		V
20th	Sun	Very strong winds with possible gale gusts in the Southeast, particularly over Mt Wellington. Unusually hot in the Burnie to Devonport districts and into the East Coast region.		
21st	Mon			3Q
22nd	Tue	Scattered frosts over the State.		
23rd	Wed			
24th	Thur			
25th	Fri	Unusually hot in the Northern district.		
26th	Sat			XhN
27th	Sun		Show ers	
28th	Mon		Rain, Th	P3 N
29th	Tue			
30th	Wed			

May 2025

Date	Day	Tasmania - May 2025	Hobart	Moon
1st	Thu		Show ers	
2nd	Fri		Show ers	^
3rd	Sat		Show ers	
4th	Sun	Very cold overnight at Mt Wellington.		
5th	Mon			1Q
6th	Tue			
7th	Wed			
8th	Thur		Show ers	
9th	Fri	Heavy rain in the East Coast region.	Show ers	XhS
10th	Sat			
11th	Sun		Show ers	A
12th	Mon	Strong damaging winds in the Burnie Northwest region, particularly around the Somerset area, which may cause some localised damage to property.		
13th	Tue	Stormy air with flash flooding in the Launceston Tamar Valley region. A line of thunderstorms with very strong wind gusts may cross Flinders Island.	Show ers	F
14th	Wed		Rain	
15th	Thur	Very heavy rain in the West Coast possible around this time.	Show ers	
16th	Fri	Strong and persistent northwest winds may prevail across the State.	Show ers	V
17th	Sat	Chance of frosts in the Central Plateau, Derwent Valley, and Midlands.	Show ers	
18th	Sun			
19th	Mon	Strong winds across the State.		
20th	Tue			3Q
21st	Wed	Unusually mild in the Southeast.		

Date	Day	Tasmania - May 2025	Hobart	Moon
22nd	Thu	Chance of frost pockets in the East Coast and Derwent Valley regions.	Fg	
23rd	Fri			XhN
24th	Sat			
25th	Sun	Warm in the Midlands, while chance of frosts around Lake Leake in the East Coast region.		
26th	Mon	Chance of frost patches on Macquarie Island.		P5
27th	Tue		Fg	N
28th	Wed		Rain, Fg	
29th	Thur	Strong and gusty northwest winds across the State.		^
30th	Fri			
31st	Sat		Show ers	

June 2025

Date	Day	Tasmania - June 2025	Hobart	Moon
1st	Sun			
2nd	Mon			
3rd	Tue		Show ers	1Q
4th	Wed			
5th	Thu			XhS
6th	Fri			
7th	Sat			A
8th	Sun	Warm on Flinders Island.		
9th	Mon	Very strong winds over Maatsuyker Island and strong northwest winds over the State.	Show ers	
10th	Tue	Very cool in the Southeast with frosts possible on Mt Wellington.	Show ers, Th	
11th	Wed	Unusually cold across the State with the chance of snow flurries in the Central Highlands.		F
12th	Thur	Very heavy rain in the Northeast region of the State, particularly around Mt Victoria.		V
13th	Fri			
14th	Sat			
15th	Sun	Milder maximum temperatures in the East Coast region over the next couple of days.		
16th	Mon		Show ers	
17th	Tue			
18th	Wed		Fg	
19th	Thu	Light snow in the Central Highlands over the next several days.		3Q XhN
20th	Fri	Chance of snowflurries on Mt Mawson. Unusually mild temperatures on Hogan Island.		
21st	Sat			
22nd	Sun		Show ers	
23rd	Mon		Show ers	P8
24th	Tue			
25th	Wed		Fg	N
26th	Thur		Fg	^
27th	Fri			
28th	Sat	Very cold across the State, especially in the southeast. Widespread scattered frosts likely over the next 3-4 days with severe frosts in the Central Plateau.	Rain	
29th	Sun			
30th	Mon	Chance of snow in the higher reaches of Hobart over the next couple of days.	Rain	

July 2025

Date	Day	Tasmania - July 2025	Hobart	Moon
1st	Tue	Light snow in the Central Highlands over the next few days.	Show ers	
2nd	Wed	Very cold across the State, particularly in the Northeast. Chance of frosts at Eddystone Point.	Fg	XhS
3rd	Thu	Light snow on Mt Mawson.		1Q
4th	Fri		Fg	
5th	Sat			A
6th	Sun			
7th	Mon	Snowflurries in the Central Highlands and on Mt Mawson. Strong and blustery winds across Hogan Island.	Show ers	
8th	Tue	Chance of snowflurries on Mt Mawson.		
9th	Wed			
10th	Thur	Snow in the Central Highlands and on Mt Mawson over the next couple of days.		V

Date	Day	Tasmania - July 2025	Hobart	Moon
11th	Fri	Chance of light snow on Mt Wellington.		F
12th	Sat		Showers	
13th	Sun	Low pressure prevails across the State over the next few days. Very strong winds across the State with possible gale gusts at times on Hogan Island.		
14th	Mon	Colder than normal maximum temperatures over most of the State over the next three weeks. Chance of thick fog in many parts of the State over the next couple of days.	Showers, Fg	
15th	Tue	Wet spell across the State. Severe frosts on Macquarie Island.	Showers	
16th	Wed		Showers	
17th	Thu	High pressure moves onto the State and lingers for around a week.	Showers	XhN
18th	Fri	Milder overnight.		3Q
19th	Sat	Light snow in the Central Highlands over the next few days.		
20th	Sun	Very cold overnight with widespread pockets of frost.		
21st	Mon		Fg	P11
22nd	Tue			
23rd	Wed	Snowflurries in the Central Highlands.	Rain	^
24th	Thur	Wet spell across the State. Very cold on Mt Wellington and in the Southeast region.	Showers	
25th	Fri	Very cold overnight with widespread scattered frosts over the next couple of days particularly in the Northern region.		N
26th	Sat		Showers	
27th	Sun	Cold change across the State over the next couple of days. Snowflurries in the Central Highlands. Heavy rain in the West Coast.	Showers	
28th	Mon	Wet spell across the State, followed by high pressure moving onto the State and lingering for another 4-5 days.		
29th	Tue			
30th	Wed			XhS
31st	Thu	Very cold overnight, with severe frosts in the Central Highlands.		

August 2025

Date	Day	Tasmania - August 2025	Hobart	Moon
1st	Fri			
2nd	Sat	Light snow in the Central Highlands and flurries on Mt Mawson. Change to milder temperatures in most parts of the State over the next week.		1Q A
3rd	Sun	Snowflurries in the Central Highlands over the next couple of days.		
4th	Mon	Mild maximum temperatures in the Northeast.		
5th	Tue	Light snow in the Central Highlands and flurries on Mt Mawson. Milder overnight elsewhere, particularly on Hogan Island.		
6th	Wed	Snow in the Central Highlands, light snow on Mt Mawson, and flurries at Ben Lomond.		V
7th	Thur	Snow in the Central Highlands and Mt Mawson.	Showers	
8th	Fri	Light snow in the Central Highlands.		
9th	Sat			F
10th	Sun			
11th	Mon	Widespread frosts most nights over the next four weeks.	Showers	
12th	Tue		Showers	
13th	Wed			XhN
14th	Thu		Showers	
15th	Fri			P12
16th	Sat			3Q
17th	Sun	Milder minimum temperatures over the next few days in many parts of the State.	Showers	
18th	Mon	An intense storm may cross the State over the next few days with the chance of damage to property and infrastructure.	Rain	
19th	Tue	Storm system may continue to bring strong winds and heavy rain across the State. Milder temperatures overnight in the Northeast.	Rain	^
20th	Wed	Widespread gale force winds and flooding level rains likely, particularly along watercourses, including on the Forth River, Huon, and South Esk river systems. Cooler spell in the west and south of the State over the next few days. Very heavy rain with chance of flash flooding in the West Coast, particularly in the Strathgordon area.	Showers	
21st	Thur	Very cold overnight with widespread frosts over most of the State for the next 7-10 nights.		
22nd	Fri	Snow in the Central Highlands over the next few days.	Showers	
23rd	Sat	Snow on Mt Mawson over the next couple of days.		N
24th	Sun	Severe frosts in the Central Plateau.		
25th	Mon	Snowflurries in the Central Highlands and on Mt Mawson.		
26th	Tue			XhS

Date	Day	Tasmania - August 2025	Hobart	Moon
27th	Wed		Showers	
28th	Thu			
29th	Fri			
30th	Sat	A cold front may cross the State around this time, bringing heavy snow in the Central Highlands and Mt Mawson, and light snow on Mt Wellington.	Showers	A
31st	Sun	Light snow in the Central Highlands and flurries on Mt Mawson.		1Q

September 2025

Date	Day	Tasmania - September 2025	Hobart	Moon
1st	Mon	Snowflurries likely on Mt Mawson. Severe frosts in the Central Plateau.		
2nd	Tue	Severe frosts in the Central Plateau.		V
3rd	Wed			
4th	Thur	Milder minimum temperatures over the next few days in many parts of the State.		
5th	Fri			
6th	Sat	Snowflurries in the Central Highlands, while milder temperatures along the East Coast.		
7th	Sun			
8th	Mon			F
9th	Tue	Snowflurries in the Central Highlands and on Mt Mawson. Strong and blustery winds across Hogan Island. Milder spell in the Southeast to South Midlands and up into the Northeast region.	Showers	XhN
10th	Wed	Light snow in the Central Highlands.	Rain	
11th	Thu			P9
12th	Fri	Wet front crosses the State.		
13th	Sat	Cold spell in the State over the next couple of days. Overnight frost on Mt Wellington.	Showers	
14th	Sun	Overnight frosts at Mt Wellington, and severe frosts in the Central Plateau. Scattered pockets of frost in sheltered parts in other regions of the State over the next couple of nights.		3Q
15th	Mon			^
16th	Tue			
17th	Wed			
18th	Thur			
19th	Fri			
20th	Sat	Light snow in the Central Highlands and on Mt Mawson.	Showers	
21st	Sun	Snowflurries in the Central Highlands.	Showers	
22nd	Mon	Light snow in the Central Highlands and on Mt Mawson.		N XhS
23rd	Tue	Snow in the Central Highlands and light snow on Mt Mawson.		
24th	Wed	Chance of snowflurries in the Central Highlands.	Showers	
25th	Thu			
26th	Fri	Unusually cold over night.		A
27th	Sat			
28th	Sun			
29th	Mon	Warmer turn over the State.		
30th	Tue		Showers	1Q V

October 2025

Date	Day	Tasmania - October 2025	Hobart	Moon
1st	Wed			
2nd	Thur	Cool spell in the State over the next couple of days. Overnight scattered frosts likely across the State over the next couple of days.		
3rd	Fri		Showers	
4th	Sat	Wet front crosses the State.	Rain	
5th	Sun		Fg	
6th	Mon			
7th	Tue	A series of wet fronts may cross the State over the next few days.	Showers	F XhN
8th	Wed	Very strong winds with possible gale gusts for a time over Hogan and Maatsuyker Islands over the next couple of days.	Showers	
9th	Thu	Strong gusty winds around the Southwest region.	Rain	P6
10th	Fri	Heavy rain in the West Coast region.	Rain	

Date	Day	Tasmania - October 2025	Hobart	Moon
11th	Sat	Light snow in the Central Highlands and on Mt Mawson. A cooler spell may cross the State over the coming week. Unsettled and blustery winds over the next 3-4 days.	Show ers	
12th	Sun		Show ers	
13th	Mon	Chance of snowflurries in the Central Highlands over the next couple of days.	Show ers	^
14th	Tue	Chance of gale gusts in the Southeast region including over Mt Wellington.	Show ers	3Q
15th	Wed			
16th	Thur			
17th	Fri			
18th	Sat	Snow in the Central Highlands likely on most days over the next two weeks. Chance of frost pockets in some Northern and Eastern districts overnight, including in the Cressy, Launceston and Fingal areas.		
19th	Sun			
20th	Mon			XhS
21st	Tue	Milder overnight.		
22nd	Wed	Cooler and unsettled southwesterly flow over the State. Winds gusty at times over the coming week.	Show ers	N
23rd	Thu	Light snow on Mt Mawson over the next few days, while continuing in the Central Highlands.		
24th	Fri			A
25th	Sat	Milder overnight. Heavy rain in the West Coast, particuarly around Mount Read.		
26th	Sun	Cooler southwesterly flow over the State.	Rain, Sn	
27th	Mon			V
28th	Tue			
29th	Wed		Show ers	
30th	Thur	Wet front crosses the State, with a generally northerly airflow prevailing.		1Q
31st	Fri		Rain	

November 2025

Date	Day	Tasmania - November 2025	Hobart	Moon
1st	Sat	Snow in the Central Highlands and light snow on Mt Mawson over the next couple of days.		
2nd	Sun	Severe frosts in the Central Plateau.		
3rd	Mon			XhN
4th	Tue	Snow in the Central Highlands over the next few days.		
5th	Wed	Light snow likely on Mt Mawson over the next couple of days. In contrast, many parts of the State may be milder overnight during the next few days, than of recent.	Show ers	
6th	Thur	Wet fronts cross the State over the next couple of days. Some shower activity in most districts, particularly along the northeast coasts. Generally warmer northerly winds prevail. Unusually warmer in the Southwest.	Show ers	F P1
7th	Fri	Unsettled and blustery winds at times n many parts of the State over the next 3-4 days.	Show ers	
8th	Sat	Cooler southwesterly flow over the State.	Show ers	
9th	Sun			^
10th	Mon			
11th	Tue	Cooler than normal overnight across the State with chance of patches of frost mostly in the Midlands, Central Plateau, Derwent Valley, and Southeast regions.		
12th	Wed			3Q
13th	Thu		Show ers	
14th	Fri	A cold southerly flow may cover the State over the next 3-4 days. Very strong and blustery winds to gale force at times on Hogan Island.	Show ers	
15th	Sat	Cooler than normal overnight across the State. Some pockets of frost likely in the Central Plateau, Derwent Valley, and Southeast regions, including on Mt Wellington. Chance of frosts on Macquarie Island.		
16th	Sun			XhS
17th	Mon			
18th	Tue			
19th	Wed			
20th	Thur			N A
21st	Fri			
22nd	Sat			
23rd	Sun			V
24th	Mon			
25th	Tue	Milder than average temperatures overnight over the next couple of weeks.		
26th	Wed	Unusually warm spell in the State over the next few days.		
27th	Thu		Show ers	
28th	Fri			1Q
29th	Sat		Fg	
30th	Sun	A band of thunderstorms may cross the State, but may only bring passing showers. Strong northerly winds prevail. Unusualy warm overnight temperatures on Flinders Island. High maximum temperatures of 30°C or more likely in a number of areas across the State.	Show ers	XhN

December 2025

Date	Day	Tasmania - December 2025	Hobart	Moon
1st	Mon	Outbreaks of rain in the West Coast, with some heavy falls at times.		
2nd	Tue			
3rd	Wed			
4th	Thur			**P2**
5th	Fri			**F**
6th	Sat			**^**
7th	Sun			
8th	Mon			
9th	Tue			
10th	Wed			
11th	Thu			
12th	Fri	Unusually warm airflow over the State.	Rain, Th	**3Q**
13th	Sat	Stormy air in the Hobart region may bring thunderstorms and hail showers and some heavy rain in places that may cause flash flooding, particularly at risk in the Rokeby and Bellerive areas.	Rain, Th	**XhS**
14th	Sun			
15th	Mon			
16th	Tue			
17th	Wed	A band of thunderstorms may cross the State with some heavy rain and hail likely. Very warm on Flinders Island overnight. Very strong winds with possible gale gusts for a time across the Southwest of the State.	Showers	**A**
18th	Thur			
19th	Fri	Heavy rain and possible flash flooding in the Hobart area.	Rain	
20th	Sat	Colder than of recent in the south and west of the State. Chance of frosts on Mt Wellington and in some parts of the Central Plateau.	Showers	**N V**
21st	Sun	Very cold overnight on the Central Plateau with chance of frost patches, particularly in the Liawenee area.	Showers	
22nd	Mon			
23rd	Tue			
24th	Wed			
25th	Thu			
26th	Fri			
27th	Sat			**XhN**
28th	Sun			**1Q**
29th	Mon	Chance of snowflurries for a short time in the Central Highlands over the next couple of days.	Showers, Th	
30th	Tue	A band of thunderstorms may cross the State with some heavy rain and hail likely. Maximum temperatures may be unusually high in many parts of the State.	Fg	
31st	Wed	An intense depression near the State may bring widespread rain over the next couple of days, with heavy falls at times.	Rain	

TASMANIA GRAPHS AND TABLES 2025

Seasonal Breakdown –	State Overall Rainfall Timing and Estimated mm Trend Graphs State Overall Maximum Temperature Trend Graphs Regional Rainfall Expectations Tables - mm and Trends Regional Maximum and Minimum Temperature Trends Tables
Monthly Breakdown -	**Rainfall:** Regional Expectations by mm Regional Expectations by mm Variance to Average Regional Variance Trend by mm Regional Estimated Expectations by % of Variance to Average Regional Estimatedl Variance Trend by % **Maximum & Minimum Temperature Trends:** Regional Estimated Average Temperature Regional Estimated Degree of Variation from the Average Regional Estimated Temperature Trend

Seasonal Breakdowns

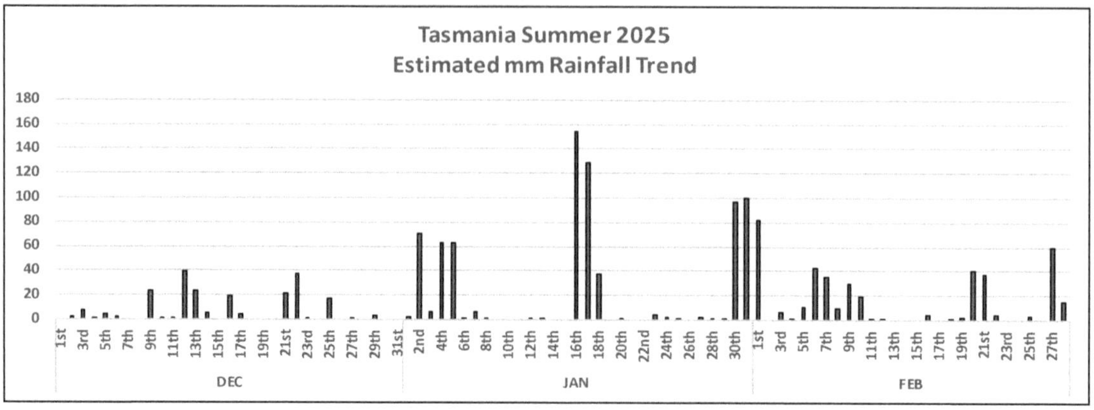

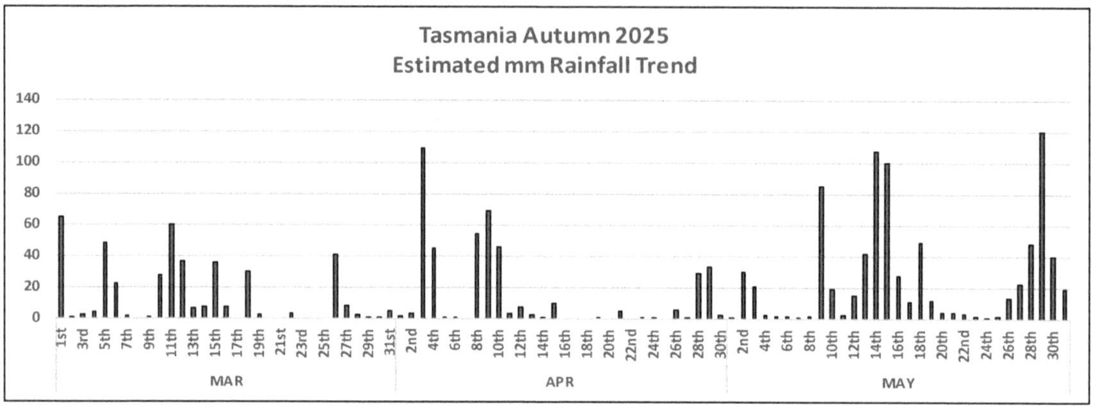

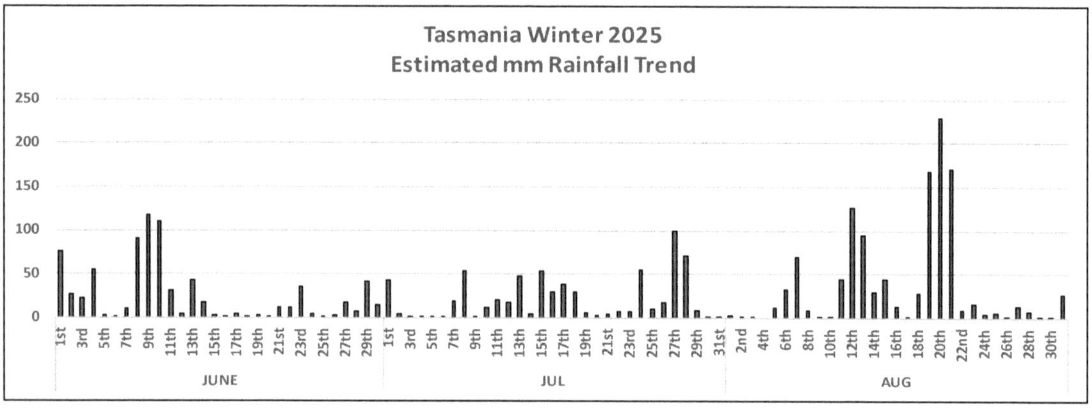

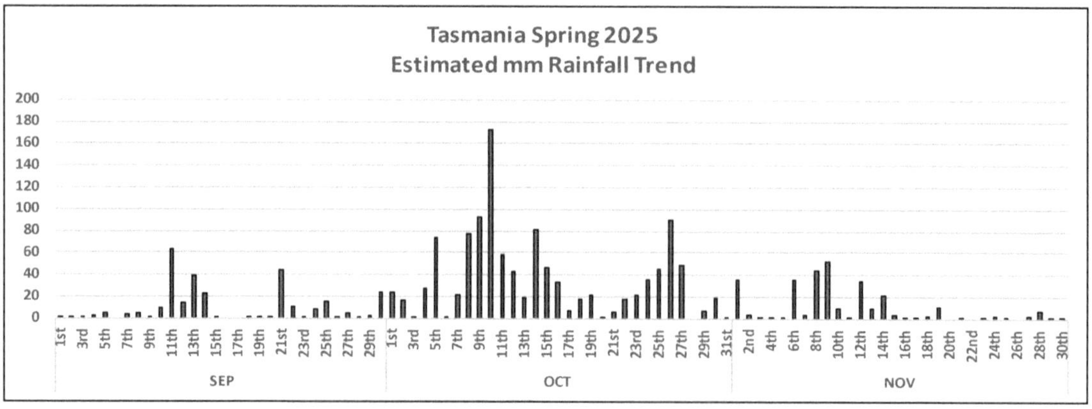

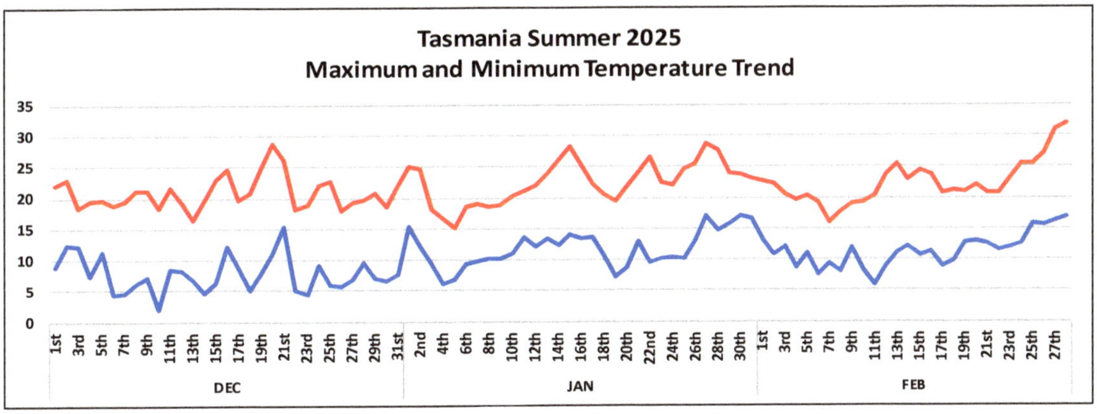
Tasmania Summer 2025
Maximum and Minimum Temperature Trend
DEC
JAN
FEB

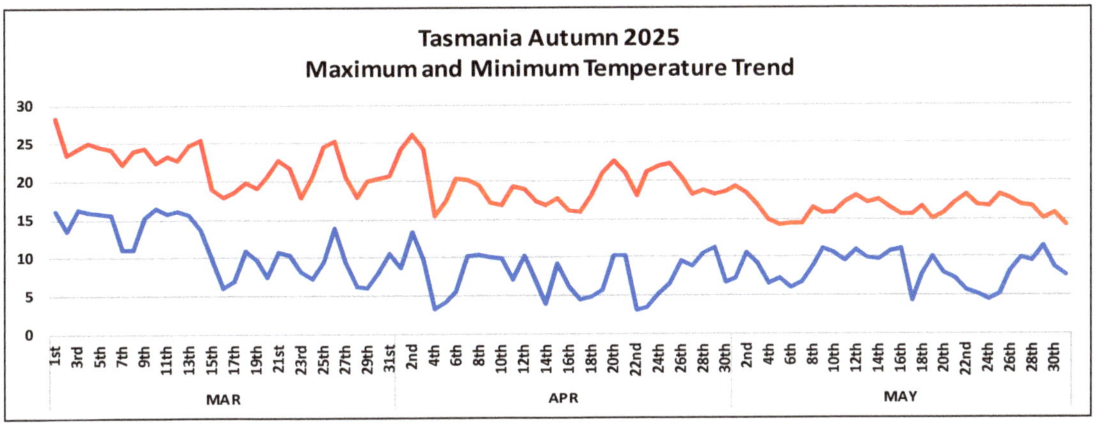
Tasmania Autumn 2025
Maximum and Minimum Temperature Trend
MAR
APR
MAY

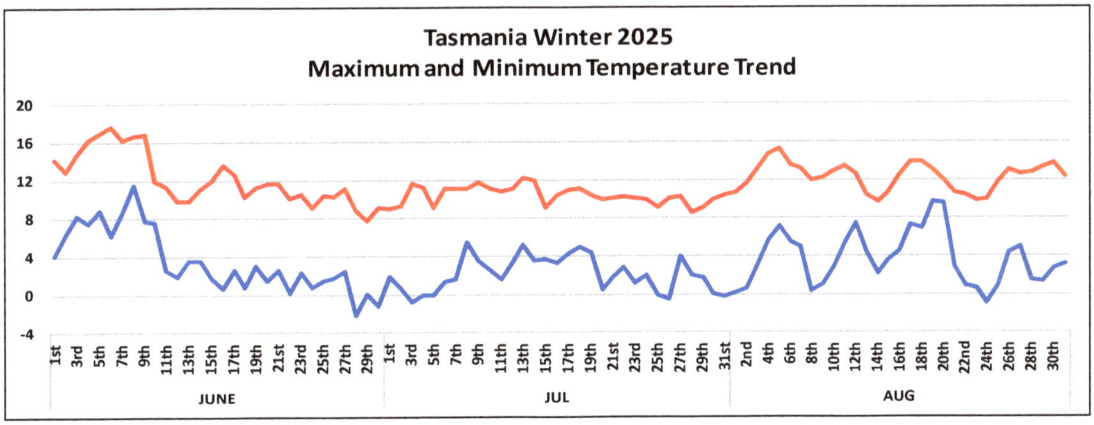
Tasmania Winter 2025
Maximum and Minimum Temperature Trend
JUNE
JUL
AUG

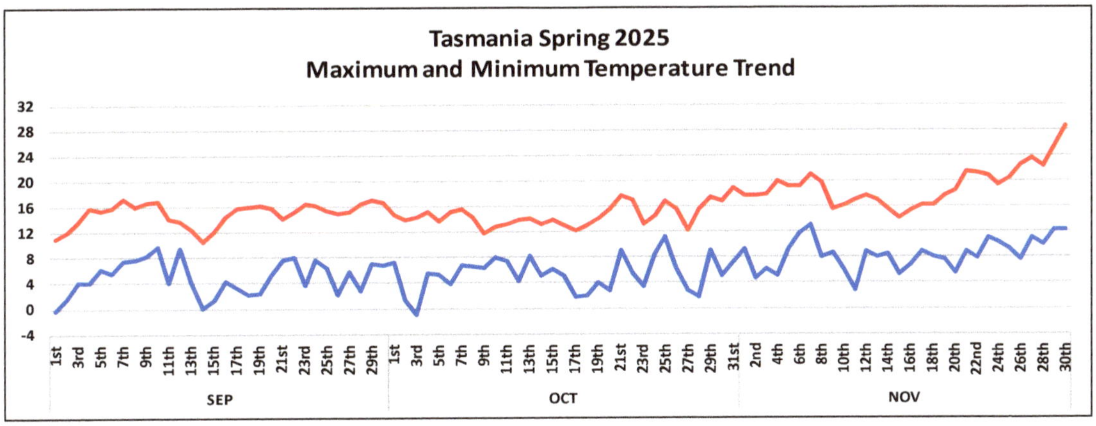
Tasmania Spring 2025
Maximum and Minimum Temperature Trend
SEP
OCT
NOV

2025 TAS - Estimated Seasonal Rainfall Expectations (mm)

REGION	NW Coast & King Is	W & S Coasts & Highlands	Central Plateau	Upper Derwent	South East	Central North	Lower Derwent	Midlands Alpine	East Coast	NE Coast & Flinders Is	TAS
Town Sample	Devonport	Strathan	Lake St Clair	Bushy Park	Hobart	Launceston	Huonville	Fingal	Swansea	Killiecrankie	Overall
Summer	73	280	296	127	160	70	148	68	69	66	1356
Autumn	107	308	360	54	86	158	91	154	129	206	1653
Winter	220	582	629	126	135	239	169	156	118	234	2607
Spring	121	331	546	131	112	87	149	67	47	90	1682
Overall	570	1526	1791	458	537	577	575	467	371	627	7499

2025 TAS - Estimated Seasonal Rainfall Variation from the Average (mm)

REGION	NW Coast & King Is	W & S Coasts & Highlands	Central Plateau	Upper Derwent	South East	Central North	Lower Derwent	Midlands Alpine	East Coast	NE Coast & Flinders Is	TAS
Town Sample	Devonport	Strathan	Lake St Clair	Bushy Park	Hobart	Launceston	Huonville	Fingal	Swansea	Killiecrankie	Overall
Summer	-51	34	-5	2	35	-51	-7	-81	-87	-67	-278
Autumn	-61	-74	-21	-77	-15	26	-53	25	9	47	-193
Winter	-32	68	80	-26	17	38	-31	-4	0	24	135
Spring	-71	-34	37	-32	-14	-72	-55	-79	-85	-74	-478
Overall	-165	18	51	-112	68	-35	-128	-117	-154	-40	-613

2025 TAS - Estimated Seasonal Rainfall Trend Variation by 10mm

REGION	NW Coast & King Is	W & S Coasts & Highlands	Central Plateau	Upper Derwent	South East	Central North	Lower Derwent	Midlands Alpine	East Coast	NE Coast & Flinders Is	TAS
Town Sample	Devonport	Strathan	Lake St Clair	Bushy Park	Hobart	Launceston	Huonville	Fingal	Swansea	Killiecrankie	Overall
Summer	drr	wtr	av	av	wtr	drr	av	drr	drr	drr	drr
Autumn	drr	drr	av	drr	av	av	drr	av	av	wtr	drr
Winter	drr	wtr	wtr	av	av	wtr	drr	av	av	av	wtr
Spring	drr	drr	wtr	drr	av	drr	drr	drr	drr	drr	drr
Overall	drr	av	av	drr	av	av	drr	drr	drr	av	drr

2025 TAS - Estimated Seasonal Rainfall Expectations by %

REGION	NW Coast & King Is	W & S Coasts & Highlands	Central Plateau	Upper Derwent	South East	Central North	Lower Derwent	Midlands Alpine	East Coast	NE Coast & Flinders Is	TAS
Town Sample	Devonport	Strathan	Lake St Clair	Bushy Park	Hobart	Launceston	Huonville	Fingal	Swansea	Killiecrankie	Overall
Summer	-41%	14%	-2%	2%	28%	-42%	-4%	-54%	-56%	-50%	-17%
Autumn	-36%	-19%	-6%	-59%	-14%	19%	-37%	19%	8%	30%	-10%
Winter	-13%	13%	15%	-17%	15%	19%	-15%	-2%	0%	11%	5%
Spring	-37%	-9%	7%	-20%	-11%	-45%	-27%	-54%	-64%	-45%	-22%
Overall	-22%	1%	3%	-20%	14%	-6%	-18%	-20%	-29%	-6%	-8%

2025 TAS - Estimated Seasonal Rainfall Trend Variation by 25%

REGION	NW Coast & King Is	W & S Coasts & Highlands	Central Plateau	Upper Derwent	South East	Central North	Lower Derwent	Midlands Alpine	East Coast	NE Coast & Flinders Is	TAS
Town Sample	Devonport	Strathan	Lake St Clair	Bushy Park	Hobart	Launceston	Huonville	Fingal	Swansea	Killiecrankie	Overall
Summer	drr	av	av	av	wtr	drr	av	drr	drr	drr	av
Autumn	drr	av	av	drr	av	av	drr	av	av	wtr	av
Winter	av	av	av	av	av	av	av	av	av	av	av
Spring	drr	av	av	av	av	drr	drr	drr	drr	drr	av
Overall	av	av	av	av	av	av	av	av	drr	av	av

2025	TAS - Estimated Seasonal Maximum Temperature Average and Trend										
Region	NW Coast & King Is	W & S Coasts & Highlands	Central Plateau	Upper Derwent	South East	Central North	Lower Derwent	Midlands Alpine	East Coast	NE Coast & Flinders Is	TAS
Town Sample	Devonport	Strathan	Lake St Clair	Bushy Park	Hobart	Launceston	Huonville	Fingal	Swansea	Killiecrankie	Overall
Summer Est	21.0	21.0	18.9	24.6	24.4	23.1	22.4	22.5	19.6	22.1	22.0
Trend	av	av	av	wmr	av	av	wmr	av	a-w	av	av
Autumn Est	19.9	18.6	15.9	20.7	21.6	19.8	20.0	19.8	16.4	20.9	19.3
Trend	wmr	wmr	wmr	wmr	wmr	wmr	wmr	wmr	wmr	wmr	wmr
Winter Est	13.0	12.1	7.2	12.1	12.9	11.9	12.9	12.0	8.6	13.8	11.7
Trend	av	av	av	av	a-w	a-c	a-w	a-c	av	av	av
Spring Est	15.5	15.9	12.8	18.4	17.5	17.4	17.6	17.2	13.6	16.8	16.3
Trend	av	av	av	a-w	av	av	a-w	av	av	av	av

2025	TAS - Estimated Seasonal Minimum Temperature Average and Trend									
Region	NW Coast & King Is	W & S Coasts & Highlands	Central Plateau	Upper Derwent	South East	Central North	Lower Derwent	Midlands Alpine	NE Coast & Flinders Is	TAS
Town Sample	Devonport	Strathan	Lake St Clair	Bushy Park	Hobart	Launceston	Huonville	Fingal	Killiecrankie	Overall
Summer Est	12.0	10.3	5.3	9.1	11.7	11.7	8.8	9.7	13.5	10.2
Trend	av	av	av	a-c	av	wmr	av	av	av	av
Autumn Est	11.0	9.7	4.5	7.7	10.7	10.0	8.0	7.6	12.1	9.0
Trend	wmr	wmr	wmr	wmr	wmr	wmr	wmr	wmr	wmr	wmr
Winter Est	4.4	4.5	-0.5	0.8	4.5	2.9	2.0	1.5	6.7	3.0
Trend	av	clr	a-c	clr	av	av	av	a-w	av	av
Spring Est	7.2	7.3	2.0	5.0	7.6	6.9	5.5	4.8	8.9	6.1
Trend	av	av	av	a-c	av	wmr	av	av	av	av

Monthly Breakdowns

2025 TASMANIA - Estimated Rainfall Expectations (mm)

REGION	NW Coast & King Is	W & S Coasts & Highlands	Central Plateau	Upper Derwent	South East	Central North	Lower Derwent	Midlands Alpine	East Coast	NE Coast & Flinders Is	TAS
Town Sample	Devonport	Strathan	Lake St Clair	Bushy Park	Hobart	Launceston	Huonville	Fingal	Swansea	Killiecrankie	Overall
January	60	168	129	91	63	44	75	32	52	30	744
February	9	57	70	21	83	22	55	33	15	36	403
March	22	78	75	13	33	35	30	55	57	22	419
April	17	106	124	22	18	26	21	37	29	30	428
May	69	125	161	19	36	97	41	62	44	153	806
June	79	201	179	30	25	58	37	42	30	89	769
July	49	137	101	19	51	48	46	74	51	97	672
August	92	245	348	77	60	132	86	40	37	48	1165
September	32	65	66	21	20	25	20	9	4	15	277
October	52	235	422	96	76	33	110	29	21	51	1124
November	37	31	58	14	16	29	19	30	22	25	281
December	53	78	58	36	58	28	35	25	10	31	411
Overall	570	1526	1791	458	537	577	575	467	371	627	7499

2025 TASMANIA - Estimated Rainfall Variation from the Average (mm)

REGION	NW Coast & King Is	W & S Coasts & Highlands	Central Plateau	Upper Derwent	South East	Central North	Lower Derwent	Midlands Alpine	East Coast	NE Coast & Flinders Is	TAS
Town Sample	Devonport	Strathan	Lake St Clair	Bushy Park	Hobart	Launceston	Huonville	Fingal	Swansea	Killiecrankie	Overall
January	19	87	26	50	22	-5	25	-22	-5	-15	182
February	-20	-13	-4	-13	51	-4	10	-2	-21	6	-9
March	-16	-28	-25	-26	-3	3	-18	16	11	-20	-107
April	-46	-10	-7	-24	-20	-21	-31	-13	-16	-23	-208
May	1	-36	10	-27	9	43	-4	22	14	90	123
June	0	46	25	-18	-11	-3	-18	-5	-7	23	31
July	-39	-43	-80	-31	14	-16	-19	25	10	22	-156
August	8	65	135	23	14	57	7	-24	-2	-21	260
September	-43	-82	-133	-32	-24	-37	-58	-43	-32	-48	-532
October	-10	114	244	39	32	-18	42	-20	-30	-2	391
November	-19	-65	-73	-38	-23	-17	-38	-17	-23	-25	-337
December	-1	-16	-67	-14	7	-18	-25	-35	-53	-28	-249
Overall	-165	18	51	-112	68	-35	-128	-117	-154	-40	-613

2025 TASMANIA - Estimated Rainfall Trend Variation by 10mm

REGION	NW Coast & King Is	W & S Coasts & Highlands	Central Plateau	Upper Derwent	South East	Central North	Lower Derwent	Midlands Alpine	East Coast	NE Coast & Flinders Is	TAS
Town Sample	Devonport	Strathan	Lake St Clair	Bushy Park	Hobart	Launceston	Huonville	Fingal	Swansea	Killiecrankie	Overall
January	wtr	wtr	wtr	wtr	wtr	av	wtr	drr	av	drr	wtr
February	drr	drr	av	drr	wtr	av	av	av	drr	av	av
March	drr	drr	drr	drr	av	av	drr	wtr	wtr	drr	drr
April	drr	av	av	drr	drr	drr	drr	drr	drr	drr	drr
May	av	drr	av	drr	av	wtr	av	wtr	wtr	wtr	wtr
June	av	wtr	wtr	drr	drr	av	drr	av	av	wtr	wtr
July	drr	drr	drr	drr	wtr	drr	drr	wtr	av	wtr	drr
August	av	wtr	wtr	wtr	wtr	wtr	av	drr	av	drr	wtr
September	drr	drr	drr	drr	drr	drr	drr	drr	drr	drr	drr
October	av	wtr	wtr	wtr	wtr	drr	wtr	drr	drr	av	wtr
November	drr	drr	drr	drr	drr	drr	drr	drr	drr	drr	drr
December	av	drr	drr	drr	av	drr	drr	drr	drr	drr	drr
Overall	drr	av	av	drr	av	av	drr	drr	drr	av	drr

2025 TASMANIA - Estimated Rainfall Expectations by %

REGION	NW Coast & King Is	W & S Coasts & Highlands	Central Plateau	Upper Derwent	South East	Central North	Lower Derwent	Midlands Alpine	East Coast	NE Coast & Flinders Is	TAS
Town Sample	Devonport	Strathan	Lake St Clair	Bushy Park	Hobart	Launceston	Huonville	Fingal	Swansea	Killiecrankie	Overall
January	45%	107%	25%	124%	53%	-10%	50%	-41%	-9%	-33%	32%
February	-68%	-18%	-5%	-38%	156%	-14%	21%	-7%	-58%	21%	-2%
March	-43%	-26%	-25%	-66%	-9%	9%	-38%	40%	23%	-48%	-20%
April	-73%	-9%	-5%	-52%	-53%	-45%	-59%	-25%	-35%	-43%	-33%
May	2%	-22%	7%	-60%	32%	81%	-9%	54%	48%	144%	18%
June	0%	29%	16%	-38%	-30%	-5%	-33%	-11%	-19%	34%	4%
July	-44%	-24%	-44%	-62%	39%	-25%	-29%	52%	24%	29%	-19%
August	9%	36%	63%	42%	29%	75%	8%	-37%	-6%	-31%	29%
September	-57%	-56%	-67%	-60%	-54%	-60%	-75%	-83%	-89%	-77%	-66%
October	-15%	94%	136%	67%	75%	-36%	60%	-40%	-58%	-4%	53%
November	-33%	-68%	-56%	-73%	-59%	-37%	-67%	-36%	-51%	-50%	-55%
December	-1%	-17%	-54%	-28%	14%	-40%	-41%	-58%	-84%	-47%	-38%
Overall	-22%	1%	3%	-20%	14%	-6%	-18%	-20%	-29%	-6%	-8%

2025	TASMANIA - Estimated Rainfall Trend Variation by 25%										
REGION	NW Coast & King Is	W & S Coasts & Highlands	Central Plateau	Upper Derwent	South East	Central North	Lower Derwent	Midlands Alpine	East Coast	NE Coast & Flinders Is	TAS
Town Sample	Devonport	Strathan	Lake St Clair	Bushy Park	Hobart	Launceston	Huonville	Fingal	Swansea	Killiecrankie	Overall
January	wtr	wtr	av	wtr	wtr	av	wtr	drr	av	drr	wtr
February	drr	av	av	drr	wtr	av	av	av	drr	av	av
March	drr	drr	av	drr	av	av	drr	wtr	av	drr	av
April	drr	av	av	drr	drr	drr	drr	av	drr	drr	drr
May	av	av	av	drr	wtr	wtr	av	wtr	wtr	wtr	av
June	av	wtr	av	drr	drr	av	drr	av	av	wtr	av
July	drr	av	drr	drr	wtr	av	drr	wtr	av	wtr	av
August	av	wtr	wtr	wtr	wtr	wtr	av	drr	av	drr	wtr
September	drr	drr	drr	drr	drr	drr	drr	drr	drr	drr	drr
October	av	wtr	wtr	wtr	wtr	drr	wtr	drr	drr	av	wtr
November	drr	drr	drr	drr	drr	drr	drr	drr	drr	drr	drr
December	av	av	drr	drr	av	drr	av	drr	drr	drr	drr
Overall	av	av	av	av	av	av	av	av	drr	av	av

2025	TASMANIA - Estimated Average Maximum Temperature										
Region	NW Coast & King Is	W & S Coasts & Highlands	Central Plateau	Upper Derwent	South East	Central North	Lower Derwent	Midlands Alpine	East Coast	NE Coast & Flinders Is	TAS
Town Sample	Devonport	Strathan	Lake St Clair	Bushy Park	Hobart	Launceston	Huonville	Fingal	Swansea	Killiecrankie	Overall
January	22	22	20	25	23	25	23	23	19	23	**23**
February	22	22	20	25	23	25	23	23	19	23	**22**
March	23	21	19	24	22	26	22	23	19	24	**22**
April	20	18	16	21	20	21	20	20	17	21	**19**
May	17	16	13	17	18	18	17	17	13	18	**16**
June	14	12	8	13	14	13	13	12	9	15	**12**
July	12	11	6	11	11	12	10	10	7	13	**10**
August	13	12	8	13	14	14	13	13	9	14	**12**
September	14	16	11	17	16	16	16	16	12	16	**15**
October	15	14	10	16	16	16	16	16	12	16	**15**
November	18	18	17	22	20	21	20	21	16	19	**19**
December	20	20	19	24	23	23	22	22	19	22	**21**
Overall	**17**	**17**	**14**	**19**	**18**	**19**	**18**	**18**	**14**	**18**	**17**

2025	TASMANIA - Estimated Average Minimum Temperature									
Region	NW Coast & King Is	W & S Coasts & Highlands	Central Plateau	Upper Derwent	South East	Central North	Lower Derwent	Midlands Alpine	NE Coast & Flinders Is	TAS
Town Sample	Devonport	Strathan	Lake St Clair	Bushy Park	Hobart	Launceston	Huonville	Fingal	Killiecrankie	Overall
January	13	12	8	11	13	13	10	11	15	**12**
February	14	11	6	10	13	13	10	11	15	**11**
March	13	11	7	10	13	13	10	11	14	**11**
April	10	9	3	6	9	8	6	6	10	**7**
May	10	9	4	7	10	8	8	6	12	**8**
June	5	5	0	1	5	4	2	2	7	**3**
July	3	3	-2	0	4	1	1	1	6	**2**
August	5	6	1	1	5	4	2	2	7	**4**
September	6	7	1	3	6	5	4	3	7	**5**
October	6	6	1	4	7	6	5	4	9	**5**
November	9	9	4	8	10	10	8	7	11	**8**
December	12	11	6	9	12	12	9	10	13	**10**
Overall	9	8	3	6	9	8	6	6	11	**7**

2025	TASMANIA - Estimated Degree of Variation from the Maximum Average										
Region	NW Coast & King Is	W & S Coasts & Highlands	Central Plateau	Upper Derwent	South East	Central North	Lower Derwent	Midlands Alpine	East Coast	NE Coast & Flinders Is	TAS
Town Sample	Devonport	Strathan	Lake St Clair	Bushy Park	Hobart	Launceston	Huonville	Fingal	Swansea	Killiecrankie	Overall
January	0.0	0.6	0.5	1.3	1.2	-0.2	-0.1	-0.9	-0.3	0.9	0.3
February	-0.1	1.2	0.2	0.9	1.0	0.6	0.1	0.5	0.2	0.0	0.5
March	2.0	1.5	2.4	2.0	1.6	3.1	0.6	1.3	1.2	2.2	1.8
April	1.9	1.6	2.4	3.3	3.1	3.1	2.4	1.6	2.6	1.9	2.4
May	2.0	2.0	2.9	2.6	3.1	3.0	2.1	1.2	2.0	2.0	2.3
June	0.2	-0.3	-0.2	1.2	1.7	1.1	0.1	-0.3	0.1	0.7	0.4
July	-0.5	-0.8	-1.1	-0.6	-0.4	0.0	-2.0	-2.0	-1.3	-0.7	-0.9
August	0.1	-0.5	-0.1	-0.1	0.6	0.7	-0.3	-0.3	0.0	0.0	0.0
September	-0.1	1.2	1.2	1.4	1.3	0.5	0.9	0.1	0.5	0.9	0.8
October	-1.2	-2.2	-3.2	-1.6	-0.7	-1.8	-1.9	-1.5	-1.6	-1.4	-1.7
November	-0.2	0.2	1.7	2.1	1.3	0.7	0.5	0.9	0.4	0.3	0.8
December	0.6	0.6	1.5	2.1	2.4	0.5	1.1	0.3	0.7	1.2	1.1
Overall	0.4	0.4	0.7	1.2	1.3	0.9	0.3	0.1	0.3	0.7	0.6

2025	TASMANIA - Estimated Degree of Variation from the Minimum Average									
Region	NW Coast & King Is	W & S Coasts & Highlands	Central Plateau	Upper Derwent	South East	Central North	Lower Derwent	Midlands Alpine	NE Coast & Flinders Is	TAS
Town Sample	Devonport	Strathan	Lake St Clair	Bushy Park	Hobart	Launceston	Huonville	Fingal	Killiecrankie	Overall
January	1.0	0.9	1.6	0.8	0.6	2.6	0.2	0.9	1.5	1.1
February	0.8	0.2	-0.3	-0.1	0.7	1.6	0.6	0.6	1.3	0.6
March	2.4	1.0	2.5	1.8	2.2	3.6	2.3	2.4	1.7	2.2
April	1.0	0.5	-0.1	-0.1	0.2	1.9	0.2	0.5	-0.6	0.4
May	3.2	1.9	1.8	2.7	2.8	4.1	3.3	3.6	2.8	2.9
June	-0.3	-0.7	-0.7	-1.3	-0.2	1.3	0.4	1.2	-0.2	0.0
July	-1.3	-2.7	-1.7	-1.4	-1.0	-0.8	-0.8	-0.1	0.0	-1.1
August	0.0	0.2	0.4	-1.0	-0.2	0.9	-0.3	0.6	0.5	0.1
September	0.3	0.5	0.3	-0.9	-0.4	1.4	-0.3	0.0	-0.1	0.1
October	-1.3	-1.1	-1.3	-1.5	-0.8	0.2	-0.7	-1.1	0.0	-0.8
November	0.1	0.3	0.2	0.1	0.4	2.1	0.4	0.2	0.4	0.5
December	0.9	0.9	0.6	0.2	0.8	2.8	0.4	0.8	1.2	1.0
Overall	0.6	0.2	0.3	0.0	0.4	1.8	0.5	0.8	0.7	0.6

2025	TASMANIA - Estimated Maximum Temperature Trend										
Region	NW Coast & King Is	W & S Coasts & Highlands	Central Plateau	Upper Derwent	South East	Central North	Lower Derwent	Midlands Alpine	East Coast	NE Coast & Flinders Is	TAS
Town Sample	Devonport	Strathan	Lake St Clair	Bushy Park	Hobart	Launceston	Huonville	Fingal	Swansea	Killiecrankie	Overall
January	av	a-w	av	wmr	wmr	av	av	a-c	av	a-w	av
February	av	wmr	av	a-w	a-w	a-w	av	av	av	av	av
March	wmr	wmr	wmr	wmr	wmr	wmr	a-w	wmr	wmr	wmr	wmr
April	wmr	wmr	wmr	wmr	wmr	wmr	wmr	wmr	wmr	wmr	wmr
May	wmr	wmr	wmr	wmr	wmr	wmr	wmr	wmr	wmr	wmr	wmr
June	av	av	av	wmr	wmr	wmr	av	av	av	a-w	av
July	av	a-c	clr	a-c	av	av	clr	clr	clr	a-c	a-c
August	av	av	av	av	a-w	a-w	av	av	av	av	av
September	av	wmr	wmr	wmr	wmr	av	a-w	av	av	a-w	a-w
October	clr	clr	clr	clr	a-c	clr	clr	clr	clr	clr	clr
November	av	av	wmr	wmr	wmr	a-w	av	a-w	av	av	a-w
December	a-w	a-w	wmr	wmr	wmr	av	wmr	av	a-w	wmr	wmr
Overall	av	av	a-w	wmr	wmr	a-w	av	av	av	a-w	a-w

2025	TASMANIA - Estimated Minimum Temperature Trend									
Region	NW Coast & King Is	W & S Coasts & Highlands	Central Plateau	Upper Derwent	South East	Central North	Lower Derwent	Midlands Alpine	NE Coast & Flinders Is	TAS
Town Sample	Devonport	Strathan	Lake St Clair	Bushy Park	Hobart	Launceston	Huonville	Fingal	Killiecrankie	Overall
January	a-w	a-w	wmr	a-w	a-w	wmr	av	a-w	wmr	wmr
February	a-w	av	av	av	a-w	wmr	a-w	a-w	wmr	a-w
March	wmr	a-w	wmr	wmr	wmr	wmr	wmr	wmr	wmr	wmr
April	a-w	av	av	av	av	wmr	av	av	a-c	av
May	wmr	wmr	wmr	wmr	wmr	wmr	wmr	wmr	wmr	wmr
June	av	a-c	a-c	clr	av	wmr	av	wmr	av	av
July	clr	clr	clr	clr	a-c	a-c	a-c	av	av	clr
August	av	av	av	a-c	av	a-w	av	a-w	av	av
September	av	av	av	a-c	av	wmr	av	av	av	av
October	clr	clr	clr	clr	a-c	av	a-c	clr	av	a-c
November	av	av	av	av	av	wmr	av	av	av	av
December	a-w	a-w	a-w	av	a-w	wmr	av	a-w	wmr	a-w
Overall	a-w	av	av	av	av	wmr	av	a-w	a-w	a-w

SOUTH AUSTRALIA 2025

GENERAL TRENDS

Year Overall:
Slightly drier but within 10% of the average
Warmer by around 0.5-1°C above the average

Summer:
Slightly wetter but within 5% of the average
Wettest – First and last few days of January
Driest – Throughout February
Warmer by around 0.5-1°C above the average
Hottest – End of third week of December and end of February
Coolest – First week of January

Autumn:
Slightly wetter but within 5-10% of the average
Wettest – Second half of May
Driest – First three weeks of March and last half of April
Warmer by up to 2-2.5°C above the average
Hottest – First half of March
Coolest – Second half of May

Winter:
Drier by around 25-30% below the average
Wettest – Second and third weeks of June
Driest – First week of August
Average temperatures overall
Hottest – First weeks of June and August and fourth week of August
Coolest – Fourth weeks of June and July

Spring:
Drier by around 20-25% below the average
Wettest –First half of November
Driest – Second half of November
Warmer by around 0.5-1°C above the average
Hottest – End of October and last week of November
Coolest – Mid-September and fourth week of October

Major Weather Events

January 2025

Date	Day	South Australia - January 2025	Adelaide	Moon
1st	Wed	Unusually cool with heavy rain at times in the Mt Lofty Ranges and in the Adelaide districts. Chance of some heavy falls of rain in the Northeast Pastoral region.	Showers	
2nd	Thur	Very cool in the Mt Lofty Ranges.		
3rd	Fri	Strong blustery winds at times around the Yorke Peninsula to Kangaroo Island.	Showers	
4th	Sat		Showers	
5th	Sun			
6th	Mon			
7th	Tue		Showers, Th	**1Q XhN**
8th	Wed	Thunderstorms with large hail like in the Mt Lofty Ranges. Very blustery at times around the Yorke Peninsula to Kangaroo Island region as well as in the Adelaide to Mt Lofty areas.	Showers, Th	**P13**
9th	Thu	Passing storms may trigger fires and cause problems with infrastructure including power outages, particularly in the Lower Southeast. Chance of lightning strikes and gusty winds causing local bushfires around Adelaide, including in the Barossa Valley and Murray regions, as well as in some parts of the Mt Lofty Ranges.		
10th	Fri	Lightning strikes may start fires, particularly along the VIC-SA border areas.	Fire	
11th	Sat	It may be hotter and wetter than average over the next 3-4 weeks compared to the norm for this time of year. Heavy rain in the Southeast to the Northeast regions, and across Eyre Peninsula.		
12th	Sun			
13th	Mon	Very hot day in the Western Agricultural region, particularly around Nullarbor, with maximum temperatures into the mid to upper 40's likely. Isolated showers in the Far Northwest.		**^**
14th	Tue			**F**
15th	Wed	Severe winds and gusts and very high maximum temperatures up into the mid 40's in the Far North, particularly around Oodnadatta. Very high temperatures into the mid 40's also extend into the Nullarbor and the Western Agricultural districts.	Showers	
16th	Thur	A band of thunderstorms with very strong winds cross the Riverland and Murraylands regions with chance of bringing damage to property and crops, particularly in the Renmark area.	Showers	
17th	Fri	There may be an extreme fire weather risk in a number of districts around this time.		
18th	Sat			
19th	Sun	Unusually cool overnight around the coastal districts from the Western Agricultural district to the Lower Southeast, particularly in the Naracoorte area, and around Ceduna, Cleve, Adelaide and Cape Jaffa, where some districts may not see double digits overnight.		
20th	Mon	There may be extreme fire weather risk alert warnings throughout the State around this time. Very hot throughout the State at this time, with many districts seeing maximum temperatures in the mid to lower 40's.		**XhS**
21st	Tue			**A**
22nd	Wed	Cooler spell on Kangaroo Island.		**3Q**
23rd	Thu			
24th	Fri	Showers possible into the western border districts over the next couple of days.		
25th	Sat	Showers spread across the Western Agricultural region and into the Southeast over the next couple of days.		
26th	Sun	There may be further extreme fire weather risk warnings in a number of districts around this time.		
27th	Mon	An intrusion of tropical air crosses the State over the next few days bringing outbreaks of rain and scattered thunderstorms in most districts, particularly in the western regions. Chance of flash-flooding in some of the Pastoral and Northern Agricultural districts during this time.		**V**
28th	Tue	Thunderstorms cross the Eyre Peninsula bringing the occasional burst of heavy rain, wind gusts, and pockets of localised flooding.	Showers	
29th	Wed	Widespread thunderstorms cross the southeast quarter of the State including over the Murraylands, Flinders Ranges, Southeast, Spencer Gulf and Yorke Peninsula regions and Kangaroo and Fleurieu Islands. Heavy rain and strong wind gusts may cause localised flooding and property damage in a number of places, including in the Flinders Ranges, Yorke Peninsula, Spencer Gulf areas. Chance of lightning strikes around the Yorke Peninsula. Thunderstorms and heavy rain also in the Far North region, particularly in the Yunta area.	Rain	

Date	Day	South Australia - January 2025	Adelaide	Moon
30th	Thur	A developing low pressure system may cross to the south of the State around this time bringing storm activity into the coastal regions and gale force gusty winds at times, particularly in the Port Lincoln area to Kangaroo Island. Very heavy rain in the Riverland region to the Flinders Ranges and over the Fleurieu and Kangaroo Islands, possibly accompanying passing thunderstorm activity, with the chance of flash flooding in some districts, including in the Hawker and Parawa areas. Heavy rain also in the Far North in the Woomera area over the next couple of days.	Showers	N
31st	Fri	Unusually cool spell in the Mt Lofty Ranges, particularly on Mt Crawford.		

February 2025

Date	Day	South Australia - February 2025	Adelaide	Moon
1st	Sat	Chance of further heavy rain and flash flooding in the Flinders Ranges, including in the Hawker area.		
2nd	Sun			P10
3rd	Mon			XhN
4th	Tue			
5th	Wed			1Q
6th	Thu	Monsoonal rain activity over the Northern Territories may briefly spread into the Northwest around this time, before becoming widespread throughout the State over the next few days.		
7th	Fri	Cool spell in the Far North to the Flinders Ranges over the next couple of days.		
8th	Sat	Heavy monsoonal rains from the tropics may spread down and throughout the State over the next few days.		
9th	Sun			^
10th	Mon			
11th	Tue	The next 3-4 weeks may continue to be hotter than is typical for this time of year, while rainfall levels may ease back to below average in the southern Agricultural districts and Far North to nearer to normal elsewhere.		
12th	Wed	Mostly dry throughout the State over the next couple of weeks.		
13th	Thur			F
14th	Fri	Very hot across much of the State with many areas seeing maximum temperatures into the mid-40's.		
15th	Sat			
16th	Sun			XhS
17th	Mon			
18th	Tue	Unusually cold overnight in the Southeast region of the State, particularly in the Robe area, with minimum temperatures not reaching into double digits in numerous areas.		A
19th	Wed			
20th	Thu	Unusually cool in the Fleurieu and Kangaroo Islands region, particularly in the Parawa district, over the next few days.		
21st	Fri			3Q
22nd	Sat			
23rd	Sun			V
24th	Mon	Thunderstorms and some flash flooding likely in the Flinders Ranges to the Mid-North regions, particularly in the Hawker and Jamestown areas, which may bring some damage to local infrastructure and roads.		
25th	Tue	Thunderstorms with bursts of heavy rain and large hail likely in the Limestone Coast area of the Southeast, near Bordertown. Chance of some damage to property and crops. Cooler overnight in the Western Agricultural.		
26th	Wed			
27th	Thur	Very hot around this time in the Western Agricultural region, with maximum temperatures likely to reach into the upper 40's, particularly in the Nullarbor area. High fire risk in numerous districts over the next couple of days.		
28th	Fri	Very warm overnight in the southern coastal regions of the State from the Western Agricultural, over the Eyre Peninsula and down into the South East region including at Mt Gambier and in Adelaide.		N

March 2025

Date	Day	South Australia - March 2025	Adelaide	Moon
1st	Sat	Thunderstorms in the Flinders Ranges over the next couple of days, with bursts of localised heavy rain likely that may cause some pockets of brief flooding.	Th	
2nd	Sun	Chance of further thunderstorms in the Flinders Ranges as well as up into the Northeast Pastoral regions, particularly in the Black Rock and Maynards areas. Thunderstorms with strong wind gusts at times in the Far North, in the Tarcooa area. Very strong gusty winds cross the Fleurieu and Kangaroo Islands, with the chance of causing some localised damage.	Showers	**P7 XhN**
3rd	Mon			
4th	Tue			
5th	Wed	Thunderstorms and squally winds in the Far North, particuarly around Oodnadatta.	Showers	
6th	Thu			
7th	Fri			**1Q**
8th	Sat			**^**
9th	Sun			
10th	Mon			
11th	Tue	Hotter than average conditions for this time of year may still continue over the coming 3-4 weeks, with rainfall amounts also above normal almost everywhere, except in some parts of the southwest, where it may be drier than what is typical for this time of year.		
12th	Wed	Very hot in the Far North region around this time with numerous areas seeing temperatures soar into the mid 40's, particularly in the Oodnadatta district.	Th	
13th	Thur			
14th	Fri	Very warm overnight in the Far North, particularly in the Moomba district.		**F**
15th	Sat			**XhS**
16th	Sun	Isolated showers over the next couple of days along the southern coasts and in the Far North.		
17th	Mon	Cold overnight in the Mt Lofty Ranges to the Adelaide area over the next few nights.		
18th	Tue			**A**
19th	Wed			
20th	Thu	Warmer than normal spell in the Spencer Gulf region to Kangaroo Island.		
21st	Fri	Light showers in the western border districts over the next couple of days.		
22nd	Sat			
23rd	Sun	Cooler spell in the Western Agricultural, particularly around the Nullarbor district.		**3Q V**
24th	Mon	Mostly dry spell over the coming week.		
25th	Tue	Warm in the Victor Harbor area to the Murray Valley and upper Southeast districts with maximum temperatures into the upper 30's.	Showers	
26th	Wed			
27th	Thur			
28th	Fri			
29th	Sat			**N XhN**
30th	Sun			**P4**
31st	Mon			

April 2025

Date	Day	South Australia - April 2025	Adelaide	Moon
1st	Tue	Showers moving into the western border districts and spreading across the Western Agricultural district over the next few days, with some heavy falls at times.		
2nd	Wed	Scattered thunderstorms and squally winds cross the Far North, particulary affecting the Oodnadatta district.	Rain	
3rd	Thu			
4th	Fri	Unusually cold overnight in the Southeast region of the State, particularly in the Coonawarra area, where minimum temperatures may even reach frost level in some sheltered parts.		**^**
5th	Sat			**1Q**
6th	Sun	Extreme fire risks likely in numerous areas across the State around this time.	Showers	
7th	Mon	Cooler spell in the Far North over the next few days, particularly in the Oodnadatta and Woomera districts.		
8th	Tue	Cool in the Mt Lofty Ranges. Outbreaks of rain activity over most of the eastern half of the State over the next couple of days.		

Date	Day	South Australia - April 2025	Adelaide	Moon
9th	Wed	A band of heavy rain crosses through the Bight bringing heavy falls at times into the southern coastal districts and light showers into the adjacent inland areas, including into the Northwest for a time.		
10th	Thur			
11th	Fri	Once again, the next few weeks may be generally wetter and warmer than normal compared to what is typical for this time of year.		XhS
12th	Sat			
13th	Sun	Wet front crosses the State bringing scattered rain activity into most districts, with heavier falls at times in the Eyre Peninsula to the Southeast regions.		F
14th	Mon	Cold overnight in the Southeast, particularly in the Coonawarra area, where sheltered areas might see some patches of frost.		A
15th	Tue			
16th	Wed	Cool overnight in Adelaide to the Mt Lofty Ranges and down into the Southeast districts over the next couple of nights with many areas not reaching up into the double digits overnight.		
17th	Thu	Heavy rain across most of the State, particularly in the Northwest and down through the Western Agricultural to the Eyre Peninsula.		
18th	Fri	Shower activity contracts out of the western half of the State and crosses the eastern half.		
19th	Sat	Some isolated showers likely to linger in the Southeast region for a short time over the next couple of days.		V
20th	Sun	Warm spell across the State over the next couple of days. Particularly warmer than normal overnight in the Adelaide region.		
21st	Mon	Warm overnight in the Spencer Gulf region to Kangaroo and Fleurieu Islands. A dry spell settles over the State for the next week to 10 days.		3Q
22nd	Tue			
23rd	Wed			
24th	Thur	Warm in the Northwest region over the next couple of days, with maximum temperatures into the mid to upper 30's in numerous districts.		
25th	Fri			
26th	Sat			XhN
27th	Sun			
28th	Mon			P3 N
29th	Tue			
30th	Wed	Very warm spell in the Far North when maximum temperatures into the upper 30's are likely, particulary in the Oodnadatta district.	Showers	

May 2025

Date	Day	South Australia - May 2025	Adelaide	Moon
1st	Thu	Some light showers activity likely in the Western Agricultural region to the Northwest.		
2nd	Fri	Cooler overnight minimum temperature settle in around this time in most parts of the State.		^
3rd	Sat	Dry spell over the week ahead.		
4th	Sun			
5th	Mon		Showers	1Q
6th	Tue	Cold overnight in the Southeast districts, particularly in the Coonawarra area, where sheltered areas might see some patches of frost.	Rain	
7th	Wed	A colder spell arrives across much of the State over the next few days.	Rain, Th	
8th	Thur	Heavy rain likely in the Adelaide region. Cold and with very strong wind gusts in the Mt Lofty Ranges.	Rain	
9th	Fri		Showers	XhS
10th	Sat			
11th	Sun	Near gale force winds ahead of a front cross the Fleurieu and Kangaroo Islands, over the Yorke Peninsula, and up into the Mount Lofty Ranges, which may cause some damage to property and infrastructure, including power outages. Some light shower activity expected in the west of the State and along the southern coasts over the next 3-4 days.		A
12th	Mon	The next 3-4 weeks may continue the wetter and warmer trend of the previous several weeks, compared to what is typical for this time of year.	Showers	
13th	Tue	Unusually warm overnight in the Flinders Ranges.	Showers	F
14th	Wed			
15th	Thur	Heavy rain through the southern coastal districts over the next couple of days, particularly over the Eyre Peninsula to the Southeast.		
16th	Fri			V

Date	Day	South Australia - May 2025	Adelaide	Moon
17th	Sat	Warm in the Far North, particularly in the Oodnadatta to Marla districts with maximum temperatures in the early 30's. Rain in the southern half of the State eases back to showers around this time, but persist for a few more days, gradually contracting to the Southeast.	Showers	
18th	Sun			
19th	Mon			
20th	Tue			3Q
21st	Wed	Low minimum temperatures in the Western Agricultural region over the next couple of days with minimum temperatures likely to consistently be in the single digits.		
22nd	Thu	Cold overnight in the southeast of the State around this time, with minimum temperatures in the single digits in many districts. Some outbreaks of rain in the far Northwest and along the northern border fringes.		
23rd	Fri	Mostly dry throughout the State over the next ten days, except for the occasional brief showery intervals, mostly in the coastal fringe districts.		XhN
24th	Sat	Heavy rain associated with a line of thunderstorms cross the Yorke Peninsula, particularly in the Kimba area. Chance of thunderstorms, heavy rain and large hail also in the Cleve area of the Western Eyre Peninsula.	Rain, Th	
25th	Sun			
26th	Mon			P5
27th	Tue			N
28th	Wed		Showers	
29th	Thur	Isolated showers in the Far North likely over the next few days.		^
30th	Fri	Strong unsettled winds across the Adelaide region over the next couple of days may turn tornadic for a time and cause damage to property.	Th	
31st	Sat		Showers	

June 2025

Date	Day	South Australia - June 2025	Adelaide	Moon
1st	Sun		Showers	
2nd	Mon			
3rd	Tue	Low minimum temperatures throughout the State over the next few days with minimum temperatures likely to consistently be in the single digits. Showers over the eastern half of the State during the next 3 days.		1Q
4th	Wed			
5th	Thu	Cold overnight, particularly in the Yongala district of the Upper North, where low temperatures may develop frosts in some areas.		XhS
6th	Fri	Cold with patches of frost over the next couple of nights in the Upper North and Northeast regions. Showers in the east of the State turn to pockets of heavy rain along the eastern border districts over the next couple of days.	Showers	
7th	Sat		Showers, Th	A
8th	Sun	A deepening low in the south of the State brings widespread showery conditions across the southern half of the State, and very strong winds to gale force at times, particularly from the Western Eyre region, across the Yorke Peninsula, over the Fleurieu and Kangaroo Islands and into the Adelaide and Mid-North Port Pirie areas. The strong winds, reaching gale force at times, may cause damage to local property and infrastructure. Cold in the Mt Lofty Ranges.	Showers	
9th	Mon	Low minimum temperatures throughout the State over the next couple of days with minimum temperatures likely to consistently be in the single digits. Cold in the Northwest over the next couple of nights with scattered frosts in some districts, including in the Nonning area.	Showers	
10th	Tue		Rain	
11th	Wed	While the next 3-4 weeks may be unusually colder than is normal for this time of year, it may also be generally a drier than normal period as well. Some shower activity in the southeast quarter of the State over the next few days.	Showers	F
12th	Thur		Showers	V
13th	Fri		Showers	
14th	Sat		Showers	
15th	Sun			
16th	Mon			
17th	Tue	Rain activity throughout the southern half of the State over the next few days, starting with showers at first, but building to heavy rain in many districts from the Western Agricultural to the Southeast over 2-3 days.		
18th	Wed	Heavy rain along the coasts to adjacent inland regions, including through the Gulfs and across the Peninsulas.	Showers	

Date	Day	South Australia - June 2025	Adelaide	Moon
19th	Thu	Rain may contract to the southeast quarter of the State around this time, easing to isolated showers over lingering over the southern coastal fringes over the next few days.		3Q XhN
20th	Fri		Showers	
21st	Sat			
22nd	Sun			
23rd	Mon			P8
24th	Tue	Light showers along the coasts from the Eyre Peninsula to the Southeast around this time. Scattered frosts across the State over the next few days.		
25th	Wed			N
26th	Thur	Dry spell in most areas over the coming week.		^
27th	Fri			
28th	Sat	Strong and blustery winds for a time around Kangaroo Island.	Showers	
29th	Sun		Showers, Fg	
30th	Mon		Showers, Fg	

July 2025

Date	Day	South Australia - July 2025	Adelaide	Moon
1st	Tue	Isolated shower activity returns around the southern coasts.		
2nd	Wed			XhS
3rd	Thu	Mostly dry over the coming week. Cold in the Northeast Pastoral region around this time, with severe frosts likely in a number of districts and scattered frosts elsewhere across the State over the next couple of days.		1Q
4th	Fri			
5th	Sat		Showers	A
6th	Sun			
7th	Mon		Showers	
8th	Tue	Warm spell over the next few days in the Western Agricultural region, including in the Nullarbor area.	Showers	
9th	Wed	Some light shower activity in the southeast quarter of the State over the next couple of days.	Showers	
10th	Thur		Showers	V
11th	Fri			F
12th	Sat	Strong gusty winds across the Fleurieu and Kangaroo Islands, with the chance of some damage resulting.	Showers, Fg	
13th	Sun	Generally, the next few weeks may be cooler and drier than normal for this time of year in most districts, with some regions, particularly in parts of the Pastoral districts, possibly not receiving any rainfall at all during the next 3-4 weeks, while in contrast, some parts in the southeast may receive above normal rainfall during this same period of time.	Showers	
14th	Mon	A fast moving front bearing a band of thunderstorms crosses the Eyre Peninsula to the Murraylands. Chance of damaging winds, and possible formations of tornado activity. Strong winds to gale force also likely throughout the south of the State across the Western Eyre to the Limestone Coast-Southeast, Yorke Peninsula and Mid-North Coast, in Adelaide to the Mt Lofty Ranges, Murray-Mallee, and Barossa, over the Fleurieu and Kangaroo Islands. Strong winds also in the Far North in the Yunta area.	Showers	
15th	Tue		Showers, Fg	
16th	Wed		Rain, Th	
17th	Thu	Chance of isolated showers in the southeast of the State.	Showers	XhN
18th	Fri		Fg	3Q
19th	Sat	Some patches of frosts likely in the Upper and Lower Southeast regions.		
20th	Sun	Isolated showers in the west and southern coastal regions over the next couple of days. Scattered frosts in the Northwest districts overnight.	Fg	
21st	Mon		Showers	P11
22nd	Tue	Some frosts in the Upper North region, particularly in the Hawker area.	Showers	
23rd	Wed	Wet spell through the southeast quarter of the State over the next 5-7 days. Some patches of frosts overnight in the Western Agricultural region, particularly in the Wudinna and Kyancutta districts.		^
24th	Thur		Showers	
25th	Fri	Scattered frosts likely in some of the Lower North districts.		N
26th	Sat		Showers	
27th	Sun	Gusty winds in the Murray-Mallee, particularly around Lameroo.	Showers	
28th	Mon			

Date	Day	South Australia - July 2025	Adelaide	Moon
29th	Tue	Mostly dry over the coming week except for the isolated showery outbreak in some southeastern parts. Very cold day in the Mt Lofty Ranges		
30th	Wed	Very cold overnight in the Northeast Pastoral region, particularly in the Yunta area, where severe frosts may occur.		XhS
31st	Thu			

August 2025

Date	Day	South Australia - August 2025	Adelaide	Moon
1st	Fri	Further frosts likely over the next couple of nights in the Northwest, Far North, Murray Valley, and Upper North regions.		
2nd	Sat			1Q A
3rd	Sun			
4th	Mon			
5th	Tue		Showers	
6th	Wed	Showers in the coastal fringes and over the southeast regions over the next couple of days.	Showers, Fg	V
7th	Thur		Rain, Fg	
8th	Fri		Fg	
9th	Sat			F
10th	Sun	Unusually hot spell in the Far North, particularly in the Marla area.	Showers, Th	
11th	Mon	Mostly dry throughout the State over the next 4-5 days.	Showers	
12th	Tue	Similar to the last few weeks, the next 3-4 weeks may also be cooler and drier than what is typical for this time of year, except in the Southeast, which may be wetter than average.	Showers	
13th	Wed		Showers	XhN
14th	Thu	Some patches of frosts likely overnight in the Northwest region. Cold in the Mt Lofty Ranges.	Showers	
15th	Fri		Fg	P12
16th	Sat	Showers in the southeast regions over the next few days.		3Q
17th	Sun		Showers, Fg	
18th	Mon		Fg	
19th	Tue		Fg	^
20th	Wed			
21st	Thur	Light showers in the southeast over the next couple of days.		
22nd	Fri			
23rd	Sat	Isolated showers activity spreads into the Northwest for a short time. Chance of isolated patches of frost in the Western Agricultural region, Southeast, and on the Adelaide Plains over the next couple of days.		N
24th	Sun	Light shower activity contracts back into the southeastern regions over the next couple of days. Some frost patches likely in the Northeast, Yorke Peninsula, Kangaroo Island, and Upper North regions over the next couple of nights.		
25th	Mon	Some shower activity in the southeast of the State every other day over the coming couple of weeks. Cold overnight in the Murraylands, with sub-zero frost patches likely, particularly in the Renmark area.	Showers	
26th	Tue		Showers	XhS
27th	Wed		Showers, Fg	
28th	Thu	Chance of frosts overnight in the Hawker district of the Upper North.		
29th	Fri	Some patches of frosts likely overnight in the Northwest region.		
30th	Sat			A
31st	Sun			1Q

September 2025

Date	Day	South Australia - September 2025	Adelaide	Moon
1st	Mon	Chance of overnight frosts in the next couple of days in the Western Agricultural regions, particularly in the Ceduna and Cummins areas.		
2nd	Tue			V
3rd	Wed			
4th	Thur			
5th	Fri	Rain in the Western Agricultural region to the Eyre Peninsula.		
6th	Sat	Rain activity eases in the west while spreading showers into the southeastern regions.		
7th	Sun	Dry spell over the next 3 weeks throughout the State.		

Date	Day	South Australia - September 2025	Adelaide	Moon
8th	Mon			F
9th	Tue	Warm spell over most of the State, particularly in the Ceduna and Tarcoola areas of the Western Agricultural region. Very strong and blustery winds for a time across the Spencer Gulf, including over Neptune Island.	Showers	XhN
10th	Wed			
11th	Thu	The next 3-4 weeks may be generally drier and hotter than what is normal for this time of year.		P9
12th	Fri	Frosts likely overnight in the Western Agricultural region, particularly in the Ceduna and Nullarbor areas, as well as in patches in the Northeast region to the Lower North.		
13th	Sat			
14th	Sun	Chance of light frosts in the Kyancutta area of the Western Agricultural region and in some parts of the Southeast.		3Q
15th	Mon	Cold overnight in the State, particularly in the Agricultural districts and Far North region, with scattered frosts likely, particularly in the Yunta and Murray Bridge areas as well as in isolated pockets in the Yorke Peninsula region to Kangaroo Island.		^
16th	Tue			
17th	Wed			
18th	Thur			
19th	Fri			
20th	Sat		Showers	
21st	Sun	Cooler spell in the Mt Lofty Ranges.	Fg	
22nd	Mon			N XhS
23rd	Tue		Showers	
24th	Wed			
25th	Thu	Possible frosts on Mt Barker in the Mt Lofty Ranges as well as in some sheltered parts of the Upper Southeast, including in the Lameroo area.		
26th	Fri	The passage of a front brings severe winds over the State for a time, particularly in the Murray-Mallee and winds possibly to gale force in the Far North, possibly with severe gusts in the Yunta district.	Showers	A
27th	Sat			
28th	Sun			
29th	Mon	Unsettled and blustery winds ahead of a front cross the Far Northwest in the Woomera area. Very strong and gusty winds also cross the Limestone Coast-Southeast region, particularly in the Naracoorte area.	Showers	
30th	Tue	Brief showery spell over the next couple of days in the southeast quarter of the State.		1Q V

October 2025

Date	Day	South Australia - October 2025	Adelaide	Moon
1st	Wed			
2nd	Thur			
3rd	Fri			
4th	Sat	Showers in the southeast regions.		
5th	Sun	Strong winds to gale force at times may cross the Far North, particularly in the Marree area, and unusually hotter than normal in the region, particularly in the Oodnadatta district. Some districts in the Pastoral regions may reach into the upper 30's to early 40's for a time.	Rain	
6th	Mon	Very warm day in the agricultural districts including in the Western Agricultural region. Very strong winds in the Far North around this time, particularly in the Marree area.		
7th	Tue	A front crossing through the Bight may bring some showers and strong winds with severe gusts, possibly briefly to gale force for a time, across the Eyre Peninsula, Fleurieu and Kangaroo Islands, Adelaide, Lower Murray, and Limestone Coast-Southeast regions.		F XhN
8th	Wed	Strong winds possibly to gale force at times may persist across the southeast quarter of the State, particularly around the Limestone Coast-Southeast region and over the Fleurieu and Kangaroo Islands.	Showers	
9th	Thu	Light shower activity in the southeast quarter of the State over the next couple of days.		P6
10th	Fri	Dry spell over the coming week.		
11th	Sat	The next 3-4 weeks may see around average to slightly below average rainfall across the State, with above normal temperatures for this time of year.		
12th	Sun			
13th	Mon			^
14th	Tue	Unusually cool overnight in the Northeast to the Far North regions, particularly in the Yunta to Marree districts, with some patches of frost possible in some districts.		3Q

Date	Day	South Australia - October 2025	Adelaide	Moon
15th	Wed			
16th	Thur	Outbreaks of rain likely in the southeast quarter of the State with light showers over the Western Agricultural coastal fringes to the Eyre Peninsula and rain into the Northeast to the Southeast regions along the VIC border.		
17th	Fri	Patchy showery spells in the southeast over the next few days.		
18th	Sat	Low minimum temperatures in the agricultural districts around this time, with chance of frost patches in some areas, incuding in the Keith district in the Southeast region and in the Mt Lofty Ranges to the Murray Bridge area.		
19th	Sun	Chance of frost patches overnight on Kangaroo Island.	Th	
20th	Mon		Showers	**XhS**
21st	Tue		Showers	
22nd	Wed			**N**
23rd	Thu			
24th	Fri	Isolated shower activity in the Adelaide region into the Flinders Ranges over the next few days.		**A**
25th	Sat			
26th	Sun			
27th	Mon	Dry spell over the next few days. Chance of frost patches overnight around the Mt Lofty Ranges to the Adelaide Plains.		**V**
28th	Tue			
29th	Wed	Chance of frosts overnight in isolated pockets in the Western Agricultural region.		
30th	Thur	Very hot day in the Western Agricultural region, particularly in the Ceduna and Nullabor districts, where maximum temperatures may reach into the low 40's for a time.		**1Q**
31st	Fri	Isolated showers in the southeast likely over the next couple of days. Very warm night over the Eyre Peninsula, and high maximum temperatures into the upper 30's during the day in many parts of the State.	Showers	

November 2025

Date	Day	South Australia - November 2025	Adelaide	Moon
1st	Sat		Rain	
2nd	Sun	Thunderstorms and pulses of heavy rain, some hail showers and pockets of flash flooding in the southeast quarter of the State from the Yorke Peninsula and Spencer Gulf region to the Mid-North Flinders regions likely. Cold spell in the Mt Lofty Ranges.		
3rd	Mon	Dry spell over the coming week.		**XhN**
4th	Tue			
5th	Wed	Very hot over the next few days in the Far North, particularly in the Oodnadatta district, where maximum temperatures may reach into the low 40's for a time.		
6th	Thur	An intense depression may cross the State bringing very strong and gusty winds to many regions over the next couple of days, including in the Far North and Northwest, Flinders Ranges, Mid-North, Yorke Peninsula, Adelaide Hills to the Mt Lofty and Flinders Ranges, Spencer Gulf, and Fleurieu and Kangaroo Islands and down into the Southeast around the Limestone Coast. Very hot spell in the Far North around this time, particularly in the Oodnadatta district, where maximum temperatures may reach into the low 40's.	Showers, Th	**F P1**
7th	Fri	A deepening low may continue to cross the State maintaining very strong winds across the southeast from the Yorke and Eyre Peninsulas and Mid-North, Adelaide and Limestone Coast regions to across the Fleurieu and Kangaroo Islands.	Showers	
8th	Sat	The gale force winds of the past couple of days may generate some storm surges along the coastal fringes around the Adelaide area.		
9th	Sun		Showers	**^**
10th	Mon	The next 3-4 weeks may be wetter and warmer than normal for this time of year.	Showers	
11th	Tue	Showers mostly in the west of the State at first, but spreading into the south and southeast over the next couple of days.	Rain	
12th	Wed		Rain	**3Q**
13th	Thu	A trough of low pressure crossing through the Bight may bring scattered thunderstorms, heavy rain, and some flash-flooding across the Eyre Peninsula, particularly around Port Augusta. Cold spell over the next couple of days in the Southeast region to the Mt Lofty Ranges.	Showers	
14th	Fri	Cold overnight in the Far North, including in the Yunta district.	Showers	
15th	Sat	Very cold overnight in the Southeast region, particularly in the Naracoorte district.		
16th	Sun			**XhS**
17th	Mon	Some showers activity over the next couple of days in the Southeast region to the Flinders Ranges.		

Date	Day	South Australia - November 2025	Adelaide	Moon
18th	Tue	Cold overnight in the Southeast of the State, with chance of frost patches in some areas, including in the Keith district.		
19th	Wed			
20th	Thur	Rain activity moving into the Northwest and Western Agricultural regions around this time may spread eastward through the northern half of the State into the Far North over the next couple of days.		N A
21st	Fri			
22nd	Sat	Rain moves into the southern half of the State during the next few days, with heavy falls in the southeast quarter for a time.		
23rd	Sun	Heavy rain across the Eyre Peninsula to the Southeast regions and up through the Flinders Ranges.		V
24th	Mon			
25th	Tue	Mostly dry over the next couple of weeks.		
26th	Wed	Very hot spell likely over the next 3-4 days across the State, with many districts seeing daily maximas into the upper 30's to low 40's.		
27th	Thu			
28th	Fri			1Q
29th	Sat	Strong and occasionally damaging winds possible in the Southeast to the Eyre Peninsula, particularly around the Cleve, Keith, and Limestone Coast areas. Very warm night in the Far North to the Flinders Ranges, particularly in the Coober Pedy district.		
30th	Sun	Srong and unsettled winds across the Far North and Far Northwest, particularly in the Tarcoola and Woomera areas. Very hot spell in the Western Agricultural region around this time, with daily maximas into the lower 40's in many areas, including in the Nullarbor district.	Showers	XhN

December 2025

Date	Day	South Australia - December 2025	Adelaide	Moon
1st	Mon			
2nd	Tue			
3rd	Wed			
4th	Thur			P2
5th	Fri			F
6th	Sat			^
7th	Sun	Very hot spell in the Far North with maximum temperatures into the low 40's likely, including in the Oodnadatta district.		
8th	Mon			
9th	Tue	A surface trough crossing the Murray-Mallee to the Riverland regions may bring extensive thunderstorm activity with chance of localised flooding, particulary in Taplan and Berri. Heavy rain and thunderstorms also in the Far North with flooding also possible in the Olary area and in the Mid-North region, especially in the Florieton area.		
10th	Wed	Spotty showers in the Western Agricultural may spread and intensify over the following day into the Eyre Peninsula and all southeastern regions of the State. Very hot in the Far North around this time, with maximum temperatures in the low to mid 40's in a number of districts.		
11th	Thu	By the end of the next 3-4 week period, rainfall amounts may be above average for this time of year, but mostly because of a single widespread rain event expected in a couple of weeks time. This Christmas into New Year period, may also be significantly warmer than average, particularly in the south of the State, and with the chance of heatwave conditions for a time early on in the New Year.		
12th	Fri	Mostly dry throughout the State over the next couple of weeks, with just a couple of exceptions.		3Q
13th	Sat			XhS
14th	Sun			
15th	Mon			
16th	Tue	A vigorous cold front crossing the Bight may extend very strong to gale force winds and scattered thunderstorms through the southern half of the State, including in the Eyre Peninsula, Spencer Gulf region, the Adelaide-Barossa area, and to the Limestone Coast-Southeast regions, as well as in the Far North, and Northwest regions around Coober Pedy, and Maree areas. Lightning strikes associated with passing storms may generate numerous fires, particularly on Kangaroo Island.	Th	
17th	Wed			A
18th	Thur	Unusually cooler over the next few nights in many parts of the State when minimum temperatures may not reach into double digits in a number of districts.		

Date	Day	South Australia - December 2025	Adelaide	Moon
19th	Fri			
20th	Sat	Unusually cold overnight in the Southeast region to the Mt Lofty Ranges over the next couple of days.		**N V**
21st	Sun			
22nd	Mon			
23rd	Tue			
24th	Wed		Showers	
25th	Thu			
26th	Fri			
27th	Sat	Brief showers in the Northwest Pastoral to the Flinders Ranges over the next few days.		**XhN**
28th	Sun		Showers, Th	**1Q**
29th	Mon	Scattered thunderstorms and gusty winds cross the Murray-Mallee and may cause some localised damage, particularly in the Pinnaroo area.		
30th	Tue	A low pressure trough may cross the Northwest over the next couple of days. Heavy rain and scattered thunderstorms with severe wind gusts and flash flooding likely, particularly in the Coober Pedy, Cleve, and Amata areas. Winds may be very gusty for a time in the Oodnadatta area.	Rain, Th	
31st	Wed	Strong unsettled winds with gusts in the Northwest, particularly around Coober Pedy. Uncomfortably warm overnight in the Far North.	Rain, Th	

SOUTH AUSTRALIA GRAPHS AND TABLES 2025

Seasonal Breakdown – State Overall Rainfall Timing and Estimated mm Trend Graphs
State Overall Maximum Temperature Trend Graphs
Regional Rainfall Expectations Tables - mm and Trends
Regional Maximum and Minimum Temperature Trends Tables

Monthly Breakdown - **Rainfall:**
Regional Expectations by mm
Regional Expectations by mm Variance to Average
Regional Variance Trend by mm
Regional Estimated Expectations by % of Variance to Average
Regional Estimatedl Variance Trend by %

Maximum & Minimum Temperature Trends:
Regional Estimated Average Temperature
Regional Estimated Degree of Variation from the Average
Regional Estimated Temperature Trend

Seasonal Breakdowns

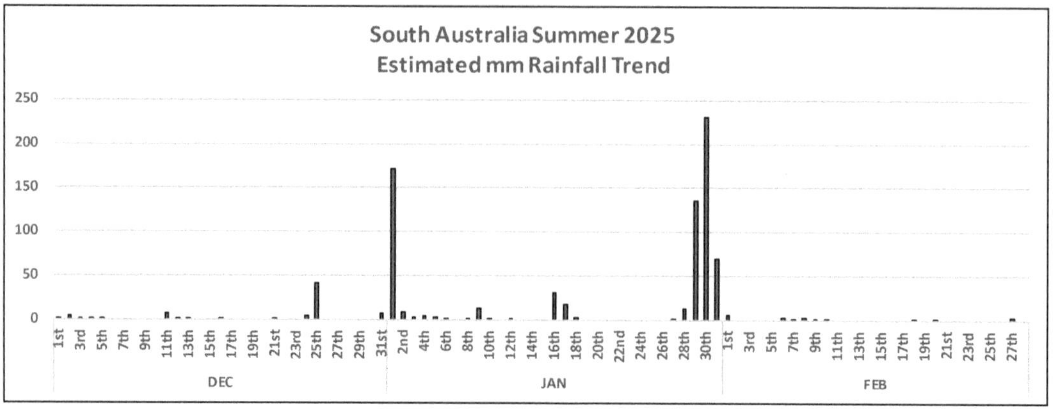

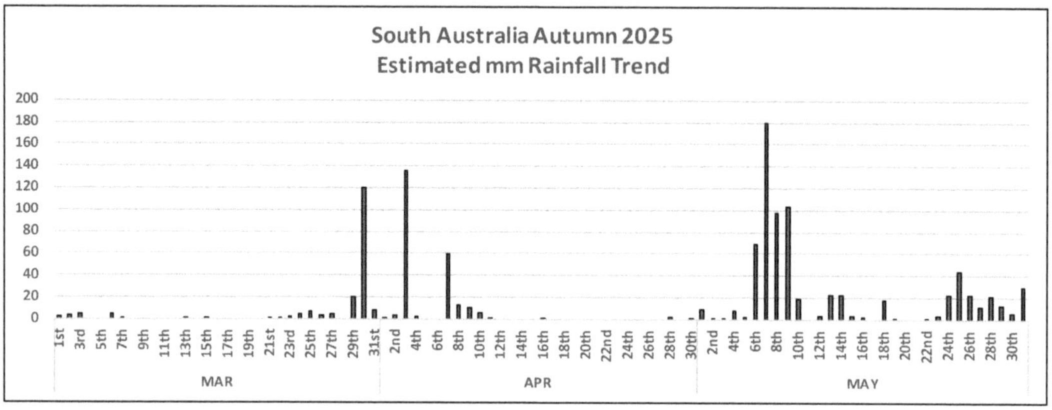

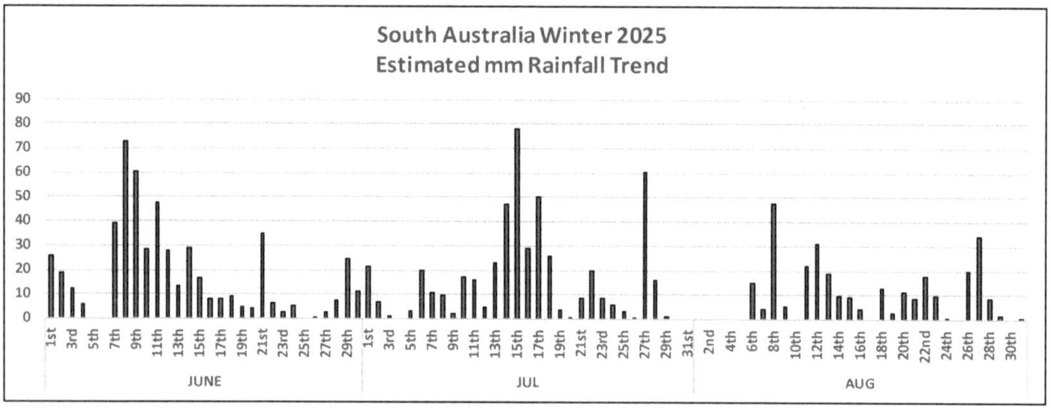

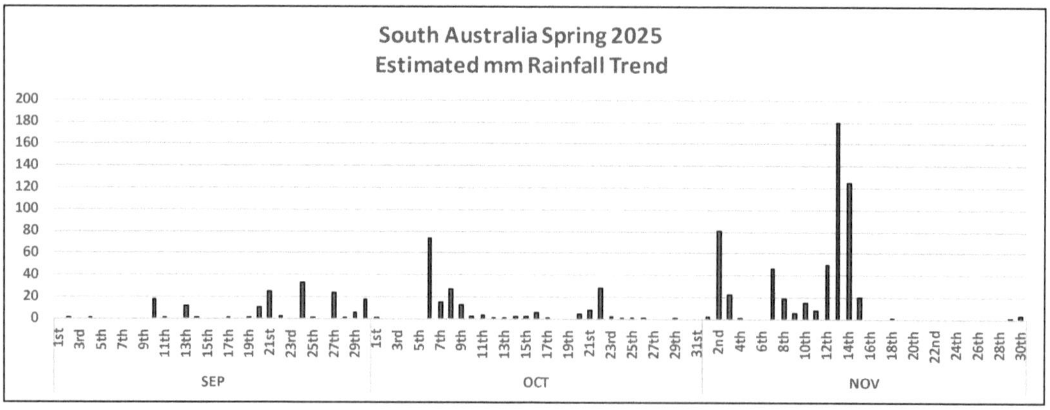

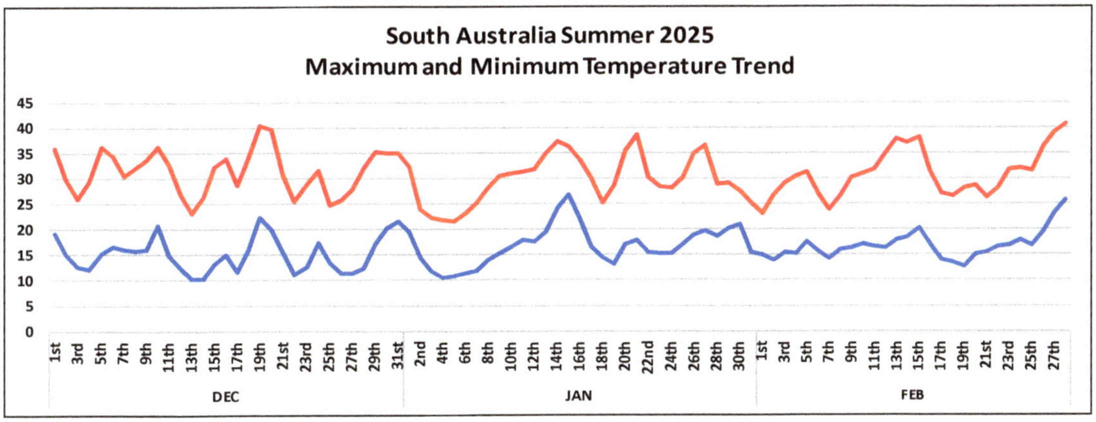
South Australia Summer 2025
Maximum and Minimum Temperature Trend
DEC
JAN
FEB

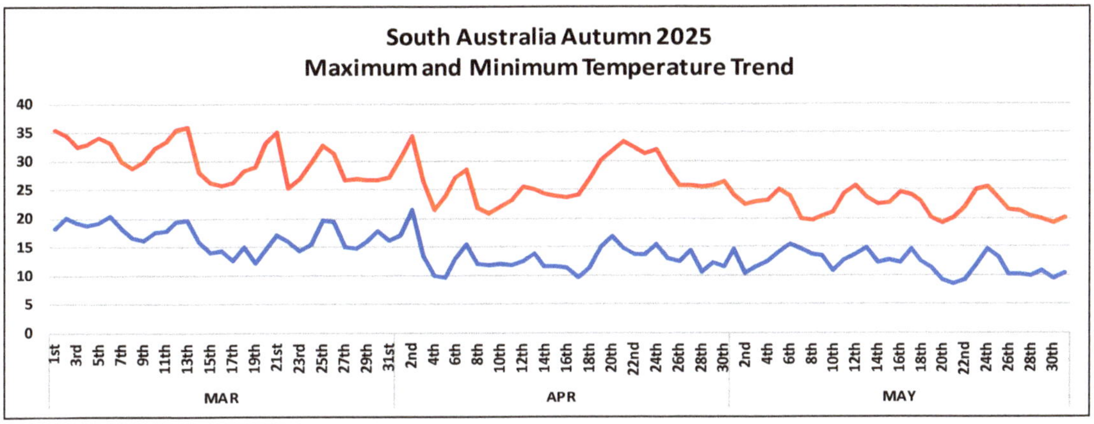
South Australia Autumn 2025
Maximum and Minimum Temperature Trend
MAR
APR
MAY

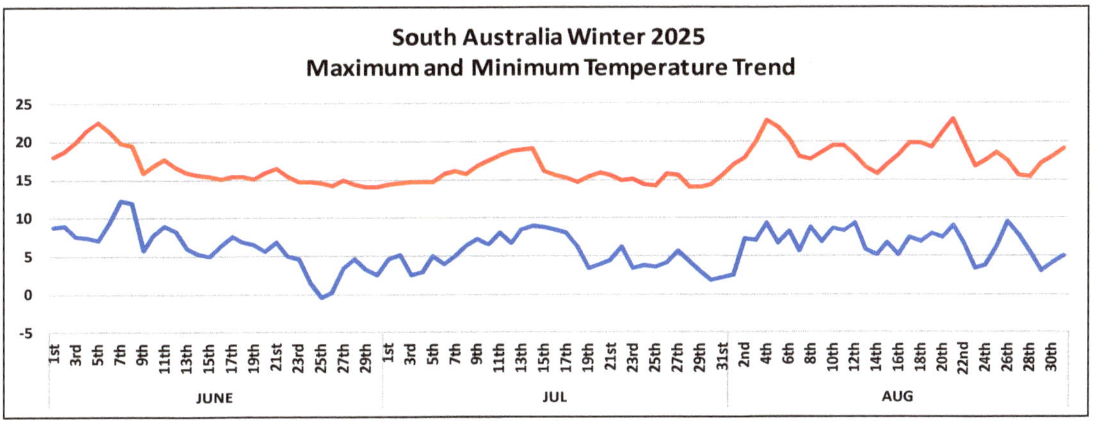
South Australia Winter 2025
Maximum and Minimum Temperature Trend
JUNE
JUL
AUG

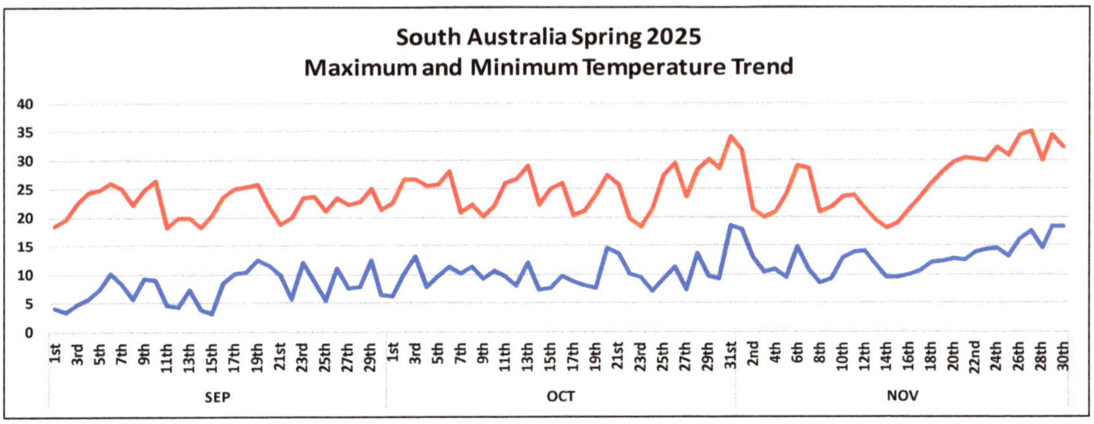
South Australia Spring 2025
Maximum and Minimum Temperature Trend
SEP
OCT
NOV

SA - Estimated Seasonal Rainfall Expectations (mm)

2025														
REGION	NW Pastoral	West Coast	Inland West Coast	Far NE	Lower Eyre Peninsula	Eastern Eyre Peninsula	Spencers Gulf	Greater Adelaide	Mid-North	Murraylands	Yorke Peninsula	Upper Southeast	Lower Southeast	SA
Town Sample	Coober Pedy	Ceduna	Wudinna	Marree	Pt Lincoln	Cleve	Pt Augusta	Adelaide	Port Pirie	Murrays Bridge	Victor Harbour	Keith	Mt Gambier	Overall
Summer	19	4	79	52	32	59	87	59	130	53	61	78	76	788
Autumn	10	127	92	24	88	91	78	115	83	120	111	88	130	1157
Winter	0	37	65	1	152	102	29	119	57	77	216	156	300	1313
Spring	5	38	21	3	67	45	55	76	74	63	164	88	226	925
Overall	44	236	296	82	357	320	257	379	348	324	566	417	743	4369

SA - Estimated Seasonal Rainfall Variation from the Average (mm)

2025														
REGION	NW Pastoral	West Coast	Inland West Coast	Far NE	Lower Eyre Peninsula	Eastern Eyre Peninsula	Spencers Gulf	Greater Adelaide	Mid North	Murraylands	Yorke Peninsula	Upper SE	Lower SE	SA
Town Sample	Coober Pedy	Ceduna	Wudinna	Marree	Pt Lincoln	Cleve	Pt Augusta	Adelaide	Port Pirie	Murrays Bridge	Victor Harbour	Keith	Mt Gambier	Overall
Summer	-44	-41	33	-5	-18	-6	32	4	71	-6	1	12	-12	20
Autumn	-24	72	36	-8	-25	1	26	13	9	41	-9	-17	-22	92
Winter	-30	-66	-40	-32	-87	-39	-38	-53	-54	-50	-11	-10	10	-501
Spring	-32	-29	-48	-37	-41	-61	-15	-31	-23	-37	39	-35	48	-302
Overall	-120	-34	19	-81	-153	-83	14	-59	7	-41	35	-44	36	-503

SA - Estimated Seasonal Rainfall Trend Variation by 10mm

2025														
REGION	NW Pastoral	West Coast	Inland West Coast	Far NE	Lower Eyre Peninsula	Eastern Eyre Peninsula	Spencers Gulf	Greater Adelaide	Mid North	Murraylands	Yorke Peninsula	Upper SE	Lower SE	SA
Town Sample	Coober Pedy	Ceduna	Wudinna	Marree	Pt Lincoln	Cleve	Pt Augusta	Adelaide	Port Pirie	Murrays Bridge	Victor Harbour	Keith	Mt Gambier	Overall
Summer	drr	drr	wtr	av	av	av	wtr	wtr	av	av	av	av	av	av
Autumn	av	wtr	wtr	av	av	av	av	av	av	av	wtr	av	av	wtr
Winter	drr	drr	drr	drr	drr	drr	drr	drr	drr	av	drr	av	av	drr
Spring	drr	av	drr	drr	drr	drr	av	av	drr	wtr	drr	drr	wtr	drr
Overall	drr	av	av	drr	av	av	av	av	av	av	av	av	av	drr

SA - Estimated Seasonal Rainfall Expectation by %

2025														
REGION	NW Pastoral	West Coast	Inland West Coast	Far NE	Lower Eyre Peninsula	Eastern Eyre Peninsula	Spencers Gulf	Greater Adelaide	Mid North	Murraylands	Yorke Peninsula	Upper SE	Lower SE	SA
Town Sample	Coober Pedy	Ceduna	Wudinna	Marree	Pt Lincoln	Cleve	Pt Augusta	Adelaide	Port Pirie	Murrays Bridge	Victor Harbour	Keith	Mt Gambier	Overall
Summer	-70%	-92%	71%	-9%	-37%	-9%	59%	7%	119%	-10%	2%	18%	-13%	3%
Autumn	-70%	131%	63%	-24%	-22%	1%	51%	12%	12%	51%	-7%	-17%	-14%	9%
Winter	-100%	-64%	-38%	-96%	-36%	-28%	-57%	-31%	-48%	-39%	-5%	-6%	3%	-28%
Spring	-86%	-44%	-70%	-92%	-38%	-57%	-21%	-29%	-24%	-37%	31%	-29%	27%	-25%
Overall	-73%	-12%	7%	-50%	-30%	-21%	6%	-13%	2%	-11%	7%	-9%	5%	-10%

2025 SA - Estimated Seasonal Rainfall Trend Variation by 25%

REGION	NW Pastoral	West Coast	Inland West Coast	Far NE	Lower Eyre Peninsula	Eastern Eyre Peninsula	Spencers Gulf	Greater Adelaide	Mid North	Murraylands	Yorke Peninsula	Upper SE	Lower SE	SA
Town Sample	Coober Pedy	Ceduna	Wudinna	Marree	Pt Lincoln	Cleve	Pt Augusta	Adelaide	Port Pirie	Murrays Bridge	Victor Harbour	Keith	Mt Gambier	Overall
Summer	drr	drr	wtr	av	drr	av	wtr	av	wtr	av	av	av	av	av
Autumn	drr	wtr	wtr	av	av	av	wtr	av	av	wtr	av	av	av	av
Winter	drr	drr	drr	drr	drr	drr	drr	drr	drr	drr	av	av	av	drr
Spring	drr	drr	drr	drr	drr	drr	av	drr	av	drr	wtr	drr	wtr	av
Overall	drr	av	av	drr	drr	av	av	av	av	av	av	av	av	av

2025 SA - Estimated Seasonal Maximum Temperature Average and Trend

Region	NW Pastoral	West Coast	Inland West Coast	Far NE	Lower Eyre Peninsula	Eastern Eyre Peninsula	Spencers Gulf	Greater Adelaide	Mid North	Murraylands	Yorke Peninsula	Upper SE	Lower SE	SA
Town Sample	Coober Pedy	Ceduna	Wudinna	Marree	Pt Lincoln	Cleve	Pt Augusta	Adelaide	Port Pirie	Murrays Bridge	Victor Harbour	Keith	Mt Gambier	Overall
Summer Est	35.9	29.4	33.4	26.5	28.6	38.2	33.7	32.5	28.8	25.2	31.2	30.3	26.3	30.8
Trend	av	wmr	a-w	a-w	av	a-w	av	wmr	wmr	a-w	wmr	a-w	wmr	a-w
Autumn Est	29.4	26.0	27.7	24.0	24.6	32.7	28.2	27.5	24.8	23.1	26.6	25.8	23.0	26.4
Trend	wmr	wmr	wmr	wmr	wmr	wmr	wmr	wmr	wmr	wmr	wmr	wmr	wmr	wmr
Winter Est	19.1	18.7	18.0	16.3	15.8	20.4	18.5	17.3	15.6	15.6	16.9	15.2	13.9	17.0
Trend	av	a-w	av	av	av	av	av	av	av	av	av	av	av	av
Spring Est	29.2	26.2	26.9	21.8	24.2	30.1	27.8	26.0	22.8	20.5	24.5	22.5	18.6	24.7
Trend	av	wmr	a-w	a-w	wmr	av	a-w	wmr	wmr	av	a-w	a-w	av	a-w

2025 SA - Estimated Seasonal Minimum Temperature Average and Trend

Region	NW Pastoral	West Coast	Inland West Coast	Far NE	Lower Eyre Peninsula	Eastern Eyre Peninsula	Spencers Gulf	Greater Adelaide	Mid North	Murraylands	Yorke Peninsula	Upper SE	Lower SE	SA
Town Sample	Coober Pedy	Ceduna	Wudinna	Marree	Pt Lincoln	Cleve	Pt Augusta	Adelaide	Port Pirie	Murrays Bridge	Victor Harbour	Keith	Mt Gambier	Overall
Summer Est	21.6	14.9	15.4	21.8	14.6	16.0	19.2	16.7	18.8	14.4	13.8	13.7	11.4	16.3
Trend	av	av	a-w	av	a-c	av	a-w	a-w	wmr	a-w	av	a-w	av	av
Autumn Est	16.8	13.5	12.9	17.5	12.9	14.3	15.2	14.6	16.1	12.3	12.8	12.0	11.1	14.0
Trend	wmr	wmr	wmr	wmr	av	wmr	wmr	wmr	wmr	wmr	wmr	wmr	wmr	wmr
Winter Est	6.2	4.8	4.0	5.3	6.9	7.7	4.8	7.3	7.5	5.4	7.3	5.0	6.1	6.0
Trend	a-c	clr	clr	av	a-c	av	av	av	a-c	a-w	av	a-c	a-w	av
Spring Est	13.5	9.2	8.1	14.8	9.4	10.1	11.5	11.0	12.5	8.8	8.8	8.5	7.8	10.3
Trend	a-c	a-c	clr	a-w	av	av	av	av	av	a-w	av	av	av	av

Monthly Breakdowns

2025 SOUTH AUSTRALIA - Estimated Rainfall Expectations (mm)

REGION	NW Pastoral	West Coast	Inland West Coast	Far NE	Lower Eyre Peninsula	Eastern Eyre Peninsula	Spencers Gulf	Greater Adelaide	Mid-North	Murraylands	Yorke Peninsula	Upper Southeast	Lower Southeast	SA
Town Sample	Coober Pedy	Ceduna	Wudinna	Marree	Pt Lincoln	Cleve	Pt Augusta	Adelaide	Port Pirie	Murrays Bridge	Victor Harbour	Keith	Mt Gambier	Overall
January	19	3	67	50	18	54	75	55	119	49	56	74	67	705
February	0	0	0	0	1	1	0	0	3	1	1	1	7	16
March	0	55	20	10	2	9	31	2	41	6	3	5	2	185
April	0	22	24	11	26	36	16	18	16	14	20	11	24	236
May	10	50	49	4	60	46	31	95	27	99	88	73	104	737
June	0	15	21	0	71	49	11	49	12	31	91	59	117	526
July	0	14	31	1	61	37	15	45	29	34	75	53	98	493
August	0	7	14	0	20	16	4	26	16	12	50	44	85	294
September	0	3	7	0	19	10	1	12	4	10	33	14	38	152
October	0	5	1	0	6	7	8	16	12	24	25	33	57	193
November	5	29	13	3	42	28	46	48	58	29	106	41	132	579
December	10	32	51	3	31	26	20	12	13	14	19	10	13	255
Overall	44	236	296	82	357	320	257	379	348	324	566	417	743	4369

2025 SOUTH AUSTRALIA - Estimated Rainfall Variation from the Average (mm)

REGION	NW Pastoral	West Coast	Inland West Coast	Far NE	Lower Eyre Peninsula	Eastern Eyre Peninsula	Spencers Gulf	Greater Adelaide	Mid North	Murraylands	Yorke Peninsula	Upper SE	Lower SE	SA
Town Sample	Coober Pedy	Ceduna	Wudinna	Marree	Pt Lincoln	Cleve	Pt Augusta	Adelaide	Port Pirie	Murrays Bridge	Victor Harbour	Keith	Mt Gambier	Overall
January	-1	-10	52	29	5	37	59	39	99	33	37	54	39	472
February	-18	-9	-16	-19	-11	-21	-17	-13	-10	-12	-13	-19	-12	-189
March	-18	39	4	-4	-24	-12	14	-21	20	-13	-24	-16	-37	-93
April	-6	5	7	5	-8	7	3	-13	-6	-13	-19	-22	-26	-84
May	0	28	25	-9	6	6	10	47	-5	67	34	21	40	270
June	-14	-17	-16	-11	-9	2	-15	-12	-28	-14	16	7	26	-83
July	-9	-25	-2	-12	-21	-9	-6	-14	-8	-5	-2	-2	-0	-114
August	-8	-24	-22	-9	-58	-32	-17	-27	-18	-31	-26	-15	-16	-304
September	-9	-23	-18	-10	-32	-32	-23	-35	-34	-29	-23	-36	-38	-341
October	-15	-21	-20	-16	-30	-30	-14	-20	-20	-9	-17	-8	-1	-219
November	-8	14	-10	-12	20	0	22	24	30	2	78	9	87	259
December	-15	8	35	-14	6	1	-1	-13	-14	-16	-9	-16	-28	-76
Overall	-120	-34	19	-81	-153	-83	14	-59	7	-41	35	-44	36	-503

2025 SOUTH AUSTRALIA - Estimated Rainfall Trend Variation by 10mm

REGION	NW Pastoral	West Coast	Inland West Coast	Far NE	Lower Eyre Peninsula	Eastern Eyre Peninsula	Spencers Gulf	Greater Adelaide	Mid North	Murraylands	Yorke Peninsula	Upper SE	Lower SE	SA
Town Sample	Coober Pedy	Ceduna	Wudinna	Marree	Pt Lincoln	Cleve	Pt Augusta	Adelaide	Port Pirie	Murrays Bridge	Victor Harbour	Keith	Mt Gambier	Overall
January	av	av	wtr	av	wtr	wtr	wtr	wtr	wtr	wtr	wtr	wtr	wtr	wtr
February	drr	av	drr	drr	drr	drr	drr	av	drr	drr	drr	drr	drr	drr
March	drr	wtr	av	drr	drr	av	wtr	wtr	drr	drr	drr	drr	drr	drr
April	av	av	av	av	av	av	av	av	drr	drr	drr	drr	drr	drr
May	av	wtr	wtr	av	av	av	av	av	wtr	wtr	wtr	wtr	wtr	wtr
June	drr	drr	drr	av	av	drr	drr	drr	drr	wtr	drr	av	wtr	drr
July	av	drr	av	drr	av	drr	av	av	drr	av	av	av	av	drr
August	av	drr	drr	drr	drr	av	drr	drr	drr	drr	drr	drr	drr	drr
September	av	drr	drr	drr	drr	av	drr	drr	drr	drr	drr	drr	drr	drr
October	drr	drr	drr	drr	drr	drr	drr	drr	drr	drr	av	av	av	drr
November	av	wtr	av	wtr	av	drr	wtr	wtr	wtr	wtr	av	av	wtr	wtr
December	drr	av	wtr	av	av	drr	av	drr	drr	av	drr	drr	drr	drr
Overall	drr	av	av	drr	av	av	av	av	av	av	av	av	av	drr

2025 SOUTH AUSTRALIA - Estimated Rainfall Expectation by %

REGION	NW Pastoral	West Coast	Inland West Coast	Far NE	Lower Eyre Peninsula	Eastern Eyre Peninsula	Spencers Gulf	Greater Adelaide	Mid North	Murraylands	Yorke Peninsula	Upper SE	Lower SE	SA
Town Sample	Coober Pedy	Ceduna	Wudinna	Marree	Pt Lincoln	Cleve	Pt Augusta	Adelaide	Port Pirie	Murrays Bridge	Victor Harbour	Keith	Mt Gambier	Overall
January	-5%	-79%	364%	135%	40%	219%	356%	228%	513%	203%	205%	283%	136%	203%
February	-100%	-100%	-100%	-100%	-90%	-94%	-100%	-100%	-79%	-91%	-90%	-94%	-63%	-92%
March	-100%	243%	25%	-29%	-92%	-57%	80%	-92%	93%	-68%	-89%	-77%	-96%	-34%
April	-100%	31%	39%	83%	-23%	25%	23%	-42%	-27%	-49%	-49%	-67%	-51%	-26%
May	3%	127%	107%	-70%	12%	14%	45%	98%	-16%	207%	62%	40%	62%	58%
June	-100%	-52%	-43%	-100%	-11%	4%	-59%	-20%	-70%	-31%	22%	14%	29%	-14%
July	-100%	-64%	-6%	-90%	-26%	-19%	-30%	-23%	-20%	-13%	-2%	-4%	0%	-19%
August	-100%	-77%	-62%	-100%	-74%	-67%	-80%	-51%	-54%	-72%	-34%	-25%	-16%	-51%
September	-100%	-87%	-73%	-100%	-62%	-76%	-94%	-74%	-89%	-74%	-40%	-72%	-51%	-69%
October	-100%	-81%	-93%	-100%	-84%	-81%	-64%	-56%	-63%	-28%	-40%	-20%	-1%	-53%
November	-61%	96%	-44%	-78%	96%	1%	95%	103%	110%	5%	284%	28%	196%	81%
December	-60%	36%	212%	-83%	26%	2%	-5%	-51%	-51%	-52%	-33%	-62%	-68%	-23%
Overall	-73%	-12%	7%	-50%	-30%	-21%	6%	-13%	2%	-11%	7%	-9%	5%	-10%

2025 SOUTH AUSTRALIA - Estimated Rainfall Trend Variation by 25%

REGION	NW Pastoral	West Coast	Inland West Coast	Far NE	Lower Eyre Peninsula	Eastern Eyre Peninsula	Spencers Gulf	Greater Adelaide	Mid North	Murraylands	Yorke Peninsula	Upper SE	Lower SE	SA
Town Sample	Coober Pedy	Ceduna	Wudinna	Marree	Pt Lincoln	Cleve	Pt Augusta	Adelaide	Port Pirie	Murrays Bridge	Victor Harbour	Keith	Mt Gambier	Overall
January	av	drr	wtr	wtr	wtr	wtr	wtr	wtr	wtr	wtr	wtr	wtr	wtr	wtr
February	drr	drr	drr	drr	drr	drr	drr	drr	drr	drr	drr	drr	drr	drr
March	drr	wtr	av	drr	drr	drr	wtr	drr	wtr	drr	drr	drr	drr	drr
April	drr	wtr	wtr	wtr	av	av	av	drr	drr	drr	drr	drr	drr	drr
May	av	wtr	wtr	drr	av	av	wtr	wtr	av	wtr	wtr	wtr	wtr	wtr
June	drr	drr	drr	drr	av	av	drr	av	drr	drr	av	av	wtr	av
July	drr	drr	av	drr	drr	av	drr	av	av	av	av	av	av	av
August	drr	drr	drr	drr	drr	drr	drr	drr	drr	drr	drr	av	av	drr
September	drr	drr	drr	drr	drr	drr	drr	drr	drr	drr	drr	drr	drr	drr
October	drr	drr	drr	drr	drr	drr	drr	drr	drr	drr	drr	av	av	drr
November	drr	wtr	drr	drr	wtr	av	wtr	wtr	wtr	av	wtr	wtr	wtr	wtr
December	drr	wtr	wtr	drr	wtr	av	av	drr	drr	drr	drr	drr	drr	av
Overall	drr	av	av	drr	drr	av	av	av	av	av	av	av	av	av

2025 SOUTH AUSTRALIA - Estimated Average Maximum Temperature

Region Town Sample	NW Pastoral Coober Pedy	West Coast Ceduna	Inland West Coast Wudinna	Far NE Marree	Lower Eyre Peninsula Pt Lincoln	Eastern Eyre Peninsula Cleve	Spencers Gulf Pt Augusta	Greater Adelaide Adelaide	Mid-North Port Pirie	Murraylands Murrays Bridge	Yorke Peninsula Victor Harbour	Upper Southeast Keith	Lower Southeast Mt Gambier	SA Overall
January	34	28	31	36	26	27	32	28	31	31	26	30	27	**30**
February	37	30	35	39	26	29	35	29	32	31	24	30	26	**31**
March	35	28	33	39	27	28	32	28	32	30	25	30	26	**30**
April	29	27	28	32	25	25	28	25	28	27	24	26	23	**27**
May	24	23	23	27	21	21	24	21	23	22	21	21	19	**22**
June	18	18	17	20	16	15	18	16	17	17	16	15	14	**17**
July	18	18	17	19	15	14	17	14	16	15	14	14	13	**16**
August	22	21	20	23	17	18	21	17	19	19	17	16	14	**19**
September	26	25	24	27	20	22	25	21	23	22	20	20	17	**23**
October	31	27	27	31	23	25	29	23	26	24	21	21	18	**25**
November	31	27	29	32	23	25	30	25	28	27	21	26	21	**26**
December	34	29	31	37	25	27	32	27	30	29	23	29	24	**29**
Overall	**28**	**25**	**26**	**30**	**22**	**23**	**27**	**23**	**26**	**25**	**21**	**23**	**20**	**25**

2025 SOUTH AUSTRALIA - Estimated Average Minimum Temperature

Region Town Sample	NW Pastoral Coober Pedy	West Coast Ceduna	Inland West Coast Wudinna	Far NE Marree	Lower Eyre Peninsula Pt Lincoln	Eastern Eyre Peninsula Cleve	Spencers Gulf Pt Augusta	Greater Adelaide Adelaide	Mid-North Port Pirie	Murraylands Murrays Bridge	Yorke Peninsula Victor Harbour	Upper Southeast Keith	Lower Southeast Mt Gambier	SA Overall
January	20	15	16	21	16	16	19	17	19	16	16	15	13	**17**
February	22	14	16	22	16	17	20	17	19	15	14	15	13	**17**
March	22	16	16	24	15	16	19	16	18	15	15	13	12	**17**
April	15	13	12	15	13	14	14	14	15	11	12	11	10	**13**
May	13	11	11	14	10	13	12	13	15	11	12	12	11	**12**
June	6	5	4	5	6	8	5	8	8	6	8	4	6	**6**
July	5	4	3	4	8	7	4	6	7	5	6	4	5	**5**
August	8	6	4	7	7	8	5	8	8	5	7	7	7	**7**
September	10	6	5	10	8	8	8	9	9	6	6	7	8	**8**
October	14	10	8	17	9	10	11	11	12	8	8	7	7	**10**
November	16	12	11	17	11	12	16	14	16	12	12	11	9	**13**
December	19	14	13	22	14	14	18	15	17	13	13	12	11	**15**
Overall	14	11	10	15	11	12	13	12	14	10	11	10	9	**12**

2025 SOUTH AUSTRALIA - Estimated Degree of Variation from the Maximum Average

Region Town Sample	NW Pastoral Coober Pedy	West Coast Ceduna	Inland West Coast Wudinna	Far NE Marree	Lower Eyre Peninsula Pt Lincoln	Eastern Eyre Peninsula Cleve	Spencers Gulf Pt Augusta	Greater Adelaide Adelaide	Mid North Port Pirie	Murraylands Murrays Bridge	Yorke Peninsula Victor Harbour	Upper SE Keith	Lower SE Mt Gambier	SA Overall
January	-3.1	-1.1	-2.2	-2.7	0.0	-1.9	-2.3	0.1	-1.0	0.0	0.8	-0.3	1.8	-0.9
February	1.5	1.8	2.0	1.4	0.3	0.5	1.2	0.8	0.5	1.0	0.0	0.4	0.7	0.9
March	2.9	1.4	3.1	5.0	2.3	1.9	1.7	3.0	2.5	2.6	1.4	3.2	3.2	2.6
April	2.1	3.1	2.3	2.9	2.1	1.7	1.5	2.8	2.7	3.3	2.2	3.4	3.6	2.6
May	1.7	1.6	1.7	3.9	1.2	1.5	2.1	2.3	2.7	3.2	1.9	2.9	3.1	2.3
June	-0.7	-0.2	0.0	0.5	-0.6	-1.1	-0.3	-0.1	-0.1	0.8	-0.2	-0.1	0.5	-0.1
July	-0.8	0.1	-0.4	-0.6	-0.8	-1.3	-1.0	-0.6	-0.5	-0.3	-1.1	-1.0	-0.2	-0.7
August	0.9	1.9	1.4	0.7	0.5	1.2	0.8	0.7	0.8	1.4	0.1	-0.1	0.3	0.8
September	0.5	3.1	1.7	0.4	1.5	2.0	1.2	2.4	2.0	2.2	1.2	1.5	1.4	1.6
October	1.5	3.1	1.0	0.8	1.6	2.2	1.5	1.8	1.8	-0.3	0.4	-0.4	-0.6	1.1
November	-1.5	0.9	-0.1	-1.6	-0.4	-0.8	-0.7	0.8	0.6	-0.1	-1.7	1.1	0.4	-0.2
December	-0.7	1.1	-0.1	0.4	0.0	-0.6	-0.3	0.9	0.5	-0.3	-1.3	0.8	1.0	0.1
Overall	0.4	1.4	0.9	0.9	0.7	0.4	0.4	1.2	1.0	1.1	0.3	1.0	1.3	0.8

2025 SOUTH AUSTRALIA - Estimated Degree of Variation from the Minimum Average

Region Town Sample	NW Pastoral Coober Pedy	West Coast Ceduna	Inland West Coast Wudinna	Far NE Marree	Lower Eyre Peninsula Pt Lincoln	Eastern Eyre Peninsula Cleve	Spencers Gulf Pt Augusta	Greater Adelaide Adelaide	Mid North Port Pirie	Murraylands Murrays Bridge	Yorke Peninsula Victor Harbour	Upper SE Keith	Lower SE Mt Gambier	SA Overall
January	-1.9	0.0	-0.1	-1.7	0.2	0.3	-0.5	1.2	1.0	1.3	0.8	1.7	1.8	0.3
February	0.3	-0.9	0.8	-0.5	-0.4	0.9	0.8	0.8	1.1	0.9	-0.1	1.5	1.1	0.5
March	3.3	3.1	3.2	4.9	0.9	1.6	2.8	1.8	2.2	2.2	2.0	1.7	1.7	2.4
April	1.0	2.1	1.5	0.7	0.8	1.7	0.8	2.3	2.2	1.7	1.4	1.9	1.6	1.5
May	3.2	2.6	2.2	4.5	-0.5	2.3	3.4	3.5	3.9	3.7	3.0	3.9	3.2	3.0
June	-1.1	-1.5	-2.1	-0.5	-2.5	0.0	-0.6	0.2	-0.5	1.0	1.2	-1.8	0.7	-0.6
July	-1.4	-1.7	-3.0	-1.3	0.5	-0.8	-0.4	-0.9	-1.1	0.1	-0.5	-1.4	0.0	-0.9
August	0.2	-0.7	-1.1	0.2	-0.3	0.8	-0.4	0.2	-0.1	1.0	0.6	0.9	1.1	0.2
September	-0.9	-1.7	-2.0	0.0	-0.1	-0.2	-0.7	-0.4	-0.3	1.1	-1.2	0.4	1.3	-0.4
October	-0.4	0.0	-1.0	2.6	-0.7	-0.2	-0.6	-0.2	0.3	0.7	-0.6	-0.7	-0.5	-0.1
November	-1.2	-0.8	-0.2	-0.4	-0.7	-0.1	0.4	0.8	1.3	1.1	0.2	0.6	0.6	0.1
December	-0.7	0.0	-0.2	1.4	-0.3	0.0	0.3	0.2	1.1	1.0	0.6	0.4	1.2	0.4
Overall	0.0	0.0	-0.2	0.8	-0.2	0.5	0.4	0.8	0.9	1.3	0.6	0.8	1.1	0.5

2025 SOUTH AUSTRALIA - Estimated Maximum Temperature Trend

Region Town Sample	NW Pastoral Coober Pedy	West Coast Ceduna	Inland West Coast Wudinna	Far NE Marree	Lower Eyre Peninsula Pt Lincoln	Eastern Eyre Peninsula Cleve	Spencers Gulf Pt Augusta	Greater Adelaide Adelaide	Mid North Port Pirie	Murraylands Murrays Bridge	Yorke Peninsula Victor Harbour	Upper SE Keith	Lower SE Mt Gambier	SA Overall
January	clr	clr	clr	clr	av	clr	clr	av	a-c	av	a-w	av	wmr	a-c
February	wmr	wmr	wmr	wmr	av	av	wmr	a-w	av	a-w	av	av	a-w	a-w
March	wmr	wmr	wmr	wmr	wmr	wmr	wmr	wmr	wmr	wmr	wmr	wmr	wmr	wmr
April	wmr	wmr	wmr	wmr	wmr	wmr	wmr	wmr	wmr	wmr	wmr	wmr	wmr	wmr
May	wmr	wmr	wmr	wmr	wmr	wmr	wmr	wmr	wmr	wmr	wmr	wmr	wmr	wmr
June	a-c	av	av	av	a-c	clr	av	av	av	a-w	av	av	av	av
July	a-c	av	av	a-c	a-c	clr	a-c	a-c	av	av	clr	a-c	av	a-c
August	a-w	wmr	wmr	a-w	av	wmr	a-w	a-w	a-w	wmr	av	av	av	a-w
September	av	wmr	wmr	av	wmr	wmr	wmr	wmr	wmr	wmr	wmr	wmr	wmr	wmr
October	wmr	wmr	a-w	a-w	wmr	wmr	wmr	wmr	wmr	av	av	av	a-c	wmr
November	clr	a-w	av	clr	av	a-c	a-c	a-w	a-w	av	clr	wmr	av	av
December	a-c	wmr	av	av	av	a-c	av	a-w	av	av	clr	a-w	a-w	av
Overall	av	wmr	a-w	a-w	a-w	av	av	wmr	a-w	wmr	av	a-w	wmr	a-w

2025 SOUTH AUSTRALIA - Estimated Minimum Temperature Trend

Region Town Sample	NW Pastoral Coober Pedy	West Coast Ceduna	Inland West Coast Wudinna	Far NE Marree	Lower Eyre Peninsula Pt Lincoln	Eastern Eyre Peninsula Cleve	Spencers Gulf Pt Augusta	Greater Adelaide Adelaide	Mid North Port Pirie	Murraylands Murrays Bridge	Yorke Peninsula Victor Harbour	Upper SE Keith	Lower SE Mt Gambier	SA Overall
January	clr	av	av	clr	av	av	av	wmr	a-w	wmr	a-w	wmr	wmr	av
February	av	a-c	a-w	av	av	a-w	a-w	a-w	wmr	a-w	av	wmr	wmr	av
March	wmr	wmr	wmr	wmr	a-w	wmr	wmr	wmr	wmr	wmr	wmr	wmr	wmr	wmr
April	a-w	wmr	wmr	a-w	a-w	wmr	a-w	wmr	wmr	wmr	wmr	wmr	wmr	wmr
May	wmr	wmr	wmr	wmr	av	wmr	wmr	wmr	wmr	wmr	wmr	wmr	wmr	wmr
June	clr	clr	clr	av	clr	av	a-c	av	av	a-w	wmr	clr	a-w	a-c
July	clr	clr	clr	clr	av	a-c	av	a-c	clr	av	av	clr	av	a-c
August	av	a-c	clr	av	av	a-w	av	av	av	a-w	a-w	a-w	wmr	av
September	a-c	clr	clr	av	av	av	a-c	av	av	wmr	clr	av	wmr	av
October	av	av	a-c	wmr	a-c	av	a-c	av	av	a-w	a-c	a-c	av	av
November	clr	a-c	av	av	a-c	av	av	a-w	wmr	wmr	av	a-w	a-w	av
December	a-c	av	av	wmr	av	av	av	av	wmr	a-w	a-w	av	wmr	av
Overall	av	av	av	a-w	av	av	av	a-w	a-w	wmr	a-w	a-w	wmr	av

WESTERN AUSTRALIA 2025

GENERAL TRENDS

Year Overall: Slightly drier but within 5% of the average
Cooler by around 2-2.5°C below the average

Summer: Drier by around 30% below the average
Wettest – Second half of January
Driest – Throughout February
Cooler by around 4.5°C below the average
Hottest – Start of second week of January and second half of February
Coolest – End of second week of January

Autumn: Wetter by up to 30% more than the average
Wettest – Second half of March to mid-April
Driest – Second half of May
Slightly cooler but within 0.5°C of the average
Hottest – Mid-March
Coolest – End of the first week of April and last week of May

Winter: Slightly drier but within 5% of the average
Wettest – Second weeks of July and August
Driest – Fourth week of June
Warmer by around 0.5-1°C above the average
Hottest – First week of June and last week of July
Coolest – Last week of June

Spring: Slightly drier but within 5% of the average
Wettest –Last week of September
Driest – Second week of October and second half of November
Cooler by around 4.5°C below the average
Hottest – Second half of November
Coolest – First week of September

Major Weather Events

January 2025

Date	Day	Western Australia - January 2025	Perth	Moon
1st	Wed	Very hot in the Gascoyne and Pilbara regions over the next couple of days. Heavy rain at times in the Kimberleys.		
2nd	Thur	Very hot in the Central West of the State. Pockets of heavy rain likely in the Kimberleys and Interior over the next few days. Very windy at times in the East Gascoyne and Lower West regions.		
3rd	Fri	Heavy outbursts of rain likely in the Halls Creek and Margaret River areas. Chance of strong wind gusts in the Lower West - Perth area. Very hot in the Pilbara and Gascoyne regions over the coming week.		
4th	Sat	Heavy rain in the West Kimberleys.		
5th	Sun	High maximum temperatures continue in the Pilbara and Gascoyne, as well as extending into the Central West. Unusually cooler overnight in the Emu Creek area.		
6th	Mon			
7th	Tue	Very hot temperatures continue in the Pilbara and Gascoyne as well as spreads into the Central Wheatbelt region.		**1Q XhN**
8th	Wed	Storms around the South Coast may cause some fires due to lightning strikes. High maximum temperatures spread into the Goldfields. Very blustery at times in the Lower West, particularly around Esperance.		**P13**
9th	Thu	High temperatures continue in the Pilbara, Gascoyne, and Interior regions continue over the next few days. Blustery at times in the West Kimberley.		
10th	Fri	Storms and showers associated with a tropical depression off the De Grey coast bring heavy falls and unstable winds over the next few days in the Gascoyne to Kimberley regions.		
11th	Sat	Many parts of the State may be slightly cooler than is normal over the next few weeks.		
12th	Sun	A band of tropical moisture associated with the tropical low off the De Grey coast brings some outbursts of heavy rain over the southeast coastal regions, and into the Goldfields and Eucla over the next few days. Chance of localised pockets of flooding.	Rain	
13th	Mon	Widespread monsoonal rains throughout most of the State over the next couple of days. Unusually cooler than normal in the southwest of the State over the next couple of days.		**^**
14th	Tue	Unusually low maximum temperatures in the Central Wheat Belt to the Goldfields regions around this time. Very strong winds with severe gusts at times in the Esperance district.		**F**
15th	Wed	Rain heavy around the South Coast with showers lingering in the Nullarbor Plains, Eucla, and into the Goldfields and Interior.		
16th	Thur			
17th	Fri			
18th	Sat			
19th	Sun		Showers	
20th	Mon		Showers	**XhS**
21st	Tue	Spells of heavy monsoonal rain over the coming week in the Kimberleys spreading down through the northern half of the State into the Pilbara and Interior.		**A**
22nd	Wed			**3Q**
23rd	Thu			
24th	Fri			
25th	Sat	Very heavy rain in the Kimberley, particularly in the Kuri Bay area. Very hot overnight in the Goldfields region.		
26th	Sun			
27th	Mon			**V**
28th	Tue			
29th	Wed			
30th	Thur	Unusually cold around the South Coast overnight.		**N**
31st	Fri			

February 2025

Date	Day	Western Australia - February 2025	Perth	Moon
1st	Sat	Mostly dry over the State during the coming week and dry in the southern half throughout the month ahead.		
2nd	Sun			**P10**
3rd	Mon	Some showers likely around the northern coastal fringes of the Kimberleys over the next couple of days.		**XhN**

Date	Day	Western Australia - February 2025	Perth	Moon
4th	Tue			
5th	Wed	Very strong and blustery winds in the Central Wheatbelt, particularly in the Gingin area.		1Q
6th	Thu	Very hot in the Pilbara over the next few days.		
7th	Fri	Occasional monsoonal rains around the northern coastal fringes of the Kimberleys over the next couple of days.	Showers, Th	
8th	Sat			
9th	Sun	Unusually warm overnight in the Esperance region.		^
10th	Mon			
11th	Tue	The following 4 weeks may be considerably hotter in many districts away from the coasts than is normal for this time of year, as well as unusually drier than average, particularly in the Wheat Belt districts over the coming week, when maximum temperatures may regularly reach into the upper 40's in many areas.		
12th	Wed	A tropical cyclone that may have developed over the NT Top End a few days earlier may cross into the Kimberleys around this time, and proceed to drift southwestward down through the coastal regions of the State over the next week to ten days, weakening and then reinvigorating several times during its transition.		
13th	Thur	Another tropical low may form off the Pilbara coast near Christmas Island in the next few days strengthening to a cyclonic system in about 3-4 days. This system may wander around the Timor Sea for at least a week before possibly making landfall near Port Hedland by around the 22nd. Maximum temperatures in some parts of the southwest, including in the Central Wheat Belt districts may reach into the upper 40's over the next couple of days.		F
14th	Fri			
15th	Sat	Pockets of rain in the Kimberleys to the Interior over the next couple of days.		
16th	Sun	Very hot overnight in the Gascoyne region.		XhS
17th	Mon	A tropical cyclone that entered the country a few days ago, may have reached the De Grey region by around this time, at full storm intensity.		
18th	Tue	Brief cool respite around the South Coast.		A
19th	Wed			
20th	Thu			
21st	Fri	Another cyclonic system that has been wandering around the Timor Sea for the past week or more, may make landfall near Port Hedland around this time, exacerbating the local conditions caused by the previous cyclonic impact only a couple of days earlier.		3Q
22nd	Sat			
23rd	Sun	Very hot in the Pilbara to the Murchison over the next couple of days with maximum temperatures likely into the upper 40's and minimum temperatures in the 30's.		V
24th	Mon			
25th	Tue		Showers	
26th	Wed		Showers	
27th	Thur		Showers	
28th	Fri	Unusually cold overnight in the southwest quarter of the State, including around the South Coast, in the Great Southern region, and some parts of the Central Wheat Belt.		N

March 2025

Date	Day	Western Australia - March 2025	Perth	Moon
1st	Sat	Blustery winds in the Kimberleys, particularly around Kununurra.		
2nd	Sun	Monsoonal showers across the Kimberleys over the week ahead.		P7 XhN
3rd	Mon	A supercell storm system crossing the Esperance region, may cause very strong and unsettled winds in the Murchison, particularly around Mt Magnet.	Fg	
4th	Tue		Fg	
5th	Wed			
6th	Thu			
7th	Fri	Very heavy rain for a time in the Northern Wheatbelt region.		1Q
8th	Sat	A thunderstorm embedded supercell storm crosses the Perth region during the next few days, bringing lightning strikes, hail, heavy rain and localised flooding in some Perth suburbs.	Th	^
9th	Sun		Showers	
10th	Mon	Unsettled blustery winds with gusts in the Margaret River region, particularly around Cape Leeuwin and Busselton. Heavy rain likely in the Goldfields. Heavy rain in the Kimberleys and showers through the Pilbara and Interior.	Showers	

Date	Day	Western Australia - March 2025	Perth	Moon
11th	Tue	The next four weeks ahead may be considerably wetter than is normal for this time of year, with the chance of up to 3 cyclonic systems bringing storms and heavy rainfall in numerous regions, including in the Kimberley, Pilbara, Interior and the Eucla. Temperatures may also be unusually higher than normal during the next week ahead, particularly in the Gascoyne, while overall the next few weeks may be cooler than average in many regions away from the coasts.	Showers	
12th	Wed		Showers	
13th	Thur	Showers in the Pilbara to the Interior, spreading into the Kimberleys over the next few days, with heavy falls at times.		
14th	Fri	Strong winds with gusts in the Pilbara, particularly around the coasts.		F
15th	Sat	Strong winds in the Pilbara, particularly around Paraburdoo.		XhS
16th	Sun	Monsoonal rains may prevail over the northern half of the State over the coming week, with heavy falls at times. Very hot in the Gascoyne with maximum temperatures possibly into the upper 40's.		
17th	Mon	Strong gusty winds at times in the Pilbara region, particularly offshore around Varanus Island. Very hot overnight in the Pilbara.		
18th	Tue	Very hot overnight from the Central Coast to the South Coast.		A
19th	Wed	Very heavy rain in the De Grey to the Fortescue regions over the next few days.		
20th	Thu			
21st	Fri	Heavy rain across the Nullabor Plains and up through the Goldfields and into the Interior over the next few days.		
22nd	Sat	Unusually cool day in the Eucla.		
23rd	Sun			3Q V
24th	Mon			
25th	Tue			
26th	Wed	Strong winds in the Pilbara, particularly around Paraburdoo. Pockets of heavy rain over the northern half of the State.		
27th	Thur	Pockets of rain activity throughout the State at first, contracting to the northern half the following day with monsoonal rains continuing in the Kimberleys on most days until the end of the month.	Th	
28th	Fri			
29th	Sat			N XhN
30th	Sun			P4
31st	Mon	Some rain activity through most of the eastern half of the State from the Eucla northwards.		

April 2025

Date	Day	Western Australia - April 2025	Perth	Moon
1st	Tue	Monsoonal rains may prevail over the northern half of the State over the next 7-10 days.		
2nd	Wed	A low off the West Kimberley coast may steadily strengthen to cyclonic conditions over the next few days as it tracks around the Timor Sea, occasionally brushing the coastal regions of the De Grey to West Kimberleys, bringing unsettled storm conditions throughout the Pilbara for much of the following week.		
3rd	Thu			
4th	Fri	Very heavy rain in the Kimberleys over he next 3-4 days.		^
5th	Sat			1Q
6th	Sun	Heavy rain spreads into the De Grey and Fortescue over the next couple of days and later into the Interior. Very heavy rain also in the Pilbara, particularly in the Pardoo area, where localised flooding may occur. Unusually cool in the Great Southern and Central Wheat Belt regions over the next couple of days.		
7th	Mon	Unsually cool spell around the South Coast, particularly in the Mandurah area.		
8th	Tue			
9th	Wed			
10th	Thur			
11th	Fri	After a brief dry start, the next few weeks may see a series of cold fronts and cloud bands bringing some rain relief to the Southwest Land Division districts, as well as in the Pilbara to the Gascoyne regions.		XhS
12th	Sat	Mostly dry spell over the coming week throughout the State.		
13th	Sun			F
14th	Mon			A
15th	Tue	Very hot in the Pilbara. Unsually warm overnight in the Wheat Belt districts, particularly in the Southern Cross area.	Showers, Th, Fg	
16th	Wed		Fg	
17th	Thu			
18th	Fri	Showers along the South Coast to the Nullarbor Plains over the next few days.	Showers	

Date	Day	Western Australia - April 2025	Perth	Moon
19th	Sat	Unusually cool respite in the Great Southern region.		V
20th	Sun			
21st	Mon			3Q
22nd	Tue			
23rd	Wed			
24th	Thur			
25th	Fri		Rain, Th	
26th	Sat	Heavy rain around the Central Coast to the Southwest, with particularly heavy falls in the Karnet, Dwellingup, and Mt William area.		XhN
27th	Sun		Fg	
28th	Mon			P3 N
29th	Tue			
30th	Wed			

May 2025

Date	Day	Western Australia - May 2025	Perth	Moon
1st	Thu	Dry start to the month over the first couple of days.		
2nd	Fri		Showers, Th	^
3rd	Sat	Scattered outbreaks of rain thorughout the southern half of the State over the next couple of days.	Fg	
4th	Sun		Rain	
5th	Mon	Outbreaks of heavy rain around the Hamersley Ranges and the Fortescue region over the next couple of days. Very heavy rain in the Gascoyne particularly in the Dairy Creek area.		1Q
6th	Tue	Very cold overnight in the South Coast region, particulary in the Jarrahwood area.		
7th	Wed	Very low minimum temperatures over the next couple of days around the South Coast and in the Goldfields.		
8th	Thur			
9th	Fri	Unsettled and blustery winds in the Margaret River region, particularly around Cape Leeuwin and Busselton.	Rain	XhS
10th	Sat	Very strong winds persist in the Margaret River region, as well as north up to Perth, and offshore on Rottnest Island and around Cape Leeuwin and Cape Naturaliste. Some showers from the Central Coast to the around the South Coast over the next few days, with chance of heavy falls for a time in the Margaret River area. Very warm overnight on Troughton Island.	Rain	
11th	Sun	The next few weeks may be fairly dry compared to the norm for this time of year, with only the Kimberleys and Interior likely to be wetter than normal. Temperatures may be fairly average.	Showers	A
12th	Mon	Showers in the Interior. Unusually hot in the Kimberley over the coming week.	Showers	
13th	Tue	Mostly dry throughout the State over the coming week ahead.		F
14th	Wed		Showers, Fg	
15th	Thur	Unusually hot in the Kimberley with maximum temperatures possibly into the low 40's.	Fg	
16th	Fri			V
17th	Sat	Chance of isolated showers in the Eucla to the Nullarbor Plains over the next couple of days.		
18th	Sun			
19th	Mon	Dry everywhere over the next few days.		
20th	Tue			3Q
21st	Wed	Unusually warm overnight on Troughton Island.	Showers	
22nd	Thu	Isolated showers briefly around the South Coast over the next couple of days.	Showers	
23rd	Fri			XhN
24th	Sat	Isolated showers along the southern coastal fringes over the next few days.		
25th	Sun			
26th	Mon			P5
27th	Tue	Rain in the Kimberleys and along the King Leopold Ranges with some heavy falls likely at first, easing to spotty showers over the next couple of days. Very heavy rain in some parts of the Pilbara around this time.	Showers	N
28th	Wed	Chance of rain around the coastal South Coast region.		
29th	Thur		Showers	^
30th	Fri			
31st	Sat	Chance of showers in the Kimberleys.		

June 2025

Date	Day	Western Australia - June 2025	Perth	Moon
1st	Sun	Dry over the first few days of the month. Unusually cold overnight in the Central Wheat Belt to the Goldfields region with chance of frosts in some areas.		
2nd	Mon			
3rd	Tue			**1Q**
4th	Wed		Showers, Fg	
5th	Thu	Some showers around the South Coast spreading into the Central Wheat Belt districts and the Pilbara over the next couple of days.	Showers	**XhS**
6th	Fri	Wind gusts in the Margaret River region, particularly around Cape Naturaliste. Unusually cool day in the South Coast to the Great Southern regions.	Rain	
7th	Sat	Rain activity in the southwest of the State with some heavy falls, particularly in the South Coast to the South Central regions.		**A**
8th	Sun			
9th	Mon	Mostly dry everywhere over the next few days.		
10th	Tue			
11th	Wed	Drier than normal conditions may continue over the next few weeks ahead, with many districts seeing well below normal rainfall than is typical for this time of year, particularly in the Southwest Land Division area. In contrast, the next few weeks may be coolder and wetter than average in the north and east of the Kimberleys, mostly due to unseasonal cloud bands crossing the regions. The Interior region may also be much cooler than normal.		**F**
12th	Thur	Unusually warm day in the Kimberley and on Troughton Island overnight.		**V**
13th	Fri			
14th	Sat	Some shower activity in the Nullarbor Plains and Eucla regions over the next couple of days.		
15th	Sun			
16th	Mon		Showers	
17th	Tue	Showers around the South Coast over the next few days. Unusually warm overnight in the Gascoyne, particularly in the Geraldton area.	Showers	
18th	Wed		Fg	
19th	Thu			**3Q XhN**
20th	Fri			
21st	Sat	Mostly dry over the State over the coming week.		
22nd	Sun			
23rd	Mon	Very warm in the Central Wheat Belt area, particularly in the Narembeen district.	Showers, Fg	**P8**
24th	Tue	Chance of some showers along the coastal areas from the Central Coast to the South Coast.	Fg	
25th	Wed			**N**
26th	Thur			**^**
27th	Fri			
28th	Sat			
29th	Sun	Light showers along the southern coastal fringes over the next couple of days, while showers develop in the Kimberleys developing into heavy rain by the end of the month.		
30th	Mon	Cooler and with heavy rain in the Kimberleys and showers into the Interior.		

July 2025

Date	Day	Western Australia - July 2025	Perth	Moon
1st	Tue	Brief rain in the Kimberleys.		
2nd	Wed	A front may cross the Margaret River region, bringing strong winds, and chance of some localised damage in the Margaret River region, including around Cape Naturaliste, Rottnest Island and Mandurah. Very cold overnight in the Murchison with chance of frosts in some areas, particularly around the Yeelirrie districts.	Rain	**XhS**
3rd	Thu	Strong winds with gusts in the Margaret River region around Cape Leeuwin. Rain moves onto the western coasts from the Gascoyne to the South Coast with showers spreading eastward over the next couple of days.	Showers	**1Q**
4th	Fri		Showers, Fg	
5th	Sat	Heavy rain briefly over the Kimberleys.	Showers, Fg	**A**
6th	Sun			
7th	Mon	Strong winds in the Perth region, including over Rottnest Island.	Rain	

Date	Day	Western Australia - July 2025	Perth	Moon
8th	Tue	A wet front may move onto the western coasts from the Gascoyne to the South Coast and spreads showers eastward through the Central Wheat Belt to the Interior over the next few days. Rain may be particularly heavy in the Paynedale area. Unusually warm night in the Murchison.	Showers	
9th	Wed	Gusty winds in the Margaret River to Perth regions, including in the Lower Southwesst to the Central Coast-Perth regions, which may cause some property damage. Gusts likely around Cape Naturaliste, Busselton, Manjimup, Mandurah, Rottnest Island, Swanbourne, and Garden Island areas. Chance of tornado formation winds developing northwest of Gnowangerup.	Rain	
10th	Thur		Showers	V
11th	Fri	The next few weeks may see a change to above average temperatures compared to the norm for this time of year. A series of cold fronts may also cross the State during the next few weeks bringing fairly average rainfall compared to the seasonal norm.	Rain	F
12th	Sat	Rain activity intensifies over the southern half of the State over the next couple of days before clearing. Very strong and blustery winds around Cape Leeuwin.	Rain, Th	
13th	Sun		Showers	
14th	Mon			
15th	Tue	Mostly dry across the State over the remainder of the month with just a few brief exceptions.		
16th	Wed	Isolated showers along the southern coastal fringes of the South Coast over the next few days.		
17th	Thu	Very cold in the Eucla overnight with frost likely in the Eyre area.		XhN
18th	Fri		Showers	3Q
19th	Sat			
20th	Sun			
21st	Mon			P11
22nd	Tue			
23rd	Wed	Chance of showers around the Fortescue coastal districts over the next few days. Chance of heavy rain for a time in the Gascoyne region, particularly in the Exmouth district.		^
24th	Thur			
25th	Fri			N
26th	Sat	Some showers possible in the West Kimberleys. Very cold in the Eucla with frosts likely in the Eyre district.		
27th	Sun	Isolated showers likely along the southern coasts.		
28th	Mon	Warm spell in the Gascoyne to the Central Wheat Belt. Warm overnight on Troughton Island.	Showers	
29th	Tue	Strong winds with gusts from the Central Coast-Perth region to the Margaret River regions, with chance of wind gusts causing some damage to property. Chance of a tornado forming in the Margaret River area.	Rain, Th	
30th	Wed	Shower activity along the western coasts from the Gascoyne to the South Coast region over the next couple of days, with heavy falls briefly around the South Coast to the Central Wheat Belt regions.	Showers	XhS
31st	Thu	Blustery winds with gusts in the Central Coast region, incuding in Perth, Mandurah, and Rottnest Island, which may cause some damage to property. Very blustery in the Margaret River region including around Cape Leeuwin, Cape Naturaliste, and Busselton.	Rain	

August 2025

Date	Day	Western Australia - August 2025	Perth	Moon
1st	Fri	Gusty winds in the Margaret River region including around Capes Leeuwin and Naturaliste and in Busselton. Rain around the South Coast and Central Wheat Belt and up along the coastal districts to the Gascoyne over the next 3-5 days.	Showers	
2nd	Sat	Very strong winds with gusts prevail across the southwest of the State. A storm crossing the Perth region to the Central Wheatbelt may generate tornado systems with the chance of wind damage to property, particularly around Rockingham and Meckering. Chance of tornado systems also crossing Rottnest Island, around Cape Leeuwin and Naturaliste, and strong gusts in Busselton.	Rain, Th	1Q A
3rd	Sun	Very cool day in the South Coast.	Rain	
4th	Mon		Showers, Fg	
5th	Tue		Showers	
6th	Wed		Rain	V
7th	Thur	Widespread rain and storms across the southern half of the State from the Fortescue to the Interior and Nullarbor Plains. Blustery around Cape Leeuwin and Naturaliste, Busselton and Rottnest Island areas in the Margaret River region over the next few days.	Showers	
8th	Fri	Gusty winds may cause some damage in the Margaret River region.	Rain	

Date	Day	Western Australia - August 2025	Perth	Moon
9th	Sat	Strong winds in the southwest of the State also bring unsettled gusts into Perth. Very hot day in the Kimberleys.	Rain	F
10th	Sun	Strong blustery winds persist in the Margaret River region, including around the Capes. Rain at first turning quickly to showers in the southwest from the North Coast to the Wheat Belt districts over the next few days. Warm overnight on Cocos Island.	Showers	
11th	Mon	The next 3-4 weeks may be unusually drier than is normal for this time of year across the State as a whole, with some large areas much drier than average, particularly in the southern half of the State. Temperatures may also be warmer than average over much of the coming few weeks.	Rain, Th, Fg	
12th	Tue			
13th	Wed			XhN
14th	Thu			
15th	Fri		Rain	P12
16th	Sat	Strong gusty winds in the Goldfields-Esperance region, including around Norseman. Storm activity with the chance of hail showers in the Perth Metropolitan region over the next couple of days. Unusually low temperatures around the South Coast region.	Rain, Hail	3Q
17th	Sun		Showers	
18th	Mon			
19th	Tue	Strong winds with gusts around Cape Leeuwin and into the Margaret River region.	Rain	^
20th	Wed	Rain at times in the South Coast at first, spreading eastward through the Central Wheat Belt disricts and north to the Central Coast over the next couple of days. Cold in the South Coast region.	Rain, Fg	
21st	Thur			
22nd	Fri	Showers in the Nullarbor Plains to the Eucla.		
23rd	Sat	Cold overnight in the Eucla with frost patches likely, particularly in the Eyre district.	Showers	N
24th	Sun	Showers in the Nullarbor Plains to the Eucla over the next few days.		
25th	Mon			
26th	Tue			XhS
27th	Wed	Dry throughout the State over the next few days.		
28th	Thu			
29th	Fri		Showers	
30th	Sat	Heavy coastal showers from the Central Coast to the South Coast.	Showers, Fg	A
31st	Sun			1Q

September 2025

Date	Day	Western Australia - September 2025	Perth	Moon
1st	Mon	Strong and blustery winds around Cape Leeuwin, and blustery into the Margaret River region. Mostly dry throughout the month in the northern half of the State, and only a few brief wet spells in some southern parts.	Rain	
2nd	Tue	Some rain activity in coastal parts from the North Coast to the South Coast.		V
3rd	Wed		Showers	
4th	Thur	Strong winds in the Perth to Margaret River regions, including strong gusts around Capes Leeuwin and Naturaliste, Busselton, Bunbury and Rottnest Island.	Showers	
5th	Fri	Strong winds with gusts in the Margaret River Region, as well as around the Capes Leeuwin and Naturaliste, Busselton, Bunbury, and Rottnest Island. Some showers around the South Coast over the next couple of days, possibly briefly extending into the Central Wheat Belt districts.	Showers	
6th	Sat		Showers	
7th	Sun		Showers	
8th	Mon	Strong blustery winds at times in the Perth region to Rottnest Island areas. Very hot day in the Pilbara.	Rain, Th	F
9th	Tue	Heavy rain for a time around the South Coast region.		XhN
10th	Wed	Very hot in the Pibara as well as around the Kimberley coasts with maximum temperatures in the upper 30's.		
11th	Thu	The next 3-4 weeks may once again be drier and see maximum temperatures generally above average over most of the State.		P9
12th	Fri	Chance of frosts overnight in the Central Wheat Belt region, particularly in the Southern Cross area.		
13th	Sat	Strong winds in the Margaret River region and later onto the South Coast-Esperance regions over the next couple of days, including around Cape Leeuwin and in Jacup.		
14th	Sun	Strong winds around the South Coast-Esperance, including in Jacup.	Showers	3Q
15th	Mon	Showers around the South Coast.		^

Date	Day	Western Australia - September 2025	Perth	Moon
16th	Tue			
17th	Wed	Cold around the South Coast.	Rain, Th, Sn	
18th	Thur	Showers around the South Coast and into the Central Wheat Belt districts. Chance of heavy rain in the Pemberton district for a time.	Showers, Fg	
19th	Fri	Showers along the southern coasts of the South Coast and Nullarbor Plains regions.		
20th	Sat			
21st	Sun	Chance of tornadic systems forming in the Perth area ahead of a front, with possible damage to property in some of the western suburbs. Strong gusty winds also across Cape Leeuwin, Cape Naturaliste, and Norseman, Busselton, Rottenest Island and Swanbourne areas.	Rain	
22nd	Mon	Storm activity with rain in the southwest quarter of the State over the next couple of days. Strong winds across the southwest of the State, including from Perth, Ocean Reef, and Rottnest Island to the Margaret River regions, Cape Naturaliste, Busselton, and Swanbourne. Gusty winds also into the Goldfields-Esperance region, particularly around Norseman. Very strong wind gusts across Cape Leeuwin.	Rain	N XhS
23rd	Tue	Further strong winds over the next couple of days in the Margaret River region, including around the Capes Naturaliste and Leeuwins and in Busselton.		
24th	Wed		Showers	
25th	Thu	A strong westerly airflow brings gusty winds over the Margaret River region including across Cape Leeuwin over the next few days. Some outbreaks of rain around the South Coast region easing to isolated showers by tomorrow.	Showers	
26th	Fri			A
27th	Sat		Showers	
28th	Sun	Isolated showers in the southwest coastal districts over the next couple of days.	Showers	
29th	Mon			
30th	Tue			1Q V

October 2025

Date	Day	Western Australia - October 2025	Perth	Moon
1st	Wed	Dry throughout the coming month over the northern half of the State, and only some passing wet spells in the southern half. Very hot in the Kimberley around this time with maximum temperatures possibly into the low 40's.	Showers	
2nd	Thur	Some showers around the South Coast.		
3rd	Fri	An approaching cold front brings very strong winds over Rottnest Island to the Perth region.	Rain	
4th	Sat	Showers along the western coasts from the North Coast to the South Coast, spreading inland across the Central Wheat Belt and into the Goldfields over the next couple of days.	Rain, Fg	
5th	Sun	Very warm overnight in the Kimberleys particularly in the Halls Creek district.	Showers	
6th	Mon	Isolated showers around the South Coast over the next few days.	Showers	
7th	Tue		Showers, Fg	F XhN
8th	Wed			
9th	Thu			P6
10th	Fri		Fg	
11th	Sat	Over the next 3-4 weeks, temperatures may be with a degree either side of the average for this period of time. Similarly, rainfall may also be near to average, with a mix of some areas, such as the Kimberley, northern Interior, Goldfields and Gascoyne regions much drier than average, in contrast to the south coastal districts and the Pilbara much wetter than average.		
12th	Sun			
13th	Mon			^
14th	Tue			3Q
15th	Wed			
16th	Thur			
17th	Fri		Rain	
18th	Sat	Rain activity in the southwest of the State from the North Coast to the South Coast spreading by tomrorow into the Central Wheat Belt and Goldfield regions.	Rain	
19th	Sun	Isolated showers spread into the Nullarbor Plains and Eucla over the next couple of days.	Showers	
20th	Mon		Showers	XhS
21st	Tue			
22nd	Wed			N
23rd	Thu	Very hot in the Kimberley with maximum temperatures around the mid-40's, particularly in the Curtin area.	Fg	
24th	Fri			A
25th	Sat			
26th	Sun			

Date	Day	Western Australia - October 2025	Perth	Moon
27th	Mon	The passage of a cold front may bring strong winds at times in the Margaret River to the Goldfields-Esperance regions, incuding around Cape Leeuwin and in Esperance.	Showers	V
28th	Tue	Isolated pockets of showers around the coastal fringes of the South Coast.		
29th	Wed			
30th	Thur	Isolated showers in the Central Wheat Belt to the Nullarbor Plains.	Showers, Fg	1Q
31st	Fri	Isolated showers along the southern coastal fringes.		

November 2025

Date	Day	Western Australia - November 2025	Perth	Moon
1st	Sat	Isolated showers along the southern coastal fringes.		
2nd	Sun	Very warm overnight in the Kimberley, particularly in the Warmun area.		
3rd	Mon			XhN
4th	Tue			
5th	Wed	A series of cold fronts may cross the southwest of the State over the next couple of days bringing unsettled and often very strong winds through the regions from Perth to Margaret Airport and later into the Goldfields-Esperance regions. Winds may be very strong for a time in the Forrest, Hopetoun, Munglinup, and Esperance areas as well as around Cape Naturaliste, Cape Leeuwin, and Ocean Reef. Chance of heavy rain in the Kimberley region, particulary in the Warmun area.	Rain	
6th	Thur	Unusually cold snap in the southwest of the State, particularly around the South Coast. Chance of snow on Mt Stirling, and possible snowflurries in Katanning for a short time. Some bursts of heavy rain along the southern coasts during the next couple of days and showers into the Central Wheatbelt districts. Very strong winds across Rottnest Island.	Showers	F P1
7th	Fri			
8th	Sat	Pockets of frost likely in the Goldfields to the Esperance regions, particularly in Norseman.		
9th	Sun			^
10th	Mon	Shower activity likely in the Interior over the next 3-4 days. Some showers in the Nullarbor Plains and along the south coasts. Some shower activity begins to return to the northern half of the State around this time, with showers in the far north of the Kimberleys.		
11th	Tue	Above average temperatures may return through much of the State over the next 3-4 weeks ahead, while rainfall expectations may be in line with what is typical for this time of year overall, with parts of the west and south well below normal rainfall, while the Kimberley, Interior, and Pilbara much wetter than average.		
12th	Wed	Strong winds and scattered thunderstorms may across the Kimberleys, and especially blustery in the Wyndham area. Unusually cold day around the South Coast, particularly in the Albany district.		3Q
13th	Thu	Some rain in the North Kimberleys over the next 4-5 days, with occasional heavy falls particularly around the coasts.		
14th	Fri			
15th	Sat			
16th	Sun			XhS
17th	Mon			
18th	Tue	Very wet for a time around the South Coast.		
19th	Wed			
20th	Thur			N A
21st	Fri	Dry over the next few days over most of the State. Very hot in the Central Wheat Belt over the next few days, with maximas into the 40's in a number of districts.		
22nd	Sat			
23rd	Sun		Fg	V
24th	Mon	Heavy rain and scattered thunderstorms with wind gusts in the Kimberleys.		
25th	Tue			
26th	Wed	Heavy rain along the northern Kimberley coasts, and chance of brief rain in the Interior.		
27th	Thu		Showers	
28th	Fri	Thunderstorms and strong winds with gusts cross the Nullarbor Plains, with very strong winds at times in the Forrest area.		1Q
29th	Sat			
30th	Sun	Very hot in the Pilbara with many districts maximum temperatures reaching into the upper 40's.		XhN

December 2025

Date	Day	Western Australia - December 2025	Perth	Moon
1st	Mon			
2nd	Tue	Some shower activity likely in the Kimberleys over the next 2-3 days. Very cold overnight in the Eucla, with chance of some frost pockets mostly in the Eyre district.		
3rd	Wed	Thunderstorms and strong wind gusts in the Kimberley, particularly in the Kununurra area. Very hot in the Pilbara, including overnight.		
4th	Thur			**P2**
5th	Fri	A band of thunderstorms may cross the Pilbara - Dampier Peninsula regions bringing strong gusty winds with damage potential, hail and some heavy rain at times.		**F**
6th	Sat			**^**
7th	Sun	Very warm in the Pilbara with temperatures unlikely to drop below the low 30's overnight and reach into the mid-40's during the day.		
8th	Mon			
9th	Tue		Showers	
10th	Wed	Thunderstorms and strong winds at times in the Kimberley, particularly around Kununurra. Further south, strong winds associated with the passage of a cold front may cross the Margaret River region, bringing very strong gusts around Cape Leeuwin.	Showers	
11th	Thu	The next 3-4 weeks may be generally wetter than normal for this time of year, resulting in generally cooler than normal temperatures.	Showers	
12th	Fri	Brief outbreaks of rain in the North Kimberleys.		**3Q**
13th	Sat	Very warm overnight in the Kimberley.		**XhS**
14th	Sun			
15th	Mon	Heavy rain along the northern Kimberley coasts, and chance of brief rain in the Interior.	Showers	
16th	Tue	Some showers around the South Coast region with unusually cold overnight temperatures in some districts, as well as in the Central Wheat Belt.		
17th	Wed	Unusually cold overnight in the Perth area.		**A**
18th	Thur	Monsoonal activity builds over the northern half of the State around this time, with shower activity becoming more frequent from the Kimberleys to the Interior over the remainder of the month. Strong winds through the Gibson Desert likely.		
19th	Fri			
20th	Sat	Outbreaks of heavy rain north of the Hamersley Ranges.		**N V**
21st	Sun	Rain activity may spread throughout the central regions of the State through the Pilbara, Gascoyne, Interior and Murchison regions over the next couple of days. Cooler temperatures prevail, particularly in the Gascyone.	Showers, Th	
22nd	Mon	Very heavy rain likely in the central parts of the State from the Gacoyne, through the Murchison and into the Intnerior, as well as spreading shower activity southward into the Central Wheat Belt and Goldfields and with isolated showers possibly reaching the southern coasts.		
23rd	Tue	Rain in the North Kimberleys.		
24th	Wed	Showers likely in the eastern half of the State over the next few days. Cooler spell in the Esperance region.		
25th	Thu	Widespread rain and storm activity srepads throughout many parts of the State over the next few days.	Showers	
26th	Fri	A large supercell system may bring thunderstorms across the Central Wheat Belt with heavy rainfall, and unsettled winds in the Esperance - Goldfield region, which may develop into tornado formations particularly in the area north of Dalwallinu.	Showers	
27th	Sat	Severe thunderstorms may cross the southern Goldfields-Esperance, Southeast Coastal regions, particularly bringing severe storms across the Norseman area. Very heavy rain in the Central Wheat Belt districts.	Showers	**XhN**
28th	Sun	Heavy rain along the southern coasts likely, with showers persisting in the Interior to the East Kimberleys, while clearing along the western coastal areas.		**1Q**
29th	Mon	Mostly dry across the State except for som eisolated showers mostly along the southern coasts and in the North Kimberley.		
30th	Tue	Heavy rain in the North Kimberleys.		
31st	Wed	Mostly dry across the State except for some isolated showers in the East Kimberleys and eastern border fringes to the Eucla, as well as around the southern coasts.		

WESTERN AUSTRALIA GRAPHS AND TABLES 2025

Seasonal Breakdown –	State Overall Rainfall Timing and Estimated mm Trend Graphs State Overall Maximum Temperature Trend Graphs Regional Rainfall Expectations Tables - mm and Trends Regional Maximum and Minimum Temperature Trends Tables
Monthly Breakdown -	**Rainfall:** Regional Expectations by mm Regional Expectations by mm Variance to Average Regional Variance Trend by mm Regional Estimated Expectations by % of Variance to Average Regional Estimatedl Variance Trend by % **Maximum & Minimum Temperature Trends:** Regional Estimated Average Temperature Regional Estimated Degree of Variation from the Average Regional Estimated Temperature Trend

Seasonal Breakdowns

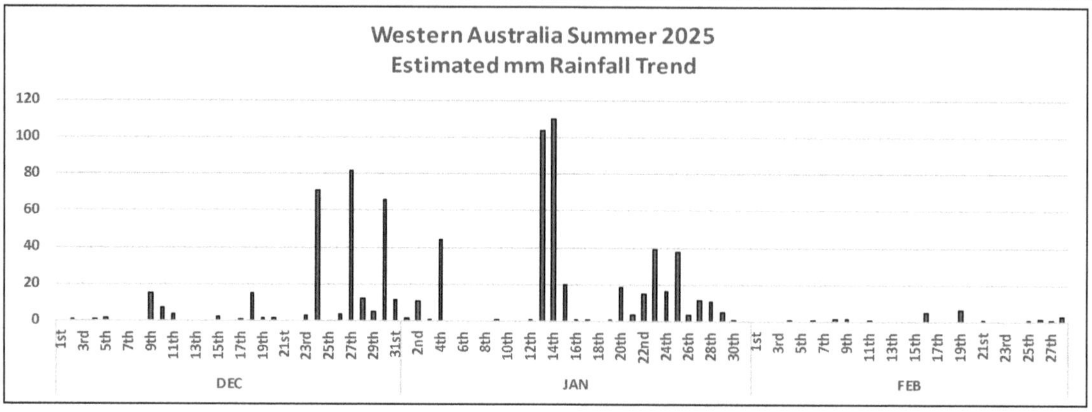

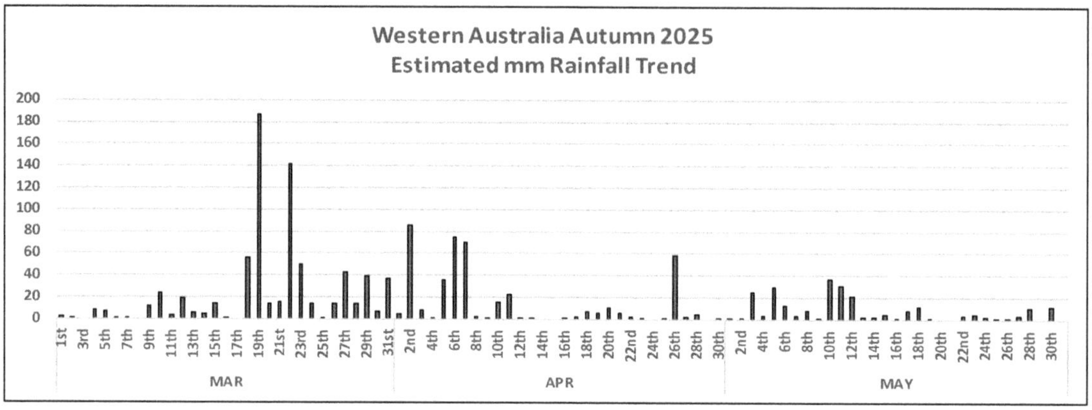

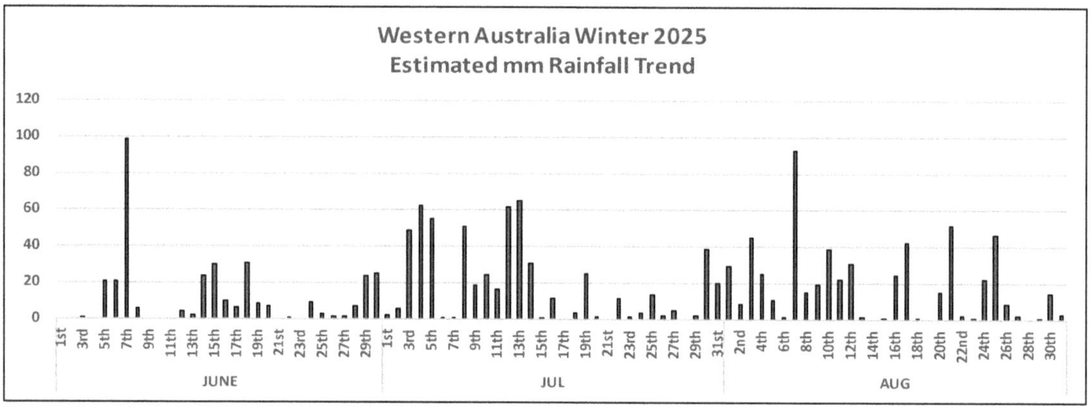

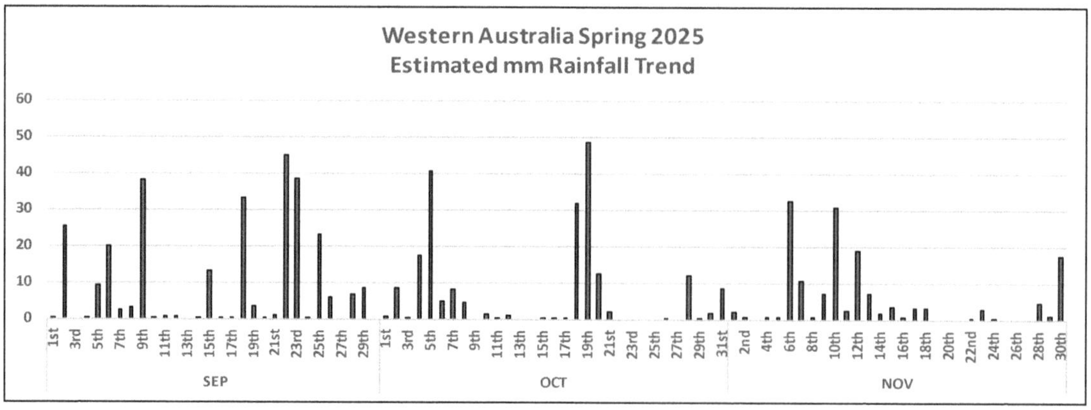

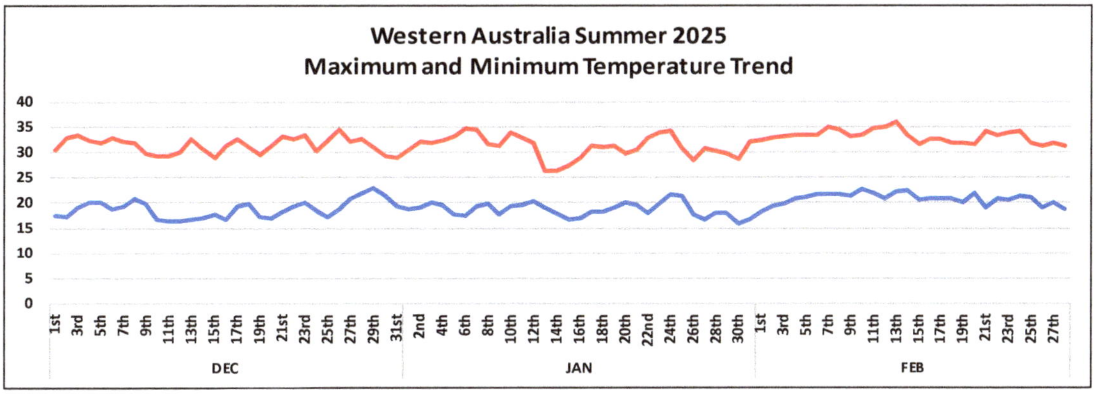

Western Australia Summer 2025
Maximum and Minimum Temperature Trend
DEC
JAN
FEB

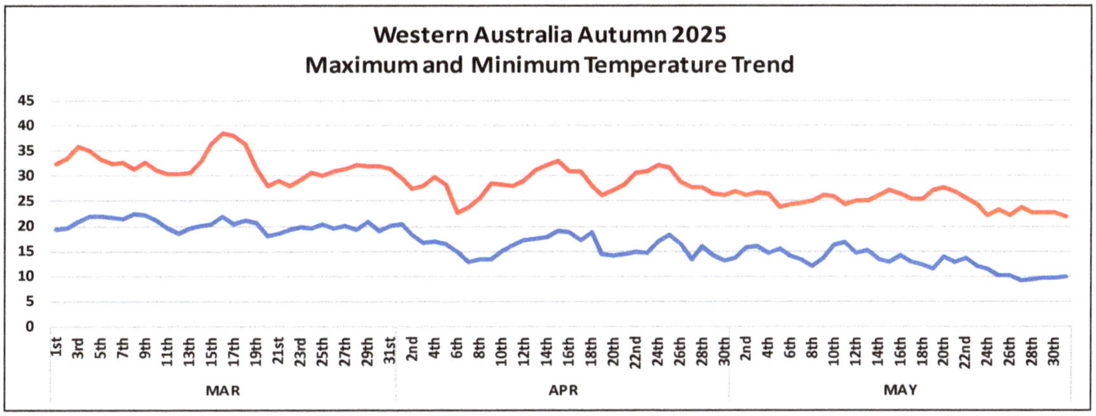

Western Australia Autumn 2025
Maximum and Minimum Temperature Trend
MAR
APR
MAY

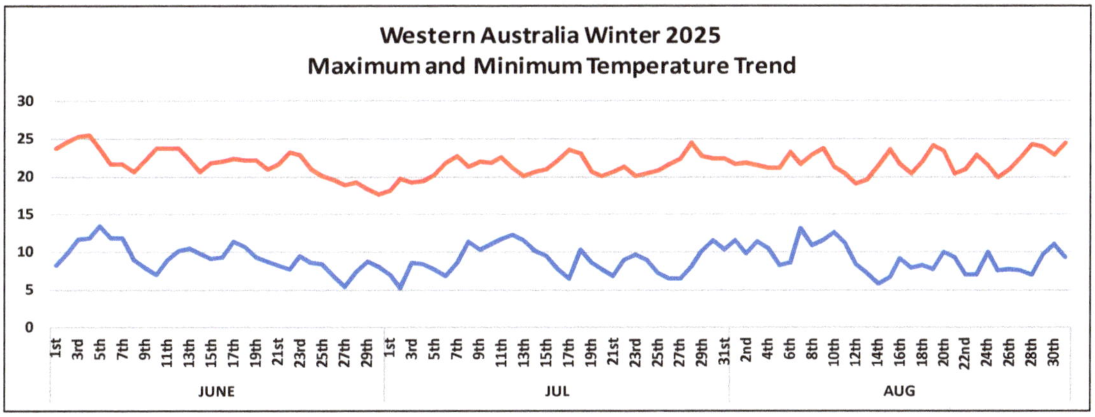

Western Australia Winter 2025
Maximum and Minimum Temperature Trend
JUNE
JUL
AUG

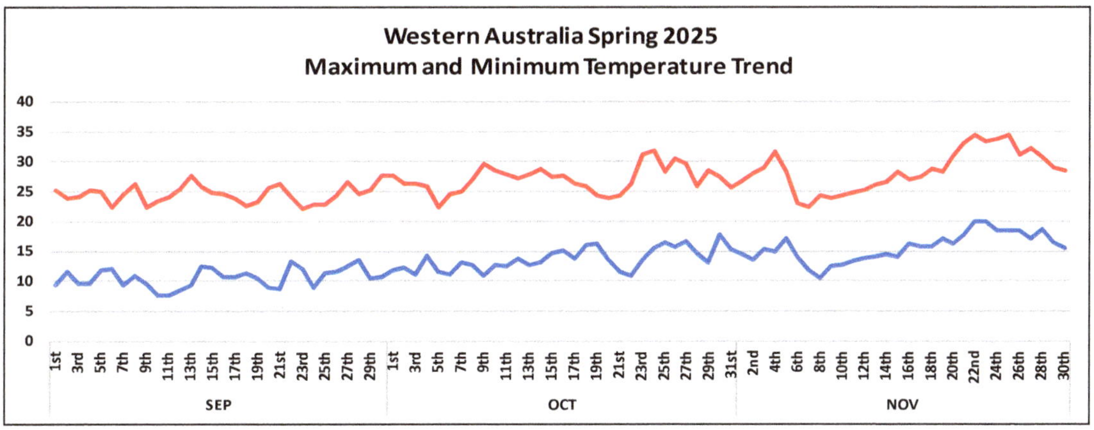

Western Australia Spring 2025
Maximum and Minimum Temperature Trend
SEP
OCT
NOV

2025 WA - Estimated Seasonal Rainfall Expectations (mm)

REGION	Kimberleys	Pilbara	Gascoyne	Central West	Lower & South West	Great Southern	Central Wheat Belt	Goldfields	Eucla		WA
Town Sample	Broome	Port Hedland	Carnarvon	Geraldton	Perth	Albany	Merredin	Kalgoorlie	Eucla	Giles	Overall
Summer	438	62	1	10	35	55	28	79	27	37	773
Autumn	367	433	23	13	109	166	13	51	99	125	1399
Winter	2	9	165	141	368	509	107	63	125	0	1488
Spring	3	1	3	59	183	265	50	13	32	35	642
Overall	640	506	191	231	703	1035	226	199	312	280	4323

2025 WA - Estimated Seasonal Rainfall Variation from the Average (mm)

REGION	Kimberleys	Pilbara	Gascoyne	Central West	Lower & South West	Great Southern	Central Wheat Belt	Goldfields	Eucla		WA
Town Sample	Broome	Port Hedland	Carnarvon	Geraldton	Perth	Albany	Merredin	Kalgoorlie	Eucla	Giles	Overall
Summer	-30	-105	-36	-11	-3	-18	-16	-0	-26	-91	-336
Autumn	214	335	-37	-91	-34	-33	-72	-26	20	42	319
Winter	-24	-33	56	-93	-35	128	-31	-8	44	-36	-33
Spring	-8	-1	-14	-2	42	47	-10	-37	-29	-24	-36
Overall	-19	196	-30	-188	-22	165	-100	-79	37	-27	-67

2025 WA - Estimated Seasonal Rainfall Trend Variation by 10mm

REGION	Kimberleys	Pilbara	Gascoyne	Central West	Lower & South West	Great Southern	Central Wheat Belt	Goldfields	Eucla		WA
Town Sample	Broome	Port Hedland	Carnarvon	Geraldton	Perth	Albany	Merredin	Kalgoorlie	Eucla	Giles	Overall
Summer	drr	drr	drr	av	av	av	av	av	av	drr	drr
Autumn	wtr	wtr	drr	drr	drr	drr	drr	av	av	wtr	wtr
Winter	av	drr	wtr	drr	drr	wtr	drr	av	wtr	drr	av
Spring	av	av	av	av	wtr	wtr	av	drr	av	av	av
Overall	av	wtr	av	drr	av	wtr	drr	av	av	av	av

2025 WA - Estimated Seasonal Rainfall Expectations by %

REGION	Kimberleys	Pilbara	Gascoyne	Central West	Lower & South West	Great Southern	Central Wheat Belt	Goldfields	Eucla		WA
Town Sample	Broome	Port Hedland	Carnarvon	Geraldton	Perth	Albany	Merredin	Kalgoorlie	Eucla	Giles	Overall
Summer	-6%	-63%	-97%	-52%	-8%	-25%	-36%	0%	-49%	-71%	-30%
Autumn	139%	341%	-61%	-87%	-24%	-16%	-85%	-34%	25%	50%	30%
Winter	-93%	-78%	52%	-40%	-9%	33%	-23%	-11%	54%	-99%	-2%
Spring	-72%	-50%	-84%	-3%	30%	22%	-16%	-75%	-48%	-41%	-5%
Overall	-3%	63%	-14%	-45%	-3%	19%	-31%	-28%	14%	-9%	-2%

2025 WA - Estimated Seasonal Rainfall Trend Variation by 25%

REGION	Kimberleys	Pilbara	Gascoyne	Central West	Lower & South West	Great Southern	Central Wheat Belt	Goldfields	Eucla		WA
Town Sample	Broome	Port Hedland	Carnarvon	Geraldton	Perth	Albany	Merredin	Kalgoorlie	Eucla	Giles	Overall
Summer	av	drr	drr	drr	av	av	drr	av	drr	drr	drr
Autumn	wtr	wtr	drr	drr	av	av	drr	drr	wtr	wtr	wtr
Winter	drr	drr	wtr	drr	av	wtr	av	av	wtr	drr	av
Spring	drr	drr	drr	av	wtr	av	av	drr	drr	drr	av
Overall	av	wtr	av	drr	av	av	drr	drr	av	av	av

2025 WA - Estimated Seasonal Maximum Temperature Average and Trend

Region	Kimberleys	Pilbara	Gascoyne	Central West	Lower & South West	Great Southern	Central Wheat Belt	Goldfields	Eucla		WA
Town Sample	Broome	Port Hedland	Carnarvon	Geraldton	Perth	Albany	Merredin	Kalgoorlie	Eucla	Giles	Overall
Summer Est	33.9	37.8	30.1	29.8	30.9	24.2	33.7	33.8	27.2	37.7	31.9
Trend	av	wmr	clr	clr	av	clr	wmr	a-w	clr	wmr	clr
Autumn Est	33.1	34.7	30.0	28.8	27.3	22.4	27.9	27.3	24.7	30.1	28.6
Trend	av	wmr	clr	av	av	clr	wmr	wmr	av	wmr	av
Winter Est	29.1	28.2	23.7	21.7	19.7	17.2	18.4	18.9	19.4	21.2	21.7
Trend	av	clr	clr	clr	av	clr	wmr	wmr	wmr	wmr	a-w
Spring Est	33.4	35.1	26.2	24.6	23.0	18.6	26.0	26.6	24.2	31.1	26.8
Trend	av	wmr	clr	clr	clr	clr	wmr	wmr	clr	wmr	clr

2025 WA - Estimated Seasonal Minimum Temperature Average and Trend

Region	Kimberleys	Pilbara	Gascoyne	Central West	Lower & South West	Great Southern	Central Wheat Belt	Goldfields	Eucla		WA
Town Sample	Broome	Port Hedland	Carnarvon	Geraldton	Perth	Albany	Merredin	Kalgoorlie	Eucla	Giles	Overall
Summer Est	26.3	25.1	21.4	17.7	16.6	13.3	16.8	18.3	14.9	23.9	19.4
Trend	av	av	a-c	av	av	av	av	a-w	clr	wmr	av
Autumn Est	22.6	23.0	19.3	16.3	13.8	12.2	12.6	13.9	12.9	18.1	16.4
Trend	av	wmr	av	a-w	a-w	av	av	wmr	av	wmr	a-w
Winter Est	12.7	13.3	11.8	11.6	9.1	8.6	6.0	7.2	4.4	6.9	9.1
Trend	clr	av	av	wmr	a-w	a-w	av	wmr	clr	a-c	av
Spring Est	22.0	17.3	16.9	11.8	11.2	9.4	9.0	11.1	9.2	16.1	13.3
Trend	av	clr	av	av	av	av	a-c	av	clr	a-c	av

Monthly Breakdowns

2025 — WESTERN AUSTRALIA - Estimated Rainfall Expectations (mm)

REGION	Kimberleys	Pilbara	Gascoyne	Central West	Lower & South West	Great Southern	Central Wheat Belt	Goldfields	Eucla		WA
Town Sample	Broome	Port Hedland	Carnarvon	Geraldton	Perth	Albany	Merredin	Kalgoorlie	Eucla	Giles	Overall
January	252	62	1	3	19	29	17	43	4	22	452
February	5	0	0	0	3	6	0	7	0	0	21
March	221	312	0	2	6	42	2	49	41	60	735
April	145	119	0	2	44	56	5	2	14	37	424
May	2	2	23	9	59	68	6	1	44	27	241
June	0	0	27	20	45	117	29	34	66	0	339
July	0	4	98	39	135	243	32	12	14	0	579
August	2	5	39	82	187	149	45	17	45	0	570
September	2	0	1	28	112	119	17	0	2	0	282
October	1	0	1	29	52	80	30	9	5	0	207
November	0	1	1	1	19	66	3	3	25	35	154
December	10	0	0	15	22	61	39	22	52	98	320
Overall	640	506	191	231	703	1035	226	199	312	280	4323

2025 — WESTERN AUSTRALIA - Estimated Rainfall Variation from the Average (mm)

REGION	Kimberleys	Pilbara	Gascoyne	Central West	Lower & South West	Great Southern	Central Wheat Belt	Goldfields	Eucla		WA
Town Sample	Broome	Port Hedland	Carnarvon	Geraldton	Perth	Albany	Merredin	Kalgoorlie	Eucla	Giles	Overall
January	62	7	-7	-2	6	3	3	18	-12	-10	66
February	-196	-95	-17	-11	-15	-17	-16	-25	-19	-39	-448
March	111	259	-16	-17	-13	4	-20	19	19	12	358
April	117	97	-16	-17	9	-1	-18	-19	-12	23	163
May	-14	-22	-5	-56	-30	-35	-34	-25	13	7	-201
June	-15	-23	-19	-63	-95	-11	-19	10	36	-16	-213
July	-10	-10	54	-49	-14	109	-18	-10	-11	-13	30
August	0	-0	21	18	73	30	6	-8	18	-8	150
September	1	-2	-7	-6	35	21	-9	-13	-20	-12	-11
October	0	-0	-4	13	16	13	11	-7	-14	-14	16
November	-10	1	-4	-9	-10	14	-12	-17	5	2	-41
December	-66	-17	-12	10	14	37	25	-0	33	42	66
Overall	-19	196	-30	-188	-22	165	-100	-79	37	-27	-67

2025 — WESTERN AUSTRALIA - Estimated Rainfall Trend Variation by 10mm

REGION	Kimberleys	Pilbara	Gascoyne	Central West	Lower & South West	Great Southern	Central Wheat Belt	Goldfields	Eucla		WA
Town Sample	Broome	Port Hedland	Carnarvon	Geraldton	Perth	Albany	Merredin	Kalgoorlie	Eucla	Giles	Overall
January	wtr	av	av	av	av	av	av	wtr	drr	av	wtr
February	drr	drr	drr	drr	drr	drr	drr	drr	drr	drr	drr
March	wtr	wtr	drr	drr	drr	av	drr	wtr	wtr	wtr	wtr
April	wtr	wtr	drr	drr	av	av	drr	drr	drr	wtr	wtr
May	drr	drr	av	drr	drr	drr	drr	drr	wtr	av	drr
June	drr	drr	drr	drr	drr	drr	drr	av	wtr	drr	drr
July	av	av	wtr	drr	drr	wtr	drr	av	drr	drr	wtr
August	av	av	wtr	wtr	wtr	wtr	av	av	wtr	av	wtr
September	av	av	av	av	wtr	wtr	av	drr	drr	drr	av
October	av	av	av	wtr	wtr	wtr	wtr	av	drr	drr	av
November	av	av	av	av	av	wtr	drr	drr	av	av	drr
December	drr	drr	drr	av	wtr	wtr	wtr	av	wtr	wtr	wtr
Overall	av	wtr	av	drr	av	wtr	drr	av	av	av	av

2025	WESTERN AUSTRALIA - Estimated Rainfall Expectations by %										
REGION	Kimberleys	Pilbara	Gascoyne	Central West	Lower & South West	Great Southern	Central Wheat Belt	Goldfields	Eucla		WA
Town Sample	Broome	Port Hedland	Carnarvon	Geraldton	Perth	Albany	Merredin	Kalgoorlie	Eucla	Giles	Overall
January	32%	13%	-88%	-40%	46%	10%	19%	68%	-73%	-32%	17%
February	-97%	-100%	-100%	-100%	-83%	-75%	-97%	-79%	-99%	-100%	-96%
March	101%	489%	-100%	-92%	-70%	10%	-92%	62%	82%	24%	95%
April	425%	444%	-100%	-89%	25%	-2%	-79%	-92%	-46%	162%	62%
May	-90%	-92%	-18%	-86%	-34%	-34%	-84%	-98%	42%	34%	-46%
June	-100%	-100%	-41%	-76%	-68%	-9%	-40%	42%	121%	-100%	-39%
July	-100%	-69%	124%	-55%	-9%	81%	-35%	-46%	-44%	-98%	6%
August	20%	-7%	114%	29%	63%	25%	14%	-31%	70%	-100%	36%
September	120%	-100%	-92%	-17%	46%	21%	-33%	-97%	-91%	-100%	-4%
October	67%	-50%	-74%	84%	45%	20%	58%	-44%	-73%	-100%	8%
November	-100%	140%	-83%	-87%	-34%	26%	-80%	-84%	23%	5%	-21%
December	-87%	-100%	-100%	180%	179%	147%	182%	-1%	178%	73%	26%
Overall	-3%	63%	-14%	-45%	-3%	19%	-31%	-28%	14%	-9%	-2%

2025	WESTERN AUSTRALIA - Estimated Rainfall Trend Variation by 25%										
REGION	Kimberleys	Pilbara	Gascoyne	Central West	Lower & South West	Great Southern	Central Wheat Belt	Goldfields	Eucla		WA
Town Sample	Broome	Port Hedland	Carnarvon	Geraldton	Perth	Albany	Merredin	Kalgoorlie	Eucla	Giles	Overall
January	wtr	av	drr	drr	wtr	av	av	wtr	drr	drr	av
February	drr	drr	drr	drr	drr	drr	drr	drr	drr	drr	drr
March	wtr	wtr	drr	drr	drr	av	drr	wtr	wtr	av	wtr
April	wtr	wtr	drr	drr	av	av	drr	drr	drr	wtr	wtr
May	drr	drr	av	drr	drr	drr	drr	drr	wtr	wtr	drr
June	drr	drr	drr	drr	drr	av	drr	wtr	wtr	drr	drr
July	drr	drr	wtr	drr	av	wtr	drr	drr	drr	drr	av
August	av	av	wtr	wtr	wtr	av	av	drr	wtr	drr	wtr
September	wtr	drr	drr	av	wtr	av	drr	drr	drr	drr	av
October	wtr	drr	drr	wtr	wtr	av	wtr	drr	drr	drr	av
November	drr	wtr	drr	drr	drr	wtr	drr	drr	av	av	av
December	drr	drr	drr	wtr	wtr	wtr	wtr	av	wtr	wtr	wtr
Overall	av	wtr	av	drr	av	av	drr	drr	av	av	av

2025	WESTERN AUSTRALIA - Estimated Average Maximum Temperature										
Region	Kimberleys	Pilbara	Gascoyne	Central West	Lower & South West	Great Southern	Central Wheat Belt	Goldfields	Eucla		WA
Town Sample	Broome	Port Hedland	Carnarvon	Geraldton	Perth	Albany	Merredin	Kalgoorlie	Eucla	Giles	Overall
January	33	38	31	31	30	24	31	31	26	36	**31**
February	33	37	30	30	33	25	37	38	29	40	**33**
March	33	37	34	33	33	26	33	32	26	36	**32**
April	33	34	30	29	27	22	28	28	26	30	**29**
May	33	33	26	25	22	19	22	22	22	25	**25**
June	30	29	25	23	20	18	19	18	18	20	**22**
July	28	27	23	22	20	17	18	19	19	20	**21**
August	29	29	23	20	19	17	18	20	21	24	**22**
September	32	33	25	22	20	18	21	24	23	29	**25**
October	34	36	26	24	22	18	26	27	25	33	**27**
November	35	36	28	28	26	19	29	28	25	31	**29**
December	34	37	29	28	27	21	29	30	26	35	**29**
Overall	**32**	**34**	**28**	**26**	**25**	**20**	**26**	**26**	**24**	**30**	**27**

2025 WESTERN AUSTRALIA - Estimated Average Minimum Temperature

Region Town Sample	Kimberleys Broome	Pilbara Port Hedland	Gascoyne Carnarvon	Central West Geraldton	Lower & South West Perth	Great Southern Albany	Central Wheat Belt Merredin	Goldfields Kalgoorlie	Eucla Eucla	Giles	WA Overall
January	26	25	21	18	16	13	16	17	14	22	**19**
February	26	26	23	19	18	15	18	21	17	25	**21**
March	25	27	23	20	18	15	18	18	16	24	**20**
April	25	23	19	15	12	12	12	14	13	17	**16**
May	17	20	16	14	11	10	8	10	10	13	**13**
June	15	15	12	11	9	9	6	6	4	6	**9**
July	11	12	13	12	9	8	6	8	4	6	**9**
August	12	13	11	11	9	9	6	7	5	9	**9**
September	17	15	13	10	10	9	6	8	7	13	**11**
October	23	17	16	12	11	9	10	12	10	17	**14**
November	26	20	18	14	13	10	12	13	11	18	**16**
December	27	24	20	15	14	12	14	15	13	21	**18**
Overall	21	20	17	14	13	11	11	12	10	16	**14**

2025 WESTERN AUSTRALIA - Estimated Degree of Variation from the Maximum Average

Region Town Sample	Kimberleys Broome	Pilbara Port Hedland	Gascoyne Carnarvon	Central West Geraldton	Lower & South West Perth	Great Southern Albany	Central Wheat Belt Merredin	Goldfields Kalgoorlie	Eucla Eucla	Giles	WA Overall
January	0.1	1.7	0.0	-1.1	-1.9	-0.7	-2.5	-2.8	0.3	-1.7	-0.9
February	0.5	1.1	-2.6	-3.0	0.7	0.5	3.7	5.5	2.9	4.0	1.3
March	-1.3	0.1	2.1	2.3	3.1	2.1	3.0	2.2	1.0	1.7	1.6
April	-0.8	-1.2	1.2	0.9	1.0	0.2	2.5	2.4	2.2	0.0	0.8
May	1.4	2.3	-0.2	0.5	0.6	0.0	1.6	1.7	1.0	1.5	1.0
June	0.7	1.3	1.7	1.7	1.0	0.8	1.2	0.3	-0.2	-0.3	0.8
July	-0.9	-0.6	1.0	2.8	1.9	1.3	2.0	2.0	0.6	-0.3	1.0
August	-1.1	-0.5	-0.6	0.1	0.5	0.8	0.7	1.2	1.8	0.8	0.4
September	0.1	0.6	0.1	0.0	-0.1	0.5	0.1	1.3	1.6	1.5	0.6
October	0.6	0.9	-0.5	-0.7	-0.3	-0.4	0.6	1.5	1.7	1.5	0.5
November	1.0	0.0	0.7	0.5	0.3	-1.5	-0.1	-0.9	0.5	-3.4	-0.3
December	0.0	-0.1	-0.3	-1.8	-2.1	-2.5	-3.1	-2.0	0.8	-1.4	-1.2
Overall	0.0	0.5	0.2	0.2	0.4	0.1	0.8	1.0	1.2	0.3	0.5

2025 WESTERN AUSTRALIA - Estimated Degree of Variation from the Minimum Average

Region Town Sample	Kimberleys Broome	Pilbara Port Hedland	Gascoyne Carnarvon	Central West Geraldton	Lower & South West Perth	Great Southern Albany	Central Wheat Belt Merredin	Goldfields Kalgoorlie	Eucla Eucla	Giles	WA Overall
January	-0.6	-0.3	-1.8	-0.8	-1.5	-0.6	-1.7	-1.6	-2.9	-1.5	-1.3
February	0.2	0.2	-0.7	-0.4	0.6	0.5	0.2	2.6	0.3	2.5	0.6
March	-0.6	2.1	0.6	2.3	2.4	1.4	1.6	1.9	-0.3	3.4	1.5
April	2.5	1.2	-0.6	-0.4	-0.6	-0.1	-1.4	0.9	-0.6	0.5	0.1
May	-0.9	2.3	1.6	0.7	0.3	0.3	-0.9	1.3	-0.5	1.9	0.6
June	-0.5	0.5	-0.2	0.6	0.1	0.8	-1.0	0.0	-3.7	-2.0	-0.6
July	-2.6	-0.2	1.9	2.7	1.0	0.4	0.9	2.8	-3.4	-1.0	0.3
August	-2.5	-0.3	-0.9	1.9	1.1	1.1	0.3	1.5	-2.6	0.0	0.0
September	-1.8	-0.9	-1.0	0.8	0.7	0.5	-1.1	0.2	-2.6	0.0	-0.5
October	0.4	-1.3	-0.3	0.6	0.6	0.2	-0.1	0.6	-1.7	0.1	-0.1
November	1.2	-1.4	-0.3	-0.1	0.2	-0.6	-1.6	-0.9	-2.3	-2.1	-0.8
December	0.0	0.0	-1.0	-1.0	-0.7	-0.8	-2.2	-1.4	-2.0	-0.8	-1.0
Overall	-0.4	0.2	-0.2	0.6	0.3	0.2	-0.6	0.7	-1.8	0.1	-0.1

2025 WESTERN AUSTRALIA - Estimated Maximum Temperature Trend

Region Town Sample	Kimberleys Broome	Pilbara Port Hedland	Gascoyne Carnarvon	Central West Geraldton	Lower & South West Perth	Great Southern Albany	Central Wheat Belt Merredin	Goldfields Kalgoorlie	Eucla Eucla	Giles	WA Overall
January	av	wmr	av	clr	clr	a-c	clr	clr	av	clr	a-c
February	av	wmr	clr	clr	a-w	av	wmr	wmr	wmr	wmr	wmr
March	clr	av	wmr	wmr	wmr	wmr	wmr	wmr	a-w	wmr	wmr
April	a-c	clr	wmr	a-w	a-w	av	wmr	wmr	wmr	av	a-w
May	wmr	wmr	av	av	a-w	av	wmr	wmr	a-w	wmr	a-w
June	a-w	wmr	wmr	wmr	a-w	a-w	wmr	av	av	av	a-w
July	a-c	a-c	a-w	wmr	wmr	wmr	wmr	wmr	a-w	av	a-w
August	clr	av	a-c	av	av	a-w	a-w	wmr	wmr	a-w	av
September	av	a-w	av	av	av	av	av	wmr	wmr	wmr	a-w
October	a-w	a-w	av	a-c	av	av	a-w	wmr	wmr	wmr	av
November	a-w	av	a-w	av	av	clr	av	a-c	av	clr	av
December	av	av	av	clr	clr	clr	clr	clr	a-w	clr	clr
Overall	av	av	av	av	av	av	a-w	wmr	wmr	av	av

2025 WESTERN AUSTRALIA - Estimated Minimum Temperature Trend

Region Town Sample	Kimberleys Broome	Pilbara Port Hedland	Gascoyne Carnarvon	Central West Geraldton	Lower & South West Perth	Great Southern Albany	Central Wheat Belt Merredin	Goldfields Kalgoorlie	Eucla Eucla	Giles	WA Overall
January	a-c	av	clr	a-c	clr	a-c	clr	clr	clr	clr	clr
February	av	av	a-c	av	a-w	av	av	wmr	av	wmr	a-w
March	a-c	wmr	a-w	wmr	wmr	wmr	wmr	wmr	av	wmr	wmr
April	wmr	wmr	a-c	av	a-c	av	clr	a-w	a-c	av	av
May	a-c	wmr	wmr	a-w	av	av	a-c	wmr	av	wmr	a-w
June	av	av	av	a-w	av	a-w	a-c	av	clr	clr	a-c
July	clr	av	wmr	wmr	a-w	av	a-w	wmr	clr	a-c	av
August	clr	av	a-c	wmr	wmr	wmr	av	wmr	clr	av	av
September	clr	a-c	a-c	a-w	a-w	av	clr	av	clr	av	av
October	av	clr	av	a-w	a-w	av	av	a-w	clr	av	av
November	wmr	clr	av	av	av	a-c	clr	a-c	clr	clr	a-c
December	av	av	a-c	a-c	a-c	a-c	clr	clr	clr	a-c	a-c
Overall	av	av	av	a-w	av	av	a-c	a-w	clr	av	av

MAPS SECTION 2025

ESTIMATED ONE MONTH / PAGE RAIN MAPS 2025

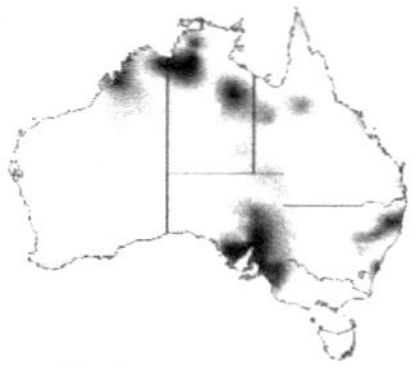
Wed 1st Jan 2025

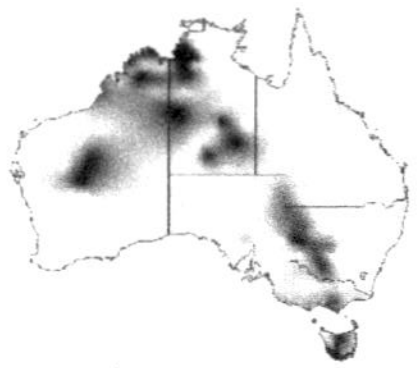
Thu 2nd Jan 2025

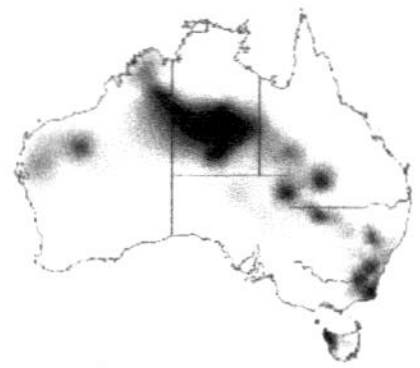
Fri 3rd Jan 2025

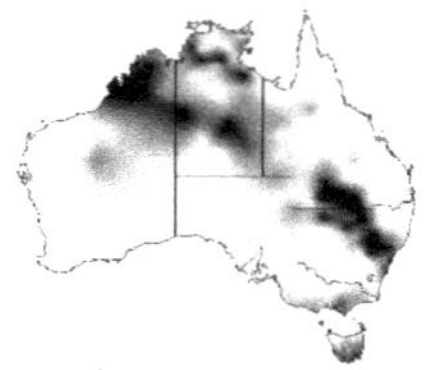
Sat 4th Jan 2025

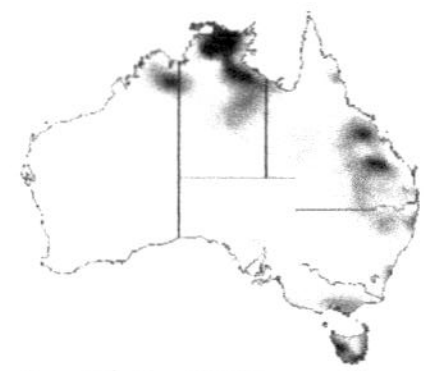
Sun 5th Jan 2025

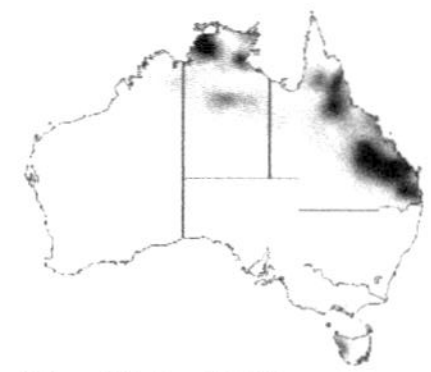
Mon 6th Jan 2025

Tue 7th Jan 2025

Wed 8th Jan 2025

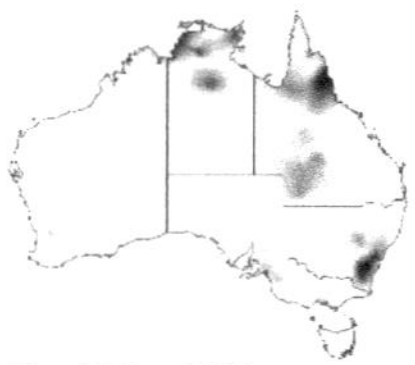
Thu 9th Jan 2025

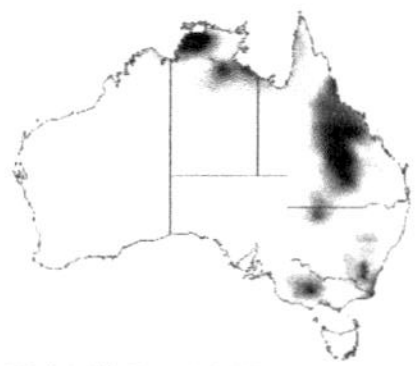
Fri 10th Jan 2025

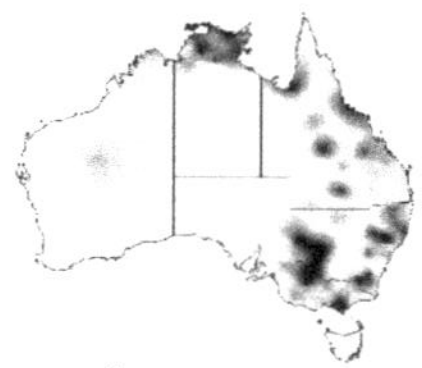
Sat 11th Jan 2025

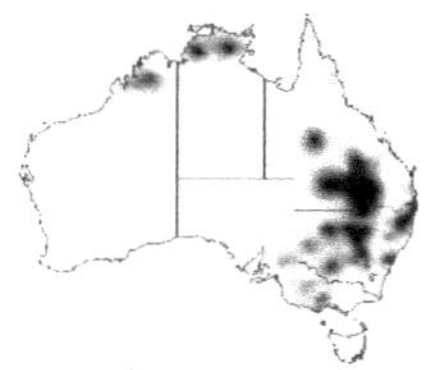
Sun 12th Jan 2025

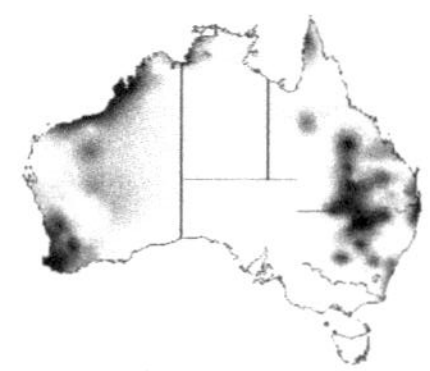
Mon 13th Jan 2025

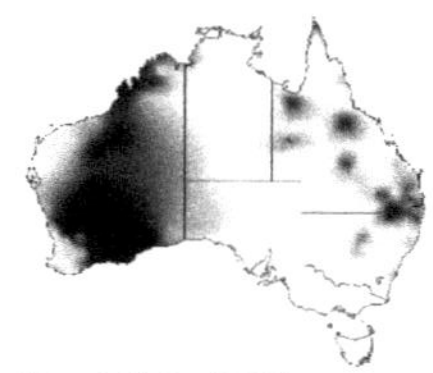
Tue 14th Jan 2025

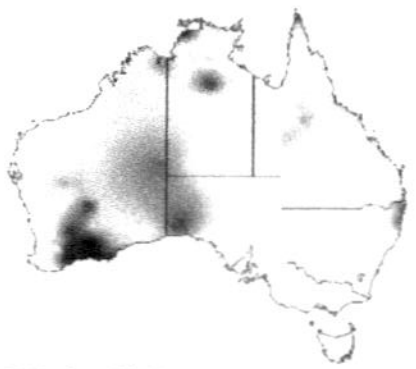
Wed 15th Jan 2025

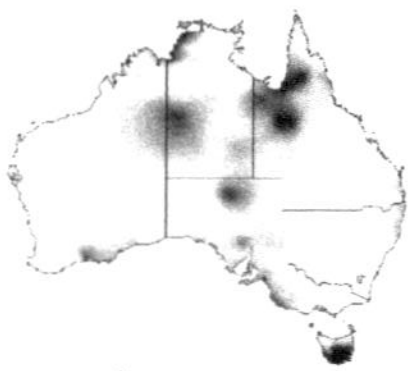
Thu 16th Jan 2025

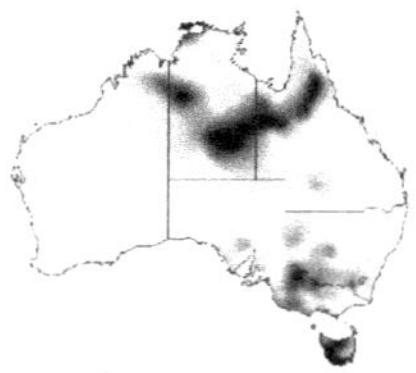
Fri 17th Jan 2025

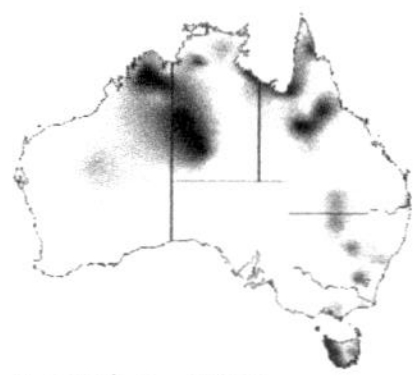
Sat 18th Jan 2025

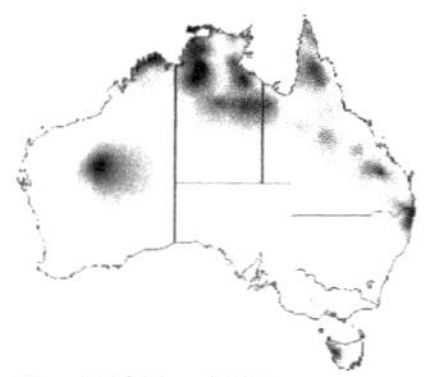
Sun 19th Jan 2025

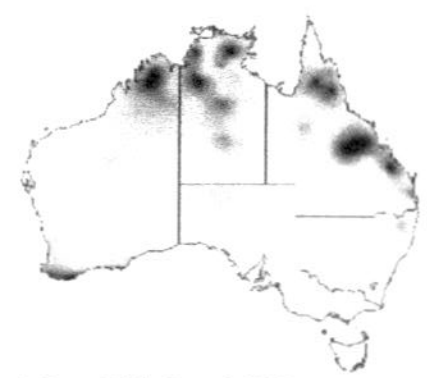
Mon 20th Jan 2025

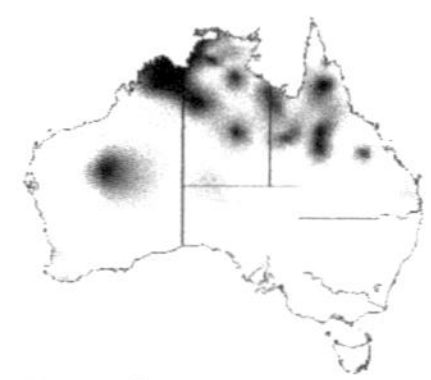
Tue 21st Jan 2025

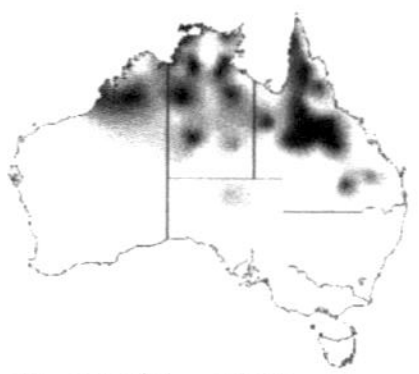
Wed 22nd Jan 2025

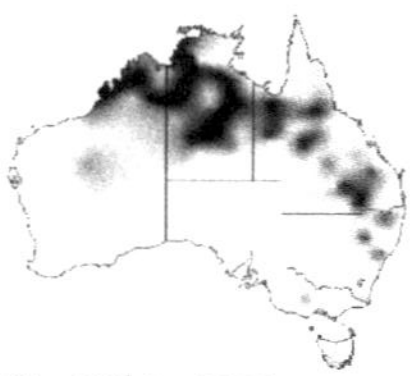
Thu 23rd Jan 2025

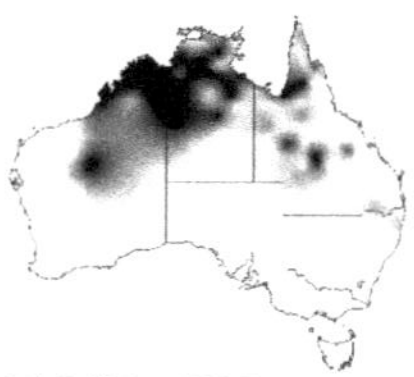
Fri 24th Jan 2025

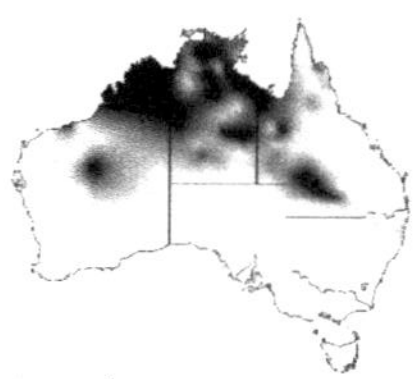
Sat 25th Jan 2025

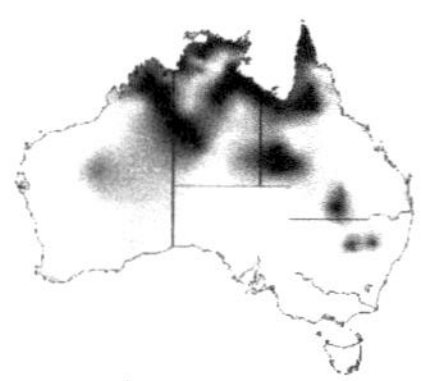
Sun 26th Jan 2025

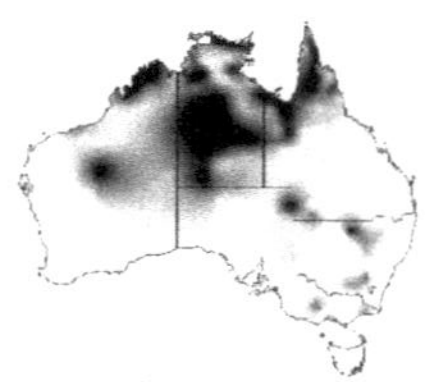
Mon 27th Jan 2025

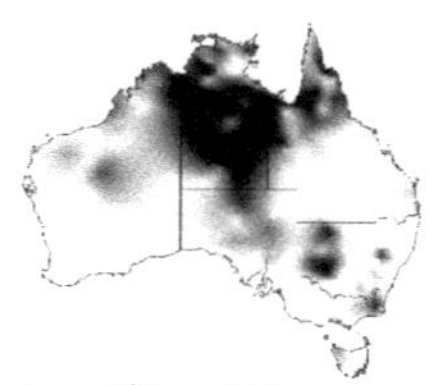
Tue 28th Jan 2025

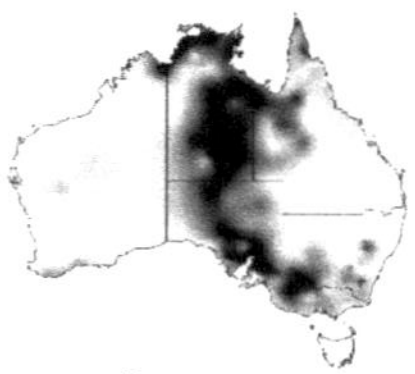
Wed 29th Jan 2025

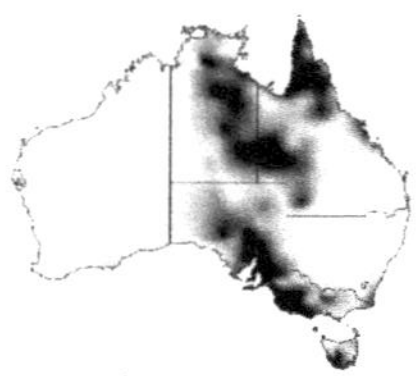
Thu 30th Jan 2025

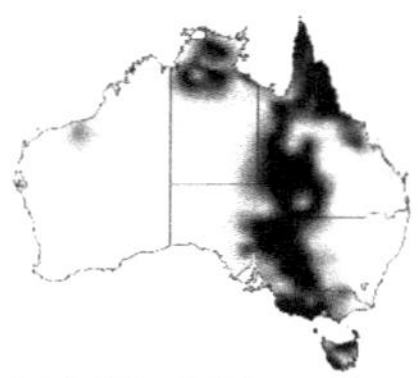
Fri 31st Jan 2025

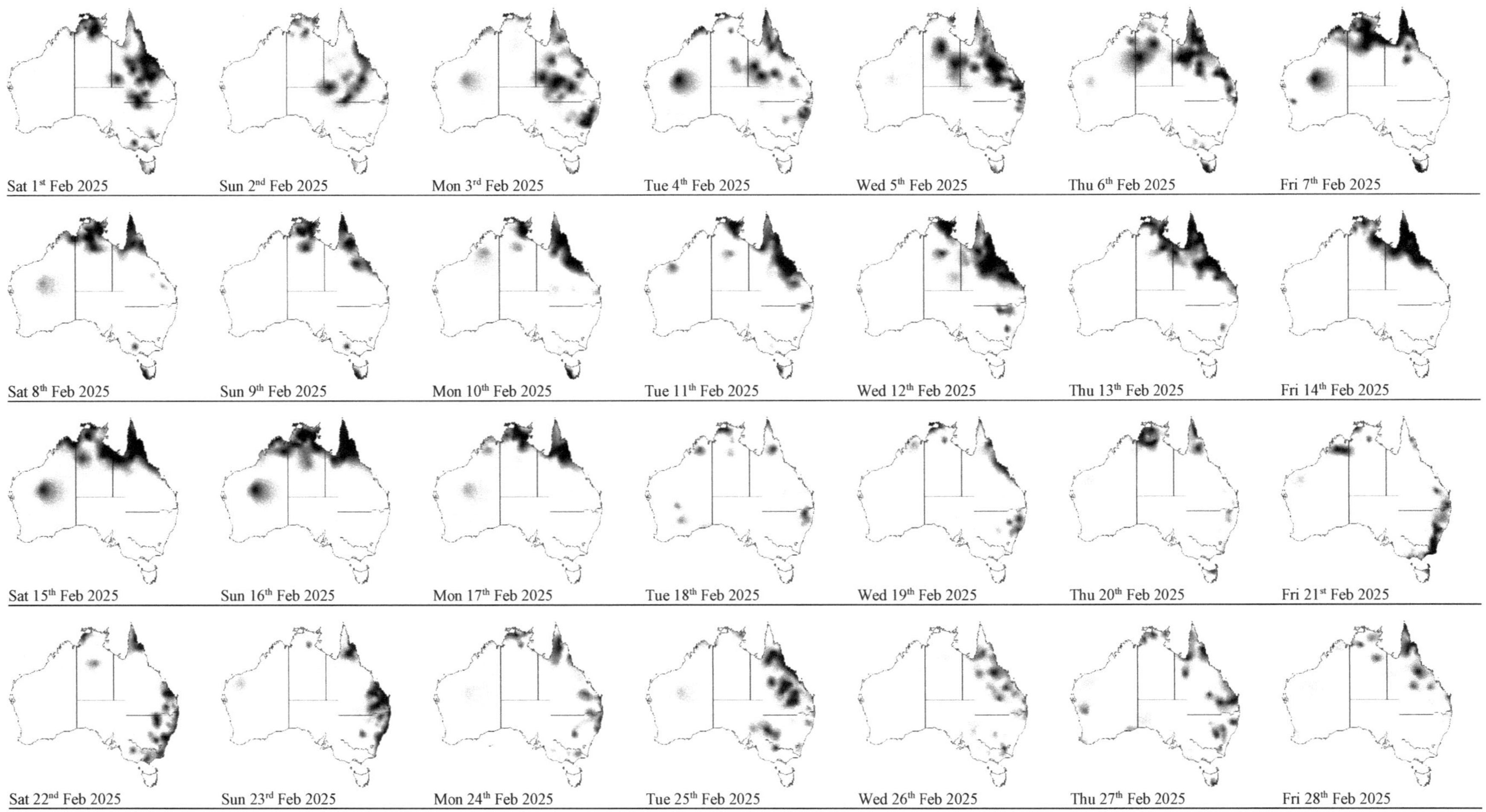

Sat 1st Feb 2025
Sun 2nd Feb 2025
Mon 3rd Feb 2025
Tue 4th Feb 2025
Wed 5th Feb 2025
Thu 6th Feb 2025
Fri 7th Feb 2025
Sat 8th Feb 2025
Sun 9th Feb 2025
Mon 10th Feb 2025
Tue 11th Feb 2025
Wed 12th Feb 2025
Thu 13th Feb 2025
Fri 14th Feb 2025
Sat 15th Feb 2025
Sun 16th Feb 2025
Mon 17th Feb 2025
Tue 18th Feb 2025
Wed 19th Feb 2025
Thu 20th Feb 2025
Fri 21st Feb 2025
Sat 22nd Feb 2025
Sun 23rd Feb 2025
Mon 24th Feb 2025
Tue 25th Feb 2025
Wed 26th Feb 2025
Thu 27th Feb 2025
Fri 28th Feb 2025

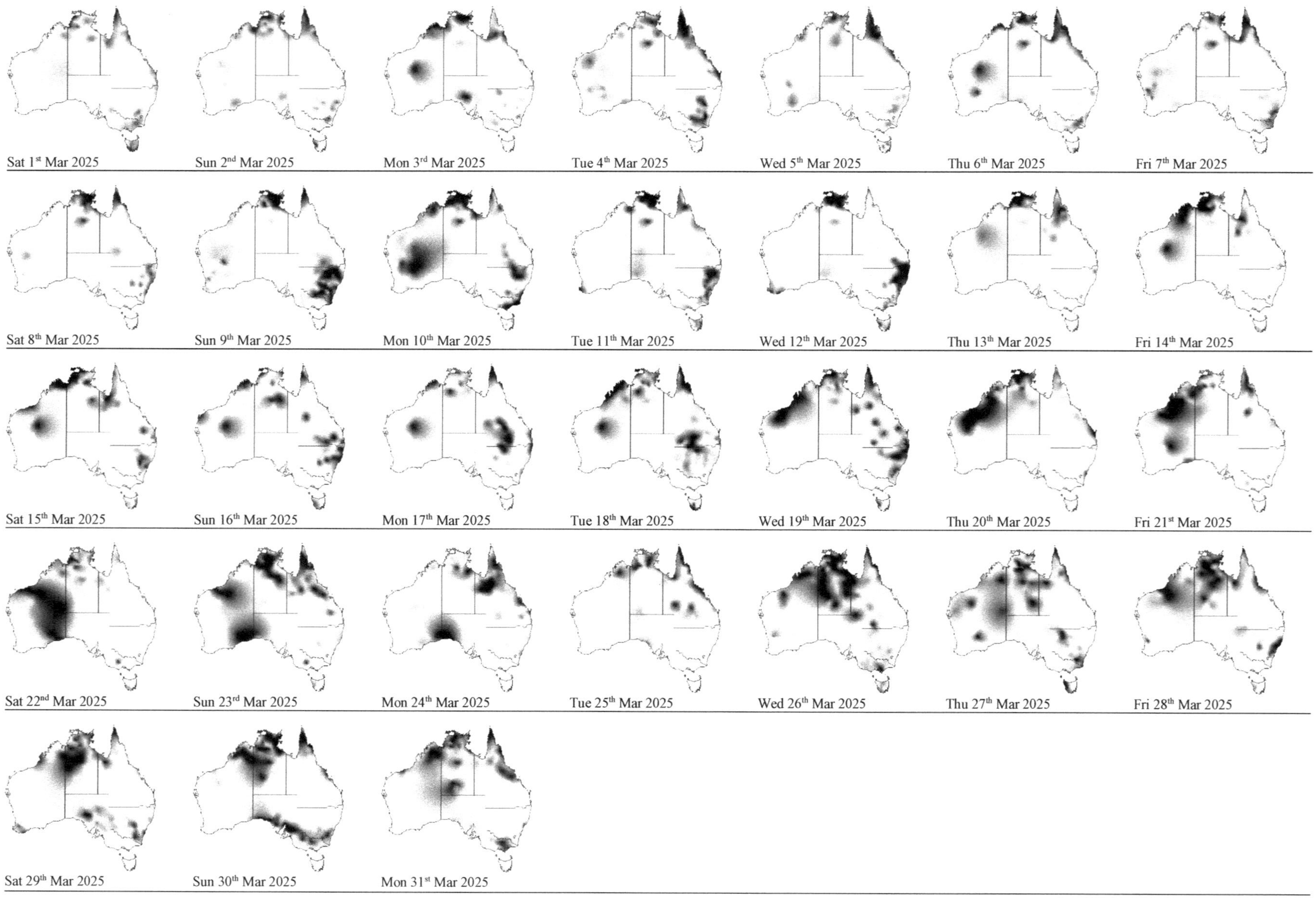
Sat 1st Mar 2025
Sun 2nd Mar 2025
Mon 3rd Mar 2025
Tue 4th Mar 2025
Wed 5th Mar 2025
Thu 6th Mar 2025
Fri 7th Mar 2025
Sat 8th Mar 2025
Sun 9th Mar 2025
Mon 10th Mar 2025
Tue 11th Mar 2025
Wed 12th Mar 2025
Thu 13th Mar 2025
Fri 14th Mar 2025
Sat 15th Mar 2025
Sun 16th Mar 2025
Mon 17th Mar 2025
Tue 18th Mar 2025
Wed 19th Mar 2025
Thu 20th Mar 2025
Fri 21st Mar 2025
Sat 22nd Mar 2025
Sun 23rd Mar 2025
Mon 24th Mar 2025
Tue 25th Mar 2025
Wed 26th Mar 2025
Thu 27th Mar 2025
Fri 28th Mar 2025
Sat 29th Mar 2025
Sun 30th Mar 2025
Mon 31st Mar 2025

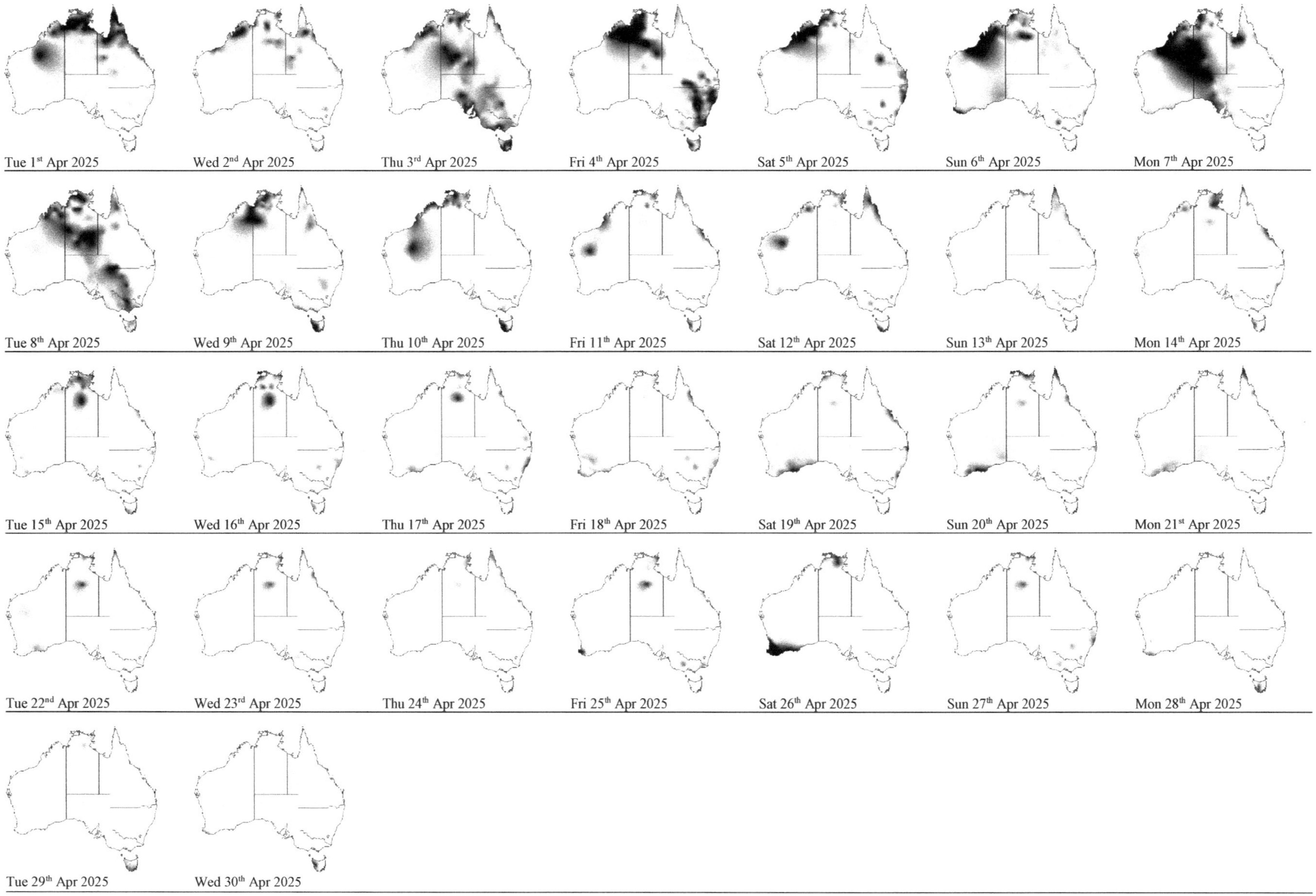
Tue 1st Apr 2025
Wed 2nd Apr 2025
Thu 3rd Apr 2025
Fri 4th Apr 2025
Sat 5th Apr 2025
Sun 6th Apr 2025
Mon 7th Apr 2025
Tue 8th Apr 2025
Wed 9th Apr 2025
Thu 10th Apr 2025
Fri 11th Apr 2025
Sat 12th Apr 2025
Sun 13th Apr 2025
Mon 14th Apr 2025
Tue 15th Apr 2025
Wed 16th Apr 2025
Thu 17th Apr 2025
Fri 18th Apr 2025
Sat 19th Apr 2025
Sun 20th Apr 2025
Mon 21st Apr 2025
Tue 22nd Apr 2025
Wed 23rd Apr 2025
Thu 24th Apr 2025
Fri 25th Apr 2025
Sat 26th Apr 2025
Sun 27th Apr 2025
Mon 28th Apr 2025
Tue 29th Apr 2025
Wed 30th Apr 2025

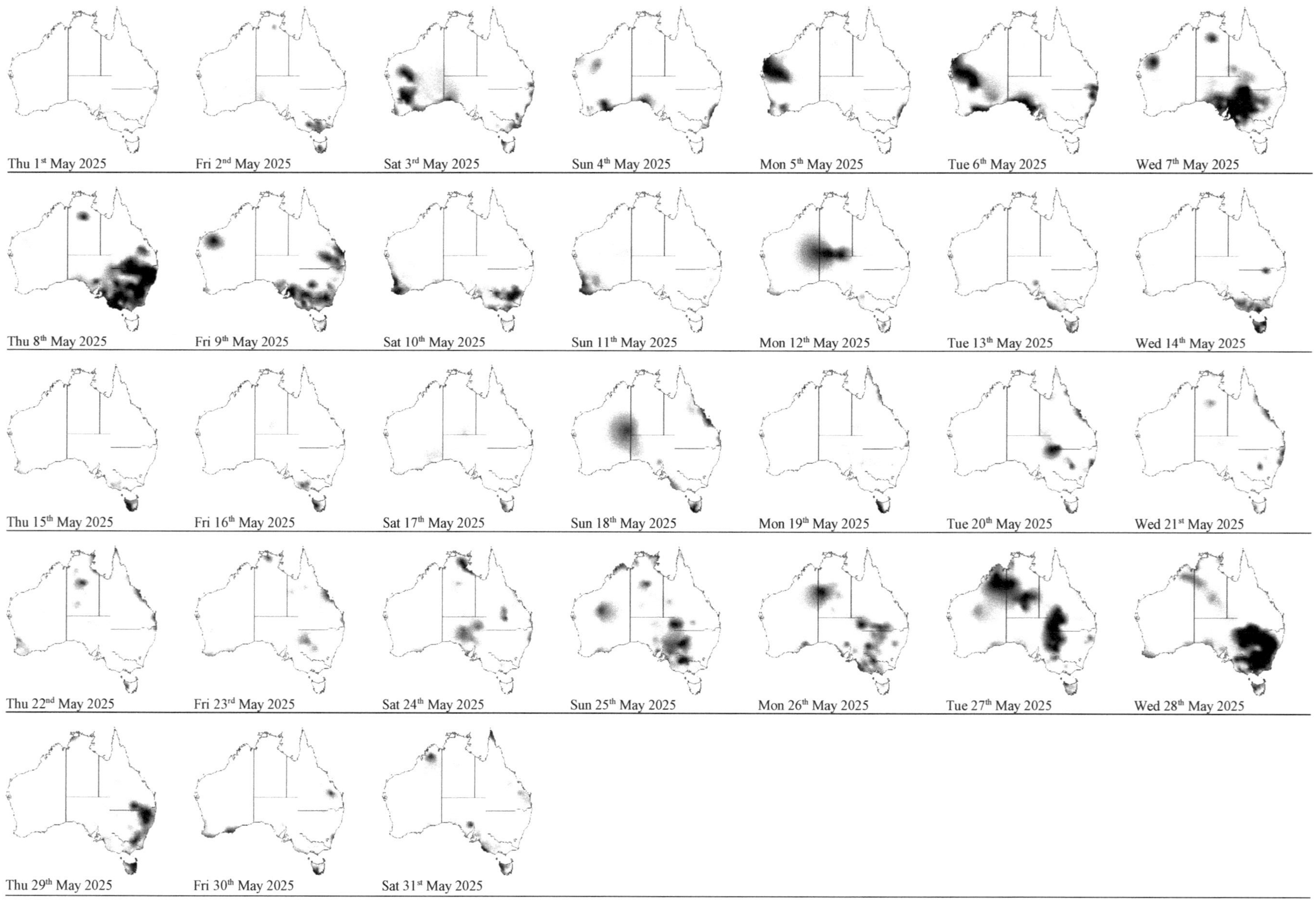

Thu 1st May 2025
Fri 2nd May 2025
Sat 3rd May 2025
Sun 4th May 2025
Mon 5th May 2025
Tue 6th May 2025
Wed 7th May 2025
Thu 8th May 2025
Fri 9th May 2025
Sat 10th May 2025
Sun 11th May 2025
Mon 12th May 2025
Tue 13th May 2025
Wed 14th May 2025
Thu 15th May 2025
Fri 16th May 2025
Sat 17th May 2025
Sun 18th May 2025
Mon 19th May 2025
Tue 20th May 2025
Wed 21st May 2025
Thu 22nd May 2025
Fri 23rd May 2025
Sat 24th May 2025
Sun 25th May 2025
Mon 26th May 2025
Tue 27th May 2025
Wed 28th May 2025
Thu 29th May 2025
Fri 30th May 2025
Sat 31st May 2025

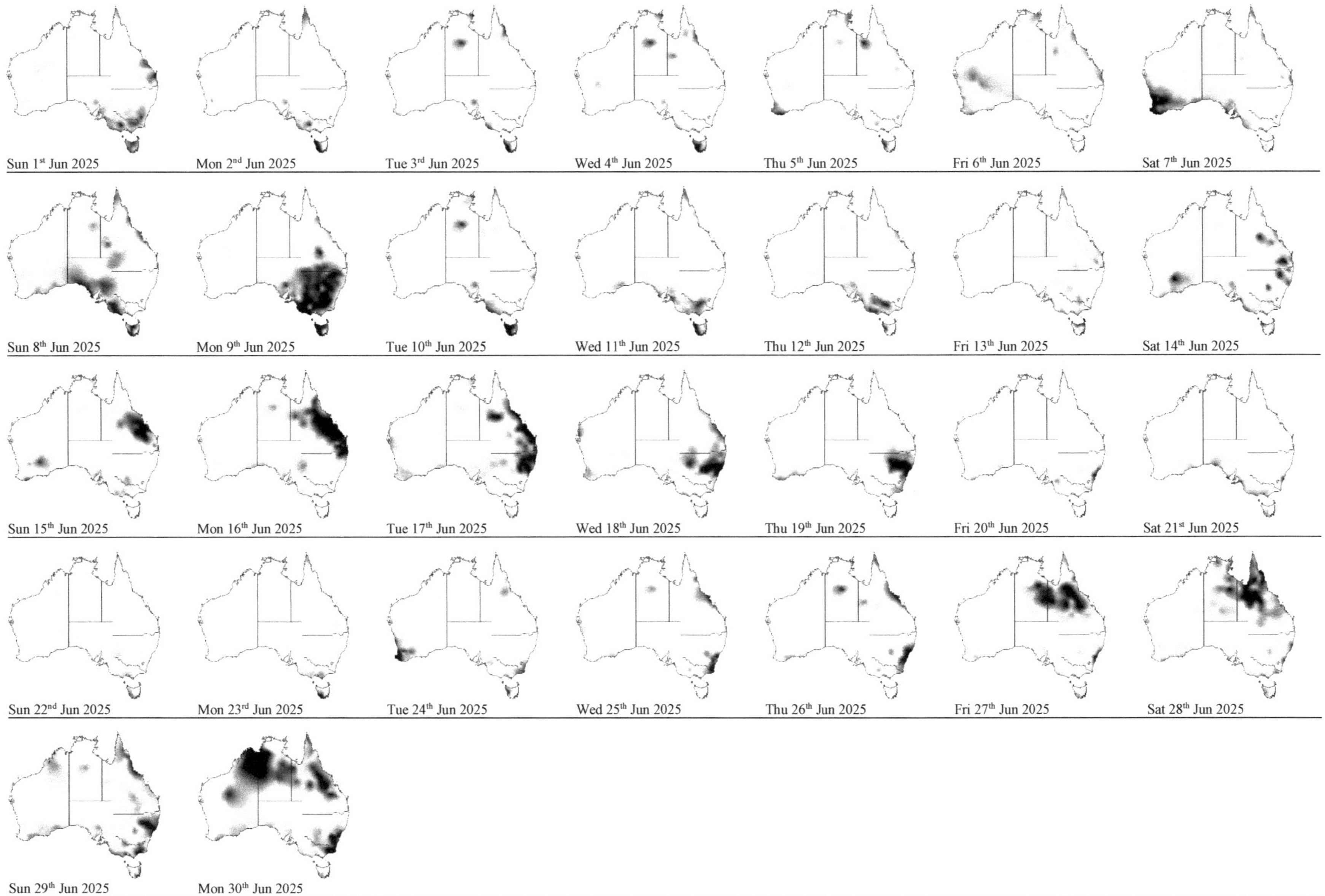

Sun 1st Jun 2025
Mon 2nd Jun 2025
Tue 3rd Jun 2025
Wed 4th Jun 2025
Thu 5th Jun 2025
Fri 6th Jun 2025
Sat 7th Jun 2025
Sun 8th Jun 2025
Mon 9th Jun 2025
Tue 10th Jun 2025
Wed 11th Jun 2025
Thu 12th Jun 2025
Fri 13th Jun 2025
Sat 14th Jun 2025
Sun 15th Jun 2025
Mon 16th Jun 2025
Tue 17th Jun 2025
Wed 18th Jun 2025
Thu 19th Jun 2025
Fri 20th Jun 2025
Sat 21st Jun 2025
Sun 22nd Jun 2025
Mon 23rd Jun 2025
Tue 24th Jun 2025
Wed 25th Jun 2025
Thu 26th Jun 2025
Fri 27th Jun 2025
Sat 28th Jun 2025
Sun 29th Jun 2025
Mon 30th Jun 2025

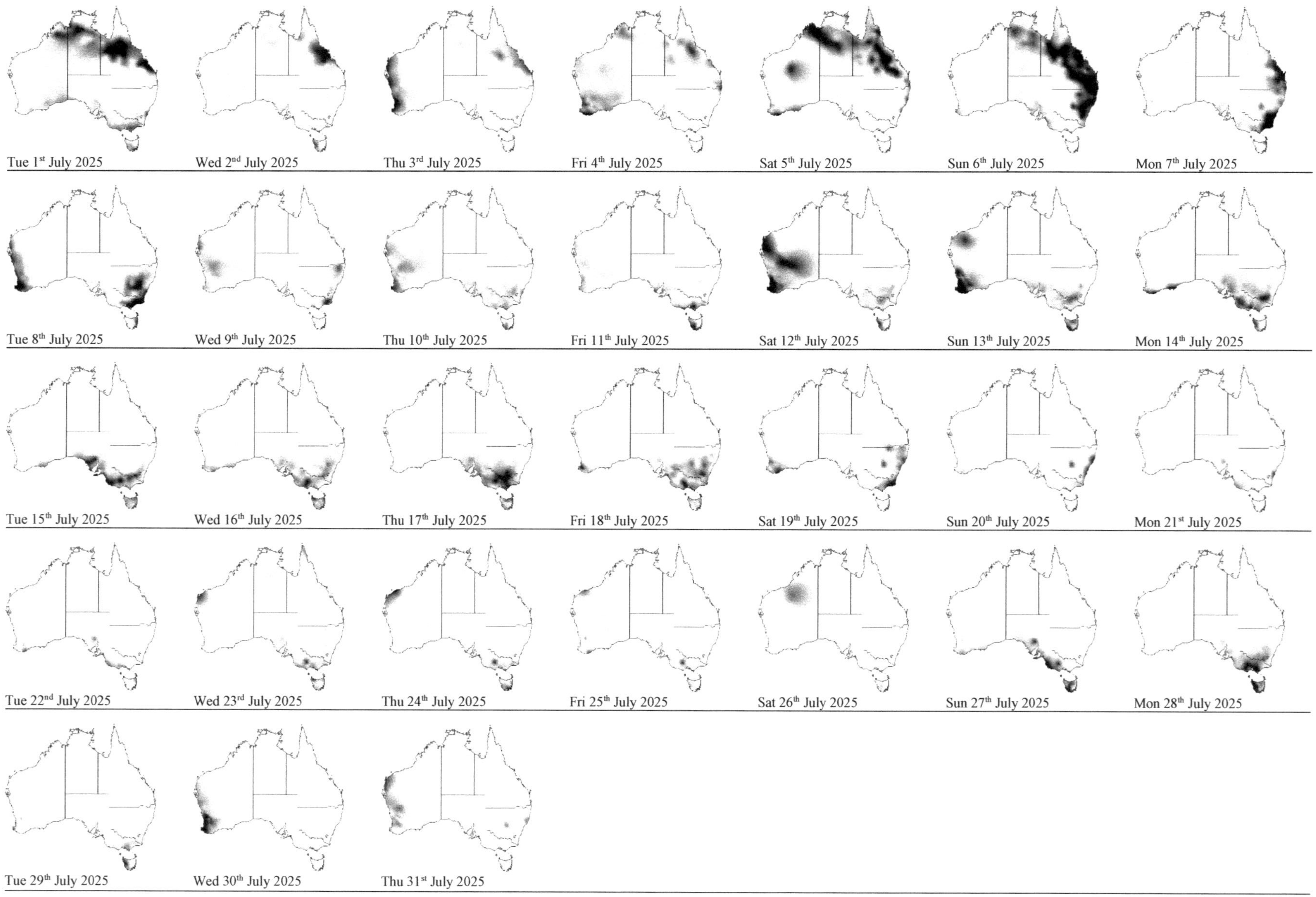
Tue 1st July 2025
Wed 2nd July 2025
Thu 3rd July 2025
Fri 4th July 2025
Sat 5th July 2025
Sun 6th July 2025
Mon 7th July 2025
Tue 8th July 2025
Wed 9th July 2025
Thu 10th July 2025
Fri 11th July 2025
Sat 12th July 2025
Sun 13th July 2025
Mon 14th July 2025
Tue 15th July 2025
Wed 16th July 2025
Thu 17th July 2025
Fri 18th July 2025
Sat 19th July 2025
Sun 20th July 2025
Mon 21st July 2025
Tue 22nd July 2025
Wed 23rd July 2025
Thu 24th July 2025
Fri 25th July 2025
Sat 26th July 2025
Sun 27th July 2025
Mon 28th July 2025
Tue 29th July 2025
Wed 30th July 2025
Thu 31st July 2025

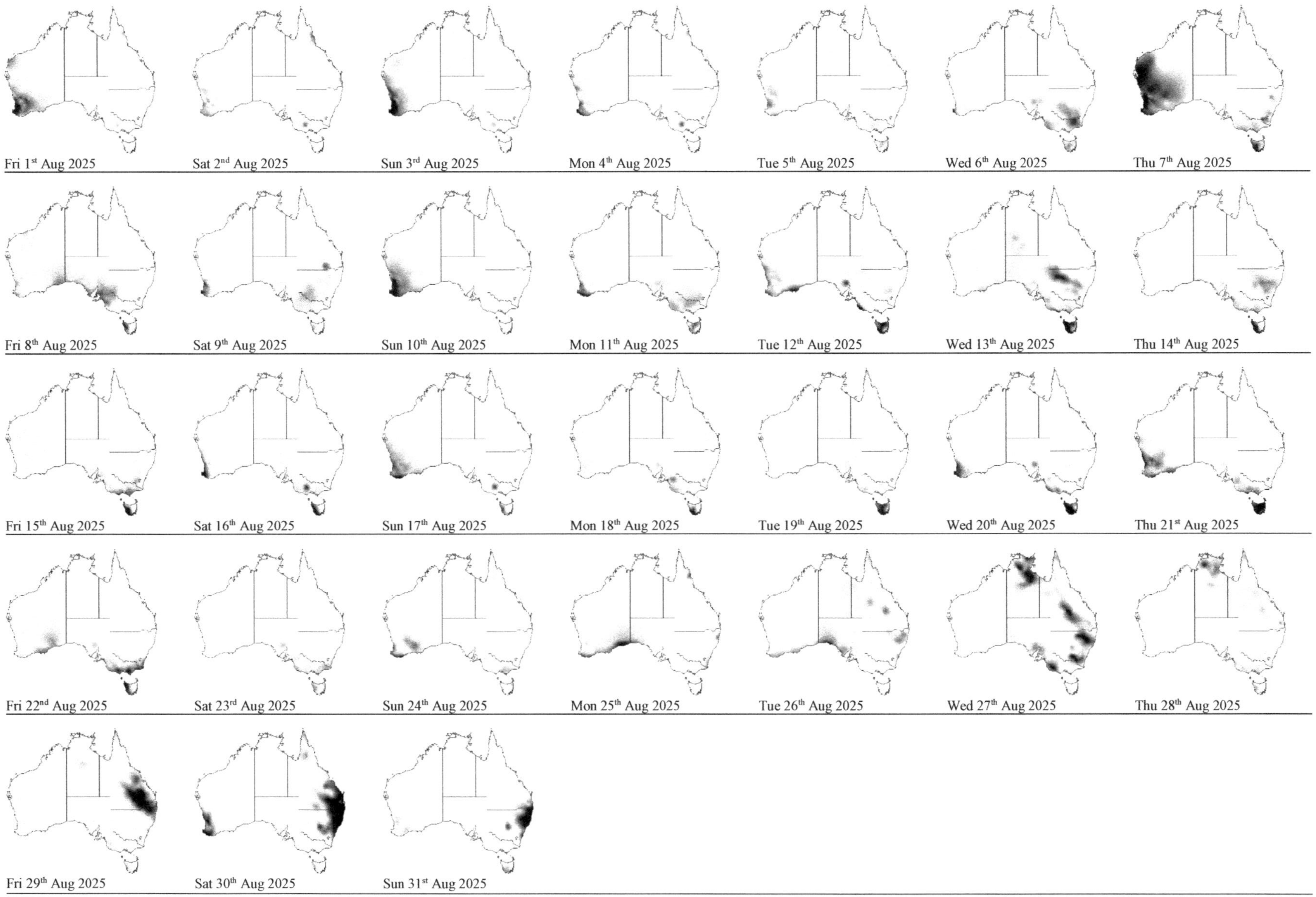
Fri 1st Aug 2025
Sat 2nd Aug 2025
Sun 3rd Aug 2025
Mon 4th Aug 2025
Tue 5th Aug 2025
Wed 6th Aug 2025
Thu 7th Aug 2025
Fri 8th Aug 2025
Sat 9th Aug 2025
Sun 10th Aug 2025
Mon 11th Aug 2025
Tue 12th Aug 2025
Wed 13th Aug 2025
Thu 14th Aug 2025
Fri 15th Aug 2025
Sat 16th Aug 2025
Sun 17th Aug 2025
Mon 18th Aug 2025
Tue 19th Aug 2025
Wed 20th Aug 2025
Thu 21st Aug 2025
Fri 22nd Aug 2025
Sat 23rd Aug 2025
Sun 24th Aug 2025
Mon 25th Aug 2025
Tue 26th Aug 2025
Wed 27th Aug 2025
Thu 28th Aug 2025
Fri 29th Aug 2025
Sat 30th Aug 2025
Sun 31st Aug 2025

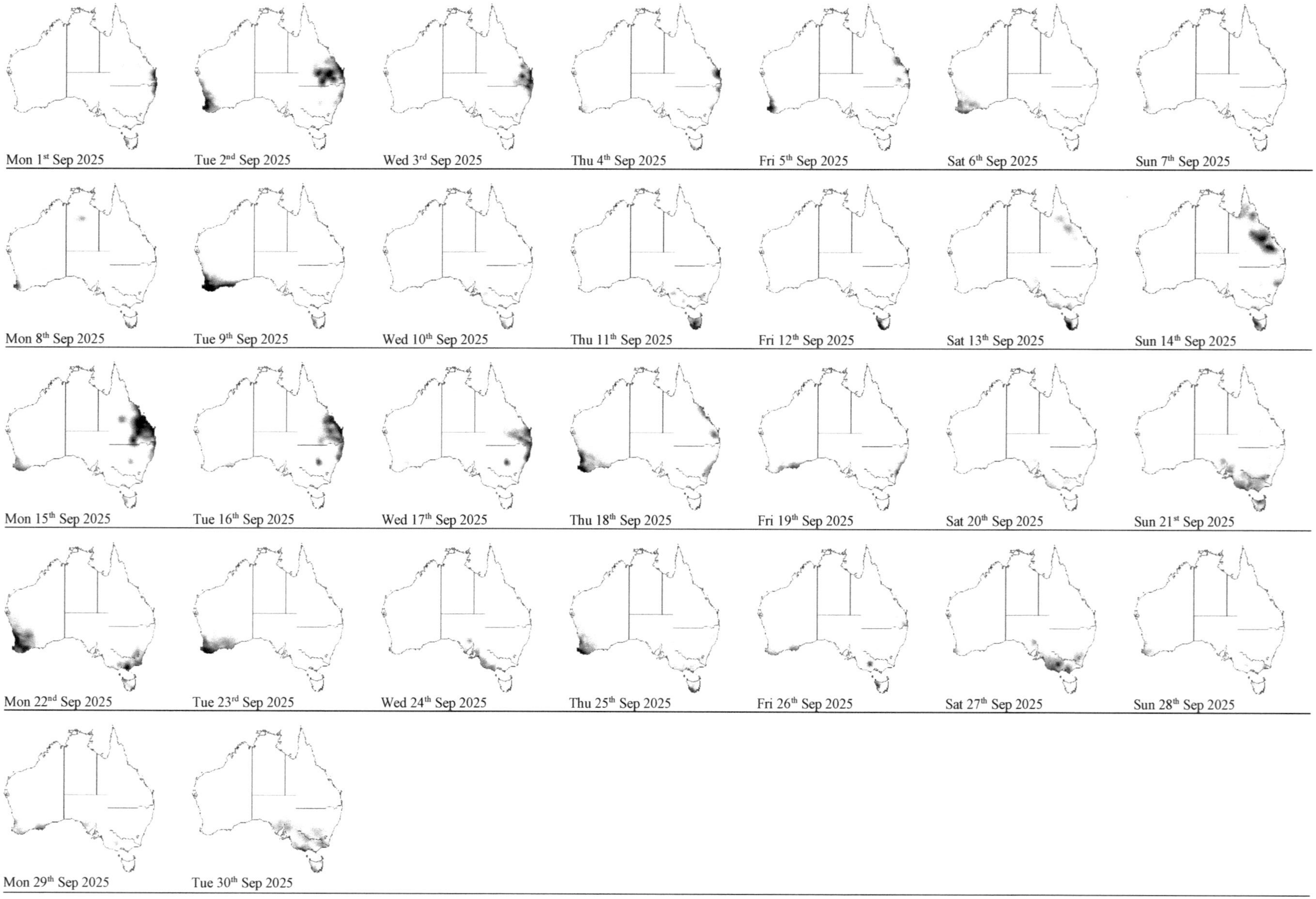

Mon 1st Sep 2025
Tue 2nd Sep 2025
Wed 3rd Sep 2025
Thu 4th Sep 2025
Fri 5th Sep 2025
Sat 6th Sep 2025
Sun 7th Sep 2025
Mon 8th Sep 2025
Tue 9th Sep 2025
Wed 10th Sep 2025
Thu 11th Sep 2025
Fri 12th Sep 2025
Sat 13th Sep 2025
Sun 14th Sep 2025
Mon 15th Sep 2025
Tue 16th Sep 2025
Wed 17th Sep 2025
Thu 18th Sep 2025
Fri 19th Sep 2025
Sat 20th Sep 2025
Sun 21st Sep 2025
Mon 22nd Sep 2025
Tue 23rd Sep 2025
Wed 24th Sep 2025
Thu 25th Sep 2025
Fri 26th Sep 2025
Sat 27th Sep 2025
Sun 28th Sep 2025
Mon 29th Sep 2025
Tue 30th Sep 2025

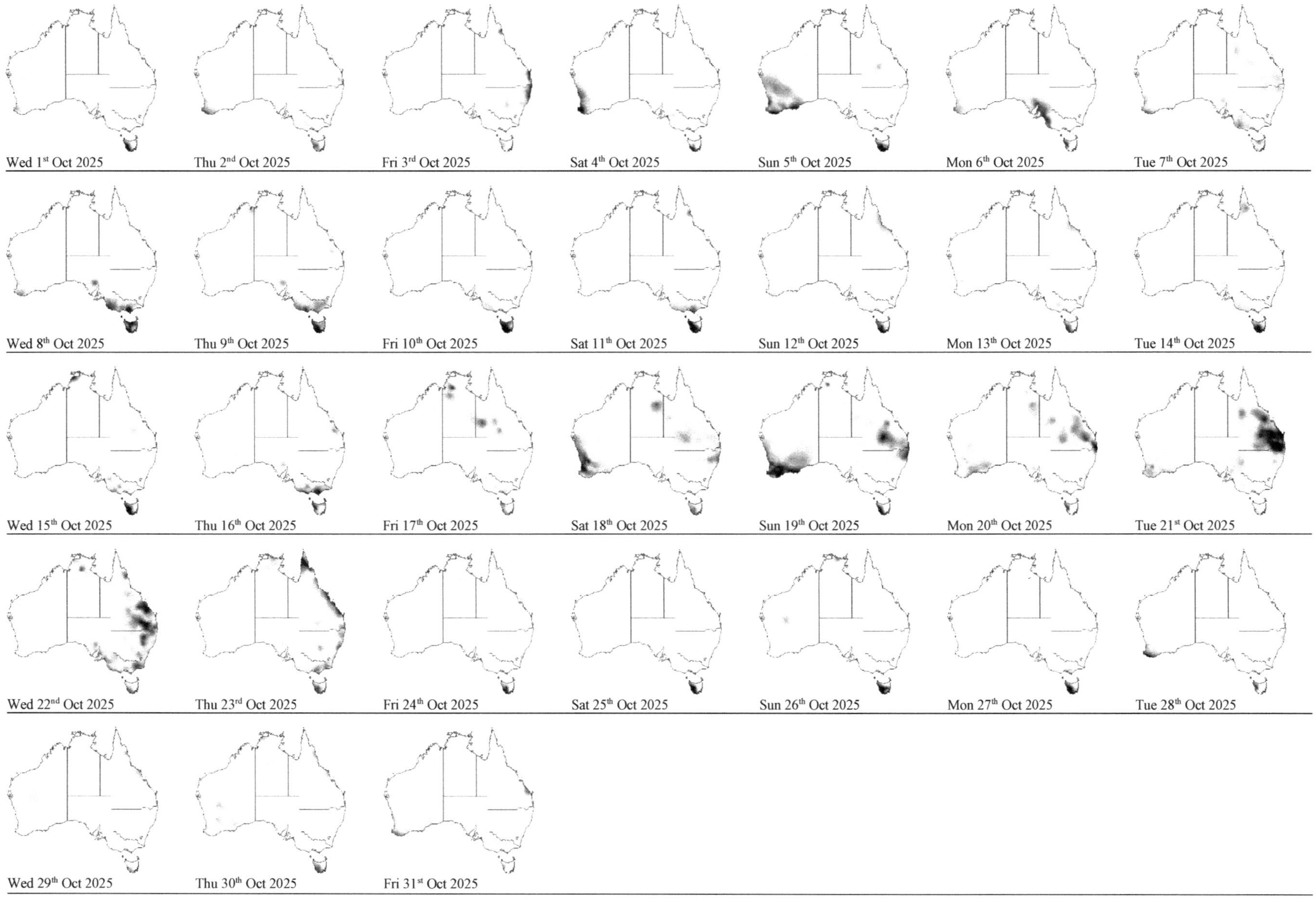
Wed 1st Oct 2025
Thu 2nd Oct 2025
Fri 3rd Oct 2025
Sat 4th Oct 2025
Sun 5th Oct 2025
Mon 6th Oct 2025
Tue 7th Oct 2025
Wed 8th Oct 2025
Thu 9th Oct 2025
Fri 10th Oct 2025
Sat 11th Oct 2025
Sun 12th Oct 2025
Mon 13th Oct 2025
Tue 14th Oct 2025
Wed 15th Oct 2025
Thu 16th Oct 2025
Fri 17th Oct 2025
Sat 18th Oct 2025
Sun 19th Oct 2025
Mon 20th Oct 2025
Tue 21st Oct 2025
Wed 22nd Oct 2025
Thu 23rd Oct 2025
Fri 24th Oct 2025
Sat 25th Oct 2025
Sun 26th Oct 2025
Mon 27th Oct 2025
Tue 28th Oct 2025
Wed 29th Oct 2025
Thu 30th Oct 2025
Fri 31st Oct 2025

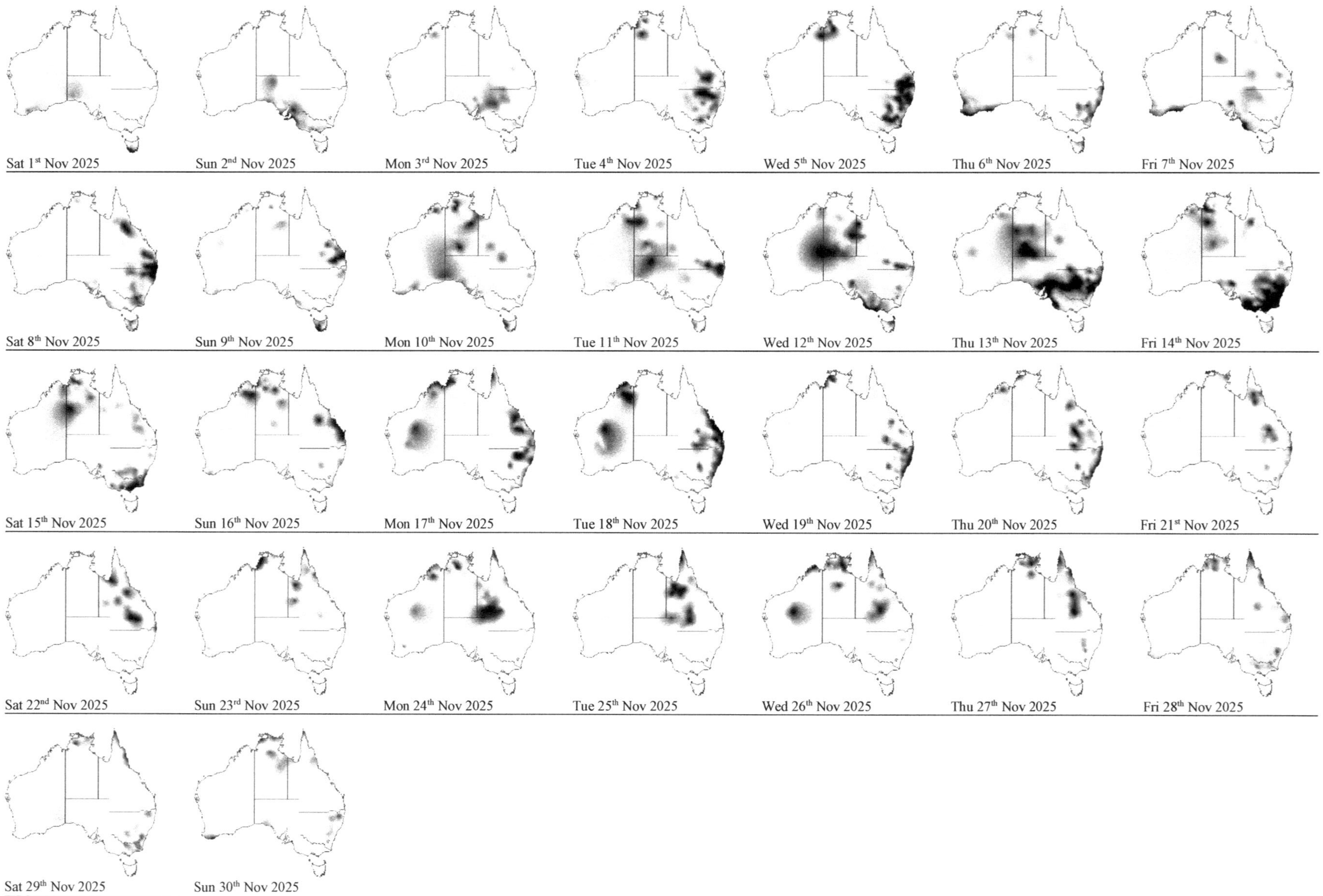

Sat 1st Nov 2025
Sun 2nd Nov 2025
Mon 3rd Nov 2025
Tue 4th Nov 2025
Wed 5th Nov 2025
Thu 6th Nov 2025
Fri 7th Nov 2025
Sat 8th Nov 2025
Sun 9th Nov 2025
Mon 10th Nov 2025
Tue 11th Nov 2025
Wed 12th Nov 2025
Thu 13th Nov 2025
Fri 14th Nov 2025
Sat 15th Nov 2025
Sun 16th Nov 2025
Mon 17th Nov 2025
Tue 18th Nov 2025
Wed 19th Nov 2025
Thu 20th Nov 2025
Fri 21st Nov 2025
Sat 22nd Nov 2025
Sun 23rd Nov 2025
Mon 24th Nov 2025
Tue 25th Nov 2025
Wed 26th Nov 2025
Thu 27th Nov 2025
Fri 28th Nov 2025
Sat 29th Nov 2025
Sun 30th Nov 2025

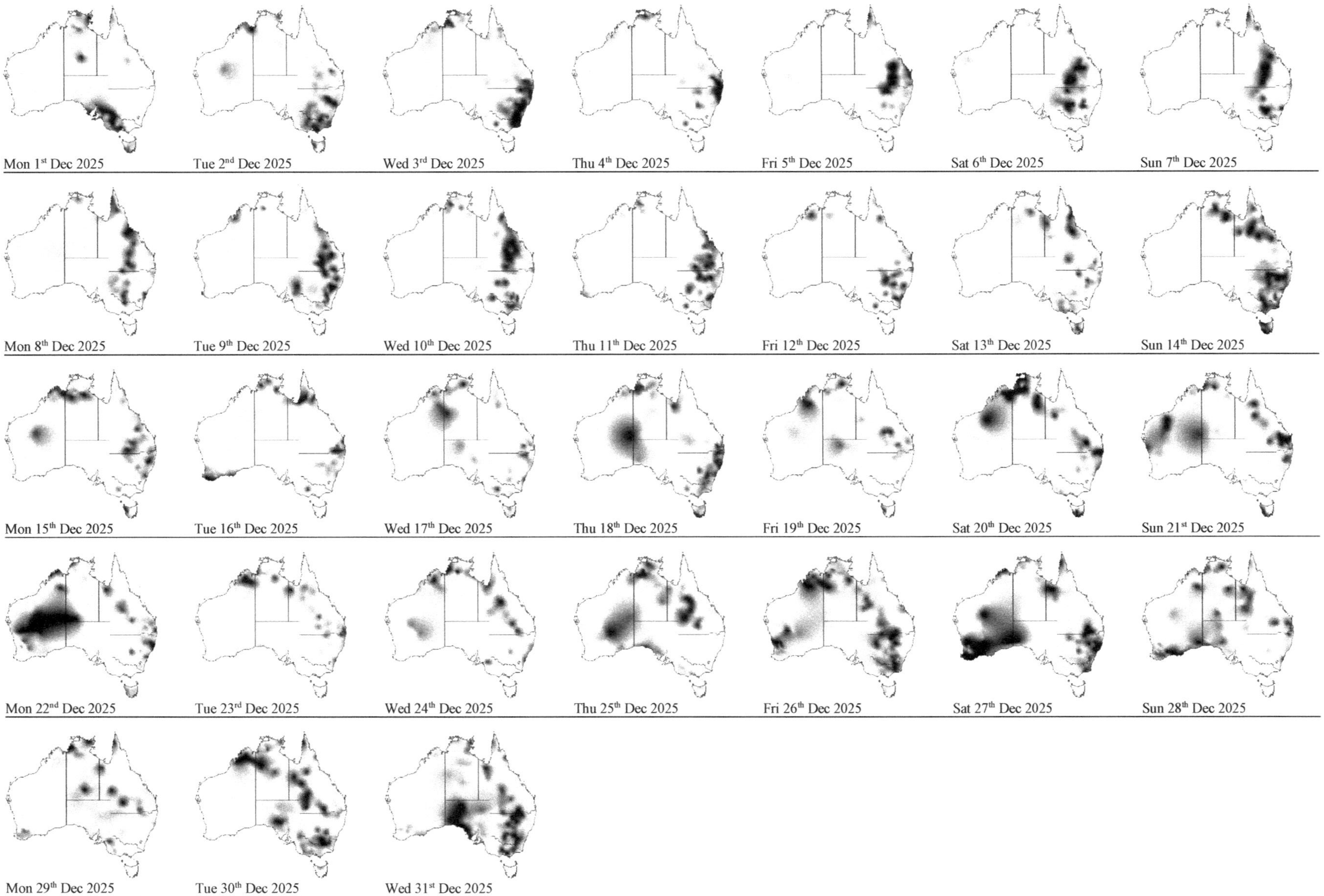
Mon 1st Dec 2025
Tue 2nd Dec 2025
Wed 3rd Dec 2025
Thu 4th Dec 2025
Fri 5th Dec 2025
Sat 6th Dec 2025
Sun 7th Dec 2025
Mon 8th Dec 2025
Tue 9th Dec 2025
Wed 10th Dec 2025
Thu 11th Dec 2025
Fri 12th Dec 2025
Sat 13th Dec 2025
Sun 14th Dec 2025
Mon 15th Dec 2025
Tue 16th Dec 2025
Wed 17th Dec 2025
Thu 18th Dec 2025
Fri 19th Dec 2025
Sat 20th Dec 2025
Sun 21st Dec 2025
Mon 22nd Dec 2025
Tue 23rd Dec 2025
Wed 24th Dec 2025
Thu 25th Dec 2025
Fri 26th Dec 2025
Sat 27th Dec 2025
Sun 28th Dec 2025
Mon 29th Dec 2025
Tue 30th Dec 2025
Wed 31st Dec 2025

ESTIMATED WEEKLY RAIN SUMMARY MAPS 2025

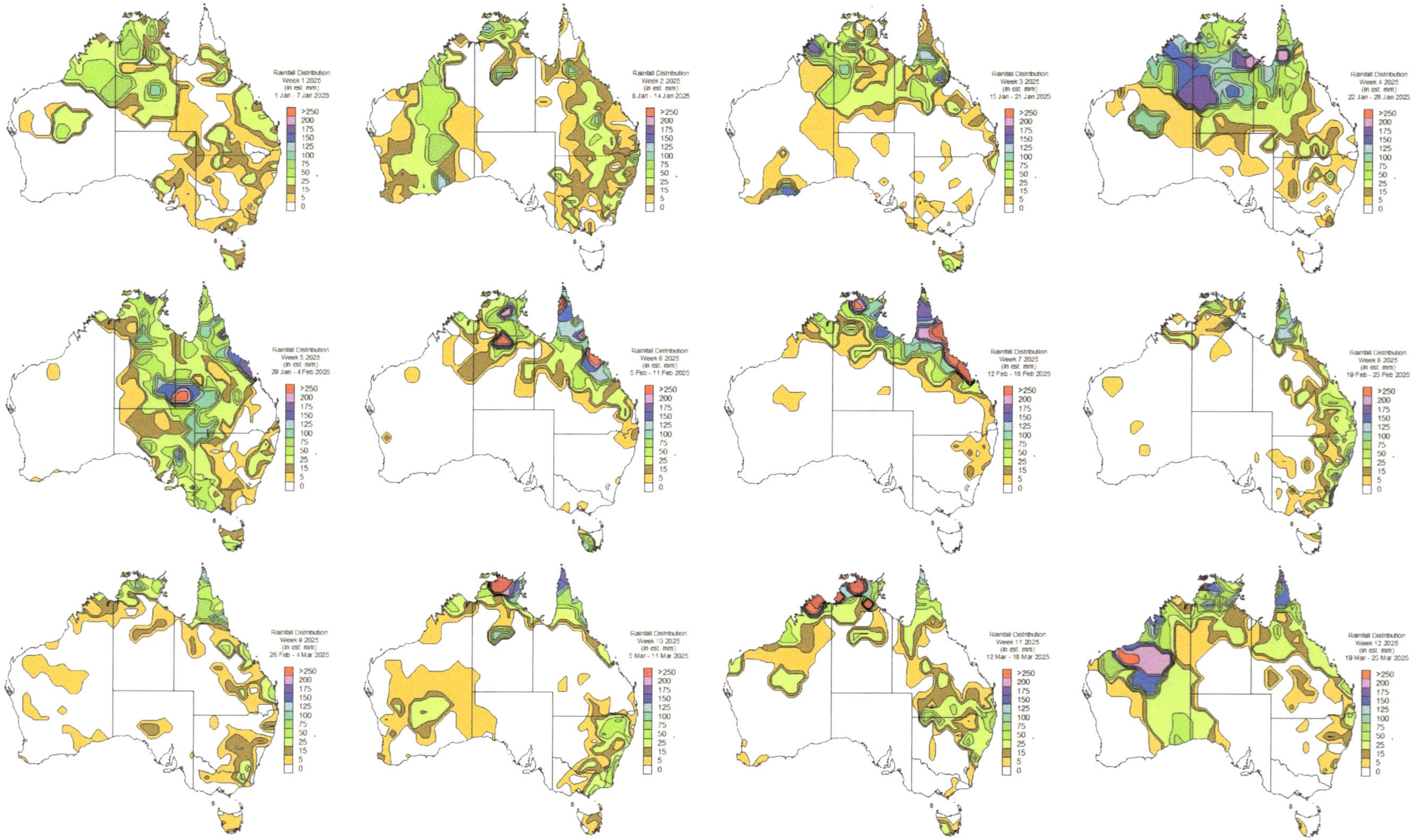

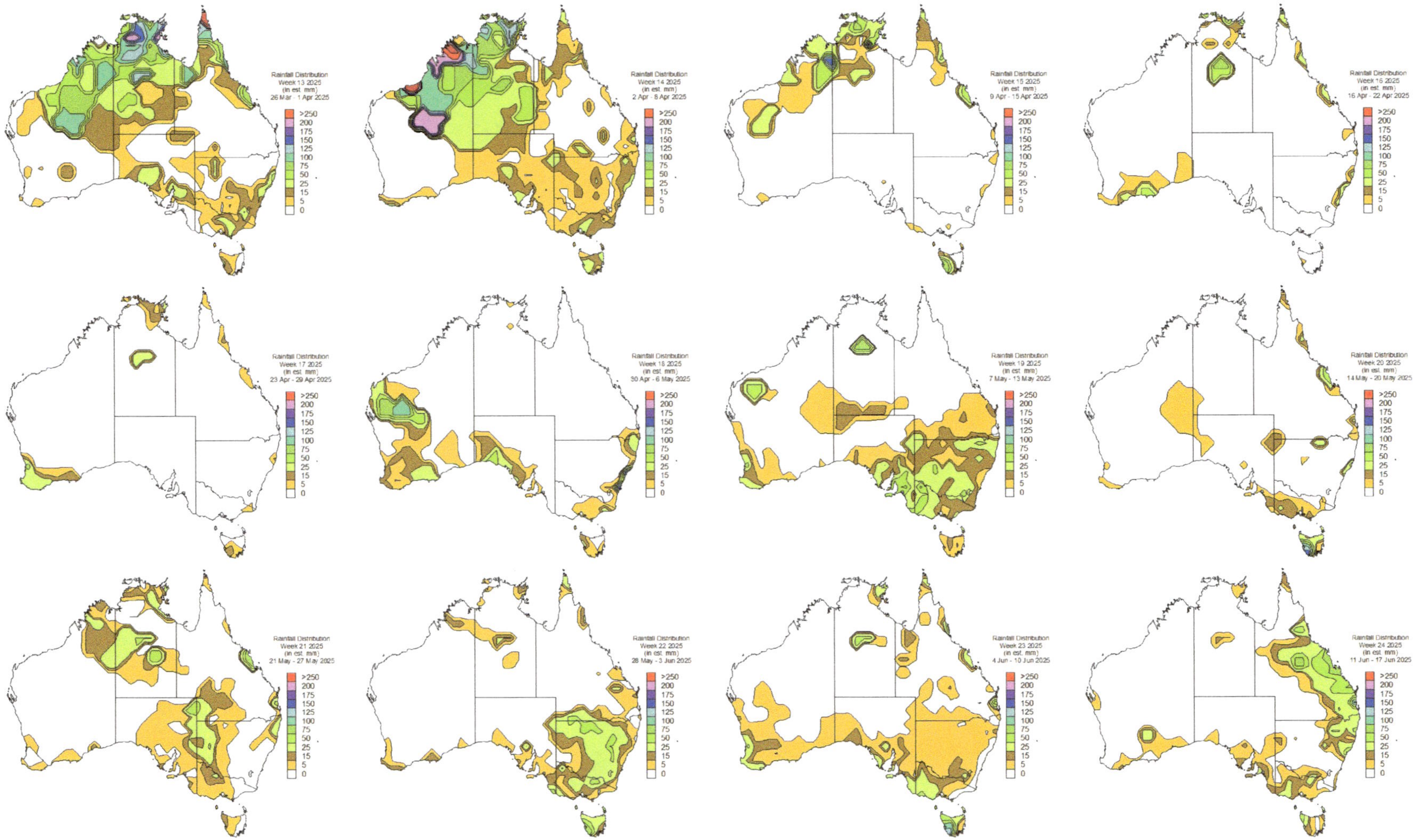
Rainfall Distribution
Week 13 2025
(in est. mm)
26 Mar - 1 Apr 2025
>250 200 175 150 125 100 75 50 25 15 5 0
Rainfall Distribution
Week 14 2025
(in est. mm)
2 Apr - 8 Apr 2025
>250 200 175 150 125 100 75 50 25 15 5 0
Rainfall Distribution
Week 15 2025
(in est. mm)
9 Apr - 15 Apr 2025
>250 200 175 150 125 100 75 50 25 15 5 0
Rainfall Distribution
Week 16 2025
(in est. mm)
16 Apr - 22 Apr 2025
>250 200 175 150 125 100 75 50 25 15 5 0
Rainfall Distribution
Week 17 2025
(in est. mm)
23 Apr - 29 Apr 2025
>250 200 175 150 125 100 75 50 25 15 5 0
Rainfall Distribution
Week 18 2025
(in est. mm)
30 Apr - 6 May 2025
>250 200 175 150 125 100 75 50 25 15 5 0
Rainfall Distribution
Week 19 2025
(in est. mm)
7 May - 13 May 2025
>250 200 175 150 125 100 75 50 25 15 5 0
Rainfall Distribution
Week 20 2025
(in est. mm)
14 May - 20 May 2025
>250 200 175 150 125 100 75 50 25 15 5 0
Rainfall Distribution
Week 21 2025
(in est. mm)
21 May - 27 May 2025
>250 200 175 150 125 100 75 50 25 15 5 0
Rainfall Distribution
Week 22 2025
(in est. mm)
28 May - 3 Jun 2025
>250 200 175 150 125 100 75 50 25 15 5 0
Rainfall Distribution
Week 23 2025
(in est. mm)
4 Jun - 10 Jun 2025
>250 200 175 150 125 100 75 50 25 15 5 0
Rainfall Distribution
Week 24 2025
(in est. mm)
11 Jun - 17 Jun 2025
>250 200 175 150 125 100 75 50 25 15 5 0

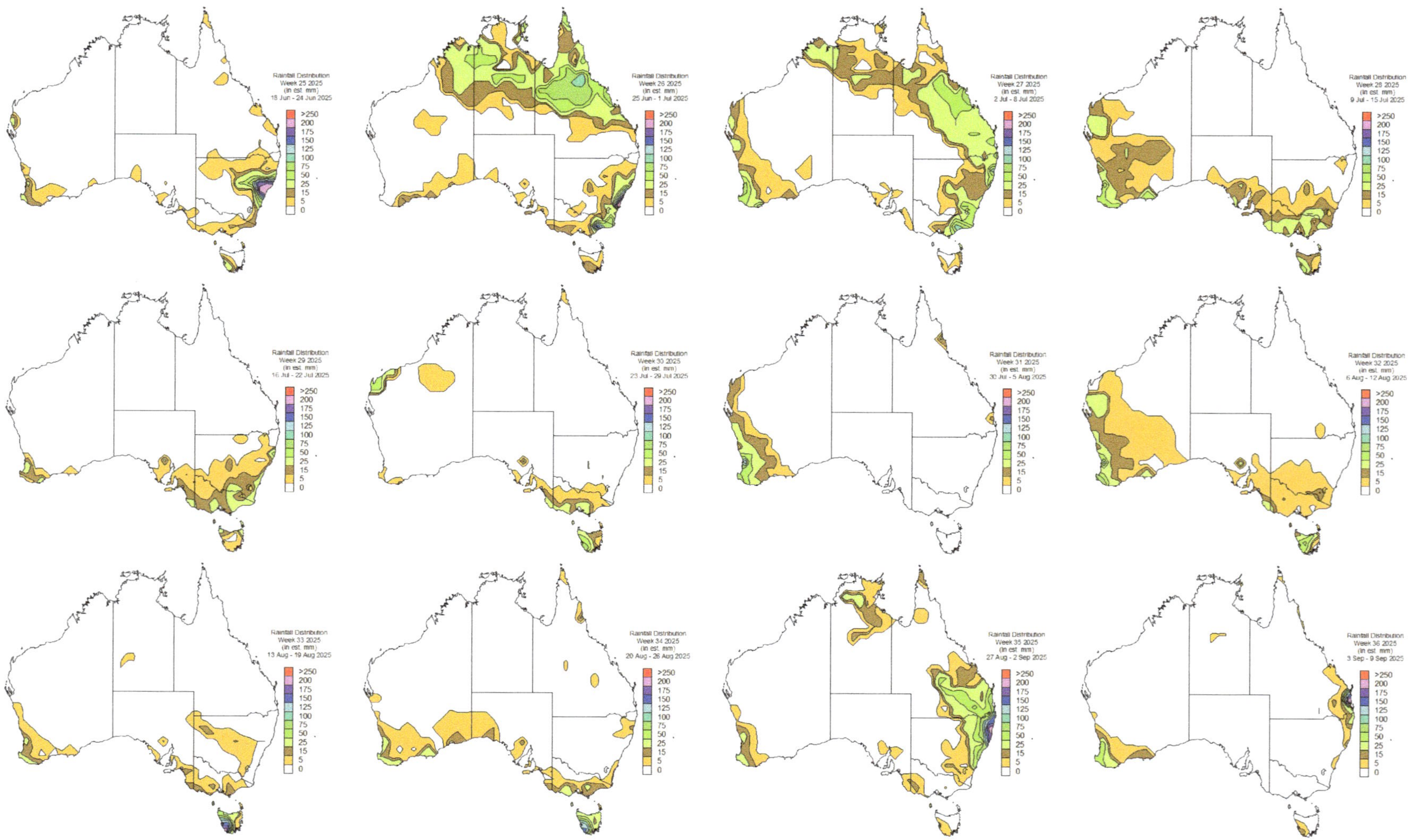
Rainfall Distribution
Week 25 2025
(in est. mm)
18 Jun - 24 Jun 2025
>250
200
175
150
125
100
75
50
25
15
5
0
Rainfall Distribution
Week 26 2025
(in est. mm)
25 Jun - 1 Jul 2025
>250
200
175
150
125
100
75
50
25
15
5
0
Rainfall Distribution
Week 27 2025
(in est. mm)
2 Jul - 8 Jul 2025
>250
200
175
150
125
100
75
50
25
15
5
0
Rainfall Distribution
Week 28 2025
(in est. mm)
9 Jul - 15 Jul 2025
>250
200
175
150
125
100
75
50
25
15
5
0
Rainfall Distribution
Week 29 2025
(in est. mm)
16 Jul - 22 Jul 2025
>250
200
175
150
125
100
75
50
25
15
5
0
Rainfall Distribution
Week 30 2025
(in est. mm)
23 Jul - 29 Jul 2025
>250
200
175
150
125
100
75
50
25
15
5
0
Rainfall Distribution
Week 31 2025
(in est. mm)
30 Jul - 5 Aug 2025
>250
200
175
150
125
100
75
50
25
15
5
0
Rainfall Distribution
Week 32 2025
(in est. mm)
6 Aug - 12 Aug 2025
>250
200
175
150
125
100
75
50
25
15
5
0
Rainfall Distribution
Week 33 2025
(in est. mm)
13 Aug - 19 Aug 2025
>250
200
175
150
125
100
75
50
25
15
5
0
Rainfall Distribution
Week 34 2025
(in est. mm)
20 Aug - 26 Aug 2025
>250
200
175
150
125
100
75
50
25
15
5
0
Rainfall Distribution
Week 35 2025
(in est. mm)
27 Aug - 2 Sep 2025
>250
200
175
150
125
100
75
50
25
15
5
0
Rainfall Distribution
Week 36 2025
(in est. mm)
3 Sep - 9 Sep 2025
>250
200
175
150
125
100
75
50
25
15
5
0

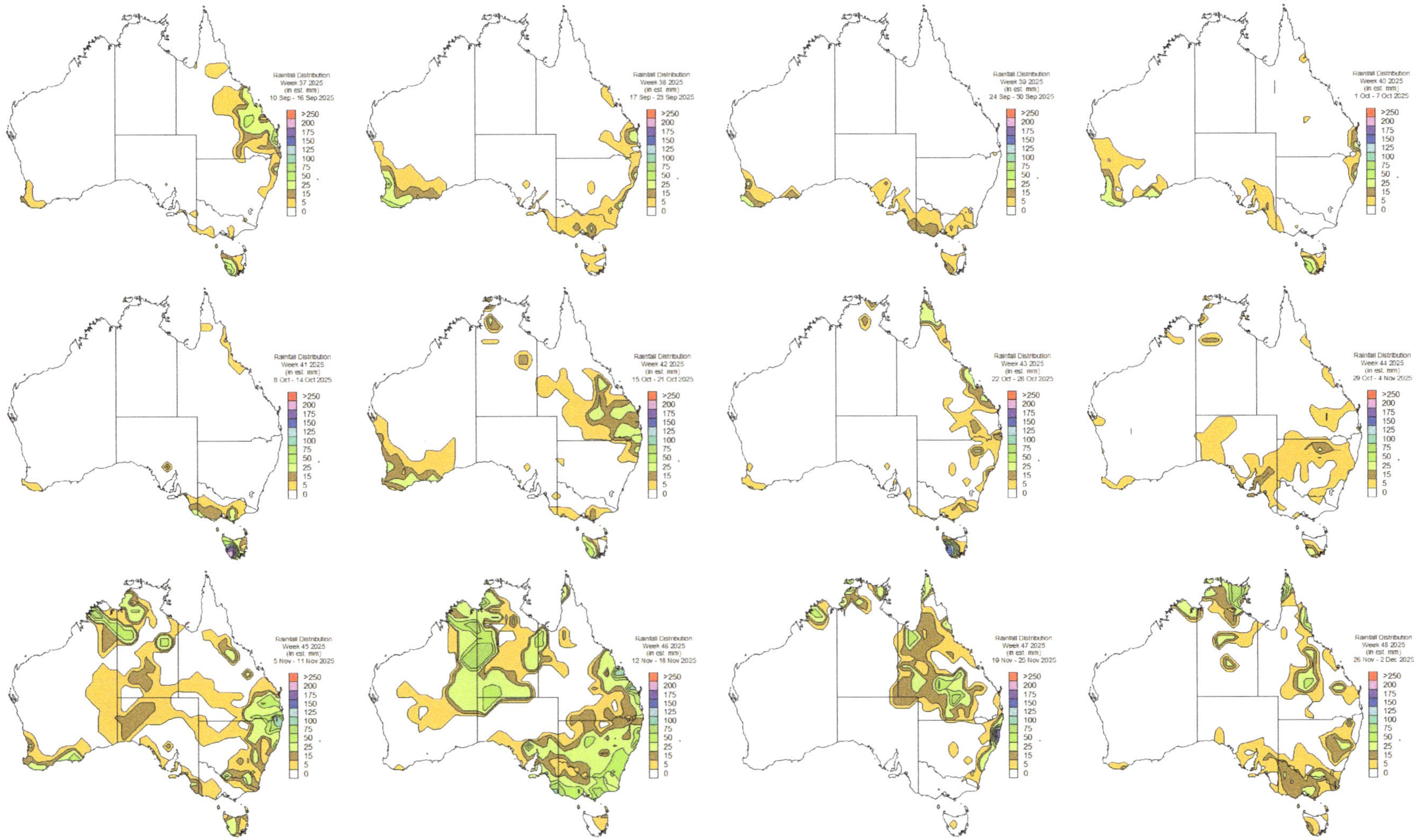
Rainfall Distribution
Week 37 2025
(in est. mm)
10 Sep - 16 Sep 2025
Rainfall Distribution
Week 38 2025
(in est. mm)
17 Sep - 23 Sep 2025
Rainfall Distribution
Week 39 2025
(in est. mm)
24 Sep - 30 Sep 2025
Rainfall Distribution
Week 40 2025
(in est. mm)
1 Oct - 7 Oct 2025
Rainfall Distribution
Week 41 2025
(in est. mm)
8 Oct - 14 Oct 2025
Rainfall Distribution
Week 42 2025
(in est. mm)
15 Oct - 21 Oct 2025
Rainfall Distribution
Week 43 2025
(in est. mm)
22 Oct - 28 Oct 2025
Rainfall Distribution
Week 44 2025
(in est. mm)
29 Oct - 4 Nov 2025
Rainfall Distribution
Week 45 2025
(in est. mm)
5 Nov - 11 Nov 2025
Rainfall Distribution
Week 46 2025
(in est. mm)
12 Nov - 18 Nov 2025
Rainfall Distribution
Week 47 2025
(in est. mm)
19 Nov - 25 Nov 2025
Rainfall Distribution
Week 48 2025
(in est. mm)
26 Nov - 2 Dec 2025
>250
200
175
150
125
100
75
50
25
15
5
0

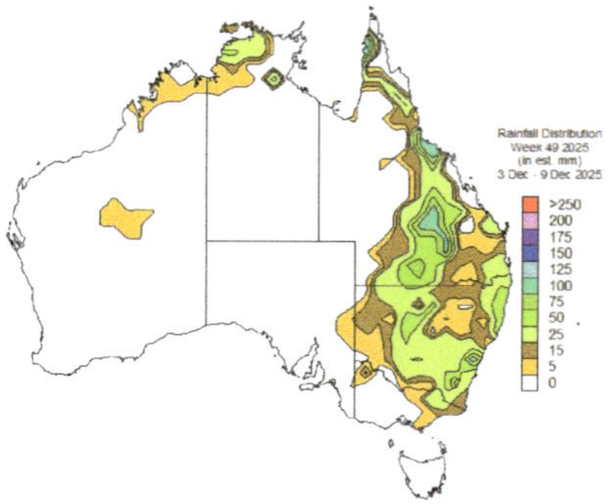
Rainfall Distribution
Week 49 2025
(in est. mm)
3 Dec - 9 Dec 2025
>250
200
175
150
125
100
75
50
25
15
5
0

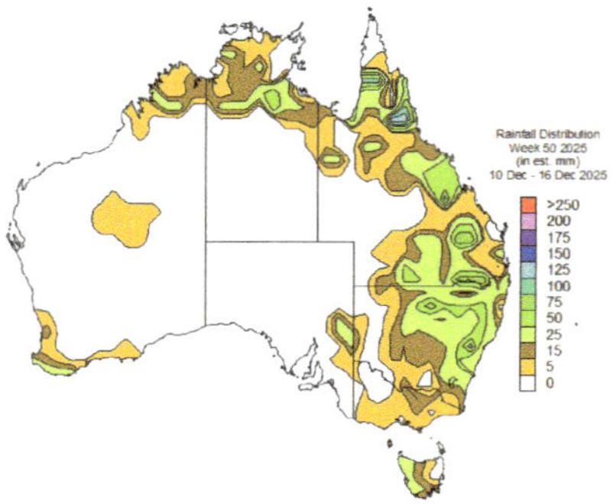
Rainfall Distribution
Week 50 2025
(in est. mm)
10 Dec - 16 Dec 2025
>250
200
175
150
125
100
75
50
25
15
5
0

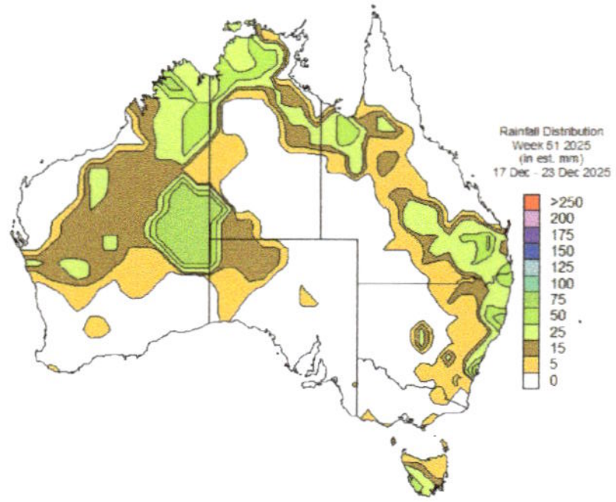
Rainfall Distribution
Week 51 2025
(in est. mm)
17 Dec - 23 Dec 2025
>250
200
175
150
125
100
75
50
25
15
5
0

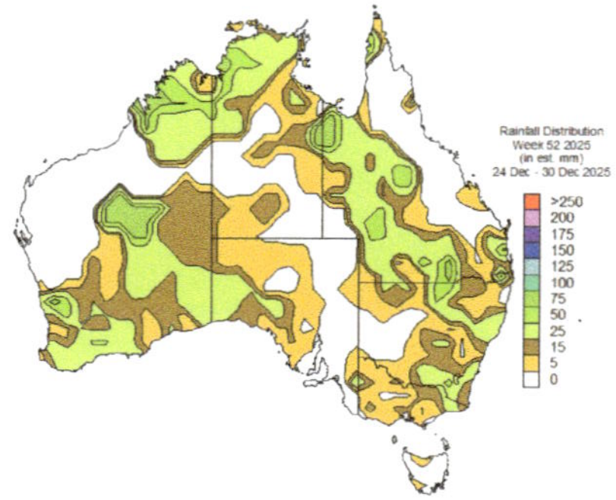
Rainfall Distribution
Week 52 2025
(in est. mm)
24 Dec - 30 Dec 2025
>250
200
175
150
125
100
75
50
25
15
5
0

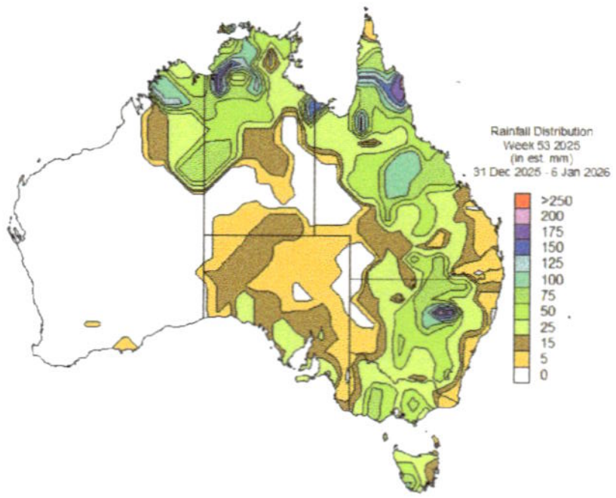
Rainfall Distribution
Week 53 2025
(in est. mm)
31 Dec 2025 - 6 Jan 2026
>250
200
175
150
125
100
75
50
25
15
5
0

ESTIMATED SEASONAL RAIN SUMMARY MAPS 2025

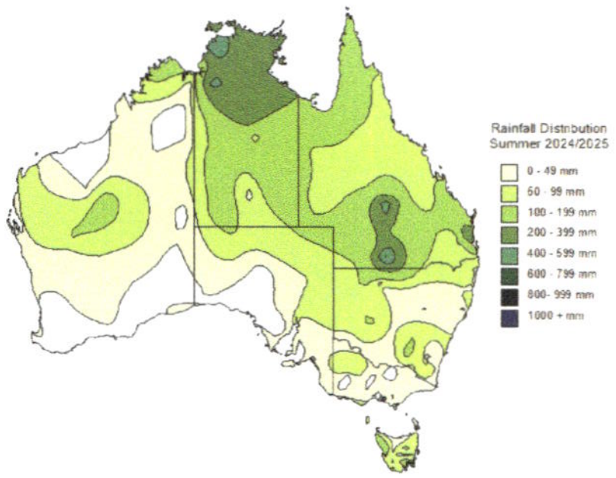

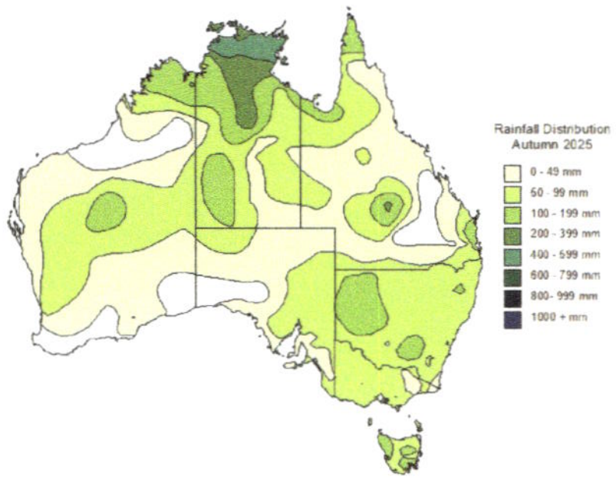

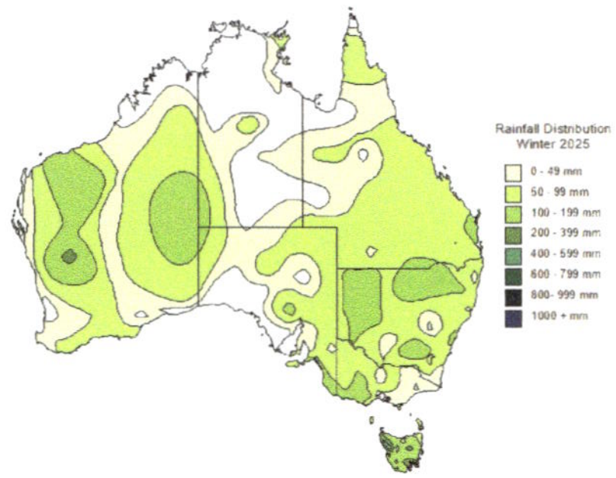

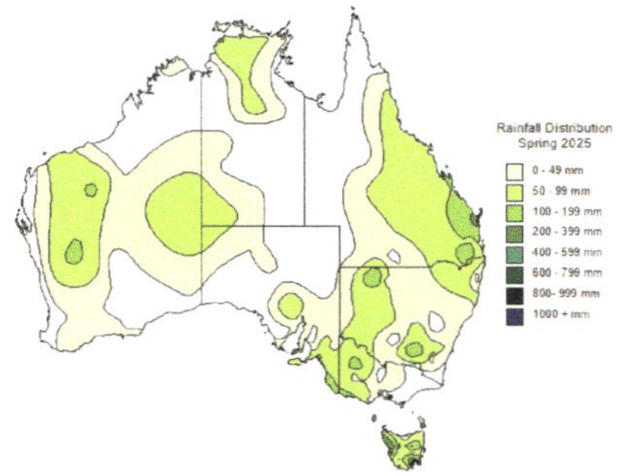

ESTIMATED MAXIMUM AND MINIMUM TEMPERATURE ANOMALIES 2025

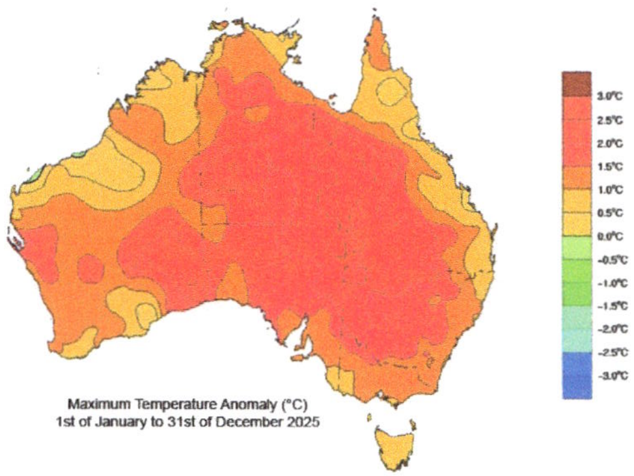

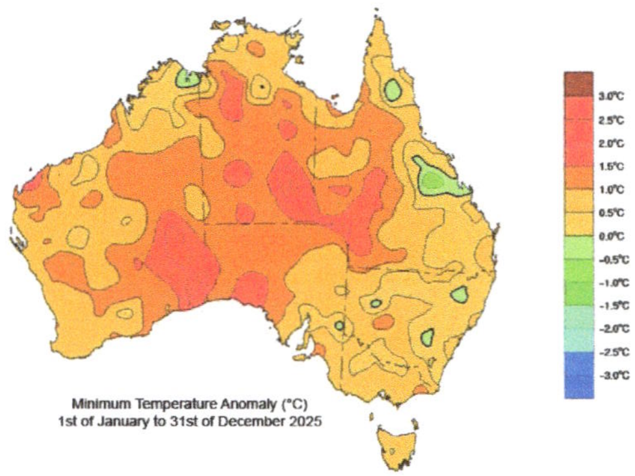

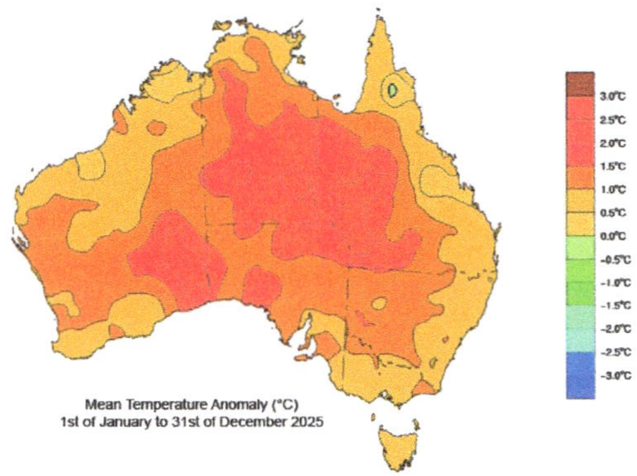

AUSTRALIAN WEATHER 2026

GENERAL OVERVIEW 2026

The coming year may bring wetter conditions across the Top End of the continent, in eastern Queensland to the northeast of New South Wales, and in the far western districts of Western Australia, with average to below average rainfall elsewhere. In particular, the drier than normal conditions throughout the southern Murray Darling Basin may be exasperated by continued below normal rainfall throughout this year. The anticipated arrival of a La Nina conditions towards the end of 2025 may be the cause of a wet start to 2026 across the Top End of the Northern Territory and over most of eastern Australia. By autumn however, La Nina may weaken resulting in the failure to produce abundant seasonal rains that usually occur in autumn. Winter rains may be mixed this year, with spring rains also being well below average in many parts of the country, particularly in the southeast. Overall, this year may not bring sufficient rainfall amounts to break the long dry spell in the southeast quarter of the country, including in Tasmania.

Temperatures will be generally warmer than normal during this coming year in most parts of the country, with only some parts of Queensland, the northeast of New South Wales, and the Kimberley region of Western Australia cooler than normal. Some parts of inland Western Australia, the Northern Territory, western Victoria and southern areas of South Australia may be particularly warmer than normal, especially during the summer months, with heatwave conditions likely in many of these areas. However, by the end of the year, the mix of warmer and cooler than normal areas may result in the mean temperatures being slightly cooler than of recent for the country as a whole. This may be the result of the anticipated arrival of a La Nina event towards the end of last year and continuing into the first part of this year.

MAJOR WEATHER EVENTS 2026

General 2026

- ***Cyclones*** - Overall, this summer season (2025-2026), may be unusually longer than normal, with the chance of up to 5 cyclonic systems making landfall in Australia, and up to 14 systems overall forming within the monsoonal trough. However, this season may have an unusually early pre-season flex with the potential for a cyclonic system to develop west of Western Australia in mid-July to early August 2025, and drift around the southwest Indian Ocean for a time. No other pre-season tropical cyclones are expected after this, with the first arrival of the traditional cyclone season starting in mid-November 2025 and the season persisting until mid-May 2026.

  During this cyclone season, the following states may be directly impacted:
  Queensland: second half of November 2025
  Northern Australia: second half of January 2026
  Western Australia: second half of February 2026
  Northern Territory: Western Australia: second half of March 2026
  Western Australia: first half of April

- ***Drought*** - Hot and dry conditions that affected much of the country last year are likely to continue throughout this year as well as possibly extending into 2027, particularly in the southeast of the country, where, despite the occasional outbreaks of heavy rain now and then, the moisture amounts may not be enough combined to break the drought. Drought conditions may finally ease around late April to early May in the Sydney-Metropolitan region, while in contrast, the Melbourne-Metropolitan region may continue to suffer very low water levels in their reservoirs. Some parts of Tasmania may also find drought conditions worsened during this year, with some parts of the State possibly not having received any significant rainfall for up to two to three years by this stage.

It may be particularly dry by May and early June in the southern half of the Northern Territory, the east of South Australia, and in some parts of Victoria and Western Australia.

Some good rain relief however, may finally arrive in November to early December when rainfall amounts may be very much above normal throughout many parts of the country, bringing the overall year rainfall amounts up to nearly average despite the drought conditions over most of the year.

Generally, 2026 may be much drier in a wide band from the interior of Western Australia, across central and southern Northern Territory and into central Queensland, as well as across most of Victoria, especially in the east of the State, and up along the New South Wales coasts to the Tablelands near Sydney. In contrast, it may be wetter than average over the west of Western Australia and the northern Kimberley, over much of the northern half of the Northern Territory, east into the Cape York Peninsula region, and along the eastern coasts of New South Wales north of Newcastle. Near to above-average rainfall may occur elsewhere, including in the northern half of the Murray-Darling Basin in northern New South Wales and into the southern inland and southeast Queensland.

- ***Heatwaves*** - Chance of heatwave temperatures in inland Western Australia to throughout the Northern Territory in January to early February.

  The middle couple of weeks in March to the end of the month may bring a very slow-moving high-pressure system into the southern Tasman Sea, directing a north to northeast flow over much of the southern half of Australia. This influence is likely to bring an exceptionally hot and prolonged heatwave throughout the Nullarbor region of Western Australia, and eastward over the southern half of South Australia and into the adjacent western half of Victoria and onto the northern half of Tasmania.

January 2026

Date	Day	Australia Overall - January 2026	Moon
1st	Thur	A cold front may cross the west of Queensland around this time bringing further rainfall and renewed river rises through most river systems in the region.	
2nd	Fri		P6
3rd	Sat	A trough over Western Queensland may cause flooding along many of the inland river systems during the next several days.	F ^
4th	Sun		
5th	Mon		
6th	Tue	Another tropical low may form in the Timor Sea around this time and track southwards towards northern Western Australia over the next couple of days, and wander along the north-western coastline of the state over the coming week, but it is not expected to make landfall. However, strong sea surges along the Kimberley and Pilbara coasts are likely.	
7th	Wed		
8th	Thu		
9th	Fri	Hot and dry conditions in the Perth to the Goldfield regions of Western Australia may see an increased risk of bush fires over the next few weeks.	XhS
10th	Sat	Further flooding possible over the next couple of weeks in the Queensland river systems around Longreach, as well as continuing along the Barcoo River and downstream to Windorah on Cooper Creek.	
11th	Sun	A low-pressure centre forms over the Northern Territory Top End region around this time and over the next few days builds into cyclonic intensity as it drifts around the Joseph Bonaparte Gulf. This system, although only likely to be of a moderate strength, is expected to make landfall near Darwin in a few days' time bringing storms and gale winds and heavy rains into Top End districts to the Katherine regions for several days, and include the risk of flooding along the local river systems.	3Q
12th	Mon	The next few weeks may be unusually hotter than what is typical for this time of year in many parts of the country.	
13th	Tue	Heat wave temperatures may prevail over much of the southern half of the Northern Territory over the next couple of weeks, with daily maximas in the early to mid 40's for several days consecutively.	
14th	Wed	Rain in the Gold Coast region of Queensland and in the northeast quarter of New South Wales over the next few days may result in some flooding along the major river systems in the regions.	A
15th	Thur	The next 3-4 weeks may be much warmer than what is typical for this time of year in most parts of the country, except for eastern and central Queensland and northeast New South Wales, which may be cooler than average. Rainfall may be above average over most of the eastern half of the continent, while below average to the west and particularly drier than normal in Tasmania.	
16th	Fri	Chance of funnel formation clouds forming in the Northern Rivers region around this time.	
17th	Sat		V
18th	Sun	Chance of flooding level rains over Cape York Peninsula around this time as a result of a decaying tropical cyclone.	
19th	Mon		N
20th	Tue	Remnants of a tropical cyclone that affected the Northern Territory over the previous week, brings significant heavy rains with flooding potential over the northern parts of Queensland, particularly affecting the Central Highlands over the next week to 10 days.	
21st	Wed		
22nd	Thu		
23rd	Fri		
24th	Sat		XhN
25th	Sun		
26th	Mon		1Q
27th	Tue		
28th	Wed		
29th	Thur		
30th	Fri	Heavy monsoonal rains of recent in Queensland may still be causing issues with flooding in a number of areas in the State over the coming week, including from over spillage from storage dams near Townsville, Charters Towers, and Emerald districts.	P10 ^
31st	Sat		

February 2026

Date	Day	Australia Overall - February 2026	Moon
1st	Sun		
2nd	Mon		F
3rd	Tue		
4th	Wed		
5th	Thu		
6th	Fri	Chance of funnel formation clouds forming in the Northern Tablelands around this time.	XhS
7th	Sat		
8th	Sun		
9th	Mon		
10th	Tue		3Q
11th	Wed	In contrast to the previous few weeks, a cooler than normal change is likely to affect most parts of the country over the next few weeks. During this cooler spell, the wet-season rains may fail to reach the interior regions of the continent, while the east and north of the country may be particularly wet.	A
12th	Thur	The next 3-4 weeks may be cooler than average in many parts of the country, particularly in the eastern states, reflecting the above average rainfall expected in the eastern parts of both New South Wales and Queensland. It may also be unusually wetter than normal in most of the northern tropics and over the southern half of Western Australia. In contrast, the central continent regions may be generally mostly dry. In contrast to the eastern states however, the northern tropics of the Northern Territory, Queensland and Western Australia may be slightly warmer than normal.	
13th	Fri	A rain-embedded trough over southern Queensland over the next couple of days may continue the flooding problems in many parts of the State, particularly in some western districts to the Darling Downs, but also spread once more into the South East regions with subsequent flooding of border river systems spreading over between Queensland and northern New South Wales.	V
14th	Sat		
15th	Sun		
16th	Mon		
17th	Tue		
18th	Wed		N
19th	Thu		
20th	Fri	An active monsoonal trough brings severe storms and heavy rain over the coming week, which may produce flash flooding in many river catchments in the Central coast to the Mackay regions of Queensland, particularly in the districts between Townsville and Bundaberg.	XhN
21st	Sat	A low-pressure system off the Kimberley coast may turn cyclonic over the next few days as it slowly tracks in a southeast direction at first before turning southwesterly and building in strength. After about a week of drifting off the western coast of Western Australia, this system may eventually make landfall near Exmouth where it may bring widespread heavy rain into the West Gascoyne and Murchison regions and east into the Goldfields and northern parts of the South West Land Division over several days.	
22nd	Sun		
23rd	Mon		
24th	Tue		
25th	Wed	Very heavy rain with major flash flooding likely around this time in the Mackay region of Queensland.	1Q P14
26th	Thur	The monsoonal trough may briefly drift northwards around this time, easing rain in the south of Queensland but intensifying rain and flooding issues northwards into Cape York Peninsula and the Gulf of Carpentaria regions over the next several days.	^
27th	Fri		
28th	Sat		

March 2026

Date	Day	Australia Overall - March 2026	Moon
1st	Sun		
2nd	Mon		
3rd	Tue		F
4th	Wed		
5th	Thu		XhS
6th	Fri		
7th	Sat		
8th	Sun		
9th	Mon		
10th	Tue		
11th	Wed	A low-pressure centre that may form over the north of the Northern Territory around this time may swiftly turn cyclonic and cross the North Kimberley over the next couple of days before tracking out to sea. It may then run parallel along the coasts west of the Kimberley and Pilbara over the coming week, before dissipating back to a low- pressure centre as it lingers offshore of Carnavon.	3Q A
12th	Thur	The next three weeks may be exceptionally hot over the southern regions of the country; from the Nullarbor in the west to central Victoria in the east, and down across Tasmania. Many parts of South Australia in particular, may swelter under heat wave temperatures for at least a fortnight in duration.	V
13th	Fri	Flooding issues over the west of Queensland of recent may begin to ease around this time as the monsoonal trough weakens and slowly continues to retreat northwards.	
14th	Sat	Very hot with heat-wave temperatures in southern Australia over the next fortnight.	
15th	Sun	Heatwave conditions that may have developed by this time may continue to affect much of southern Australia over the next 3 weeks before a colder change is anticipated in about 4 weeks' time. The next 3-4 weeks may also be mostly dry, except also, for a wetter change by the 4th week, particularly in the southwest and southeast areas of the country. Overall, the next 3-4 weeks may see below average rainfall almost everywhere, and particularly drier than normal conditions in South Australia. Only small parts of the Pilbara and Gascoyne regions and the eastern coasts of northern Queensland and the Top End of the Northern Territory may be wetter than normal for this time of year.	
16th	Mon		
17th	Tue		
18th	Wed		
19th	Thu		N XhN
20th	Fri		
21st	Sat		
22nd	Sun		P11
23rd	Mon		
24th	Tue	There may be a heightened bush fire risk in the Adelaide region and adjacent ranges around this time.	
25th	Wed		
26th	Thur		1Q ^
27th	Fri		
28th	Sat	Chance of some late flooding in the coastal river systems between Townsville and Cooktown in northern Queensland around this time.	
29th	Sun	The next couple of months may see warmer than average daytime temperatures and cooler than average nighttime temperatures for the remainder of the autumn season, in most parts of the country, with particularly unusually low minimum temperatures at times in Queensland and the northern tropical districts.	
30th	Mon		
31st	Tue		

April 2026

Date	Day	Australia Overall - April 2026	Moon
1st	Wed		
2nd	Thu		F XhS

Date	Day	Australia Overall - April 2026	Moon
3rd	Fri	A low-pressure system in the Indian Ocean may swiftly turn cyclonic as it moves in a south to southeasterly tract towards the Pilbara coastline of Western Australia over the coming few days. However, although it may bring heavy rain into the West Pilbara and Gascoyne, and unseasonal heavy rain into the southwest of the State, including in the Perth area, it is not expected to make landfall before dissipating back into the monsoonal trough.	
4th	Sat		
5th	Sun		
6th	Mon		
7th	Tue		A
8th	Wed		
9th	Thur		V
10th	Fri	Chance of rare sub-zero overnight minimum temperatures occurring around this time in the Darling Downs of Queensland.	3Q
11th	Sat		
12th	Sun	By around this time, the wet season over the northern half of the country may have ended, setting into place a drier than normal period to come over the next 3-4 weeks, as well as unusually cooler than normal in the east and north of the continent, while closer to average temperatures likely elsewhere. Generally, while Tasmania may be around 30% drier than what is typical for this time of year, other states, like Queensland may be up to 80% drier than average. Only some parts of Western Australia, particularly in the Esperance-Kalgoorlie-Karratha area, and the northeast coasts of New South Wales, may see above average rainfall.	
13th	Mon	Severe winds likely to cross the southeast quarter of the country around this time, particularly affecting many districts in Victoria, Tasmania, and the southeast quarter of South Australia with the potential to cause damage to infrastructure and power outages.	
14th	Tue		
15th	Wed		
16th	Thu		XhN
17th	Fri		N
18th	Sat		
19th	Sun		P7
20th	Mon		
21st	Tue		
22nd	Wed		^
23rd	Thur		
24th	Fri		1Q
25th	Sat		
26th	Sun		
27th	Mon		
28th	Tue		
29th	Wed		XhS
30th	Thur		

May 2026

Date	Day	Australia Overall - May 2026	Moon
1st	Fri		
2nd	Sat		F
3rd	Sun		
4th	Mon		
5th	Tue		A
6th	Wed		V
7th	Thu		
8th	Fri	Cold outbreak over the next couple of days with widespread snow likely in the highlands of New South Wales and Victoria.	
9th	Sat		
10th	Sun		3Q
11th	Mon		

Date	Day	Australia Overall - May 2026	Moon
12th	Tue	The next 3-4 weeks may be considerably drier throughout most of the country than what is normal for this time of year, with some parts of the east and southern coasts possibly completely rain free. Consequently, accompanying the drier than normal conditions, temperatures, maximums may be above normal almost everywhere, while overnight minimums may be generally below average, particularly in the northern tropics regions and along the east coast of Queensland and New South Wales.	
13th	Wed		**XhN**
14th	Thur		
15th	Fri		
16th	Sat		
17th	Sun		**N**
18th	Mon		**P3**
19th	Tue		**^**
20th	Wed		
21st	Thu		
22nd	Fri		
23rd	Sat		**1Q**
24th	Sun		
25th	Mon		
26th	Tue		**XhS**
27th	Wed		
28th	Thur		
29th	Fri		
30th	Sat		
31st	Sun		**F**

June 2026

Date	Day	Australia Overall - June 2026	Moon
1st	Mon		**A**
2nd	Tue		**V**
3rd	Wed		
4th	Thu	Extremely cold in the higher reaches of New South Wales and Victoria, particularly in the Charlottes Pass district.	
5th	Fri		
6th	Sat		
7th	Sun		
8th	Mon		**3Q**
9th	Tue		**XhN**
10th	Wed	Heavy rain around the Sunshine Coast in Queensland may cause some flooding along local river systems over the coming week.	
11th	Thur		
12th	Fri	The winter season across the country may start milder than what is normal for the season, and then steadily get cooler as the season progresses, with mid-winter having typical temperatures for the time of year, but the end month of winter being colder than is normal for that time of year.	
13th	Sat	The next 3-4 weeks may return near to average rainfall for much of the northern half of the country, but continue to be very dry over much of the southern half of the continent, particularly in the Wheatbelt and South Coast regions of Western Australia. Similarly, to the previous 3-4 weeks, the next 3-4 weeks may also be warmer than what is normal for this time of year, both with regard to maximum and minimum temperatures, particularly in the eastern half of the continent.	
14th	Sun		
15th	Mon	Heavy rain in the far west of Queensland may cause some brief flooding along the Bulloo River system over the next few days.	**N P2**
16th	Tue		**^**
17th	Wed		
18th	Thu		

Date	Day	Australia Overall - June 2026	Moon
19th	Fri	A wide band of cloud may cross the country around this time, from the northern Western Australia region to the northern New South Wales and southern Queensland coasts, bringing widespread rain, and some localised heavy falls over the next few days.	
20th	Sat	Unstable winds in the Central Coast region of Western Australia may produce tornadoes, particularly in the Perth area.	
21st	Sun		
22nd	Mon		1Q XhS
23rd	Tue		
24th	Wed		
25th	Thur		
26th	Fri		
27th	Sat		
28th	Sun		A
29th	Mon		V
30th	Tue		F

July 2026

Date	Day	Australia Overall - July 2026	Moon
1st	Wed		
2nd	Thu		
3rd	Fri		
4th	Sat		
5th	Sun		
6th	Mon		
7th	Tue		XhN
8th	Wed		3Q
9th	Thur		
10th	Fri		
11th	Sat		
12th	Sun	Chance of storm surges along the southeast coasts of Tasmania around this time.	
13th	Mon	The next 3-4 weeks may be typical for this time of year, with rainfall amounts and minimum temperatures close to normal in most districts, and maximum temperatures slightly warmer than average. Only the eastern regions of Queensland, in a broad band along the tropical coasts from the North Coast to the Mackay regions may be wetter than average, with up to twice as much rainfall compared to the average for this time of year; with most of this falling in the last week of this look-ahead period, during the first part of August. In contrast, some parts of northern Western Australia, the Northern Territory and into the western regions of Queensland may not see any precipitation over the coming weeks.	P4 ^
14th	Tue	Unusually hot and dry conditions may prevail over many parts of the southeast of the continent over the coming 3-4 months.	N
15th	Wed		
16th	Thu		
17th	Fri		
18th	Sat		
19th	Sun	Temperatures may fail to rise above zero in some of the higher reaches of the Victorian ranges over the next 4-5 weeks.	
20th	Mon		XhS
21st	Tue		1Q
22nd	Wed		
23rd	Thur		
24th	Fri		
25th	Sat		
26th	Sun		A
27th	Mon		V
28th	Tue		

Date	Day	Australia Overall - July 2026	Moon
29th	Wed	Chance of storm surges and coastal flooding over the next couple of days around the Western Australian Central Coast, including in the Perth region.	
30th	Thu		F
31st	Fri		

August 2026

Date	Day	Australia Overall - August 2026	Moon
1st	Sat		
2nd	Sun	Snow may fall to low levels on several occasions over the coming weeks, particularly in Victoria and southern New South Wales.	
3rd	Mon		XhN
4th	Tue		
5th	Wed		
6th	Thur		3Q
7th	Fri		
8th	Sat		
9th	Sun		^
10th	Mon		P8
11th	Tue		
12th	Wed	Overall, most of the coming 3-4 weeks may be cooler and drier than normal over most of the country. However, early in September, a significant rain event may occur bringing heavy rains over many parts of the continent, and subsequently, a warmer change.	
13th	Thu		N
14th	Fri		
15th	Sat		
16th	Sun		XhS
17th	Mon		
18th	Tue		
19th	Wed		
20th	Thur		1Q
21st	Fri	Chance of widespread snow falling to low levels in Victoria around this time, particularly near Melbourne and in southern New South Wales.	
22nd	Sat		A
23rd	Sun		V
24th	Mon		
25th	Tue		
26th	Wed		
27th	Thu		
28th	Fri	Very cold temperatures with severe frosts likely in the Eyre district of Western Australia around this time.	F
29th	Sat		
30th	Sun		XhN
31st	Mon		

September 2026

Date	Day	Australia Overall - September 2026	Moon
1st	Tue		
2nd	Wed		
3rd	Thur		
4th	Fri	A significant rain event may occur during the coming week, bringing a wide band of rain extending from the Kimberley of Western Australia, through the southern half of the Northern Territory and into northern South Australia and the far southwest of Queensland, with heavy falls possible in many districts along this line.	3Q
5th	Sat		

Date	Day	Australia Overall - September 2026	Moon
6th	Sun		^
7th	Mon		P12
8th	Tue		
9th	Wed		
10th	Thu		
11th	Fri		N
12th	Sat	The next couple of months may be warmer than is normal for this time of year in most parts of the country, with only some pockets in the northern parts of Victoria slightly cooler than normal. Accompanying this warmer than normal trend, early spring may also be unusually drier than is typical for this time of year, resulting in low inflows into the Murray-Darling Basin, with the potential to cause trouble with cropping later in the year in the south-eastern quarter of the continent.	XhS
13th	Sun	Generally, the next 3-4 weeks may be drier than normal in South Australia, Victoria, southern New South Wales and the southeast of Western Australia, while the remainder of the country may see nearer to average rainfall to what is typical for this time of year.	
14th	Mon		
15th	Tue		
16th	Wed		
17th	Thur		
18th	Fri		
19th	Sat		1Q A V
20th	Sun		
21st	Mon		
22nd	Tue		
23rd	Wed		
24th	Thu		
25th	Fri		
26th	Sat		XhN
27th	Sun		F
28th	Mon		
29th	Tue		
30th	Wed		

October 2026

Date	Day	Australia Overall - October 2026	Moon
1st	Thur	Thunderstorm activity may prevail through the Northern Rivers region of New South Wales around this time and possibly produce some funnel cloud formations and hailstorms.	
2nd	Fri		P13
3rd	Sat	Late season frosts likely over the next couple of days in southern Western Australia and northern and western parts of Victoria with chance of damaging early spring crops.	
4th	Sun		3Q ^
5th	Mon		
6th	Tue		
7th	Wed		
8th	Thu		
9th	Fri		
10th	Sat		XhS
11th	Sun		N
12th	Mon	The next 3-4 weeks may be considerably drier than normal throughout central and southeastern parts of the country, including in Tasmania, while rainfall may be closer to normal elsewhere. Generally, this look ahead period may be warmer almost everywhere for both daytime maximum temperatures and overnight minimum temperatures.	
13th	Tue		
14th	Wed		
15th	Thur		
16th	Fri		
17th	Sat		A V

Date	Day	Australia Overall - October 2026	Moon
18th	Sun		
19th	Mon		1Q
20th	Tue		
21st	Wed		
22nd	Thur		
23rd	Fri		
24th	Sat		XhN
25th	Sun		
26th	Mon		F
27th	Tue		
28th	Wed		
29th	Thu		P9
30th	Fri		^
31st	Sat		

November 2026

Date	Day	Australia Overall - November 2026	Moon
1st	Sun		
2nd	Mon		3Q
3rd	Tue	Chance of late season damaging frosts in some parts of Victoria around this time.	
4th	Wed		
5th	Thur		
6th	Fri		XhS
7th	Sat		
8th	Sun		
9th	Mon		N
10th	Tue		
11th	Wed		
12th	Thu	Around this time, there may be a marked change in the climate patterns, with moist air finally moving into central Australia and remaining there for the coming month, with a resulting very wet spell throughout much of the country during this period, and long awaited rains in the Alice Springs region finally arriving. Generally, the next 3-4 weeks may be wetter than normal in most parts of the country, and temperatures cooler than average.	
13th	Fri	Unstable winds north of Adelaide South Australia may briefly cause a tornado.	V
14th	Sat		A
15th	Sun		
16th	Mon		
17th	Tue		
18th	Wed		1Q
19th	Thur		
20th	Fri		XhN
21st	Sat		
22nd	Sun		
23rd	Mon		
24th	Tue		
25th	Wed		F
26th	Thu		P5
27th	Fri	Severe storms across the southeast of Queensland and into the northern New South Wales regions over the next 3-4 days. Gale force winds and downbursts of large hail may cause damage to property and infrastructure during this stormy period, and heavy rains bring localised flooding in some districts.	^
28th	Sat		
29th	Sun		
30th	Mon	Chance of flooding in the southeast of Queensland around this time because of earlier heavy rain.	

December 2026

Date	Day	Australia Overall - December 2026	Moon
1st	Tue		3Q
2nd	Wed		
3rd	Thur	Chance of late season snow on the New South Wales Central and Southern Tablelands, with possible damage occurring to nearby crops.	XhS
4th	Fri		
5th	Sat		
6th	Sun		
7th	Mon		
8th	Tue		
9th	Wed	Heavy rain and chance of flooding in the northeast of New South Wales region, particularly in the Tamworth area.	N
10th	Thu		V
11th	Fri		A
12th	Sat	Wetter than normal conditions may continue to persist over the coming weeks in most parts of the country, except in Queensland, which may be drier and warmer than is typical for this time of year, despite still seeing regular storm activity with thunderstorms, flash flooding, damaging winds, and large hail.	
13th	Sun	Thunderstorm activity may prevail through the Northern Rivers region of New South Wales around this time and possibly produce some funnel cloud formations.	
14th	Mon		
15th	Tue		
16th	Wed		
17th	Thur		1Q XhN
18th	Fri		
19th	Sat		
20th	Sun		
21st	Mon		
22nd	Tue		
23rd	Wed		
24th	Thu		P1 F ^
25th	Fri		
26th	Sat		
27th	Sun		
28th	Mon		
29th	Tue	Chance of tornado activity in southern Queensland around this time.	
30th	Wed	A cyclonic system may form near Port Keats in the Northern Territory around this time and cross the northern coast of the Kimberley over the next couple of days bringing storms and heavy rain into the regions. Thunderstorm activity in the northeast of New South Wales around this time may produce funnel formation clouds briefly in the Northern Rivers region.	XhS
31st	Thur		3Q

NORTHERN TERRITORY 2026

GENERAL TRENDS

Year Overall: Slightly drier but within 5% of the average
Slightly warmer but within 0.5°C of the average

Summer: Drier by around 10-15% below the average
Wettest – Second half of February
Driest – First half of December
Slightly warmer but within 0.5°C of the average
Hottest – First half of December and second half of January
Coolest – First half of January

Autumn: Slightly wetter but within 5-10% of the average
Wettest – Second to third weeks of March
Driest – Second to third weeks of May
Warmer by up to 0.5-1°C above the average
Hottest – Last week of March to first week of April
Coolest – Last half of May

Winter: Drier by around 20-25% below the average
Wettest – Last week of June
Driest – All of July and August
Slightly cooler but within 0.5°C of the average
Hottest –Fourth week of June, end of July and mid-August
Coolest – First week of August

Spring: Drier by around 35-40% below the average
Wettest –Mid-November to end of the month
Driest – Second half of September to first half of October
Slightly cooler but within 0.5°C of the average
Hottest – Second half of October to mid-November
Coolest – First half of September

Major Weather Events

January 2026

Date	Day	Northern Territory - January 2026	Darwin	Moon
1st	Thur	Widespread heavy monsoonal rains throughout the State over the next couple of days.	Showers, Th	
2nd	Fri			P6
3rd	Sat		Th	F ^
4th	Sun	Heavy rain around the Timber Creek area.	Rain, Th	
5th	Mon	Chance of flooding in the Adelaide River catchment.	Rain, Th	
6th	Tue	Northwesterly trade winds begin to stabilise over the Top End around this time. Heavy rain over the Top End over the coming week. Chance of localised flooding in the Arnhem Lands to the Roper-McArthur regions.	Showers, Th	
7th	Wed		Rain, Th	
8th	Thu	Cool day in the Jabiru area.	Showers	
9th	Fri	Heavy rain in the Darwin-Daly region with chance of some brief flooding in the Darwin suburbs.	Rain, Th	XhS
10th	Sat	Heavy rain persists over the Top End over the coming week.	Rain, Th	
11th	Sun	A tropical low may build to cyclonic strength over the next few days as it drifts around the Top End regions.	Showers	3Q
12th	Mon	The next few weeks may be exceptionally hotter than is typical for this time of year, particularly in the next couple of weeks.	Rain	
13th	Tue		Rain	
14th	Wed	A full strength tropical cyclone may continue to bring disruption over the Top End regions around this time and persist for the next few days as it meanders over the far north of the State before moving offshore into the Gulf of Carpentaria and weakening as it melds back into the monsoonal trough. Chance of wind damage in many Top End to the Gulf, including in the Katherine area and widespread and heavy rain and localised flooding through the region over the next couple of days.	Rain, Th	A
15th	Thur	A complex of thunderstorms and strong blustery winds cross the East Arnhem region over the next couple of days.	Rain	
16th	Fri		Showers	
17th	Sat	Rain eases to isolated showers over the Top End for the next few days.		V
18th	Sun			
19th	Mon	Very hot in the Victoria River region.		N
20th	Tue	Very hot across the State over the next couple of days, particularly in the Alice Springs region, where maximum temperatures may reach into the mid-40's.	Rain, Th	
21st	Wed			
22nd	Thu	Storm lines cross the Top End over the next couple of days.		
23rd	Fri	Lines of storms continue to cross the Top End bringing heavy rain and scattered thunderstorms throughout the region including in Darwin, with the chance of flash flooding in some places.	Rain, Th	
24th	Sat		Showers	XhN
25th	Sun	Thunderstorms with accompanying strong wind gusts in the East Arnhem region.		
26th	Mon			1Q
27th	Tue	Further storms over the Top End with heavy rain in places,particularly in the Cox Peninsula area. Chance of localised flash flooding.		
28th	Wed	Strong winds and hail in the Katherine region accompanying thunderstorm activity. Cooler spell overnight in the Victoria River region.		
29th	Thur	Cooler overnight in the southeast of the Alice Springs region.	Rain	
30th	Fri	Strong blustery winds in the East Arnhem region, particularly gusty around the McArthur River area. The monsoonal trough may slip into temporary inactivity around this time and remain fairly inactive for the next week to ten days.		P10 ^
31st	Sat	Very hot in the Tamani Desert region.		

February 2026

Date	Day	Northern Territory - February 2026	Darwin	Moon
1st	Sun	Only light monsoonal activity over the coming week.		
2nd	Mon	A cloud band over the Alice Springs region brings some shower activity on most days in the week ahead.	Rain, Th	F
3rd	Tue		Rain, Th	

Date	Day	Northern Territory - February 2026	Darwin	Moon
4th	Wed		Rain, Th	
5th	Thu			
6th	Fri		Showers	XhS
7th	Sat		Showers, Th	
8th	Sun	Sporadic monsoonal activity over the Top End over the coming week. Outbreaks of rain in the Alice Springs region over the next couple of days.	Showers, Th	
9th	Mon		Showers, Th	
10th	Tue		Th	3Q
11th	Wed	Very high maximum temperatures in the upper 30's to early 40's in the Victoria River region.	Rain, Th	A
12th	Thur		Showers, Th	
13th	Fri	Outbreaks of monsoonal rains across the Territory over the coming couple of days.	Showers, Th	V
14th	Sat		Showers, Th	
15th	Sun	Widespread heavy monsoonal rains throughout the State over the next several days.	Showers	
16th	Mon	Very high maximum temperatures in the upper 30's to early 40's in the Roper-McArthur region.	Showers	
17th	Tue	The monsoonal trough reinvigorates over the northern half of the State around this time bringing regular outbreaks of heavy rain over the coming 2-3 weeks, but the southern half of the State may remain relatively dry with only a few brief spells of rain.	Showers	
18th	Wed	Very heavy rain continues over the Top End region, particularly around the Bradshaw area around this time.	Rain, Th	N
19th	Thu	Spell of very heavy rain over the next 3-4 days over the northern half of the Territory, particularly in the Darwin-Daly region and adjacent coastal areas.	Rain	
20th	Fri	Cooler spell in the Roper-McArthur region.	Rain	XhN
21st	Sat		Rain	
22nd	Sun	Very heavy monsoonal rain over the Top End over most of the coming week.	Rain, Th	
23rd	Mon	Very heavy rain in the Katherine region.	Rain	
24th	Tue		Rain	
25th	Wed		Rain	1Q P14
26th	Thur		Showers	^
27th	Fri		Showers	
28th	Sat	Very hot and humid in the Alice Springs region.	Rain	

March 2026

Date	Day	Northern Territory - March 2026	Darwin	Moon
1st	Sun	The monsoonal trough begins to contract northwards, bringing cooler temperatures and clearing most rain activity from the southern half of the State over the few weeks, and only maintaining its influence mostly over the Top End.	Rain	
2nd	Mon	Very hot from the Barkly Tablelands and southward through the Alice Springs region over the next couple of days where many places may see maximum temperatures in the mid-40's.	Rain	
3rd	Tue		Rain	F
4th	Wed		Showers	
5th	Thu		Showers	XhS
6th	Fri	Unusually cool spell in the Alice Springs region, particulary around the Watarrka to Curtin Springs districts.		
7th	Sat	Heavy monsoonal rains over the next week in the Top End regions.	Rain	
8th	Sun		Rain	
9th	Mon	A band of very heavy rain crosses through the south of the Top End districts over the next couple of days from the Katherine region in the west to the Roper-McArthur region in the east.	Rain	
10th	Tue		Showers	
11th	Wed	Storms and heavy rain in the Darwin-Daly region over the next few days.	Showers	3Q A
12th	Thur		Showers	V
13th	Fri		Showers	
14th	Sat		Rain, Th	
15th	Sun			
16th	Mon		Rain, Th	
17th	Tue		Rain, Th	
18th	Wed		Showers, Th	
19th	Thu		Showers	N XhN
20th	Fri	Heavy rain in the Top End districts to the Roper-McArthur region over the next 2-3 days.		
21st	Sat			

Date	Day	Northern Territory - March 2026	Darwin	Moon
22nd	Sun	Severe thunderstorms likely in the Darwin-Daly region over the next couple of days.	Showers, Th	P11
23rd	Mon		Rain, Th	
24th	Tue	Showers over the Top End over the next 3-4 days.	Rain, Th	
25th	Wed	Rain in the north of the Top End, with some heavy falls around the coastal areas north of Darwin.	Th	
26th	Thur	Very warm overnight in the Jervios district.	Rain	1Q ^
27th	Fri	Very hot in the Alice Springs region with maximum temperatures into the early-40's in a number of districts.	Showers, Th	
28th	Sat	Storm activity over the Top End region, with scattered thunderstorms and bursts of heavy rain in places, particularly in Darwin and near Batchelor. Chance of shower activity in the western half of the Alice Springs region.	Rain, Th	
29th	Sun	Isolated showers over the Top End over the next few days.	Rain, Th	
30th	Mon			
31st	Tue		Rain	

April 2026

Date	Day	Northern Territory - April 2026	Darwin	Moon
1st	Wed	Mostly dry over the Top End over the next couple of days.		
2nd	Thu			F XhS
3rd	Fri	Showers in the Barkly Tablelands and into the southwest of the Alice Springs region over the next few days.	Rain, Th	
4th	Sat		Showers	
5th	Sun		Rain, Th	
6th	Mon			
7th	Tue	Cooler turn in the Alice Springs region, particularly in the Yuendumu area.	Rain, Th	A
8th	Wed		Showers, Th	
9th	Thur	Unusually cool day in the West Arnhem region. Showers around the Top End coasts over the next couple of days.	Rain, Th	V
10th	Fri	Cold overnight in the Alice Springs region, particularly in the Arltunga district, where minimas may not reach into double digits for a time.	Rain, Th	3Q
11th	Sat	Unusually cool in the Katherine to Daly Waters districts.	Showers	
12th	Sun	The monsoonal trough continues to contract and disippate over the next week or so, with southeast to northeast winds becoming established and bringing drier than normal conditions over most of the State during the next 3-4 weeks, especially in areas south of the Barkly Tablelands, but still with regular passages of showers over the Top End.		
13th	Mon	Very hot day in the Barkly to the Arnhem regions. Daily maximas may still reach into the upper 30's in the Alice Springs region to the Victoria River districts over the next week. Showers in the West Arnhem over the next couple of days.		
14th	Tue			
15th	Wed	Showers over the Top End.	Showers, Th	
16th	Thu	Heavy rain over the Top End.	Showers, Th	XhN
17th	Fri			N
18th	Sat			
19th	Sun			P7
20th	Mon		Showers	
21st	Tue	Very little rainfall can be expected in the Darwin area from around this time until the end of September.		
22nd	Wed			^
23rd	Thur			
24th	Fri	Very strong gusty winds at times in the Roper-McArthur region.		1Q
25th	Sat			
26th	Sun			
27th	Mon	Overnight minimums may struggle to reach double digits in the Alice Springs region over the coming couple of weeks.		
28th	Tue			
29th	Wed			XhS
30th	Thur			

May 2026

Date	Day	Northern Territory - May 2026	Darwin	Moon
1st	Fri			
2nd	Sat	Unusually cool in the far northwest of the Top End.		F
3rd	Sun	Isolated showers in the Arnhem Lands.		
4th	Mon			
5th	Tue			A
6th	Wed			V
7th	Thu			
8th	Fri	Unusually cold overnight in the Alice Springs region.		
9th	Sat	Unusually cold overnight over the Top End particularly in the West Arnhem region, and around Daly Waters.		
10th	Sun			3Q
11th	Mon			
12th	Tue			
13th	Wed	High temperatures in the upper 30's in the Barkly region including over the Tablelands. Very hot overnight in the West Arnhem district.		XhN
14th	Thur			
15th	Fri			
16th	Sat			
17th	Sun			N
18th	Mon			P3
19th	Tue			^
20th	Wed			
21st	Thu			
22nd	Fri			
23rd	Sat			1Q
24th	Sun			
25th	Mon			
26th	Tue			XhS
27th	Wed			
28th	Thur	Unusually low maximum temperatures in the southern half of the Alice Springs region.		
29th	Fri	Minimum overnight temperatures may not reach double digits in the Alice Springs region from around this time forward through the remainder of the upcoming winter season.		
30th	Sat	Showers possible in the Arnhem Land.		
31st	Sun			F

June 2026

Date	Day	Northern Territory - June 2026	Darwin	Moon
1st	Mon			A
2nd	Tue	Strong and blustery winds likely for a time across the Tamani Desert, including in Lajamanu.		V
3rd	Wed			
4th	Thu	Very cold overnight in the south of the Alice Springs region with chance of frost patches being observed, particularly in the Arltunga district.		
5th	Fri	Chance of isolated showers in the Arnhem Lands over the next couple of days.		
6th	Sat			
7th	Sun			
8th	Mon	Very cold on the Barkly Tablelands, particularly in the Brunette Downs area.		3Q
9th	Tue	Strong winds with occasional gusts for a time in the Victoria-Daly region, including in the Douglas River area.		XhN
10th	Wed	Very cold in the Victoria-Daly region.		
11th	Thur			
12th	Fri			
13th	Sat			
14th	Sun			
15th	Mon			N P2
16th	Tue	A cloudband crosses the central parts of the State around this time, bringing the chance of some unseasonal outbreaks of rain in some districts ove the next few days.		^
17th	Wed	Unusually high maximas in the Darwin-Daly region.		

Date	Day	Northern Territory - June 2026	Darwin	Moon
18th	Thu	Unusually cold overnight in the Alice Springs region, with chance of some patches of frost, particularly in the Arltunga area. Outbreaks of rain in the southern half of the State south of the Barkly Tablenads over the next couple of days.		
19th	Fri	Shower activity may briefly spread into the northern half of the State, particularly in a line the Victoria River region to the Roper-McArthur, with isolated showers extending into the Top End.		
20th	Sat			
21st	Sun			
22nd	Mon	Warmer spell over much of the State over the next couple of days, particularly in the Alice Springs region.		1Q XhS
23rd	Tue		Fg	
24th	Wed			
25th	Thur			
26th	Fri	Chance of isolated showers in the West Arnhem coastal fringes over the next couple of days.		
27th	Sat			
28th	Sun		Fg	A
29th	Mon		Fg	V
30th	Tue			F

July 2026

Date	Day	Northern Territory - July 2026	Darwin	Moon
1st	Wed	Chance of showers in the Victoria River to Katherine districts.		
2nd	Thu			
3rd	Fri			
4th	Sat			
5th	Sun			
6th	Mon	Pockets of frost in the Alice Springs region over the next few days.		
7th	Tue	Chance of showers in the Victoria River to Katherine districts.		XhN
8th	Wed			3Q
9th	Thur			
10th	Fri		Fg	
11th	Sat		Fg	
12th	Sun		Fg	
13th	Mon		Fg	P4 ^
14th	Tue			N
15th	Wed			
16th	Thu			
17th	Fri			
18th	Sat	Showers in the south of the Alice Springs region.	Fg	
19th	Sun			
20th	Mon			XhS
21st	Tue			1Q
22nd	Wed			
23rd	Thur			
24th	Fri	Chance of some frost pockets in the Alice Springs region, particularly around Yulara.	Fg	
25th	Sat			
26th	Sun			A
27th	Mon		Fg	V
28th	Tue	High maximum temperatures in the West Arnhem region.		
29th	Wed			
30th	Thu	Unusually high maximas over the next couple of days in the Roper-McArthur region, particularly in the Bulman and Ngukurr districts.		F
31st	Fri			

August 2026

Date	Day	Northern Territory - August 2026	Darwin	Moon
1st	Sat			
2nd	Sun	Chance of rain in the Kaduku area, particularly in the Jabiru district.		
3rd	Mon	Cold spell in the south of the Alice Springs region over the next few days with patches of overnight frost likely in numerous districts.		XhN
4th	Tue			
5th	Wed	Chance of frosts in the south of the Alice Springs region, particularly in the Arltunga district.		
6th	Thur	Unusually cold in the coastal areas of the Darwin-Daly region.		3Q
7th	Fri			
8th	Sat		Fg	
9th	Sun		Fg	^
10th	Mon	Unsually hot day in the Yulara district.	Fg	P8
11th	Tue		Fg	
12th	Wed	The next 3-4 weeks may be mostly dry and generally cooler than is typical for this time of year, particularly in the southern half of the State, while some parts over the Top End may be slightly warmer than average.		
13th	Thu			N
14th	Fri			
15th	Sat		Fg	
16th	Sun		Fg	XhS
17th	Mon		Fg	
18th	Tue			
19th	Wed	Pockets of frost in the Alice Springs region over the next few days.		
20th	Thur			1Q
21st	Fri			
22nd	Sat	Very cold overnight in the Alice Springs region, with chance of frost pockets, including in the Kulgera area.		A
23rd	Sun			V
24th	Mon			
25th	Tue			
26th	Wed			
27th	Thu			
28th	Fri			F
29th	Sat			
30th	Sun			XhN
31st	Mon			

September 2026

Date	Day	Northern Territory - September 2026	Darwin	Moon
1st	Tue			
2nd	Wed	Unusually cool in the south of the Alice Springs region, particularly in the Arltunga district.	Fg	
3rd	Thur	Cold day around Lajamanu to Rabbit Flat districts.		
4th	Fri			3Q
5th	Sat			
6th	Sun			^
7th	Mon	A cloud band from Western Australia extends into the southwest of the State bringing some shower activity in the Alice Springs and Tanami Desert regions over the next several of days.		P12
8th	Tue			
9th	Wed			
10th	Thu	A trough over Western Australia around this time may extend some showers into the southern Alice Springs region, with chance of brief heavy falls in the Jervois to Curtin Springs areas over the next couple of days. Unusually hot in the Barkly region with maximas into the high 30's likely.	Fg	
11th	Fri	Chance of rain in the Atula and Jervois districts. Very warm overnight in the Roper-McArthur region.	Fg	N

Date	Day	Northern Territory - September 2026	Darwin	Moon
12th	Sat	Chance of frosts in the Alice Springs region, particularly in the Kulgera area. Overnight minimum temperatures may struggle to reach into double digits over most days in the next 3-4 weeks.		**XhS**
13th	Sun	A band of rain brings some shower activity across the State from Victoria River to the Roper-McArthur regions with showers spreading later into the Alice Springs region by tomorrow.	Fg	
14th	Mon			
15th	Tue			
16th	Wed			
17th	Thur			
18th	Fri		Fg	
19th	Sat		Fg	**1Q A V**
20th	Sun			
21st	Mon		Fg	
22nd	Tue		Fg	
23rd	Wed	Rain in the Katherine region with some heavy falls in the Upper Waterhouse River catchment.	Fg	
24th	Thu		Fg	
25th	Fri		Fg	
26th	Sat	Very hot in the Barkly with maximum temperatures possibly into the low 40's, including in the Elliott area.	Fg	**XhN**
27th	Sun	Hot spell through many districts north of the Barkly Tablelands over the next 3-4 days. Isolated showers around the Barkly Tablelands.		**F**
28th	Mon			
29th	Tue	Chance of brief rain in the Darwin-Daly region.		
30th	Wed			

October 2026

Date	Day	Northern Territory - October 2026	Darwin	Moon
1st	Thur	Very warm overnight in the Victoria River region. Some showers around the Barkly Tablelands.		
2nd	Fri			**P13**
3rd	Sat	A band of thunderstorms cross the Alice Springs region, accompanied by strong blustery winds. Some showers cross the Top End around this time breaking the long dry spell, including some showers into the Darwin district over the next couple of days.		
4th	Sun	Cooler in the southwest of the Alice Springs region, particularly around Arltunga.		**3Q ^**
5th	Mon		Showers	
6th	Tue	Warm in the Arnhem Lands.		
7th	Wed			
8th	Thu			
9th	Fri			
10th	Sat	Very hot over the northern half of the State with maximum temperatures in the upper 30's in many districts. Chance of isolated showers around the West Arnhem coastal fringes.		**XhS**
11th	Sun			**N**
12th	Mon	The next few weeks may be hotter than is typical for this time of year, particularly in the Darwin region, where it may also be the wetter than normal.		
13th	Tue			
14th	Wed			
15th	Thur		Rain	
16th	Fri	Very high maximum temperatures in the Roper-McArthur region over the coming week with temperatures regularly into the low 40's in a number of places.		
17th	Sat	Very cool overnight in the Kulgera area, with overnight temperatures possibly below 10°C.		**A V**
18th	Sun	Unusually cool in the Arnhem Lands, particulary in the Gove district. Also cooler than normal in the southern Alice Springs region, including in the Rabbit Flat and Arltunga districts, with maximum temperatures possibly not reaching into the 20's. Showers in the Roper-McArthur region to the Barkly Tablelands, with isolated showers spreading into the Alice Springs region in the next couple of days.		
19th	Mon		Rain, Th	**1Q**
20th	Tue			
21st	Wed		Th	

Date	Day	Northern Territory - October 2026	Darwin	Moon
22nd	Thur			
23rd	Fri	Very warm overnight in the Victoria River region.		
24th	Sat	Very hot in the Barkly with maximum temperatures possibly into the low 40's, including in the Elliott area. In contrast, cold overnight in the Arltunga area.	Showers, Th	XhN
25th	Sun		Th	
26th	Mon	Chance of stormy conditions across the Top End with thunderstorms, heavy rain, and strong gusty winds, particularly in the Darwin-Daly region, particularly in the Darwin suburbs. Possible showers around the central parts of the State either side of the Barkly Tablelands.	Rain, Th	F
27th	Tue	Some brief shower activity across the Top End over the next 2-3 days.	Rain, Th	
28th	Wed		Showers, Th	
29th	Thu		Rain, Th	P9
30th	Fri	Very warm overnight in the Victoria River region.	Th	^
31st	Sat			

November 2026

Date	Day	Northern Territory - November 2026	Darwin	Moon
1st	Sun			
2nd	Mon	Very hot in the Darwin-Daly region with maximum temperatures in the upper 30's in many districts over the next couple of days.		3Q
3rd	Tue			
4th	Wed			
5th	Thur			
6th	Fri	Cold overnight in the Alice Springs region, particularly in the Kulgera district where temperatures may not reach double digits.		XhS
7th	Sat	Very hot in the Alice Springs region with maximum temperatures in the lower 40's, particularly in the Rabbit Flat area.		
8th	Sun			
9th	Mon			N
10th	Tue	Very hot in the Alice Springs region with maximum temperatures into the early-40's in a number of districts, including in the Jervois district and very warm overnight.		
11th	Wed			
12th	Thu	Hot over the next 3-4 weeks in the Darwin-Daly region with fewer rain days than are typical for this time of year. No regions are likely to see minimum temperatures drop into single digits from now until the end of the year.		
13th	Fri	Cooler over the next couple of nights in the Alice Springs region.	Showers	V
14th	Sat	Brief rain activity in the Victoria River to Katherine regions.		A
15th	Sun	Isolated rain activity south of the Barkly Tablelands.		
16th	Mon			
17th	Tue			
18th	Wed	Rain in the Alice Springs region, with some heavy falls around the Barkly Tablelands over the next couple of days.		1Q
19th	Thur	Cooler temperatures in some southern parts of the Alice Springs region over the next few days.		
20th	Fri	Isolated showers around the Top End coastal fringes over the next couple of days.	Th	XhN
21st	Sat		Showers, Th	
22nd	Sun		Showers, Th	
23rd	Mon		Th	
24th	Tue	Very hot in the Victoria River region with maximums possibly reaching the mid to upper 40's.	Th	
25th	Wed		Th	F
26th	Thu	Bands of thunderstorms and heavy rain cross the State over the next 3-4 days.		P5
27th	Fri	Cooler temperatures in the Alice Springs region, particularly in the southwest.		^
28th	Sat	Heavy rain and scattered thunderstorms in the Alice Springs region over the next 3 days.		
29th	Sun	Very strong winds in the Yulara area.	Th	
30th	Mon	Widespread heavy rain throughout many parts of the State, possibly accompanying thunderstorm activity.	Rain, Th	

December 2026

Date	Day	Northern Territory - December 2026	Darwin	Moon
1st	Tue	Scattered rain activity in most parts of the State.	Rain, Th	3Q
2nd	Wed	Very hot in the Barkly over the next few days with maximum temperatures in the mid-40's.	Th	
3rd	Thur		Showers, Th	XhS
4th	Fri	Very warm overnight in the Barkly to the Arnhem regions.		
5th	Sat	Outbreaks of monsoonal rains across the Territory over the coming few days.	Th	
6th	Sun	Very high minimum temperatures overnight in the Top End region, particularly north of Darwin.		
7th	Mon		Th	
8th	Tue		Th	
9th	Wed		Th	N
10th	Thu		Showers, Th	V
11th	Fri	Very heavy monsoonal rains and thunderstorms over the Top End and with pockets of heavy rain possible in the southwest of the Alice Springs region.	Rain, Th	A
12th	Sat	The next few weeks until the end of the year may see above average temperatures almost everywhere, particularly over the Top End, and especially around the time of the onset of the monsoonal season, which may be earlier than is typical.		
13th	Sun	Very hot in the Alice Springs region over the next 3-4 days with maximum temperatures in the early 40's.	Showers, Th	
14th	Mon		Showers, Th	
15th	Tue	Monsoonal activity builds in most regions throughout the State over the next week.	Th	
16th	Wed	Very hot in the Barkly.	Rain, Th	
17th	Thur	Cool spell in the southern Alice Springs region with maximum temperatures only into the mid 20's likely.		1Q XhN
18th	Fri		Showers, Th	
19th	Sat		Th	
20th	Sun		Th	
21st	Mon	Very warm over the next couple of nights around the Roper-McArthur region including on Centre Island.	Rain, Th	
22nd	Tue	A wide cloudband over much of the State brings showers at first into the Alice Springs region turning to heavy rain by tomorrow and briefly spreading into the Top End regions.	Showers	
23rd	Wed	The monsoonal trough reactivates around this time with strong early signals, showing the development of tropical lows with cyclonic potentials.	Showers, Th	
24th	Thu	Heavy rain around the Darwin-Daly and in the southeast of the Alice Springs region.	Showers, Th	P1 F ^
25th	Fri	Cooler overnight in the Alice Springs region over the next few nights. Monsoonal rain throughout the Top End regions daily for the remainder of the year.	Rain	
26th	Sat	A tropical low may form in the Timor Sea to the northwest of Darwin around this time, but move away in an southwest trajectory over the coming couple of weeks.	Rain, Th	
27th	Sun		Rain, Th	
28th	Mon	The tropical low that formed near Darwin a couple of days ago, may be drifting around the Josephine Bonaparte Gulf around this time and intensifying as it heads towards the north of WA.	Rain	
29th	Tue	Scattered outbreaks of monsoonal rain activity in the Alice Springs region over the next few days.	Rain	
30th	Wed	Cooler overnight in the Kulgera district.	Rain, Th	XhS
31st	Thur		Rain, Th	3Q

NORTHERN TERRITORY GRAPHS AND TABLES 2026

Seasonal Breakdown – State Overall Rainfall Timing and Estimated mm Trend Graphs
State Overall Maximum Temperature Trend Graphs
Regional Rainfall Expectations Tables - mm and Trends
Regional Maximum and Minimum Temperature Trends Tables

Monthly Breakdown - **Rainfall:**
Regional Expectations by mm
Regional Expectations by mm Variance to Average
Regional Variance Trend by mm
Regional Estimated Expectations by % of Variance to Average
Regional Estimatedl Variance Trend by %

Maximum & Minimum Temperature Trends:
Regional Estimated Average Temperature
Regional Estimated Degree of Variation from the Average
Regional Estimated Temperature Trend

Seasonal Breakdowns

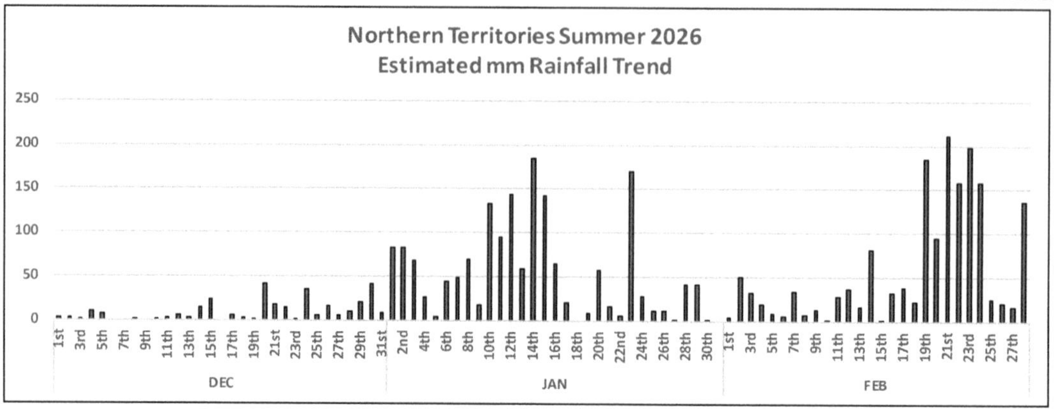

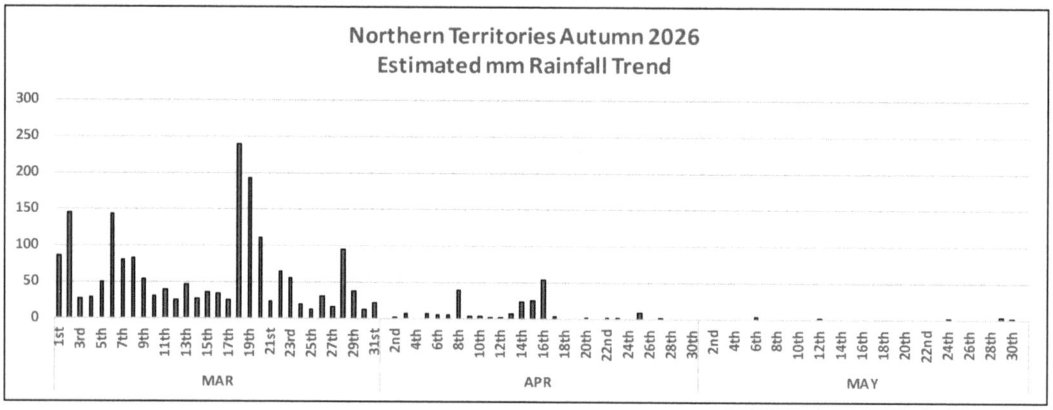

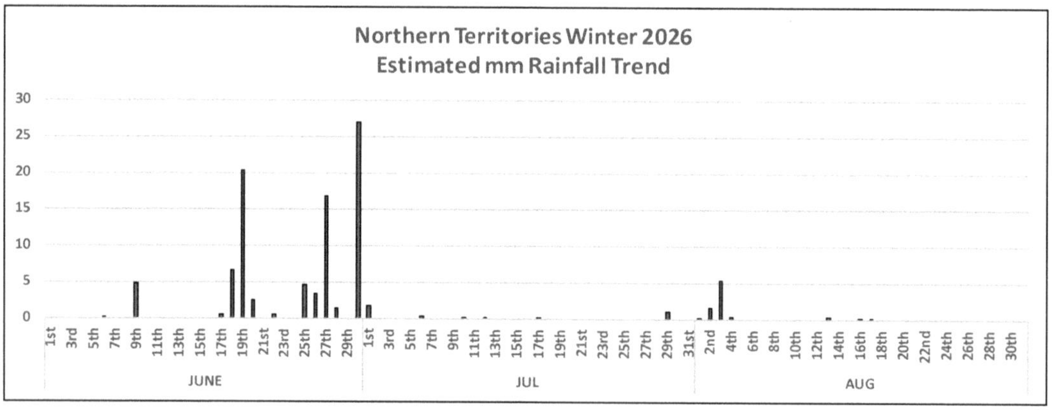

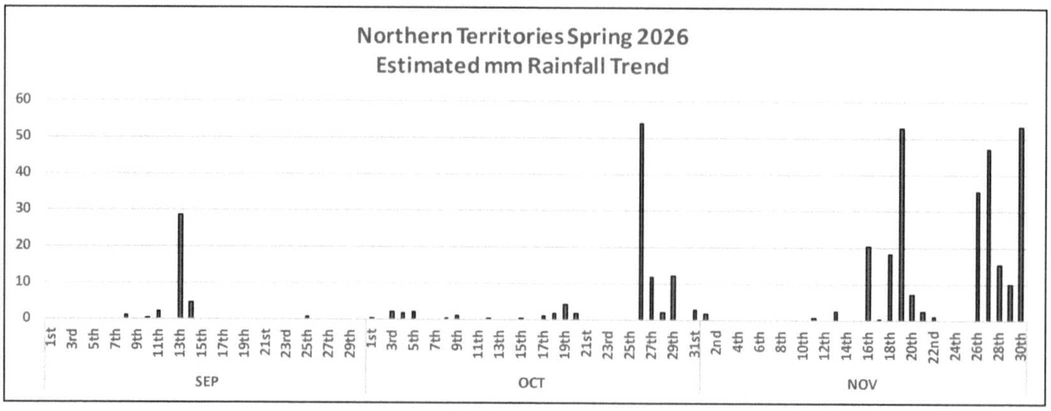

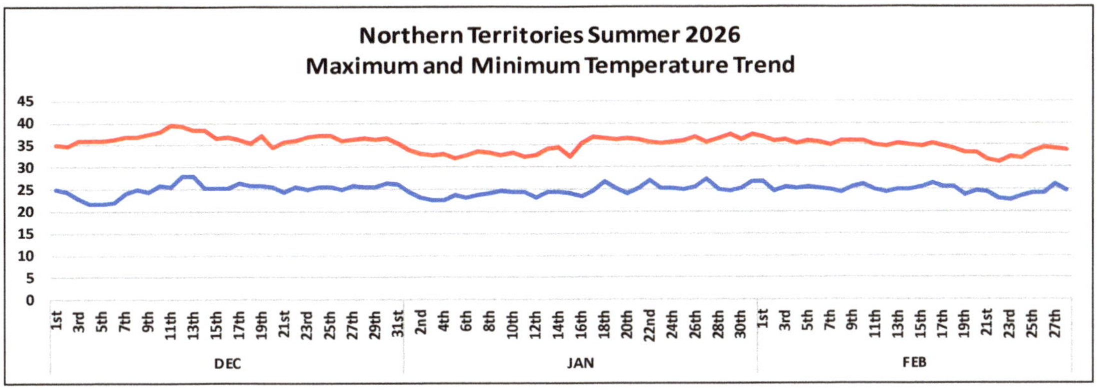
Northern Territories Summer 2026
Maximum and Minimum Temperature Trend
DEC
JAN
FEB

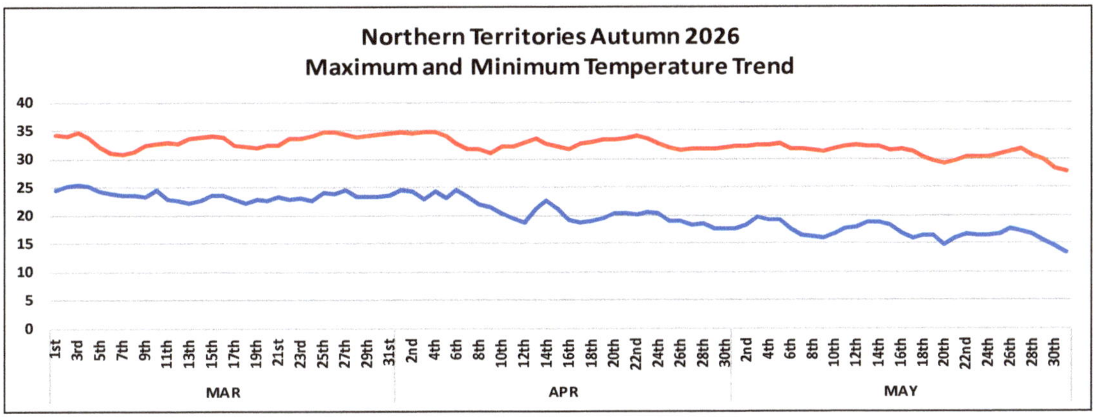
Northern Territories Autumn 2026
Maximum and Minimum Temperature Trend
MAR
APR
MAY

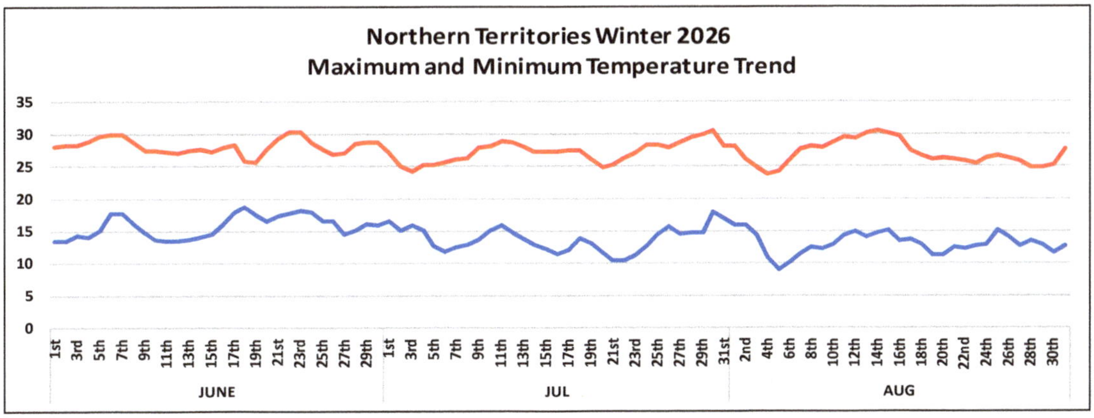
Northern Territories Winter 2026
Maximum and Minimum Temperature Trend
JUNE
JUL
AUG

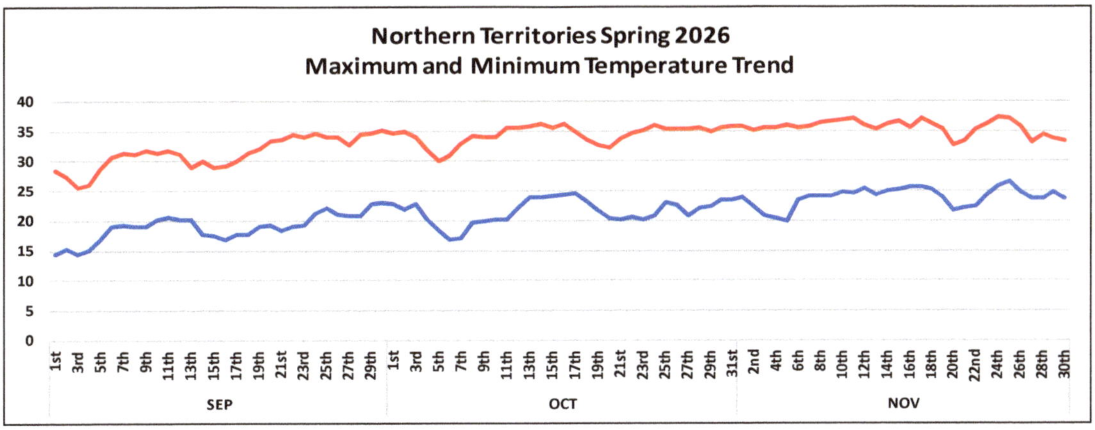
Northern Territories Spring 2026
Maximum and Minimum Temperature Trend
SEP
OCT
NOV

2026	NT - Estimated Seasonal Rainfall Expectations (mm)							
REGION	Arnhem Lands	Darwin Daly	Katherine	Roper-McArthur	Barkly	Alice Springs	Victoria River	NT
Town Sample	Gove	Darwin	Katherine	Centre Island	Tennant Ck	Alice Springs	Victoria River	Overall
Summer	564	879	876	457	88	34	716	3614
Autumn	495	580	341	457	8	0	210	2093
Winter	72	0	0	2	11	16	0	101
Spring	10	114	30	2	83	154	12	405
Overall	1207	1792	1293	958	230	220	991	6690

2026	NT - Estimated Seasonal Rainfall Variation from the Average (mm)							
REGION	Arnhem Lands	Darwin Daly	Katherine	Roper-McArthur	Barkly	Alice Springs	Victoria River	NT
Town Sample	Gove	Darwin	Katherine	Centre Island	Tennant Ck	Alice Springs	Victoria River	Overall
Summer	-205	-236	74	-164	-226	-87	278	-566
Autumn	-101	113	114	120	-72	-74	74	175
Winter	15	-9	-3	-8	2	-19	-6	-28
Spring	-49	-106	-106	-42	20	93	-72	-262
Overall	-274	-20	124	-54	-236	-72	328	-204

2026	NT - Estimated Seasonal Rainfall Trend Variation by 30mm							
REGION	Arnhem Lands	Darwin Daly	Katherine	Roper-McArthur	Barkly	Alice Springs	Victoria River	NT
Town Sample	Gove	Darwin	Katherine	Centre Island	Tennant Ck	Alice Springs	Victoria River	Overall
Summer	drr	drr	wtr	drr	drr	drr	wtr	drr
Autumn	drr	wtr	wtr	wtr	drr	drr	wtr	wtr
Winter	av	av	av	av	av	av	av	av
Spring	drr	drr	drr	drr	av	wtr	drr	drr
Overall	drr	av	wtr	av	drr	av	wtr	drr

2026	NT - Estimated Seasonal Rainfall Expectations by %							
REGION	Arnhem Lands	Darwin Daly	Katherine	Roper-McArthur	Barkly	Alice Springs	Victoria River	NT
Town Sample	Gove	Darwin	Katherine	Centre Island	Tennant Ck	Alice Springs	Victoria River	Overall
Summer	-27%	-21%	9%	-26%	-72%	-72%	64%	-14%
Autumn	-17%	24%	50%	35%	-90%	-100%	55%	9%
Winter	26%	-98%	-100%	-84%	26%	-53%	-100%	-22%
Spring	-82%	-48%	-78%	-96%	32%	151%	-86%	-39%
Overall	-19%	-1%	11%	-5%	-51%	-25%	49%	-3%

2026	NT - Estimated Seasonal Rainfall Trend Variation by 25%							
REGION	Arnhem Lands	Darwin Daly	Katherine	Roper-McArthur	Barkly	Alice Springs	Victoria River	NT
Town Sample	Gove	Darwin	Katherine	Centre Island	Tennant Ck	Alice Springs	Victoria River	Overall
Summer	drr	av	av	drr	drr	drr	wtr	av
Autumn	av	av	wtr	wtr	drr	drr	wtr	av
Winter	wtr	drr	drr	drr	wtr	drr	drr	av
Spring	drr	drr	drr	drr	wtr	wtr	drr	drr
Overall	drr	av	drr	drr	drr	av	wtr	drr

2026	NT - Estimated Seasonal Maximum Temperature Average and Trend							
Region Town Sample	Arnhem Lands Gove	Darwin Daly Darwin	Katherine Katherine	Roper-McArthu Centre Island	Barkly Tennant Ck	Alice Springs Alice Springs	Victoria River Victoria River	NT Overall
Summer Est	32.7	32.6	34.9	34.6	38.9	36.7	37.2	35.4
Trend	av	a-w	av	a-w	wmr	a-w	a-w	a-w
Autumn Est	30.8	32.5	33.3	31.9	33.8	30.4	34.2	32.4
Trend	av	av	av	av	wmr	wmr	av	a-w
Winter Est	28.0	31.5	30.4	27.0	25.1	20.6	29.8	27.5
Trend	av	a-w	av	av	av	av	av	av
Spring Est	31.3	33.8	37.4	32.4	33.8	30.2	37.9	33.8
Trend	av	a-w	av	av	a-c	a-c	av	av

2026	NT - Estimated Seasonal Minimum Temperature Average and Trend							
Region Town Sample	Arnhem Lands Gove	Darwin Daly Darwin	Katherine Katherine	Roper-McArthu Centre Island	Barkly Tennant Ck	Alice Springs Alice Springs	Victoria River Victoria River	NT Overall
Summer Est	24.9	25.2	24.4	27.3	26.2	21.7	24.8	24.9
Trend	av	av	av	av	wmr	a-w	av	av
Autumn Est	23.0	23.2	18.9	24.7	21.3	13.4	18.3	20.4
Trend	av	av	clr	av	wmr	a-w	clr	av
Winter Est	20.7	18.9	12.5	19.4	13.0	4.8	9.9	14.2
Trend	av	a-c	clr	av	av	av	clr	a-c
Spring Est	21.9	24.3	22.5	24.3	21.3	14.7	21.6	21.5
Trend	av	av	av	av	av	av	av	av

Monthly Breakdowns

2026	NORTHERN TERRITORY - Estimated Rainfall Expectations (mm)							
REGION	Arnhem Lands	Darwin Daly	Katherine	Roper-McArthur	Barkly	Alice Springs	Victoria River	NT
Town Sample	Gove	Darwin	Katherine	Centre Island	Tennant Ck	Alice Springs	Victoria River	Overall
January	373	412	419	292	17	3	162	1678
February	190	382	343	153	51	13	502	1635
March	374	538	301	457	4	0	208	1883
April	110	42	41	0	4	0	2	199
May	11	0	0	0	0	0	0	11
June	61	0	0	0	11	16	0	89
July	3	0	0	1	0	0	0	4
August	7	0	0	1	0	0	0	8
September	0	0	0	0	33	3	1	37
October	6	81	5	1	1	4	1	100
November	4	33	25	0	49	147	10	269
December	66	303	159	52	60	33	105	778
Overall	1207	1792	1293	958	230	220	991	6690

2026	NORTHERN TERRITORY - Estimated Rainfall Variation from the Average (mm)							
REGION	Arnhem Lands	Darwin Daly	Katherine	Roper-McArthur	Barkly	Alice Springs	Victoria River	NT
Town Sample	Gove	Darwin	Katherine	Centre Island	Tennant Ck	Alice Springs	Victoria River	Overall
January	67	-53	106	57	-107	-37	5	38
February	-89	9	114	-99	-71	-28	340	176
March	99	203	120	210	-49	-35	100	648
April	-112	-66	0	-82	-15	-20	-20	-314
May	-88	-25	-6	-9	-8	-18	-6	-159
June	31	-2	-1	-5	8	3	-3	30
July	-16	-1	0	-2	-4	-17	-3	-43
August	1	-6	-2	-1	-1	-5	-1	-15
September	-6	-17	-4	-1	27	-5	-3	-9
October	-8	16	-32	-5	-17	-18	-19	-82
November	-35	-104	-71	-37	10	115	-50	-172
December	-118	27	-102	-81	-9	-8	-14	-303
Overall	-274	-20	124	-54	-236	-72	328	-204

2026	NORTHERN TERRITORY - Estimated Rainfall Trend Variation by 10mm							
REGION	Arnhem Lands	Darwin Daly	Katherine	Roper-McArthur	Barkly	Alice Springs	Victoria River	NT
Town Sample	Gove	Darwin	Katherine	Centre Island	Tennant Ck	Alice Springs	Victoria River	Overall
January	wtr	drr	wtr	wtr	drr	drr	av	wtr
February	drr	av	wtr	drr	drr	drr	wtr	wtr
March	wtr	wtr	wtr	wtr	drr	drr	wtr	wtr
April	drr	drr	av	drr	drr	drr	drr	drr
May	drr	drr	av	av	av	drr	av	drr
June	wtr	av	av	av	av	av	av	wtr
July	drr	av	av	av	av	drr	av	drr
August	av	av	av	av	av	av	av	av
September	av	drr	av	av	wtr	av	av	av
October	av	wtr	drr	av	drr	drr	drr	drr
November	drr	drr	drr	drr	av	wtr	drr	drr
December	drr	wtr	drr	drr	av	av	drr	drr
Overall	drr	av	wtr	av	drr	av	wtr	drr

2026	NORTHERN TERRITORY - Estimated Rainfall Expectations by %							
REGION	Arnhem Lands	Darwin Daly	Katherine	Roper-McArthur	Barkly	Alice Springs	Victoria River	NT
Town Sample	Gove	Darwin	Katherine	Centre Island	Tennant Ck	Alice Springs	Victoria River	Overall
January	22%	-11%	34%	24%	-86%	-92%	3%	2%
February	-32%	2%	50%	-39%	-58%	-68%	211%	12%
March	36%	61%	66%	85%	-92%	-99%	93%	53%
April	-50%	-61%	1%	-100%	-79%	-100%	-90%	-61%
May	-89%	-100%	-100%	-98%	-100%	-100%	-100%	-93%
June	99%	-100%	-100%	-96%	238%	21%	-100%	51%
July	-84%	-100%	0%	-74%	-100%	-100%	-100%	-92%
August	10%	-97%	-100%	-64%	-100%	-100%	-100%	-64%
September	-100%	-100%	-100%	-64%	408%	-63%	-77%	-19%
October	-56%	24%	-86%	-77%	-94%	-82%	-97%	-45%
November	-89%	-76%	-74%	-100%	26%	350%	-83%	-39%
December	-64%	10%	-39%	-61%	-12%	-18%	-12%	-28%
Overall	-19%	-1%	11%	-5%	-51%	-25%	49%	-3%

2026	NORTHERN TERRITORY - Estimated Rainfall Trend Variation by 25%							
REGION	Arnhem Lands	Darwin Daly	Katherine	Roper-McArthur	Barkly	Alice Springs	Victoria River	NT
Town Sample	Gove	Darwin	Katherine	Centre Island	Tennant Ck	Alice Springs	Victoria River	Overall
January	av	av	wtr	av	drr	drr	av	av
February	drr	av	wtr	drr	drr	drr	wtr	av
March	wtr	wtr	wtr	wtr	drr	drr	wtr	wtr
April	drr	drr	av	drr	drr	drr	drr	drr
May	drr	drr	drr	drr	drr	drr	drr	drr
June	wtr	drr	drr	drr	wtr	av	drr	wtr
July	drr	drr	av	drr	drr	drr	drr	drr
August	av	drr	drr	drr	drr	drr	drr	drr
September	drr	drr	drr	drr	wtr	drr	drr	av
October	drr	av	drr	drr	drr	drr	drr	drr
November	drr	drr	drr	drr	wtr	wtr	drr	drr
December	drr	av	drr	drr	av	av	av	drr
Overall	drr	av	drr	drr	drr	av	wtr	drr

2026	NORTHERN TERRITORIES - Estimated Average Maximum Temperatures							
Region	Arnhem Lands	Darwin Daly	Katherine	Roper-McArthu	Barkly	Alice Springs	Victoria River	NT Estimates
Town Sample	Gove	Darwin	Katherine	Centre Island	Tennant Ck	Alice Springs	Victoria River	Overall
January	32	32	33	34	39	38	37	**35**
February	32	32	34	34	39	37	35	**35**
March	30	31	33	32	37	35	34	**33**
April	31	33	34	33	34	29	35	**33**
May	31	33	33	31	30	27	33	**31**
June	29	32	31	28	26	21	31	**28**
July	28	31	30	27	25	21	29	**27**
August	27	32	31	26	24	20	29	**27**
September	30	33	36	30	30	26	36	**31**
October	32	34	38	33	35	32	38	**34**
November	33	35	39	35	37	33	40	**36**
December	34	33	37	36	38	34	38	**36**
Overall	**31**	**33**	**34**	**31**	**33**	**29**	**35**	**32**

2026	NORTHERN TERRITORY - Estimated Average Minimum Temperatures							
Region	Arnhem Lands	Darwin Daly	Katherine	Roper-McArthu	Barkly	Alice Springs	Victoria River	NT Estimates
Town Sample	Gove	Darwin	Katherine	Centre Island	Tennant Ck	Alice Springs	Victoria River	Overall
January	25	25	24	26	26	21	25	**25**
February	25	25	24	27	26	23	24	**25**
March	24	24	23	26	25	18	24	**24**
April	23	24	19	26	22	14	18	**21**
May	22	22	14	22	17	9	12	**17**
June	22	20	14	21	15	6	12	**16**
July	21	19	12	19	12	4	9	**14**
August	19	18	11	18	12	4	8	**13**
September	21	23	20	22	18	11	19	**19**
October	22	24	23	25	22	15	21	**22**
November	23	26	25	27	24	19	24	**24**
December	26	26	26	28	26	21	26	**25**
Overall	23	23	20	24	20	14	19	**20**

2026	NORTHERN TERRITORY - Estimated Degree of Variation from the Maximum Average							
Region	Arnhem Lands	Darwin Daly	Katherine	Roper-McArthu	Barkly	Alice Springs	Victoria River	NT
Town Sample	Gove	Darwin	Katherine	Centre Island	Tennant Ck	Alice Springs	Victoria River	Overall
January	-0.4	-0.1	-1.1	0.2	2.0	1.1	0.4	0.3
February	0.4	0.5	0.4	0.8	2.8	1.6	-0.7	0.8
March	-0.9	-0.7	-1.7	-0.9	2.7	2.6	-1.4	-0.1
April	0.3	0.3	0.3	0.9	2.0	0.8	0.6	0.7
May	0.7	1.1	0.5	1.2	2.6	3.3	1.5	1.5
June	0.6	1.0	0.6	1.3	1.4	1.5	1.2	1.1
July	-0.1	0.6	-0.3	0.3	0.2	0.9	-0.2	0.2
August	-1.4	0.3	-1.8	-1.7	-3.1	-3.1	-2.4	-1.9
September	-0.8	0.6	-0.3	-0.9	-1.9	-1.0	0.1	-0.6
October	-0.2	0.5	0.2	-0.2	-0.1	0.4	0.4	0.1
November	-0.5	1.2	1.1	-0.1	0.2	-1.3	1.0	0.2
December	1.0	0.5	0.9	0.8	0.5	-1.6	0.3	0.4
Overall	-0.1	0.4	-0.1	0.1	0.8	0.4	0.0	0.2

2026	NORTHERN TERRITORY - Estimated Degree of Variation from the Minimum Average							
Region	Arnhem Lands	Darwin Daly	Katherine	Roper-McArthu	Barkly	Alice Springs	Victoria River	NT
Town Sample	Gove	Darwin	Katherine	Centre Island	Tennant Ck	Alice Springs	Victoria River	Overall
January	0.5	0.1	-0.3	-0.4	1.3	-1.0	0.2	0.1
February	-0.1	0.1	0.4	0.5	1.6	2.6	-0.1	0.7
March	0.0	-0.2	0.3	-0.5	1.9	0.5	0.7	0.4
April	-0.1	-0.5	-1.5	0.4	1.0	0.9	-1.3	-0.1
May	-0.8	-0.4	-3.3	-0.1	0.8	0.4	-3.5	-1.0
June	0.5	-0.1	0.0	1.4	2.0	1.3	0.2	0.8
July	0.8	-0.4	-1.2	0.3	0.1	-0.1	-1.9	-0.4
August	0.2	-2.2	-3.8	-1.2	-2.4	-1.8	-4.7	-2.3
September	1.2	-0.4	0.0	-0.4	-0.6	0.5	1.0	0.2
October	0.1	-0.5	-1.1	-0.3	0.0	-0.1	-1.3	-0.5
November	-0.5	0.5	-0.2	-0.3	0.3	0.8	-0.2	0.1
December	1.1	0.5	0.7	0.6	0.7	0.3	0.8	0.7
Overall	0.2	-0.3	-0.8	0.0	0.6	0.4	-0.8	-0.1

2026	NORTHERN TERRITORIES - Estimated Maximum Temperature Trend							
Region	Arnhem Lands	Darwin Daly	Katherine	Roper-McArthu	Barkly	Alice Springs	Victoria River	NT
Town Sample	Gove	Darwin	Katherine	Centre Island	Tennant Ck	Alice Springs	Victoria River	Overall
January	av	av	clr	av	wmr	wmr	av	av
February	av	av	av	a-w	wmr	wmr	a-c	a-w
March	a-c	a-c	clr	a-c	wmr	wmr	clr	av
April	av	av	av	a-w	wmr	a-w	a-w	a-w
May	a-w	wmr	av	wmr	wmr	wmr	wmr	wmr
June	a-w	a-w	a-w	wmr	wmr	wmr	wmr	wmr
July	av	a-w	av	av	av	a-w	av	av
August	clr	av	clr	clr	clr	clr	clr	clr
September	a-c	a-w	av	a-c	clr	av	av	a-c
October	av	av	av	av	av	av	av	av
November	av	wmr	wmr	av	av	clr	a-w	av
December	a-w	av	a-w	a-w	av	clr	av	av
Overall	av	av	av	av	a-w	av	av	av

2026	NORTHERN TERRITORY - Estimated Minimum Temperature Trend							
Region	Arnhem Lands	Darwin Daly	Katherine	Roper-McArthu	Barkly	Alice Springs	Victoria River	NT
Town Sample	Gove	Darwin	Katherine	Centre Island	Tennant Ck	Alice Springs	Victoria River	Overall
January	av	av	av	av	wmr	a-c	av	av
February	av	av	av	av	wmr	wmr	av	a-w
March	av	av	av	av	wmr	av	a-w	av
April	av	av	clr	av	a-w	a-w	clr	av
May	a-c	av	clr	av	a-w	av	clr	a-c
June	av	av	av	wmr	wmr	wmr	av	a-w
July	a-w	av	clr	av	av	av	clr	av
August	av	clr	clr	clr	clr	clr	clr	clr
September	wmr	av	av	av	a-c	av	a-w	av
October	av	av	clr	av	av	av	clr	av
November	av	av	av	av	av	a-w	av	av
December	wmr	av	a-w	a-w	a-w	av	a-w	a-w
Overall	av	av	a-c	av	a-w	av	a-c	av

QUEENSLAND 2026

GENERAL TRENDS

Year Overall: Wetter by up to 15-20% more than the average
Slightly cooler but within 0.5°C of the average

Summer: Wetter by up to 35-40% more than the average
Wettest – Second halves of January and February
Driest – Mid-December
Warmer by around 1°C above the average
Hottest – Second half of December
Coolest – Second week of January and second half of February

Autumn: Drier by around 20-25% below the average
Wettest – First half of March
Driest – Second and third weeks of May
Slightly warmer but within 0.5°C of the average
Hottest – First week of March
Coolest – Second half of May

Winter: Wetter by up to 60-65% more than the average
Wettest – Second and third weeks of June and first week of August
Driest – First half of July and last half of August
Slightly cooler but within 0.5°C of the average
Hottest – Second half of June and July and second week of August
Coolest – First half of August

Spring: Drier by around 20-25% below the average
Wettest – Third and fourth weeks of September
Driest – First half of October and second week of November
Cooler by up to 0.5-1°C below the average
Hottest – Second half of November
Coolest – First half of September

Major Weather Events

January 2026

Date	Day	Queensland - January 2026	Brisbane	Moon
1st	Thur	Very wet throughout the State over the coming week under a strong monsoonal influence.		
2nd	Fri		Showers	P6
3rd	Sat	Chance of heavy rain in the Mackay to the adjacent inland Highland areas.		F ^
4th	Sun	Brief respite from the monsoonal rains around this time before they return with vigour for the remainder of the week.		
5th	Mon	Heavy rain likely on the North Coast, with the chance of patches of flooding along the local river systems, including in the Bulgun and Jarra Creeks. Outbreaks of heavy rain also in the southeast in the Moreton and Gold Coast regions over the next 2-3 days.	Showers	
6th	Tue	Cool day in the Darling Downs, particularly in the Toowoomba district, where maximum temperatures may not reach into the 20's.	Showers	
7th	Wed	Mild overnight temperatures over much of the State over the next few days, with minimum temperatures dropping into the mid-teens in many districts. Unusually cold overnight in the Bedourie area of the Lower Western with minimum temperatures dropping into the single digits in some areas. Very heavy rain around this time in the Herbert-North Coast to the Cassowary Coast regions with chance of flash-flooding in some districts including in the Daintree area.	Showers	
8th	Thu	Heavy rain over Cape York Peninsula over the next couple of days, with widespread showers through the eastern and southern parts of the State. Mostly dry in the Central West to the Gulf Country.		
9th	Fri	Mostly dry spell throughout the State except for over Cape York Peninsula.	Showers	XhS
10th	Sat	Showers along the eastern coastal regions to the adjacent inland districts over the next week, while mostly dry in the west of the State.	Showers	
11th	Sun	The next 3-4 weeks may be generally wetter than average, particularly in the southeast of the State, and warmer than normal over the east of Cape York Peninsula, on the Barron North Coast, in the Lower Carpentaria, and in the far west of the State, with most other regions cooler than is normal for this time of year. Chance of strong winds over the South Coast-Moreton around this time, with chance of gale gusts.	Showers	3Q
12th	Mon	A low pressure system off the southeast of the State brings widespread rain activity over the coming week, particularly into the southeastern regions.	Showers	
13th	Tue		Rain	
14th	Wed	Thunderstorms and heavy rain in the North Tropical Coast and adjacent Tablelands. Chance of brief flooding in the Herbert River catchment. Mild temperatures in the Darling Downs to the Maranoa likely, with maximum temperatures only into the low 20's in many districts.	Rain	A
15th	Thur	A line of thunderstorms and strong gusty winds cross from the Central Coast, up through the Northern Goldfields and Upper Flinders and into the Gulf Country. Very heavy rain in the Gold Coast to the Tropical North around this time, with the chance of local flooding in some districts, including in the Proserpine and Glengaven districts. Warm spell throughout the State over the next 3-4 days, particularly in the southeast quarter and along the eastern coasts.		
16th	Fri	Severe thunderstorms cross through the southeast quarter of the State, with chance of flash flooding including in the Brisbane region.	Rain, Th	
17th	Sat	Isolated thunderstorms and some outbreaks of heavy rain cross the Darling Downs to the Granite Belt regions. Very hot in the Carpentaria to the Central Lowlands, with maximum temperatures reaching into the low 40's in some districts.		V
18th	Sun	Monsoonal activity invigorates over the North Tropical Coasts to the adjacent Tablelands bringing scattered thunderstorms and heavy rain. Chance of a tropical cyclone tracking to the Gulf of Carpentaria and across Cape York and into the central parts of the State over the coming week to 10 days. Very hot in the Lower Western around this time, particularly in the Birdsville district, with maximum temperatures into the mid-40's likely.	Showers	
19th	Mon	Showers develop over the southern half of the State around this time and intensify to heavy rain over the next 3-4 days.	Showers	N
20th	Tue			
21st	Wed	Thunderstorms and heavy rain at times in the Herbert to the Lower Burdekin regions. Mild maximum temperatures in some parts of the Maranoa and Mackay regions.	Showers	
22nd	Thu	Outbreaks of heavy rain in the Channel Country around this time and over the coming week.		
23rd	Fri	Heavy rain associated with an ex-tropical cyclone may bring some flash flooding in areas in the Herbert to the Lower Burdekin, over the next few days, particularly south of Townsville.		

Date	Day	Queensland - January 2026	Brisbane	Moon
24th	Sat	Heavy rain associated with a decaying cyclonic system continues throughout the Herbert to Lower Burdekin regions. Very heavy falls likely in the Whitsunday-Tropical North region particularly in the Proserpine district once again. Very hot in the Maranoa, Warrego, Far Southwest and Lower Carpentaria around this time with maximum temperatures into the low 40's likely in many districts.	Rain	XhN
25th	Sun	Heavy rain and very strong and blustery winds associated with an ex-tropical cyclone continues to bring flooding risks in the Herbert and Lower Burdekin regions, especially along the Burdekin River.	Rain	
26th	Mon	Hot overnight in the Lower Western and North Peninsula regions, particularly in the Birdsville district, with overnight minimums possibly into the high 20's to early 30's in some areas. Widespread rain activity throughout the State over the next 4-5 days.	Showers	1Q
27th	Tue	Widespread rain and scattered thunderstorms within the monsoonal trough brings heavy falls at times in the Central Coast - Whitsundays regions, Central Highlands and Lowlands and into the Central West. Chance of brief flooding in the Pioneer River catchment.	Showers	
28th	Wed	Heavy rain in the Central West around this time, particularly in the Listowel Downs district as well as in some parts of the southeast, including in the Wallen area.		
29th	Thur			
30th	Fri	Heavy rain in the Central Coast-Whitsunday districts, with chance of flooding along the Pioneer River and Bakers Creek catchments. Thunderstorms with severe wind gusts in the Northwest, particularly around Julia Creek. Very hot in the Lower Carpentaria, particularly in the Mt Isa district.	Showers	P10 ^
31st	Sat	Scattered showers and thunderstorms in the Southeast Coast districts, with some heavy falls at times. Very hot in the Upper Western with maximum temperatures into the low 40's in a number of districts. Rain activity in the inland parts clears to mostly dry through the Western districts as well as Carpentaria around this time.		

February 2026

Date	Day	Queensland - February 2026	Brisbane	Moon
1st	Sun	Scattered rain activity in the southeast quarter of the State while mostly dry elsewhere.		
2nd	Mon	Rain clears in the southeast, but showers spread elsewhere over the next couple of days.		F
3rd	Tue			
4th	Wed	Scattered severe thunderstorms and damaging winds in the Channel Country. Warm overnight in the southeast of the State over the next few days. Unusually cool overnight in the Darling Downs, particularly in the Applethorpe district, with minimum temperatures barely rising into the double digits in some areas.		
5th	Thu	Heavy rain and monsoonal storms over the Gulf Country and Cape York Peninsula, easing to showers by tomorrow.		
6th	Fri	Mostly dry in the southern half of the State.		XhS
7th	Sat	Brief showery spell through the western half of the State, while clearing to mostly dry over Cape York Peninsula and in the southeast quarter of the State.	Showers	
8th	Sun	Mostly dry except for some spotty showers over Cape York Peninsula.	Rain	
9th	Mon	Further thunderstorms along the east coast of the State over the next few days with heavy falls likely in the Herbert and Lower Burdekin areas, with flooding risks still lingering in the Pioneer River catchment. Heavy rain over Cape York Peninsula over the next few days.	Showers	
10th	Tue	Widespread shower activity spreads over the southern half of the State around this time.		3Q
11th	Wed	The next 3-4 weeks may bring above normal rainfall and cooler temperatures due to frequent cloud cover accompanying an active monsoonal trough, including flash flooding events, particularly in some parts of the Central, North Coastal, and Capricornia regions. In contrast, some parts in the west and northwest may be drier and warmer than is typical for this time of year.		A
12th	Thur	Widespread monsoonal rain activity throughout most regions of the State over the coming week.	Rain	
13th	Fri	Pockets of heavy rain in the southeast quarter of the State including on the Darling Downs and Granite Belt regions. Chance of heavy rain in the Mt Tamborine area that may cause damage to local roads.	Rain	V
14th	Sat	An inland trough brings widespread thunderstorms across the Darling Downs and Granite Belt regions, with some heavy downpours at times, as well as very heavy rain for a time around the Gold Coast hinterland which may cause some localised flooding.	Rain	
15th	Sun	Thunderstorms and heavy rain over the northern coasts including in the Wide Bay and Burnett districts, as well as in many up into the adjacent inland regions from the Carpentaria to the Warrego. Hot in the Lower Carpentaria region over the next couple of days, particularly in the Julia Creek and Mt Isa districts with maximum temperatures in the low 40's likely.	Showers, Fg	

Date	Day	Queensland - February 2026	Brisbane	Moon
16th	Mon	A surface trough crosses the southeast of Queensland triggering widespread showers and scattered thunderstorms from the Darling Downs and Granite Belt, in the Southeast Coastal regions, up through the Wide Bay Burnett areas, and into the Central Highlands, Northern Goldfields, Upper Flinders and Northern Tropical Coasts and Tablelands.	Rain, Th	
17th	Tue	Heavy rain in parts of the Wide Bay and Burnett districts as well as down through the Granite Belt and into the Darling Downs. Very warm spell in the eastern coastal regions over the next couple of days.		
18th	Wed	Scattered severe thunderstorms and strong gusty winds within the monsoon trough continue to bring heavy rain in the Herbert and Lower Burdekin regions to the Southeast Coast. Over the coming week, the monsoonal trough may become particularly active over the northern tropical regions bringing scattered showers and storms across the Gulf and Tropics north of the Mackay, as well as around the Capricornia coast.	Show ers	N
19th	Thu	Severe thunderstorms within a trough line continues crossing the southeast, through the Herbert and Lower Burdekin and Southeast Coastal regions. Similarly, scattered storms prevail over the north of the State for a time.		
20th	Fri	Cool overnight in the Darling Downs with minimum temperatures dropping into the single digits in a number of districts.		XhN
21st	Sat	The monsoonal trough maintains severe thunderstorms and outbeaks of heavy rain throughout the eastern coastal regions of the State from the Herbert and Lower Burdekin, Wide Bay and Burnett, and Central Coast-Whitsundays areas, with rain activity elsewhere. Winds may be very strong with damage potential at times, particularly in the Wide Bay, Burnett, and Central Coast regions.	Rain	
22nd	Sun	An inland tropical low may form in the monsoonal trough over the southeast quarter of the State bringing further widespread rain, damaging winds, and thunderstorms throughout the State over the next 3-4 days, and especially from the Central Coast-Whitsunday areas, and including in the Mackay region. Strong stormy winds at times in the Island off the Capricornia, Wide Bay and Burnett, and Central Coast-Whitsundays coasts. Chance of heavy rain causing localised flooding in the Channel Country.	Rain	
23rd	Mon	While storm activity eases in the south of the State, a band of thunderstorms within the monsoonal trough moves through the Central Coast districts, including up into the Northern Goldfields and Upper Flinders.	Rain	
24th	Tue	Scattered thunderstorms and outbreaks of heavy rain along the Central Coast, including in the Herbert and Lower Burdekin.		
25th	Wed	Thunderstorms and heavy rain in the Herbert and Lower Burdekin regions, and through the Central Coast-Whitsundays areas including in the Mackay. Very heavy rain in the Mackay around this time may see flash flooding in numerous parts of the region.		1Q P14
26th	Thur	Thunderstorms and occasional heavy falls in the Herbert and Lower Burdekin persist. Strong winds on the Central Coast-Whitsundays regions.	Show ers	^
27th	Fri	Some shower activity in most districts.	Show ers	
28th	Sat	Mild maximum temperatures in some parts of the Mackay region around this time, with daily maximas only reaching into the mid-20's likely in a number of districts. Mostly dry through the central parts of the State from the Gulf Country to the Carpentaria and Mackay with shower activity elsewhere and heavy rain persisting in the central parts of Cape York Peninsula.	Rain	

March 2026

Date	Day	Queensland - March 2026	Brisbane	Moon
1st	Sun	Mostly dry through the central parts of the State over the next 4-5 days from the Central West to the central eastern coasts, with showers over the Cape York Peninsula and in the southern regions from the Warrego to the Fraser Coast.		
2nd	Mon			
3rd	Tue	Showers clearing to some isolated showers in the Warrego to the Central Highlands, while mostly dry elsewhere.		F
4th	Wed	Mostly dry and very warm spell across the State over the next couple of days, particularly in the eastern coastal regions from Port Curtis to the South Coast-Moreton and in the Darling Downs regions. Very hot day in the Lower Western with maximum temperatures into the mid-40's likely in a number of areas, including in the Birdsville district.	Fg	
5th	Thu	A hot westerly wind passes over the southeast of the State bringing very high temperatures into the Brisbane to the Gold Coast areas, with maximums up to the 40's likely. Very high overnight temperatures into the low 30's in the Lower Western, particularly in the Boulia area.		XhS
6th	Fri	Very hot in the Upper Western around this time, with maximum temperatures into the low 40's possible, particularly in the Winton area. Some showers return in the northern half of the State.		

Date	Day	Queensland - March 2026	Brisbane	Moon
7th	Sat	Isolated thunderstorms in the Capricornia and heavy monsoonal rains in the region may bring some flash flooding in some parts, including in the Rockhamptom district. The Fitzroy River may be in flood around this time. Cool spell in the Maranoa with some districts maximum temperatures not rising into the early 20's.	Show ers	
8th	Sun	Unusually cool and showery spell in the Darling Downs region about this time, particularly in the Toowoomba district, where maximum temperatures may not reach into the 20's in a number of areas. Heavy rain throughout Cape York Peninsula over the next several days.	Rain	
9th	Mon	Falls of heavy rain within the monsoon trough over the North Tropical Coast and Tablelands. Rain activity also intensifies over the southern half of the State briefly before easing back to showers over the next couple of days.		
10th	Tue			
11th	Wed	Heavy rain along the coastal fringes from the Fraser Coast to the Moreton over the next couple of days.		3Q A
12th	Thur	Further thunderstorms and heavy rain within the monsoonal trough over the North Tropical Coasts and into the adjacent Tablelands, including in Cairns. Chance of flooding in the Johnstone River catchment over the next few days. Despite this spell, overall, the next 3-4 weeks may see a return to average rainfall across the State in most regions, except on the Northern Tropical Coast and adjacent inland parts, which may continue to be wetter than is normal for this time of year. Temperatures during the next few weeks may also continue to be cooler than normal, except in the west, which may be slightly warmer overall.		V
13th	Fri	Scattered thunderstorms and heavy rain on the North Tropical Coast and adjacent Tablelands over the coming week. Mostly dry elsewhere.		
14th	Sat	Showers in the southeast from the Fraser Coast to the Darling Downs and South Coast-Moreton over the next couple of days.		
15th	Sun			
16th	Mon	Very heavy rain in the Tropical North may see localised flooding in the Port Douglas district around this time. Mostly dry elsewhere.		
17th	Tue	Brief showers possible in the southeast quarter of the State around this time.		
18th	Wed	Rain persists over Cape York Peninsula over the next couple of days while mostly dry elsewhere.		
19th	Thu			N XhN
20th	Fri	Showers over Cape York Peninsula as well as in the South Coast-Moreton to the Darling Downs over the next couple of days.	Show ers	
21st	Sat	Very strong winds in the Moreton region. Rain activity briefly spreads up along the eastern coasts to the Mackay and Central Coasts.	Show ers	
22nd	Sun	Regular outbreaks of monsoonal rains over Cape York Peninsula over the coming week, while mostly dry with just the odd showery spell elsewhere.		P11
23rd	Mon	Chance of some showers briefly in the eastern regions tablelands from the Mackay to the Darling Downs and South Coast-Moreton.		
24th	Tue	Mostly dry over the next few days throughout the State except over Cape York Peninsula.		
25th	Wed	Hot spell in the southern regions of the State over the next couple of days, with a number of districts possibly seeing maximum temperatures into the upper 30's to the low 40's.		
26th	Thur	Very hot day in the Lower Western region, with maximums into the low 40's likely, particularly in the Bedourie district. Chance of some isolated showers around the Fraser Coast to the South Coast over the next few days.	Show ers	1Q ^
27th	Fri		Show ers	
28th	Sat	Cool day with passing showers in the Darling Downs, particularly in the Toowoomba district, where maximum temperatures may not reach into the 20's.	Show ers	
29th	Sun	Brief shower activity in the southeast quarter of the State over the next couple of days.	Show ers	
30th	Mon	Some showers from the Mackay to the South Coast over the next couple of days.	Show ers	
31st	Tue	Mostly dry on Cape York Peninsula and through the western half of the State.	Show ers	

April 2026

Date	Day	Queensland - April 2026	Brisbane	Moon
1st	Wed	Mostly dry day throughout the State.		
2nd	Thu	Some showers in the east of the State from the Mackay to the South Coast over the next couple of days. Isolated showers over Cape York Peninsula.		F XhS
3rd	Fri	A tropical cyclone may have formed off the QLD coast and drift around the Coral Sea over the next few days, but isn't expected to make landfall in the State.	Show ers	
4th	Sat	Isolated showers in the Channel Country and on Cape York Peninsula. Mostly dry elsewhere.	Show ers	

Date	Day	Queensland - April 2026	Brisbane	Moon
5th	Sun	Unusually hot spell over much of the State during the next week, with a number of districts seeing maximum temperatures into the upper 30's at times. Hot overnight in the Carpentaria with a number of districts likely to see minimum temperatures in the upper 20's. Some shower activity in the southeast quarter of the State over the next couple of days, dry elsewhere.		
6th	Mon	Isolated showers in the Gulf Country to the Carpentaria.	Showers	
7th	Tue	Heavy rain at times on the Darling Downs to the Granite Belt regions. Cooler overnight temperatures become established over the coming week in many districts, particularly in inland parts.	Rain	A
8th	Wed	A band of thunderstorms, rain, some hail showers, and strong gusty winds cross the Southeast regions of the State over the next couple of days. Chance of storm damage in some areas, including in the Mango Hill district. Strong gusty winds may generate high seas offshore particularly at Cape Moreton.	Showers, Th	
9th	Thur	Overnight minimum temperatures likely to drop into single digits in some parts of the Darling Downs, Maranoa, Warrego, and Western districts around this time, and spread into the Central Lowlands and Upper Western the following night. Cooler spell in the Port Curtis to the Mackay regions with maximum temperatures only into the early 20's likely in a number of districts. Outbreaks of heavy rain likely from the North Coast-Herbert to the Central West and Central Lowlands to the Port Curtis regions.		V
10th	Fri	Chance of patches of frost in the Stanthorpe region of the Darling Downs around this time. Chance of heavy rain for a time over the next couple of days in northern parts of Cape York Peninsula. Some lingering showers in the South Coast-Moreton region.		3Q
11th	Sat	Very cold overnight in the Darling Downs. Mostly dry throughout the State over the next week, except for showers at times in the far north of the Cape York Peninsula.		
12th	Sun	Overall, the next 3-4 weeks may be drier than is typical for this time of year, as well as minimum temperatures continuing to be cooler than average, but maximum temperatures may be closer to the norm. Warm spell over the eastern half of the State during the next few days with daily maximum temperatures in the high 20's to mid 30's in a number of districts.		
13th	Mon			
14th	Tue			
15th	Wed	Very heavy rain in the South Peninsula region, with chance of flooding along the Lockhart River. Dry elsewhere throughout the State over the next couple of days.		
16th	Thu			XhN
17th	Fri	Scattered thunderstorms on the Sunshine Coast and Hervey Bay regions, with very strong winds at times, particularly around Double Island Point over the next couple of days. Dry in the western half of the State for most of the remainder of the month.	Showers	N
18th	Sat			
19th	Sun	Isolated showers in Cape York Peninsula over the next couple of days.	Showers	P7
20th	Mon	Cool day on the Darling Downs, particularly in the Applethorpe district, where maximum temperatures may sit around the mid-teens. Brief showers likely in the southeast quarter of the State.		
21st	Tue	Mostly dry throughout the State over the next 4-5 days.		
22nd	Wed			^
23rd	Thur	Very hot day in the Boulia area of the Lower Western with maximum temperatures in the upper 30's to low 40's possible.		
24th	Fri	Warm over the next couple of days in the eastern half of the State.		1Q
25th	Sat			
26th	Sun	Some showers likely in the southeast of the State from the Port Curtis region to the Darling Downs and adjacent coastal districts.	Th	
27th	Mon	Dry spell throughout most of the State over the next few days.		
28th	Tue	Chance of isolated showers in Cape York Peninsula over the next couple of days.	Showers	
29th	Wed	Isolated shower activity likely in the southeast of the State over the next couple of days.	Showers	XhS
30th	Thur	Brief showers may spread into the Central Highlands and Port Curtis regions around this time.	Showers	

May 2026

Date	Day	Queensland - May 2026	Brisbane	Moon
1st	Fri	Mostly dry throughout the coming month in the western half of the State. Isolated showers over the eastern regions from the Mackay to the South Coast-Moreton at first, becoming more frequent by tomorrow.		
2nd	Sat	Scattered thunderstorms and possible hail showers in the southeast of the State including up into the Capricornia region and along the Southeast Coastal regions. Winds may be very strong at times in the Capricornia.		F

Date	Day	Queensland - May 2026	Brisbane	Moon
3rd	Sun	Very warm overnight in the Lower Carpentaria with minimum temperatures still lingering in the upper 20's in many districts, particularly in the Mornington Island district. Dry throughout the State.		
4th	Mon	Wet with storms and the chance of lightning strikes in the Darling Downs and Granite Belt regions. Very strong and blustery winds for a time off the Capricornia coast, particularly in the Rundle Island area. Some rain around the North Coast.		
5th	Tue	Heavy rain likely in the coastal districts of the Wide Bay-Burnett region, with isolated showers in the Central Lowlands and Highlands and Warrego. Dry elsewhere.		A
6th	Wed	Mostly dry over the next 2 weeks throughout the State with only a couple of isolated showery patches.		V
7th	Thu			
8th	Fri	Warm spell over the coming 2-3 days across the State, particularly in the southeast quarter regions. In contrast, minimum overnight temperatures may be unusually low. Chance of some isolated showers along the southern border fringes in the Darling Downs.		
9th	Sat	Cooler spell over the next few days over much of the State. Cool spell in the Darling Downs to the Southeast regions over the next couple of days, particularly in the Applethorpe district, where maximum temperatures may sit around the mid-teens.		
10th	Sun	Cold overnight in the Central Lowlands and Lower Western, as well as in some parts of the Moreton-South Coast regions, with overnight minimum temperatures dropping in the single digits in many districts. Chance of frost patches developing in sheltered parts in the Amberley area as well as in a number of districts in the Darling Downs.		3Q
11th	Mon	Chance of light frost in the Kingaroy, Applethorpe and Stanthorpe districts as well as more widespread in the Maranoa.		
12th	Tue	Overall, the next 3-4 weeks may be see a mix of wetter and drier than normal regions, for example, above normal rainfall is expected in the Wide Bay-Burnett districts, while in contrast, some parts of the nearby Herbert and Lower Burdekin may be much drier than is normal for this time of year. Overnight minimum temperatures may be lower than normal, while maximum temperatures may be closer to their norms. Warm spell throughout most of the State over the coming week.		
13th	Wed	Hot day in the Carpentaria region, particularly in the Century Mine district, with maximum temperatures into the upper 30's. Very warm overnight in the North Peninsula, particularly in the Horn Island district, with minimum temperatures still lingering around the mid-20's.		XhN
14th	Thur	Chance of some isolated showers in the far north of Cape York Peninsula over the next few days.		
15th	Fri			
16th	Sat			
17th	Sun	Cold overnight in the Darling Downs, with frost patches in some districts, including in the Stanthorpe district.		N
18th	Mon			P3
19th	Tue			^
20th	Wed			
21st	Thu	Some light showers likely around the Fraser Coast and Wide Bay-Burnett region.		
22nd	Fri			
23rd	Sat	Very warm overnight in the North Peninsula, particularly in the Horn Island district, with minimum temperatures still lingering around the mid-20's. Chance of some further light showers likely around the Fraser Coast and Wide Bay-Burnett region, as well as briefly over the northern half of Cape York Peninsula.		1Q
24th	Sun	Some isolated showers likely in the southeast of the State over the next couple of days.		
25th	Mon			
26th	Tue	Very heavy rain on the Herbert-North Coast region and brief showers over Cape York Peninsula over the next couple of days.		XhS
27th	Wed	Spell of warmer maximum temperatures over the next few days in many districts in the central and eastern regions. Some isolated showers likely in the Central West.	Fg	
28th	Thur	A storm embedded front crosses the southeast of the State over the next couple of days, with very gusty winds at times, particularly around Moreton Bay and some showers likely.	Showers, Th	
29th	Fri	Strong winds persist in the Moreton Bay region and Southeast as the storm embedded front moves offshore. Cold day in some parts of the Darling Downs region, with maximum temperatures barely into the double digits in some places, particularly in the Applethorpe district.		
30th	Sat	Cold day in some parts of the Carpentaria region, with maximum temperatures dropping into the single digits in some districts, including in the Kowanyama area. Dry throughout the State over the next couple of days and into early June.		
31st	Sun			F

June 2026

Date	Day	Queensland - June 2026	Brisbane	Moon
1st	Mon	Dry throughout the State over the next couple of days.		A
2nd	Tue	Dry throughout the western half of the State over the coming week.		V
3rd	Wed	Some showers in the southeast of the State over the next couple of days, and over Cape York Peninsula over the following 3-4 days.		
4th	Thu			
5th	Fri			
6th	Sat			
7th	Sun			
8th	Mon	Widespread showers in the Central West to the Central Coast regions, while dry to the north and south.	Showers	3Q
9th	Tue	Showers clear in the west, but spread in the east from the Mackay to the South Coast-Moreton regions, and become more intense over the next few days.	Rain	XhN
10th	Wed	Stormy in the eastern half of the State over the next 3-4 days. Strong gusty winds at times in the Wide Bay-Burnett to the South Coast-Moreton regions. Heavy rain in the Port Curtis and Wide Bay-Burnett regions, particularly in the Lady Elliot Island and Coringa areas. Heavy rain in the Sunshine Coast to the South Coast-Moreton at this time may cause flash flooding in some parts of the region, including around Hervey Bay, Noosa Heads, and Maryborough.	Rain	
11th	Thur	Strong winds in the Capricornia region, particularly around Rundle Island. Very heavy rain on the Fraser Coast likely.	Showers	
12th	Fri	The coming 3-4 weeks may be wetter than average in the southwest of the State, while drier than average in the tropics, and most districts may be generally warmer than normal for this time of year for both maximum and minimum temperatures, with overnight temperatures in some central districts, much warmer than average. Mild temperatures overnight in the Cape York Peninsula.	Rain	
13th	Sat	Widespread rain activity spreads throughout the State over the next couple of days. Chance of heavy rain causing the risk of localised flooding along the Noosa River catchment in the southeast of the State. Winds may be very strong for a time in the coastal fringes from Cape Moreton to Double Island Point in the Wide Bay and Burnett areas.	Rain, Fg	
14th	Sun	Warm spell over the next few days over the eastern half of the State with maximum temperatures rising into the upper 20's to early 30's in a number of districts. Very heavy rain around the Gold Coast at this time, with chance of localised flooding possible, including in the Springbrook district.	Fg	
15th	Mon	Storm activity eases throughout the State, and contracts to the southeast quarter, with most other regions returning to mostly dry conditions for the next few days.		N P2
16th	Tue	Unusually warm day in the Lower Carpentaria region, with some districts maximum temperatures reaching up into the mid-30's, particularly in the Kowanyama area. Some lingering showers likely around the Fraser Coast, while mostly dry elsewhere.		^
17th	Wed	Unusually warm overnight in the North Peninsula region, with minimum temperatures in the mid-20's likely in some places, particularly in the Horn Island district. Isolated showers in the southeast.	Showers	
18th	Thu	Showery in the southeast quarter of the State over the next 3-4 days.		
19th	Fri	Showers spread into the Darling Downs to the west and up along the coasts to the North Coast-Herbert region during the next couple of days.	Showers	
20th	Sat	Mostly dry in the Gulf Country with shower activity elsewhere over the next couple of days.	Showers	
21st	Sun			
22nd	Mon	Very warm in the east of the State from the North Coast to the South Coast and into the adjacent inland regions, with daily maximas into the upper 20's to early 30s likely. Some showers around the Central Coast to the adjacent inland regions, with isolated showers into the South Coast.		1Q XhS
23rd	Tue	Brief showers in the northern half of Cape York Peninsula while mostly dry elsewhere for most of the remainder of this month.	Fg	
24th	Wed			
25th	Thur			
26th	Fri	Cooler spell over the eastern coastal regions over the next couple of days, with minimum temperatures dropping back into the single digits in many areas. Chance of patches of frost in the Port Curtis region particularly in the Thangool district.		
27th	Sat			
28th	Sun	Dry day everywhere.	Fg	A
29th	Mon	Showers return to Cape York Peninsula during the next couple of days.	Showers, Fg	V
30th	Tue	Warm day over the North Peninsula. Showers throughout the southeast quarter of the State, with isolated showers into the Central Lowlands.	Rain, Th	F

July 2026

Date	Day	Queensland - July 2026	Brisbane	Moon
1st	Wed	Mostly dry throughout the State over the next fortnight with just a few brief exceptions.		
2nd	Thu	Cold day in some parts of the Darling Downs around this time, with some areas maximum temperatures only reaching into the low teens, including very cool day in the Applethorpe area. Chance of isolated showers in the southeast quarter of the State.		
3rd	Fri	Cold minimum temperatures throughout much of the State over the coming week.		
4th	Sat			
5th	Sun			
6th	Mon	Chance of frost patches in the Central Highlands and Maranoa.		
7th	Tue	Chance of some isolated showers over Cape York Peninsula.		XhN
8th	Wed			3Q
9th	Thur	Scattered frosts in the Darling Downs, Maranoa, and Warrego regions around this time. Chance of some patches of severe frosts in the Stanthorpe district.		
10th	Fri	Chance of some isolated showers over Cape York Peninsula over the next couple of days.		
11th	Sat	Warm day in the Lower Carpentaria, Central Lowlands, Upper Western and Far Southwest around this time.	Fg	
12th	Sun	Warm over the next couple of days in the southeast quarter of the State, including north into the Mackay and North Coast regions with maximum temperatures in the mid-to-upper 20's in many districts.		
13th	Mon	The next 3-4 weeks may be considerably drier than normal in the far southwest of the State, while considerably wetter than average in the northeastern regions and generally slightly wetter elsewhere. In contrast, temperatures, both maximum and minimum averages, may be within 1-2°C of what is typical for this time of year.		P4 ^
14th	Tue			N
15th	Wed	Some brief showers in the southeast quarter of the State over the next couple of days. Dry elsewhere.		
16th	Thu			
17th	Fri	Rain activity intensifies in the eastern half of the State over the coming days.	Rain	
18th	Sat	An upper trough may bring isolated thunderstorms along the Southeast Coasts. Bursts of heavy rain at times may cause some brief flooding issues along the Cooyar Creek and Brisbane River areas.	Rain	
19th	Sun	Very heavy rain throughout the southern half of the State and up through the eastern regions and onto Cape York Peninsula, while mostly dry in the Gulf Country to the western border districts.	Show ers	
20th	Mon	Rain activity swiftly eases to showers and contracts to the southeast of the State around this time.		XhS
21st	Tue	Dry spell over the next 3-4 days. Cold over the next few nights in many parts of the State with scattered patches of frosts in many districts, particularly in the Central Highlands, Maranoa, and Warrego, as well as in some sheltered parts in the Port Curtis and Moreton-South Coast regions.		1Q
22nd	Wed			
23rd	Thur			
24th	Fri		Fg	
25th	Sat	Warmer spell in many parts of the State over the next couple of days. Very warm overnight in the Lower Carpentaria region with minimum temperatures still in the mid-20's likely in some districts. Showers return to the southeast of the State from the Fraser Coast to the Darling Downs and South Coast-Moreton regions over the next couple of days.	Fg	
26th	Sun		Show ers	A
27th	Mon	Mostly dry in the Gulf Country with widespread rain activity elsewhere. Heavy rain in the Barron-North Coast to the Cassowary Coast regions as well as in central parts of the Cape York Peninsula.	Show ers, Fg	V
28th	Tue	Rain clears to spotty shower activity as it contracts into the southern half of the State.		
29th	Wed	Warmer and mostly dry spell in many parts of the State over the next 2-3 days.		
30th	Thu			F
31st	Fri	Hot day in the Lower Carpentaria region, particularly in the Century Mine area, with some districts maximum temperatures into the mid-30's likely. Chanc eof some isolated showers around the Central Coast.		

August 2026

Date	Day	Queensland - August 2026	Brisbane	Moon
1st	Sat	Unusually mild overnight in the North Peninsula region, with minimum temperatures lingering around the mid-20's in a number of districts, including at Coconut Island. Widespread scattered shower activity in most regions of the State.		
2nd	Sun	Isolated showers persist in the east and south of the State.		
3rd	Mon	Widespread rain throughout the State over the next couple of days, with heavy rain down through the eastern coastal regions from the Peninsula to the South Coast-Moreton. Cool day in the Central Lowlands to the Lower Carpentaria with maximum temperatures barely into the low double digits to mid-teens.	Rain	XhN
4th	Tue	Very strong and blustery winds in the South Coast-Moreton region around this time, with gale gusts possible. Rain and storm activity may be very intense for a time in the southern half of the State over the next couple of days.	Rain	
5th	Wed		Show ers	
6th	Thur	Rain activity clears in most parts of the State although may linger in the southeast quarter.		3Q
7th	Fri	Warmer spell in many parts of the State over the next couple of days. Mostly dry day.	Show ers	
8th	Sat	Very cold spell over the next couple of days in the Darling Downs region, particularly in the Applethorpe area. Chance of isolated outbreaks of snow in some parts of the Darling Downs around this time, particularly in the Stanthorpe, Cambooya, and Toowoomba districts and showers from the North Coast to the Central West and Warrego.		
9th	Sun	Cold overnight temperatures in the low single digits in many eastern regions of the State during the next few nights.		^
10th	Mon	Some patches of frost likely in the Darling Downs over the next couple of days, particularly in the Stanthorpe district. Mostly dry throughout the State over the coming week.		P8
11th	Tue	Warm day in the Upper and Lower Western, Far Southwest, and Lower Carpentaria, with maximum temperatures into the low 30's in many districts. In contrast, chance of patches of frost in some South Coast-Moreton districts, and more widespread across the Darling Downs.		
12th	Wed	Very warm over the next couple of days in the east of the State from the Central Highlands to the South Coast-Moreton and into the Darling Downs with daily maximas into the upper 20's to early 30's in many districts. Blustery on the Darling Downs around this time.		
13th	Thu	Overall, the next 3-4 weeks may be cooler and drier than what is typical for this time of year, mostly due to a strong high pressure system directing an almost continuous cool and dry south to southwest flow over the State.		N
14th	Fri			
15th	Sat			
16th	Sun			XhS
17th	Mon	Some brief showers likely around the southeast quarter of the State, including along the border districts of the Warrego and Darling Downs over the next couple of days.		
18th	Tue			
19th	Wed	Mostly dry throughout the State for the remainder of the month.		
20th	Thur	Low minimum temperatures across most of the State during the coming week, with many districts dropping into the low single digits. Patches of frost in the Darling Downs and Warrego around this time.		1Q
21st	Fri	Patches of frost in the Central Highlands, Port Curtis, South Coast-Moreton, and Darling Downs, with severe frost patches likely in the Stanthorpe district.		
22nd	Sat	Cold day in the Darling Downs, particularly in the Applethorpe area.		A
23rd	Sun	Patches of frost in the Central Highlands, Port Curtis, South Coast-Moreton, Darling Downs and Warrego.		V
24th	Mon			
25th	Tue			
26th	Wed	Some patches of frost likey in the Darling Downs.		
27th	Thu			
28th	Fri			F
29th	Sat	Chance of some isolated showers around the Gulf Country.		
30th	Sun	Patches of frost in the Maranoa and Central Highlands over the next couple of days. Possible isolated showers around the Mackay to the North Coast-Herbert regions.		XhN
31st	Mon			

September 2026

Date	Day	Queensland - September 2026	Brisbane	Moon
1st	Tue	Dry day throughout the State.		
2nd	Wed	Very warm in the Mackay to the Port Curtis regions around this time, with maximum temperatures into the upper 20's likely. Chance of some isolated showers around the North Coast to the Cape York Peninsula over the next couple of days.		
3rd	Thur	Very overnight in the North Peninsula region, particularly in the Coconut to Horn islands areas. Outbreaks of showers in the southeast quarter of the State.	Rain	
4th	Fri	Dry throughout the State over the next 4-5 days.		3Q
5th	Sat			
6th	Sun	Heavy rain in the Atherton Tablelands.	Fg	^
7th	Mon	Very warm overnight in the Lower Carpentaria region, particularly in the Burketown area. Chance of some showers in the North Coast-Herbert region.		P12
8th	Tue			
9th	Wed	Some shower activity around the North Coast-Herbert region and isolated showers in the southeast over the next couple of days.	Showers	
10th	Thu	Very warm over the next couple of days in many parts of the State, with many districts seeing maximum temperatures in the high 20's to the mid-30's likely, and into the upper 30's in some parts of the Upper Western.	Showers	
11th	Fri	Widespread thunderstorm activity in the western districts around this time may produce some heavy falls in the region. Warm overnight in the Darling Downs, particularly in the Texas district, with overnight minimum temperatures in the mid-20's likely in a number of places. Very warm day in the Lower Carpentaria, with some districts possibly seeing maximum temperatures up into the high-30's.	Showers	N
12th	Sat	The next 3-4 weeks may be wetter than normal in the central and western parts of the State, while closer to average rainfall elsewhere, and with average temperatures, both maximum and minimum, generally warmer than normal for this time of year. Shower activity moves into the eastern half of the State around this time, while clearing in the west.		XhS
13th	Sun	Mostly dry day throughout the State.		
14th	Mon	A band of rain crosses the southeast of the State over the next few days bringing moderate to heavy falls in many districts. Outbreaks of rain at first in the northern half of the State before spreading throughout over the next couple of days.	Showers	
15th	Tue	Cold day in the Maranoa. Heavy rain along the coasts and in the Central West to the Gulf Country with showers elsewhere.	Rain	
16th	Wed	Low single digit minimas over the coming week days in many districts of the eastern half of the State. Heavy rain in the eastern half of the State spreads from the coast into the adjacent inland regions, including in some parts of the Darling Downs and Tablelands.	Showers	
17th	Thur	Cooler spell in the Darling Downs around this time, with some maximum temperatures only reaching into the low teens, particularly in the Applethorpe district. Rain activity clears throughout the State with only a few showers lingering around the Gold Coast-Moreton likely.		
18th	Fri	Mostly dry over the next few days throughout the State except over Cape York Peninsula.		
19th	Sat	Mild overnight in the Barron-North Coast region with overnight minimas in the upper teens to the early 20's possible. Chance of some showers over Cape York Peninsula over the next 4-5 days.		1Q A V
20th	Sun	Chance of patches of frost overnight in the Darling Downs, particularly in the Stanthorpe district.		
21st	Mon	Chance of some isolated showers around the Port Curtis region.		
22nd	Tue	Showers persist over the Cape York Peninsula over the next couple of days with mostly dry conditions elsewhere.		
23rd	Wed	Strong wind gusts in the Channel Country.		
24th	Thu	Isolated showers in the Northwest of the State to the southern Peninsula over the next couple of days.		
25th	Fri	Chance of isolated thunderstorms along the southern border fringes around the South Coast-Moreton over the next couple of days.	Showers, Th	
26th	Sat	Showers briefly spread through the southeast quarter of the State around this time.	Showers	XhN
27th	Sun	Hot spell over the next 4-5 days in the Carpentaria, Upper and Lower Western, and Peninsula regions, with daily maximas into the upper 30's in many districts. Mostly dry throughout the State over the next couple of days.		F
28th	Mon			
29th	Tue	Isolated showers likely around the North Coast-Herbert. Dry elsewhere.		
30th	Wed	Chance of brief outbursts of heavy rain, wind gusts, and possible hail in isolated pockets in the southeast of the State, particularly in the Brisbane area. Dry in the west and north.	Showers, Fg	

October 2026

Date	Day	Queensland - October 2026	Brisbane	Moon
1st	Thur	Chance of flash flooding from some bursts of heavy rain in the Brisbane area and possible flooding risks along the Upper Logan River. Hail showers possible. Very hot in the Lower Western around this time, with daily maximas into the high 30's likely in some districts, particularly in the Boulia area.	Show ers	
2nd	Fri	Strong gusty winds and thunderstorms with hail likely in the Southeast Coastal areas, particularly around Amberley to Rathdowney, Ipswich and Lake Wivenhoe. Rain around the North Coast and Fraser Coast over the next couple of days.	Th	P13
3rd	Sat	High daily maximas into the upper 20's to mid 30's in many eastern parts of the State over the next few days. Very warm overnight in the Lower Western, particularly in the Birdsville district, with minimum temperatures into the mid-to-upper 20's likely.		
4th	Sun	Chance of some rain in the Central Lowlands around this time and isolated showers on the North Coast-Herbert and South Coast-Moreton regions. Dry elsewhere.		3Q ^
5th	Mon	Mostly dry throughout the State over the coming week with just a few exceptions.		
6th	Tue	Heavy rain around the Gold Coast.		
7th	Wed			
8th	Thu			
9th	Fri	Very hot spell throughout most of the State over the next couple of days with daily maximas in the mid-to-upper 30's in many districts. Showers over the northern half of Cape York Peninsula.	Fg	
10th	Sat			XhS
11th	Sun			N
12th	Mon	Overall, the nex 3-4 weeks may be warmer than normal in the west of the State, while cooler in the east, with close to average rainfall almost everywhere.		
13th	Tue			
14th	Wed	Heavy rain at times in the northern half of Cape York Peninsula over the next couple of days. Dry conditions continue elsewhere.		
15th	Thur	Some showers likely in the southeast of the State over the next couple of days.		
16th	Fri	Warm spell over the next 3-4 days throughout most of the State, with daily maximas into the 30's in many districts. Isolated showers likely in the Channel Country.		
17th	Sat	Showers spread from the Channel Country eastward to the Central Coast while dry to the north and south, except for chance of some showers around the Moreton.		A V
18th	Sun	Mostly dry throughout the State over the next couple of days.		
19th	Mon			1Q
20th	Tue	Heavy rain likely in the Somerset - Port Curtis region. A cold front crosses the State over the next few days bringing outbreaks of rain from the eastern coasts into the central regions as far north as the North Coast-Barron and South Coast-Moreton, with heavy falls at times, particularly around the Darling Downs and South Coast-Moreton.	Rain	
21st	Wed	Cool day on the Darling Downs, particularly in the Toowoomba and Oakey districts, with maximum temperatures only reaching into the mid-teens likely.	Rain	
22nd	Thur	Rain activity may be fairly widespread over the next couple of days.	Show ers	
23rd	Fri		Show ers	
24th	Sat	Showers linger for the next couple of days over Cape York Peninsula and in the southern regions of the State from the Warrego to the South Coast-Moreton, while clearing to mostly dry through central regions.		XhN
25th	Sun			
26th	Mon	Some pockets of heavy rain in the southeast of the State, particularly around Bracken Ridge. Chance of showers of large hail, particularly around Gatton and Esk. Strong winds may cause some damage in the Southeast Coast region to the Capricornia. Showers in the Gulf Country, Carpentaria, and over Cape York Peninsula.	Rain, Th	F
27th	Tue	Wet over the next couple of days along the eastern coasts from the North Coast-Herbert to the South Coast-Moreton. Heavy rain in the southeast, particularly around the Sunshine Coast, may see some localised flooding along the Noosa River catchment to headwaters.		
28th	Wed			
29th	Thu	Showers over Cape York Peninsula over the remainder of the month, while mostly dry elsewhere.		P9
30th	Fri			^
31st	Sat	Heavy rain likely in the Heathlands National Park area of Cape York Peninsula.		

November 2026

Date	Day	Queensland - November 2026	Brisbane	Moon
1st	Sun	Strong winds with gusts and showers of hail in the Darling Downs and Granite Belt and down through the Southeast Coasts. Hail stones may be particularly large in the Toowoomba, Grantham, Yamnto, and Ipswich areas.	Show ers, Th	
2nd	Mon	Large hail likely in some parts of the Southeast Coast districts. In contrast, very warm day in the Mackay and Lower Carpentaria with maximum temperatures in the low 30's in many districts.		3Q
3rd	Tue	Cool over the next few nights throughout most of the State with many regions minimum overnight temperatures dropping into the single digits. Chance of patches of frost in the Stanthorpe district of the Darling Downs. Mostly dry throughout the State over the next 7-10 days.		
4th	Wed			
5th	Thur	Storms may bring hail showers in the Southeast Coastal districts, including on the Gold Coast.	Th	
6th	Fri			XhS
7th	Sat			
8th	Sun			
9th	Mon	Hot spell throughout most of the State over the next couple of days.		N
10th	Tue	Maximum temperatures into the low 40's likely in many parts of the Carpentaria, Far Southwest, and Western regions.		
11th	Wed	Warm overnight in the Lower Western region, with minimum temperatures into the low 30's in a number of districts, including in the Birdsville area.		
12th	Thu	Hot in the Lower Western region over the next couple of days with maximum temperatures into the low 40's. Some sowers develop in the Channel Country around this time, and isolated showers in the far north of Cape York Peninsula.		
13th	Fri	Over the next 3-4 weeks, thunderstorm activity may bring higher than normal rainfall into the inland tropics and in the southwest of the State, while most eastern parts may be drier than what is typical for this time of year. Temperatures may be slightly cooler than normal in the far south and west, mostly due to increased cloud bearing airmasses, while most other parts of the State may be slightly warmer than what is typical for this time of year.		V
14th	Sat	Widespread rain over New South Wales may briefly extend some showers and thunderstorm activity into the southern border regions of the State.	Show ers, Th	A
15th	Sun	Hot spell in the Carpentaria, Central Highland regions over the next 4-5 days with maximum temperatures into the low 40's. Showers in the southeast quarter of the State and around the North Coast-Herbert regoin, while mosty dry elsewhere.		
16th	Mon	Rain develops throughout the southeast quarter of the State over the next couple of days with heavy falls at times, particularly in districts immediately north of the Fraser Coast and in the Brisbane area.	Rain	
17th	Tue	Rain in the North Coast.	Fg	
18th	Wed	Heat wave temperatures in the low 40's spread in to the Western districts around this time and possibly briefly in the Central Lowlands, Maranoa, and Warrego. Brief dry respite throughout the State around this time.		1Q
19th	Thur	Strong winds with high gusts in the Northern Goldfields and Upper Flinders regions. Widespread rain and scattered storm activity throughout the State over the next couple of days.	Rain, Th	
20th	Fri	Very heavy rain on the Atherton Tablelands, particularly in the Herberton district.		XhN
21st	Sat	Isolated lingering showers around the South Coast and far north of Cape York Peninsula, while clearing to mostly dry elsewhere.		
22nd	Sun	Dry throughout the western half of the State over the next 4-5 days.		
23rd	Mon	Some shower activity in the southeastern regions over the next few days.	Show ers	
24th	Tue	Showers may be more persistent and heavier in the Southeast for a time.	Rain	
25th	Wed	Isolated showers persist briefly along the eastern coastal fringes from the South Coast to the Mackay, while mostly dry elsewhere.	Show ers	F
26th	Thu	Dry day throughout the State.		P5
27th	Fri	Widespread rain activity moves into the State over the next few days. Heavy rain in the southeast of the State, particularly along the coastal fringes, as well as the chance of hail showers and strong winds which may cause some localised damage. The passage of severe thunderstorms with possible destructive microbursts in the Brisbane area around this time may cause some damage to infrastructure and property.	Rain, Th	^
28th	Sat	Cool day in the Darling Downs with maximum temperatures possibly only reaching into the mid teens in a number of areas, particularly in the Applethorpe district.	Rain	
29th	Sun	Heavy rain likely in the Maranoa and Warrego.	Rain	
30th	Mon	Heavy rain at times in the southeast quarter of the State, including in the Capricornia, Wide Bay - Burnett, Maranoa, Warrego, Darling Downs, Granite Belt, and the Brisbane districts, and along the Lockyer and Myall Creeks, the Paradise Lagoon catchment, and the Bremer, Logan and Condamine Rivers. Chance of hail on the Darling Downs to the Granite Belt, particularly in the Toowoomba district. Strong gusty winds at times along the Southeast Coasts.	Rain, Th	

December 2026

Date	Day	Queensland - December 2026	Brisbane	Moon
1st	Tue	Widespread rain and storm activity continues to prevail throughout the State over the next couple of days. Stormy air with thunderstorms, heavy rain, occasional bursts of hail and strong gusty winds in the Southeast Coastal regions. Hail stones may be particularly large in the Blackwater area of the Central Highlands-Coal fields which may bring some damage to property. Strong winds at times may bring some localised damage in the Darling Downs and Granite Belt regions. Heavy rain may cause some flash flooding, particularly in the Ipswich district, Lockyer Valley district, and Moreton.	Rain, Th, Fg	3Q
2nd	Wed	Heavy rain in the Wide Bay-Burnett region.		
3rd	Thur	Windy with strong gusts at times in the Central Coast - Whitsundays region.	Showers	XhS
4th	Fri	Very hot around the Cape York Peninsula to the North Coast and Mackay over the next couple of days, with maximum temperatures peaking into the low 40's in a number of districts, while it may be unusually hot overnight around the South Coast region with minimum temperatures in the upper 20's to early 30's. In contrast, unusually cool overnight in some parts of the Darling Downs with minimum temperatures only in the single digits possible, particularly in the Stanthorpe district. Showers linger in the Channel Country to the Central West, while clearing to dry elsewhere.		
5th	Sat	Mostly dry day throughout the State.		
6th	Sun	Some rain in the Carpentaria to the Gulf Country, and chance of rain developing in the South Coast-Moreton region. Dry elsewhere.	Rain	
7th	Mon	Heavy rain in the Herbert and Lower Burdekin, may see some flash flooding along the Herbert River catchment. Windy with gusts in the Darling Downs and Granite Belt regions.	Rain, Th	
8th	Tue	Heavy rain and blustery winds in the Central Highlands - Coalfields, and Herbert and Lower Burdekin areas. Chance of flooding along the Sutton and Ross Rivers. Hot in the Upper Western with maximum temperatures in the low 40's likely.	Showers	
9th	Wed	Heavy rain likely in the Maranoa and Warrego with chance of flooding along the Warrego River. Heavy rain also in the Herbert and Lower Burdekin with chance of localised flash flooding, as well as the overflowing of the some creeks and Ross River. Strong unsettled winds with severe gusts at times in the Central West. Very hot in the western half of the State with maximum temperatures in the low 40's in many districts and overnight minimums only dropping into the low 30's.	Rain	N
10th	Thu	Widespread rain prevails throughout the State over the next couple of days. Heavy rain at times on the North Tropical Coast, Tablelands and Central Lowlands and strong winds in the Southeast Coast region.	Showers, Th	V
11th	Fri			A
12th	Sat	Over the next 3-4 weeks, while the Central Highlands and the far north of the State and central Peninsula regions may be wetter than is typical for this time of year, most other regions may be drier. Temperatures may be cooler than normal in the Gulf Country, but above average elsewhere, and particularly warmer than normal in the Wide Bay-Burnett region and adjacent inland parts. Cold overnight in some parts of the Darling Downs, with single digit minimum temperatures likely in a number of districts. Some rain around the Far Southwest-Channel Country around this time, while mostly dry elsewhere.		
13th	Sun	Strong winds may cause some damage in the Northern Goldfields and Upper Flinders. Showers over Cape York Peninsula, while dry elsewhere.		
14th	Mon	Very hot in the Upper Western, particularly in the Urandangi district, with maximum temperatures into the mid-40's possible. Showers and storm activity return through the southern regions of the State over the coming week.	Showers, Th	
15th	Tue	Thunderstorms, hail and strong gusty wnds likely in the Southeast Coast region to the Wide Bay and Burnett districts. Very hot in the Western districts over the next couple of days, with maximum temperatures into the low-to-mid-40's possible. It may be particularly hot around this time in the Lower Western with some districts possibly rising into the upper 40's for a time.	Rain, Th	
16th	Wed	Warm overnight in the Upper Wester with minimum temperatures still in the low 30's likely in some places, including in the Urandangi district.	Rain	
17th	Thur	Scattered storms likely on the Southeast Coast, with chance of hail showers.	Rain, Th	1Q XhN
18th	Fri	A broad trough brings storms and heavy rain over the next few days, with hail showers at times on the Southeast Coast, as well as in the Herbert, Lower Burdekin, Darling Downs and Granite Belt areas. Chance of brief flooding along the Ross and Macintyre Rivers. Winds may be very strong at times in the Southeast Coast areas, and particularly strong with chance of bringing widespread damage in the Central Highlands - Coalfields area. Very hot day in the Clermont district of the Central Highlands.	Th	
19th	Sat			
20th	Sun	Very hot over the next few days in the Carpentaria, with many districts maximum temperatures reaching into the low-to-mid 40's.		

Date	Day	Queensland - December 2026	Brisbane	Moon
21st	Mon	Strong winds at times in the Maranoa and Warrego. Unusually warm over the next couple of nights in the Lower Carpentaria and Lower Western with minimum temperatures still remaining in the low 30's in some places, particularly in the Cloncurry district. Brief dry respite throughout the State over the next couple of days.		
22nd	Tue	Very hot in the Maranoa, Warrego, and Far Southwest around this time, with maximum temperatures in up to the mid-40's possible.		
23rd	Wed	Widespread rain and storm activity returns across the State over the coming few days. Hail likely in the Capricornia region. Pockets of heavy rain in the Southeast Coastal regions. Cool spell over the next few days on the Darling Downs, with many districts maximum temperatures only reaching into the low 20's, particularly in the Applethorpe area.	Rain, Th	
24th	Thu		Showers	**P1 F ^**
25th	Fri	Unsettled blustery winds at times in the Central Coast-Whitsundays region with accompanying heavy rain from the Central Coast to the Fraser Coast.		
26th	Sat	Milder day on the Darling Downs. Mostly dry in the Carpentaria and Gulf Country, with scattered rain activity elsewhere.		
27th	Sun	Cold overnight in some parts of the Darling Downs region, with minimum temperatures dropping into the single digits, particularly in the Applethorpe districts. Mostly dry throughout the State over the next few days except for some showers over the far north of Cape York Peninsula.		
28th	Mon			
29th	Tue			
30th	Wed	More persistent showers over Cape York around this time, and outbreaks of rain in the Central West and chance of brief outbreaks of rain around the South Coast-Moreton. Heavy rain possible in the Wide Bay and Burnett districts. Chance of hail in the Capricornia region.	Rain, Th	**XhS**
31st	Thur	Cooler spell in the Darling Downs over the next couple of days, particularly in the Stanthorpe and Applethorpe districts, with maximum temperatures only reaching into the low 20's in a number of places.		**3Q**

QUEENSLAND GRAPHS AND TABLES 2026

Seasonal Breakdown – State Overall Rainfall Timing and Estimated mm Trend Graphs
State Overall Maximum Temperature Trend Graphs
Regional Rainfall Expectations Tables - mm and Trends
Regional Maximum and Minimum Temperature Trends Tables

Monthly Breakdown - **Rainfall:**
Regional Expectations by mm
Regional Expectations by mm Variance to Average
Regional Variance Trend by mm
Regional Estimated Expectations by % of Variance to Average
Regional Estimatedl Variance Trend by %

Maximum & Minimum Temperature Trends:
Regional Estimated Average Temperature
Regional Estimated Degree of Variation from the Average
Regional Estimated Temperature Trend

Seasonal Breakdowns

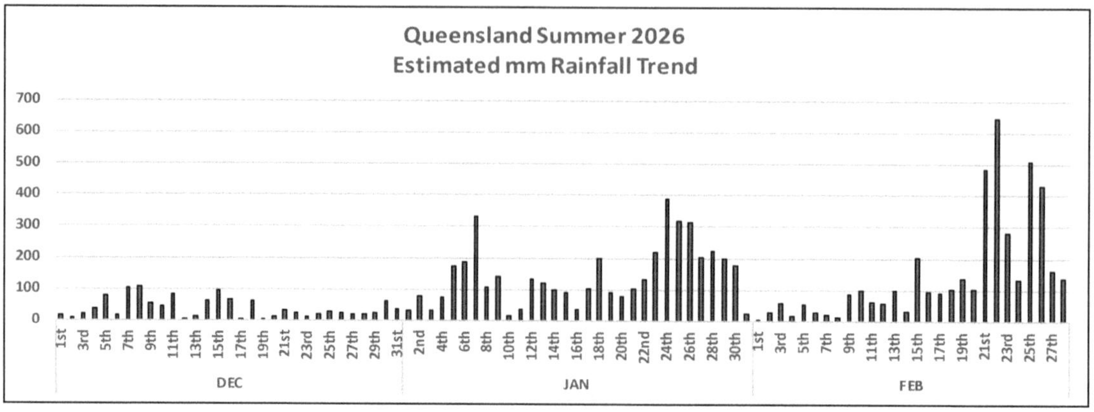

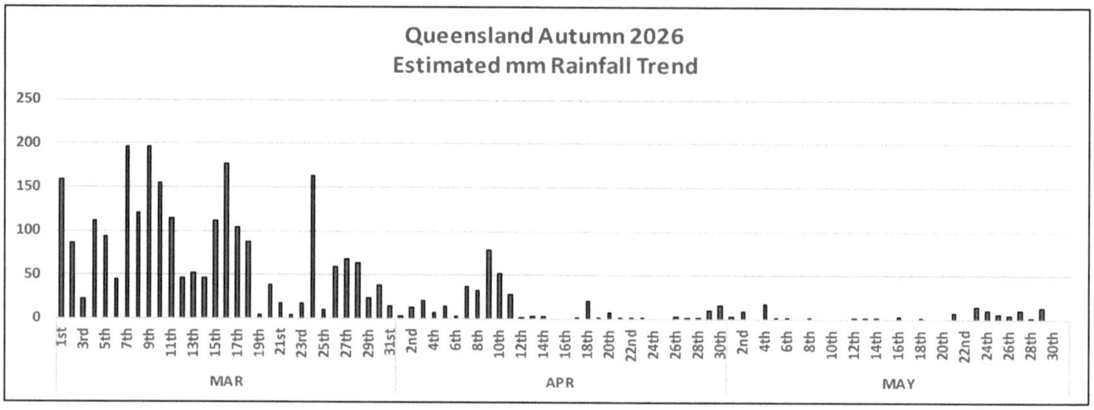

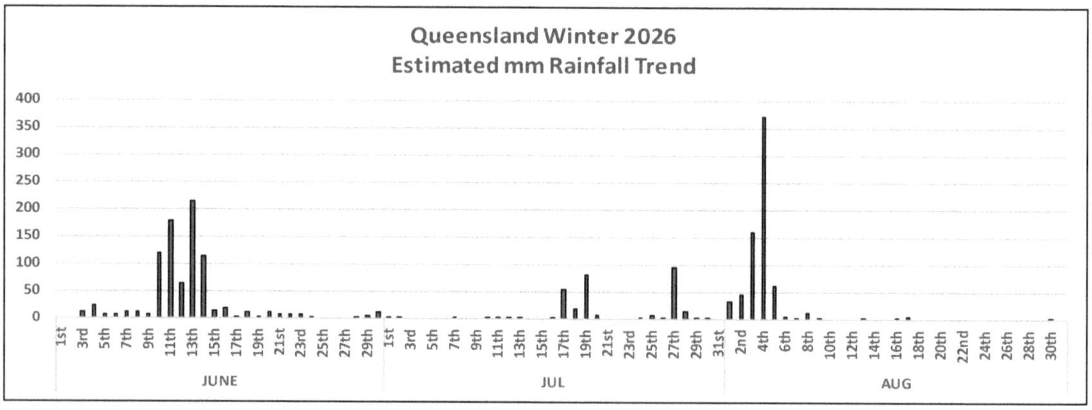

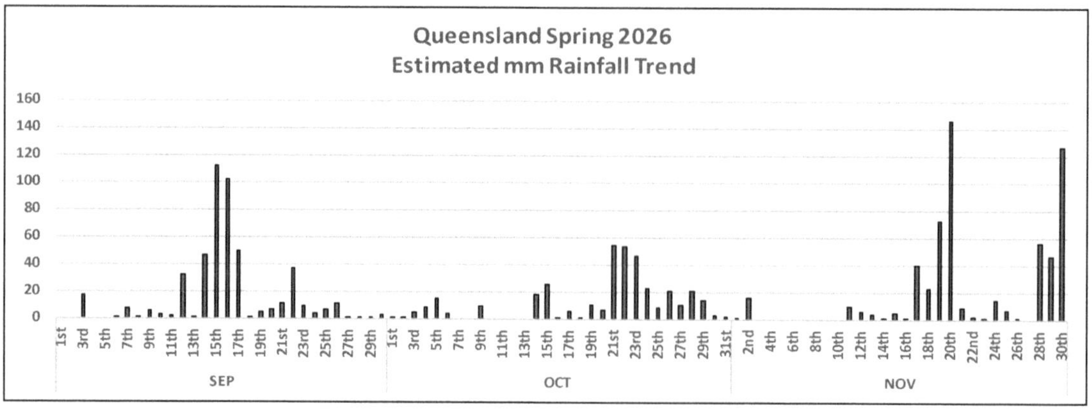

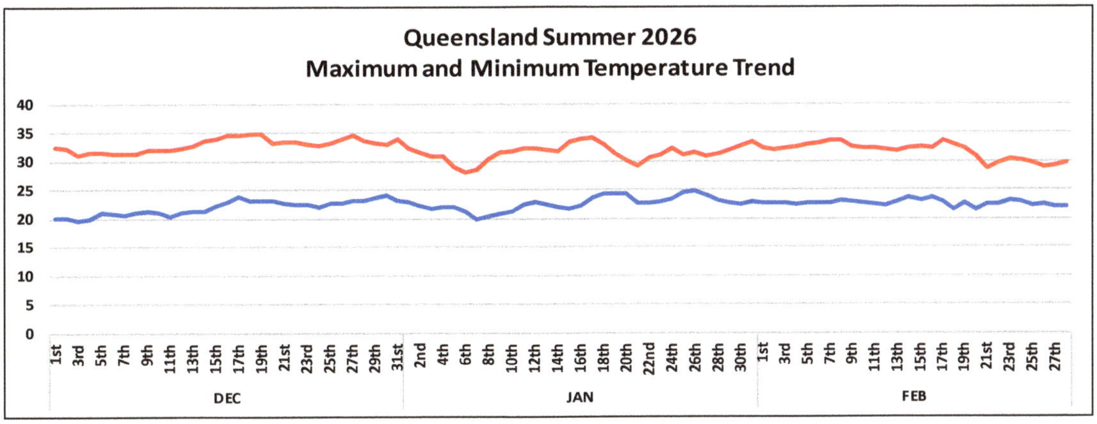
Queensland Summer 2026
Maximum and Minimum Temperature Trend
40
35
30
25
20
15
10
5
0
DEC
JAN
FEB

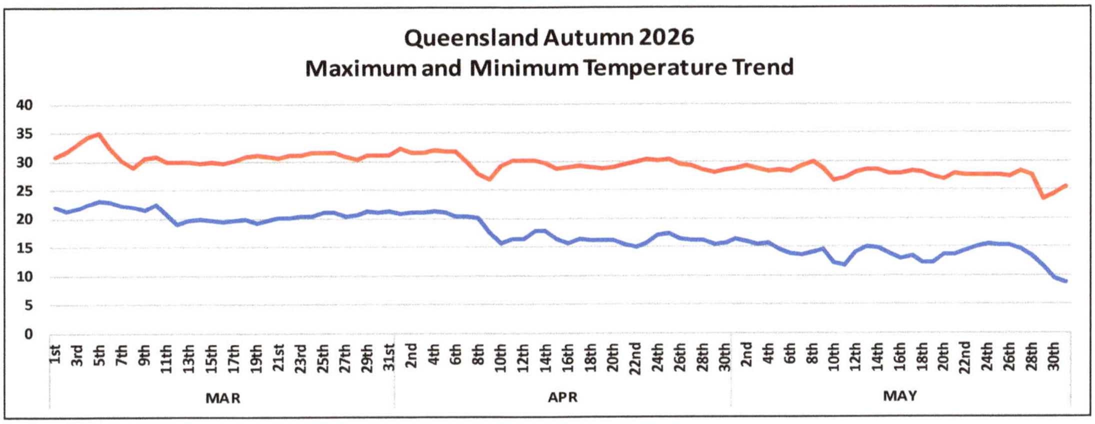
Queensland Autumn 2026
Maximum and Minimum Temperature Trend
40
35
30
25
20
15
10
5
0
MAR
APR
MAY

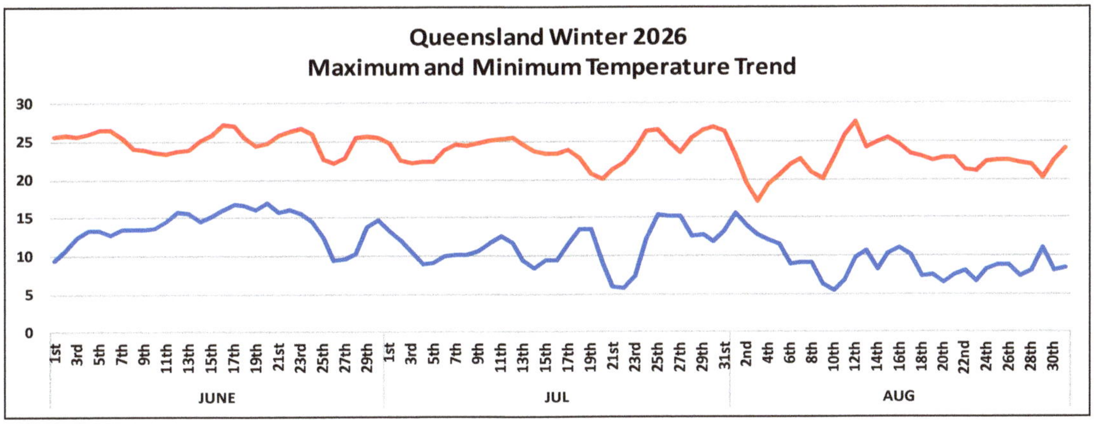
Queensland Winter 2026
Maximum and Minimum Temperature Trend
30
25
20
15
10
5
0
JUNE
JUL
AUG

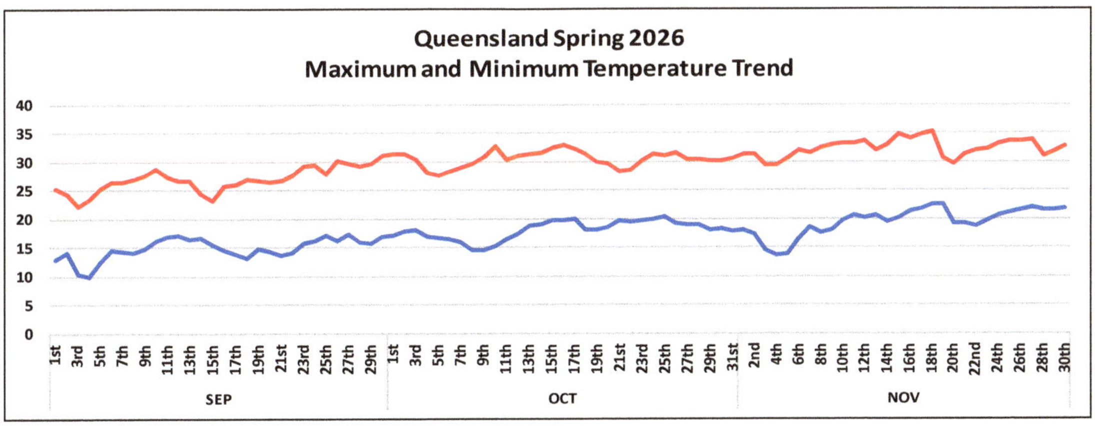
Queensland Spring 2026
Maximum and Minimum Temperature Trend
40
35
30
25
20
15
10
5
0
SEP
OCT
NOV

2026 QLD - Estimated Seasonal Rainfall Expectations (mm)

REGION	Gulf Country	Northwest	Central West	Channel Country	Peninsula	Tropical Coast & Tablelnds	Goldfields / Upper Flinders	Herbet / Lower Burdekin	Central Coast-Whitsundays	Central Highlands	Capricornia	Maranoa / Warrego	Darling Dwns / Granite Belt	Wide Bay / Burnett	Southeast	QLD
Town Sample	Normanton	Mt Isa	Longreach	Birdsville	Weipa	Cairns	Charters Twrs	Townsville	Mackay	Emerald	Rockhampton	Charleville	Toowoomba	Maryborough	Brisbane	Overall
Summer	556	132	236	41	1195	1045	810	1186	1359	978	514	472	203	702	382	9811
Autumn	148	7	13	0	756	1018	46	118	211	41	230	24	97	88	110	2907
Winter	5	2	63	11	5	97	149	96	188	137	150	48	122	441	345	1858
Spring	20	23	103	38	59	140	36	38	51	149	95	125	134	104	308	1422
Overall	693	172	399	114	2153	2339	1118	1531	1849	1264	1131	598	566	1347	1241	16513

2026 QLD - Estimated Seasonal Rainfall Variation from the Average (mm)

REGION	Gulf Country	Northwest	Central West	Channel Country	Peninsula	Tropical Coast & Tablelnds	Goldfields / Upper Flinders	Herbet / Lower Burdekin	Central Coast-Whitsundays	Central Highlands	Capricornia	Maranoa / Warrego	Darling Dwns / Granite Belt	Wide Bay / Burnett	Southeast	QLD
Town Sample	Normanton	Mt Isa	Longreach	Birdsville	Weipa	Cairns	Charters Twrs	Townsville	Mackay	Emerald	Rockhampton	Charleville	Toowoomba	Maryborough	Brisbane	Overall
Summer	-97	-168	35	-28	-74	40	438	502	671	725	206	278	-134	307	1	2702
Autumn	-48	-70	-88	-38	222	325	-88	-100	-148	-150	44	-83	-117	-198	-267	-803
Winter	-10	-15	16	-17	-5	-4	92	43	64	54	51	-26	4	299	180	726
Spring	-38	-48	28	6	-59	-55	-53	-78	-100	13	-54	12	-67	-91	98	-486
Overall	-229	-292	-26	-51	222	346	465	459	527	601	390	109	-304	330	109	2655

2026 QLD - Estimated Seasonal Rainfall Trend Variation by 10mm

REGION	Gulf Country	Northwest	Central West	Channel Country	Peninsula	Tropical Coast & Tablelnds	Goldfields / Upper Flinders	Herbet / Lower Burdekin	Central Coast-Whitsundays	Central Highlands	Capricornia	Maranoa / Warrego	Darling Dwns / Granite Belt	Wide Bay / Burnett	Southeast	QLD
Town Sample	Normanton	Mt Isa	Longreach	Birdsville	Weipa	Cairns	Charters Twrs	Townsville	Mackay	Emerald	Rockhampton	Charleville	Toowoomba	Maryborough	Brisbane	Overall
Summer	drr	drr	wtr	av	drr	wtr	wtr	wtr	wtr	wtr	wtr	wtr	drr	wtr	av	wtr
Autumn	drr	drr	drr	drr	wtr	wtr	drr	drr	drr	drr	wtr	drr	drr	drr	drr	drr
Winter	av	av	av	av	av	av	wtr	wtr	wtr	wtr	wtr	av	av	wtr	drr	wtr
Spring	drr	drr	av	av	drr	drr	drr	drr	drr	av	drr	av	drr	drr	wtr	drr
Overall	drr	drr	av	av	wtr	wtr	wtr	wtr	wtr	wtr	wtr	wtr	drr	wtr	wtr	wtr

2026 QLD - Estimated Seasonal Rainfall Expectations by %

REGION	Gulf Country	Northwest	Central West	Channel Country	Peninsula	Tropical Coast & Tablelnds	Goldfields / Upper Flinders	Herbet / Lower Burdekin	Central Coast-Whitsundays	Central Highlands	Capricornia	Maranoa / Warrego	Darling Dwns / Granite Belt	Wide Bay / Burnett	Southeast	QLD
Town Sample	Normanton	Mt Isa	Longreach	Birdsville	Weipa	Cairns	Charters Twrs	Townsville	Mackay	Emerald	Rockhampton	Charleville	Toowoomba	Maryborough	Brisbane	Overall
Summer	-15%	-56%	17%	-41%	-6%	4%	118%	73%	97%	286%	67%	143%	-40%	78%	0%	38%
Autumn	-25%	-90%	-87%	-100%	42%	47%	-66%	-46%	-41%	-79%	24%	-77%	-55%	-69%	-71%	-22%
Winter	-67%	-88%	33%	-61%	-51%	-4%	160%	80%	52%	64%	52%	-35%	3%	210%	109%	64%
Spring	-65%	-68%	37%	20%	-50%	-28%	-60%	-67%	-66%	10%	-36%	10%	-33%	-47%	47%	-25%
Overall	-25%	-63%	-6%	-31%	11%	17%	71%	43%	40%	91%	53%	22%	-35%	32%	10%	19%

2026 QLD - Estimated Seasonal Rainfall Trend Variation by 25%

REGION	Gulf Country	Northwest	Central West	Channel Country	Peninsula	Tropical Coast & Tablelnds	Goldfields / Upper Flinders	Herbet / Lower Burdekin	Central Coast-Whitsundays	Central Highlands	Capricornia	Maranoa / Warrego	Darling Dwns / Granite Belt	Wide Bay / Burnett	Southeast	QLD
Town Sample	Normanton	Mt Isa	Longreach	Birdsville	Weipa	Cairns	Charters Twrs	Townsville	Mackay	Emerald	Rockhampton	Charleville	Toowoomba	Maryborough	Brisbane	Overall
Summer	av	drr	av	drr	av	av	wtr	wtr	wtr	wtr	wtr	wtr	drr	wtr	av	wtr
Autumn	av	drr	drr	drr	wtr	wtr	drr	drr	drr	drr	av	drr	drr	drr	drr	av
Winter	drr	drr	wtr	drr	drr	av	wtr	wtr	wtr	wtr	wtr	drr	av	wtr	wtr	wtr
Spring	drr	drr	wtr	av	drr	drr	drr	drr	drr	av	drr	av	drr	drr	wtr	av
Overall	drr	av	av	wtr	av	drr	wtr	wtr	drr	av	wtr	av	drr	drr	wtr	av

2026 QLD - Estimated Seasonal Maximum Temperature Average and Trend

Region	Gulf Country	Northwest	Central West	Channel Country	Peninsula	Tropical Coast & Tablelnds	Goldfields / Upper Flinders	Herbet / Lower Burdekin	Central Coast-Whitsundays	Central Highlands	Capricornia	Maranoa / Warrego	Darling Dwns / Granite Belt	Wide Bay / Burnett	Southeast	QLD
Town Sample	Normanton	Mt Isa	Longreach	Birdsville	Weipa	Cairns	Charters Twrs	Townsville	Mackay	Emerald	Rockhampton	Charleville	Toowoomba	Maryborough	Brisbane	Overall
Summer Est	32.4	34.5	33.0	31.9	31.1	29.6	30.3	31.5	35.1	38.0	38.9	31.2	25.8	29.2	27.9	32.0
Trend	av	av	a-c	av	av	av	clr	clr	clr	wmr	a-c	clr	clr	clr	a-c	a-c
Autumn Est	31.8	33.6	29.8	29.4	29.9	26.7	28.9	29.4	32.1	33.3	32.1	28.2	23.1	27.7	25.9	29.5
Trend	av	av	av	av	av	av	av	av	a-w	wmr	a-w	av	av	a-w	av	av
Winter Est	31.3	29.7	24.7	26.2	25.5	21.8	23.5	23.0	24.2	25.8	21.7	20.9	16.9	22.3	21.1	23.9
Trend	av	a-c	a-c	av	av	av	av	clr	av	av	a-c	av	a-c	av	av	av
Spring Est	35.5	36.2	31.0	29.7	29.0	26.6	28.8	29.7	32.7	34.6	32.3	29.0	23.1	27.1	24.5	30.0
Trend	av	av	clr	av	av	a-c	a-c	clr	a-c	av	a-c	a-c	clr	av	clr	a-c

2026 QLD - Estimated Seasonal Minimum Temperature Average and Trend

Region	Gulf Country	Northwest	Central West	Channel Country	Peninsula	Tropical Coast & Tablelnds	Goldfields / Upper Flinders	Herbet / Lower Burdekin	Central Coast-Whitsundays	Central Highlands	Capricornia	Maranoa / Warrego	Darling Dwns / Granite Belt	Wide Bay / Burnett	Southeast	QLD
Town Sample	Normanton	Mt Isa	Longreach	Birdsville	Weipa	Cairns	Charters Twrs	Townsville	Mackay	Emerald	Rockhampton	Charleville	Toowoomba	Maryborough	Brisbane	Overall
Summer Est	25.5	23.6	22.2	25.2	24.3	24.0	22.4	24.5	23.4	21.2	21.9	20.2	16.8	20.7	20.7	22.4
Trend	av	av	av	av	av	av	av	av	av	a-c	av	clr	a-c	av	av	av
Autumn Est	21.3	18.0	15.4	16.8	22.3	21.1	17.6	19.9	19.5	15.4	17.4	12.3	12.4	15.8	15.4	17.4
Trend	a-c	av	a-c	av	av	av	a-c	a-c	av	clr	av	clr	a-c	a-c	clr	a-c
Winter Est	16.2	9.8	8.2	7.7	19.4	17.6	11.9	14.1	14.0	9.4	11.0	4.9	7.0	9.7	9.7	11.4
Trend	av	av	av	av	av	av	av	av	av	av	av	av	av	av	av	av
Spring Est	23.0	18.5	17.0	16.4	22.4	21.3	17.1	20.4	18.7	15.6	16.6	14.5	12.1	14.3	15.2	17.5
Trend	av	av	av	a-c	a-w	a-w	a-c	av	a-c	clr	av	av	a-c	a-c	av	av

Monthly Breakdowns

2026 — QUEENSLAND - Estimated Rainfall Expectations (mm)

REGION	Gulf Country	Northwest	Central West	Channel Country	Peninsula	Tropical Coast & Tablelnds	Goldfields / Upper Flinders	Herbet / Lower Burdekin	Central Coast-Whitsundays	Central Highlands	Capricornia	Maranoa / Warrego	Darling Dwns / Granite Belt	Wide Bay / Burnett	Southeast	QLD
Town Sample	Normanton	Mt Isa	Longreach	Birdsville	Weipa	Cairns	Charters Twrs	Townsville	Mackay	Emerald	Rockhampton	Charleville	Toowoomba	Maryborough	Brisbane	Overall
January	143	37	168	7	521	535	447	600	459	623	165	281	52	275	134	4447
February	318	22	10	32	466	459	331	468	848	202	324	58	72	379	185	4172
March	148	6	10	0	659	953	31	114	168	15	198	22	39	37	47	2446
April	0	2	1	0	98	48	15	3	14	24	31	2	35	40	46	357
May	0	0	3	0	0	18	0	1	28	2	2	1	23	11	17	104
June	0	0	31	11	1	42	4	1	69	60	57	40	49	274	234	872
July	0	0	0	0	2	39	22	20	13	11	21	1	36	62	65	292
August	5	2	32	0	2	16	123	75	106	66	71	8	37	105	46	694
September	0	6	70	1	31	33	9	1	9	43	29	93	36	31	84	474
October	17	6	23	1	10	89	1	4	15	37	45	15	33	22	47	366
November	3	11	9	36	18	18	26	33	27	69	20	17	65	52	177	582
December	59	81	42	27	345	90	109	210	94	112	167	61	89	61	160	1707
Overall	693	172	399	114	2153	2339	1118	1531	1849	1264	1131	598	566	1347	1241	16513

2026 — QUEENSLAND - Estimated Rainfall Variation from the Average (mm)

REGION	Gulf Country	Northwest	Central West	Channel Country	Peninsula	Tropical Coast & Tablelnds	Goldfields / Upper Flinders	Herbet / Lower Burdekin	Central Coast-Whitsundays	Central Highlands	Capricornia	Maranoa / Warrego	Darling Dwns / Granite Belt	Wide Bay / Burnett	Southeast	QLD
Town Sample	Normanton	Mt Isa	Longreach	Birdsville	Weipa	Cairns	Charters Twrs	Townsville	Mackay	Emerald	Rockhampton	Charleville	Toowoomba	Maryborough	Brisbane	Overall
January	-117	-81	95	-18	35	156	293	350	242	553	87	209	-48	160	21	1939
February	69	-83	-62	4	-35	11	192	175	519	121	214	-10	-44	221	40	1333
March	-10	-49	-24	-17	252	564	-43	4	28	-43	120	-19	-28	-77	-74	584
April	-31	-11	-42	-9	-9	-168	-15	-72	-133	-29	-25	-31	-36	-37	-111	-757
May	-8	-10	-22	-11	-21	-71	-30	-32	-44	-78	-51	-34	-54	-84	-82	-630
June	-9	-8	12	0	-2	1	-15	-18	4	33	15	8	7	217	171	415
July	-3	-6	-17	-11	0	8	-2	7	-12	-30	-4	-25	-9	19	-3	-87
August	3	-1	21	-6	-3	-13	109	54	72	51	40	-9	6	63	11	398
September	-3	0	59	-5	29	-2	-0	-11	-26	26	2	69	-2	-8	55	183
October	7	-14	1	-11	-10	41	-18	-22	-13	-17	-6	-17	-34	-52	-25	-189
November	-42	-35	-32	22	-78	-94	-35	-46	-61	4	-50	-40	-32	-31	68	-480
December	-86	4	-14	11	62	-88	30	69	-49	10	48	7	-32	-63	37	-55
Overall	-229	-292	-26	-51	222	346	465	459	527	601	390	109	-304	330	109	2655

2026 QUEENSLAND - Estimated Rainfall Trend Variation by 10mm

REGION Town Sample	Gulf Country Normanton	Northwest Mt Isa	Central West Longreach	Channel Country Birdsville	Peninsula Weipa	Tropical Coast & Tablelnds Cairns	Goldfields / Upper Flinders Charters Twrs	Herbet / Lower Burdekin Townsville	Central Coast-Whitsundays Mackay	Central Highlands Emerald	Capricornia Rockhampton	Maranoa / Warrego Charleville	Darling Dwns / Granite Belt Toowoomba	Wide Bay / Burnett Maryborough	Southeast Brisbane	QLD Overall
January	drr	drr	wtr	drr	wtr	wtr	wtr	wtr	wtr	wtr	wtr	wtr	drr	wtr	wtr	wtr
February	wtr	drr	drr	av	drr	wtr	wtr	wtr	wtr	wtr	wtr	av	drr	wtr	wtr	wtr
March	av	drr	drr	drr	wtr	wtr	drr	av	wtr	drr	wtr	drr	drr	drr	drr	wtr
April	drr	drr	drr	av	av	drr	drr	drr	drr	drr	drr	drr	drr	drr	drr	drr
May	av	av	drr	drr	drr	drr	drr	drr	drr	drr	drr	drr	drr	drr	drr	drr
June	av	av	wtr	av	av	av	drr	drr	av	wtr	wtr	av	av	wtr	wtr	wtr
July	av	av	drr	drr	av	av	av	av	drr	drr	av	drr	av	wtr	av	drr
August	av	av	wtr	av	av	drr	wtr	wtr	wtr	wtr	wtr	av	av	wtr	wtr	wtr
September	av	av	wtr	av	wtr	av	av	drr	drr	wtr	av	wtr	av	av	wtr	wtr
October	av	drr	av	drr	av	wtr	drr	drr	drr	drr	av	drr	drr	drr	drr	drr
November	drr	drr	drr	wtr	drr	drr	drr	drr	drr	av	drr	drr	drr	drr	wtr	drr
December	drr	av	drr	wtr	wtr	drr	wtr	wtr	drr	av	wtr	av	drr	drr	wtr	drr
Overall	drr	drr	av	av	wtr	wtr	wtr	wtr	wtr	wtr	wtr	wtr	drr	wtr	wtr	wtr

2026 QUEENSLAND - Estimated Rainfall Expectations by %

REGION Town Sample	Gulf Country Normanton	Northwest Mt Isa	Central West Longreach	Channel Country Birdsville	Peninsula Weipa	Tropical Coast & Tablelnds Cairns	Goldfields / Upper Flinders Charters Twrs	Herbet / Lower Burdekin Townsville	Central Coast-Whitsundays Mackay	Central Highlands Emerald	Capricornia Rockhampton	Maranoa / Warrego Charleville	Darling Dwns / Granite Belt Toowoomba	Wide Bay / Burnett Maryborough	Southeast Brisbane	QLD Overall
January	-45%	-69%	130%	-73%	7%	41%	191%	140%	112%	790%	112%	289%	-48%	140%	19%	77%
February	28%	-79%	-86%	14%	-7%	2%	138%	60%	158%	151%	195%	-15%	-38%	141%	28%	47%
March	-6%	-90%	-72%	-100%	62%	145%	-58%	4%	20%	-74%	152%	-46%	-42%	-67%	-61%	31%
April	-100%	-86%	-99%	-100%	-8%	-78%	-51%	-95%	-90%	-55%	-44%	-94%	-51%	-48%	-71%	-68%
May	-100%	-100%	-89%	-100%	-100%	-80%	-100%	-98%	-61%	-97%	-97%	-98%	-70%	-89%	-83%	-86%
June	-100%	-100%	62%	3%	-60%	2%	-79%	-94%	6%	121%	37%	26%	16%	379%	272%	91%
July	-100%	-100%	-100%	-100%	0%	25%	-7%	51%	-47%	-74%	-14%	-96%	-21%	44%	-4%	-23%
August	171%	-23%	196%	-100%	-62%	-43%	756%	259%	207%	340%	125%	-55%	20%	152%	33%	135%
September	-100%	4%	519%	-85%	1532%	-6%	-3%	-93%	-74%	161%	8%	286%	-4%	-20%	188%	63%
October	62%	-68%	5%	-92%	-50%	85%	-95%	-84%	-47%	-32%	-12%	-53%	-51%	-71%	-34%	-34%
November	-92%	-77%	-77%	162%	-82%	-84%	-57%	-58%	-69%	6%	-71%	-70%	-33%	-37%	62%	-45%
December	-59%	6%	-25%	69%	22%	-49%	37%	49%	-35%	9%	40%	13%	-27%	-51%	30%	-3%
Overall	-25%	-63%	-6%	-31%	11%	17%	71%	43%	40%	91%	53%	22%	-35%	32%	10%	19%

2026 QUEENSLAND - Estimated Rainfall Trend Variation by 25%

REGION Town Sample	Gulf Country Normanton	Northwest Mt Isa	Central West Longreach	Channel Country Birdsville	Peninsula Weipa	Tropical Coast & Tablelnds Cairns	Goldfields / Upper Flinders Charters Twrs	Herbet / Lower Burdekin Townsville	Central Coast-Whitsundays Mackay	Central Highlands Emerald	Capricornia Rockhampton	Maranoa / Warrego Charleville	Darling Dwns / Granite Belt Toowoomba	Wide Bay / Burnett Maryborough	Southeast Brisbane	QLD Overall
January	drr	drr	wtr	drr	av	wtr	wtr	wtr	wtr	wtr	wtr	wtr	drr	wtr	av	wtr
February	wtr	drr	drr	av	av	av	wtr	wtr	wtr	wtr	wtr	av	drr	wtr	wtr	wtr
March	av	drr	drr	drr	wtr	wtr	drr	av	av	drr	wtr	drr	drr	drr	drr	wtr
April	drr	drr	drr	drr	av	drr	drr	drr	drr	drr	drr	drr	drr	drr	drr	drr
May	drr	drr	drr	drr	drr	drr	drr	drr	drr	drr	drr	drr	drr	drr	drr	drr
June	drr	drr	wtr	av	drr	av	drr	drr	av	wtr	wtr	wtr	av	wtr	wtr	wtr
July	drr	drr	drr	drr	av	av	av	wtr	drr	drr	av	drr	av	wtr	av	av
August	wtr	av	wtr	drr	drr	drr	wtr	wtr	wtr	wtr	wtr	drr	av	wtr	wtr	wtr
September	drr	av	wtr	drr	wtr	av	av	drr	drr	wtr	av	wtr	av	av	wtr	wtr
October	wtr	drr	av	drr	drr	wtr	drr	drr	drr	drr	av	drr	drr	drr	drr	drr
November	drr	drr	drr	wtr	drr	drr	drr	drr	drr	av	drr	drr	drr	drr	wtr	drr
December	drr	av	av	wtr	av	drr	wtr	wtr	drr	av	wtr	av	drr	drr	wtr	av
Overall	drr	av	av	wtr	av	drr	wtr	wtr	drr	av	wtr	av	drr	drr	wtr	av

2026 QUEENSLAND - Estimated Average Maximum Temperatures

Region	Gulf Country	Northwest	Central West	Channel Country	Peninsula	Tropical Coast & Tablelnds	Goldfields / Upper Flinders	Herbet / Lower Burdekin	Central Coast-Whitsundays	Central Highlands	Capricornia	Maranoa / Warrego	Darling Dwns / Granite Belt	Wide Bay / Burnett	Southeast	QLD Estimates
Town Sample	Normanton	Mt Isa	Longreach	Birdsville	Weipa	Cairns	Charters Twrs	Townsville	Mackay	Emerald	Rockhampton	Charleville	Toowoomba	Maryborough	Brisbane	Overall
January	34	37	32	40	31	32	32	31	29	30	30	31	26	29	28	**31**
February	33	38	35	38	31	31	32	31	29	31	30	31	26	29	28	**32**
March	34	36	35	37	31	30	31	30	28	31	30	31	25	29	27	**31**
April	35	34	32	32	32	30	30	31	27	29	29	28	24	28	26	**30**
May	32	30	29	27	32	29	29	29	26	27	28	25	21	27	25	**28**
June	31	27	26	24	32	27	26	27	23	24	24	23	19	24	22	**26**
July	30	26	25	21	31	26	25	26	22	23	24	21	17	22	21	**24**
August	29	24	22	20	31	26	23	24	21	21	22	19	15	21	20	**23**
September	34	31	28	27	34	28	28	27	24	26	26	24	20	25	22	**27**
October	37	35	34	34	36	30	32	30	27	30	29	30	24	27	25	**31**
November	37	38	37	35	37	31	33	30	29	33	31	33	26	29	26	**32**
December	38	38	39	37	34	33	36	32	33	35	33	35	27	32	29	**34**
Overall	**34**	**33**	**31**	**31**	**33**	**29**	**30**	**29**	**26**	**29**	**28**	**28**	**22**	**27**	**25**	**29**

2026 QUEENSLAND - Estimated Average Minimum Temperatures

Region	Gulf Country	Northwest	Central West	Channel Country	Peninsula	Tropical Coast & Tablelnds	Goldfields / Upper Flinders	Herbet / Lower Burdekin	Central Coast-Whitsundays	Central Highlands	Capricornia	Maranoa / Warrego	Darling Dwns / Granite Belt	Wide Bay / Burnett	Southeast	QLD Estimates
Town Sample	Normanton	Mt Isa	Longreach	Birdsville	Weipa	Cairns	Charters Twrs	Townsville	Mackay	Emerald	Rockhampton	Charleville	Toowoomba	Maryborough	Brisbane	Overall
January	25	24	22	25	25	25	22	24	24	21	22	20	17	21	21	**23**
February	25	25	22	26	24	24	23	25	24	21	22	20	17	21	21	**23**
March	24	23	20	23	24	23	20	23	22	19	21	17	15	19	19	**21**
April	22	18	16	16	22	21	18	20	19	16	18	12	13	17	16	**18**
May	18	14	11	12	21	19	15	16	17	12	14	7	9	11	12	**14**
June	18	12	12	11	20	19	14	16	17	12	14	9	10	13	13	**14**
July	16	9	7	7	20	18	12	14	14	9	11	3	7	10	10	**11**
August	15	8	6	6	19	16	10	12	11	7	8	2	4	7	7	**9**
September	21	15	13	11	22	20	15	19	16	13	14	11	10	12	13	**15**
October	23	20	18	18	22	22	18	21	20	16	17	15	13	15	15	**18**
November	25	21	20	20	23	22	18	22	20	18	18	18	14	17	17	**20**
December	27	24	22	22	25	25	23	26	24	21	22	20	16	20	20	**22**
Overall	22	18	16	16	22	21	17	20	19	15	17	13	12	15	15	**17**

2026 QUEENSLAND - Estimated Degree of Variation from the Maximum Average

Region	Gulf Country	Northwest	Central West	Channel Country	Peninsula	Tropical Coast & Tablelnds	Goldfields / Upper Flinders	Herbet / Lower Burdekin	Central Coast-Whitsundays	Central Highlands	Capricornia	Maranoa / Warrego	Darling Dwns / Granite Belt	Wide Bay / Burnett	Southeast	QLD
Town Sample	Normanton	Mt Isa	Longreach	Birdsville	Weipa	Cairns	Charters Twrs	Townsville	Mackay	Emerald	Rockhampton	Charleville	Toowoomba	Maryborough	Brisbane	Overall
January	-0.3	0.4	-4.9	-1.1	-0.7	0.6	-1.9	-0.3	-0.7	-4.1	-2.2	-4.7	-2.8	-1.9	-1.4	-1.7
February	-0.5	2.5	-0.9	-0.5	-0.3	0.2	-0.9	-0.3	-0.4	-2.6	-1.5	-2.7	-1.6	-1.0	-0.8	-0.8
March	-0.5	1.9	0.5	1.4	-0.7	-0.9	-1.3	-0.3	-0.6	-1.4	-0.7	-1.3	-1.6	-0.3	-0.7	-0.4
April	0.5	1.6	1.0	-0.5	-0.2	0.6	0.1	0.8	-0.2	-0.4	0.3	0.2	0.3	0.3	-0.1	0.3
May	0.5	2.1	1.6	1.5	0.4	1.0	1.1	1.0	1.3	1.1	1.7	1.9	1.1	1.8	0.9	1.3
June	0.9	2.3	1.7	2.2	0.6	0.7	1.0	1.3	0.9	1.0	0.9	2.7	1.7	1.1	1.1	1.3
July	0.0	1.0	1.3	-0.8	-0.1	0.3	0.1	0.4	0.4	0.0	0.6	1.5	0.4	0.3	0.2	0.4
August	-3.0	-3.0	-4.0	-4.4	-0.7	-1.1	-3.8	-2.2	-2.1	-4.2	-2.7	-3.2	-4.0	-2.6	-2.2	-2.9
September	-0.4	-0.7	-2.4	-2.2	-0.3	0.3	-2.4	-1.0	-1.7	-2.7	-1.3	-2.3	-2.7	-0.9	-1.9	-1.5
October	0.2	0.5	-0.4	0.7	-0.1	0.1	-1.4	0.2	-0.7	-1.8	-1.0	-0.2	-0.8	-0.5	-0.6	-0.4
November	0.2	1.3	0.3	-1.3	1.1	0.3	-0.7	-0.4	-0.3	-0.6	0.2	0.5	-0.7	0.0	-0.7	-0.1
December	1.3	1.1	0.9	-2.4	0.3	1.4	1.3	0.5	2.6	0.6	1.2	0.0	-0.6	1.4	0.5	0.7
Overall	-0.1	0.9	-0.4	-0.7	-0.1	0.3	-0.7	-0.1	-0.1	-1.3	-0.4	-0.6	-1.0	-0.2	-0.5	-0.3

2026 QUEENSLAND - Estimated Degree of Variation from the Minimum Average

Region	Gulf Country	Northwest	Central West	Channel Country	Peninsula	Tropical Coast & Tablelnds	Goldfields / Upper Flinders	Herbet / Lower Burdekin	Central Coast-Whitsundays	Central Highlands	Capricornia	Maranoa / Warrego	Darling Dwns / Granite Belt	Wide Bay / Burnett	Southeast	QLD
Town Sample	Normanton	Mt Isa	Longreach	Birdsville	Weipa	Cairns	Charters Twrs	Townsville	Mackay	Emerald	Rockhampton	Charleville	Toowoomba	Maryborough	Brisbane	Overall
January	0.0	0.0	-1.5	-1.2	0.3	0.8	0.0	0.1	0.0	-0.9	0.2	-1.9	-0.4	0.6	0.1	-0.3
February	0.1	1.4	-0.5	1.0	0.0	0.4	0.3	0.4	0.4	-0.5	0.2	-1.4	-0.9	0.8	-0.6	0.1
March	-0.4	1.2	-0.8	0.7	-0.1	0.3	-0.7	0.1	-0.1	-1.9	-0.2	-1.6	-1.3	0.0	-1.1	-0.4
April	-0.9	-1.0	-0.9	-1.1	-0.4	-0.3	-0.7	-0.6	-0.8	-1.3	-0.2	-1.5	-0.8	-0.1	-0.8	-0.8
May	-1.6	-0.4	-1.2	-0.1	-0.8	-1.4	-0.6	-1.2	-0.4	-1.3	-0.6	-1.9	-0.5	-1.7	-1.3	-1.0
June	1.2	2.4	3.3	2.8	-0.5	0.8	1.6	1.4	2.5	2.0	3.1	3.4	2.2	2.4	1.9	2.0
July	0.0	-0.1	-0.2	-0.4	0.8	1.3	0.3	0.4	0.9	0.0	1.2	-0.9	0.3	1.1	0.4	0.3
August	-2.2	-1.8	-2.4	-2.7	0.3	-1.7	-2.7	-2.5	-2.9	-3.2	-2.9	-3.4	-3.2	-2.8	-3.0	-2.5
September	0.9	0.5	0.2	-1.9	1.7	1.6	0.3	1.6	-0.6	-0.7	0.5	0.7	-0.8	-0.5	0.0	0.2
October	-0.2	1.3	1.3	0.3	0.6	1.0	-0.3	0.1	0.2	-0.8	-0.1	0.4	-0.2	-0.9	-0.3	0.1
November	-0.4	-0.4	-0.5	-0.8	-0.1	-0.3	-2.4	-1.3	-1.7	-1.9	-1.2	0.0	-1.1	-1.0	-1.1	-0.9
December	0.5	0.8	-0.5	-2.0	0.6	1.4	0.7	1.4	0.9	-0.6	0.5	-0.9	-0.3	0.0	-0.5	0.1
Overall	-0.2	0.3	-0.3	-0.5	0.2	0.3	-0.3	0.0	-0.1	-0.9	0.0	-0.8	-0.6	-0.2	-0.5	-0.2

2026 QUEENSLAND - Estimated Maximum Temperature Trend

Region	Gulf Country	Northwest	Central West	Channel Country	Peninsula	Tropical Coast & Tablelnds	Goldfields / Upper Flinders	Herbet / Lower Burdekin	Central Coast-Whitsundays	Central Highlands	Capricornia	Maranoa / Warrego	Darling Dwns / Granite Belt	Wide Bay / Burnett	Southeast	QLD
Town Sample	Normanton	Mt Isa	Longreach	Birdsville	Weipa	Cairns	Charters Twrs	Townsville	Mackay	Emerald	Rockhampton	Charleville	Toowoomba	Maryborough	Brisbane	Overall
January	av	av	clr	clr	a-c	a-w	clr	av	a-c	clr	clr	clr	clr	clr	clr	clr
February	av	wmr	a-c	av	av	av	a-c	av	av	clr	clr	clr	clr	a-c	a-c	a-c
March	av	wmr	av	wmr	a-c	a-c	clr	av	a-c	clr	a-c	clr	clr	av	a-c	av
April	av	wmr	a-w	av	av	a-w	av	a-w	av	av	av	av	av	av	av	av
May	av	wmr	wmr	wmr	av	a-w	wmr	a-w	wmr	wmr	wmr	wmr	wmr	wmr	a-w	wmr
June	a-w	wmr	wmr	wmr	a-w	a-w	a-w	wmr	a-w	a-w	a-w	wmr	wmr	wmr	wmr	wmr
July	av	a-w	wmr	a-c	av	av	av	av	av	av	a-w	wmr	av	av	av	av
August	clr	clr	clr	clr	a-c	clr	clr	clr	clr	clr	clr	clr	clr	clr	clr	clr
September	av	a-c	clr	clr	av	av	clr	a-c	clr	clr	clr	clr	clr	a-c	clr	clr
October	av	av	av	a-w	av	av	clr	av	a-c	clr	a-c	av	a-c	av	a-c	av
November	av	wmr	av	clr	wmr	av	a-c	av	av	a-c	av	av	a-c	av	a-c	av
December	wmr	wmr	a-w	clr	av	wmr	wmr	av	wmr	a-w	wmr	av	a-c	wmr	av	a-w
Overall	av	a-w	av	a-c	av	av	a-c	av	av	clr	av	a-c	a-c	av	av	av

2026 QUEENSLAND - Estimated Minimum Temperature Trend

Region	Gulf Country	Northwest	Central West	Channel Country	Peninsula	Tropical Coast & Tablelnds	Goldfields / Upper Flinders	Herbet / Lower Burdekin	Central Coast-Whitsundays	Central Highlands	Capricornia	Maranoa / Warrego	Darling Dwns / Granite Belt	Wide Bay / Burnett	Southeast	QLD
Town Sample	Normanton	Mt Isa	Longreach	Birdsville	Weipa	Cairns	Charters Twrs	Townsville	Mackay	Emerald	Rockhampton	Charleville	Toowoomba	Maryborough	Brisbane	Overall
January	av	av	clr	clr	av	a-w	av	av	av	a-c	av	clr	av	a-w	av	av
February	av	wmr	av	a-w	av	av	av	av	av	av	av	clr	av	a-w	a-c	av
March	av	wmr	a-c	a-w	av	av	a-c	av	av	clr	av	clr	clr	av	clr	av
April	a-c	a-c	a-c	clr	av	av	a-c	a-c	a-c	clr	av	clr	a-c	av	a-c	a-c
May	clr	av	clr	av	a-c	clr	a-c	clr	av	clr	a-c	clr	av	clr	clr	a-c
June	wmr	wmr	wmr	wmr	av	a-w	wmr	wmr	wmr	wmr	wmr	wmr	wmr	wmr	wmr	wmr
July	av	av	av	av	a-w	wmr	av	av	a-w	av	wmr	a-c	av	wmr	av	av
August	clr	clr	clr	clr	av	clr	clr	clr	clr	clr	clr	clr	clr	clr	clr	clr
September	a-w	av	av	clr	wmr	wmr	av	wmr	a-c	a-c	av	a-w	a-c	av	av	av
October	av	wmr	wmr	av	a-w	a-w	av	av	av	a-c	av	av	av	a-c	av	av
November	av	av	av	a-c	av	av	clr	clr	clr	clr	clr	av	clr	a-c	clr	a-c
December	av	a-w	av	clr	a-w	wmr	a-w	wmr	a-w	a-c	av	a-c	av	av	av	av
Overall	av	av	av	av	av	av	av	av	av	a-c	av	a-c	a-c	av	av	av

NEW SOUTH WALES / ACT 2026

GENERAL TRENDS

Year Overall: Slightly drier but within 5% of the average
Slightly cooler but within 0.5°C of the average

Summer: Wetter by up to 60-65% more than the average
Wettest – Throughout December and second and third weeks of February
Driest – Second and fourth weeks of January
Cooler by around 1.5-2°C below the average
Hottest –Mid January and second week of February
Coolest – First week of January and last few days of January and mid-February

Autumn: Drier by around 40-45% below the average
Wettest – First half of March and first ten days of May
Driest – Second half of May
Warmer by around 0.5-1°C above the average
Hottest – Second half of March
Coolest – Second week of May to end of the month

Winter: Drier by around 20-25% below the average
Wettest – Second half of June and first week of August
Driest – First half of July
Slightly warmer but within 0.5°C of the average
Hottest – First half of June
Coolest – Mid-July and second week of August

Spring: Slightly drier but within 5% of the average
Wettest –Second to third weeks of September and last week of November
Driest – Second week of October
Cooler by around 0.5-1°C below the average
Hottest – Mid-October and second and fourth weeks of November
Coolest – First three weeks of September

Major Weather Events

January 2026

Date	Day	New South Wales / ACT - January 2026	Sydney	Canberra	Moon
1st	Thur	Widespread storm activity throughout the State over the next couple of days. Thunderstorms and rain with occasional heavy falls in the Riverina region, particularly in the Hillston area, where hail showers may occur.		Rain, Th	
2nd	Fri	Further thunderstorms with heavy rain in the Riverina, with the chance of flash flooding, damaging wind gusts, and hail possible, particularly in the Deniliquin district. High maximum temperatures likely over the next few days in the Far Northwest and Darling regions.			P6
3rd	Sat	Occasional thunderstorm activity in the Central West Plains, with heavy falls at times and chance of hail showers. Unstable winds in the Riverina may generate tornado activity near Hay. Scattered storm activity through the Lower and Upper Western regions, with damaging wind gusts likely in a number of areas including around Broken Hill, Wilcannia, Ivanhoe, and Fowlers Gap. Winds may also be extremely gusty and occasionally to gale force, on the Southern Tablelands to the Snowy Mountains regions.	Rain		F ^
4th	Sun	Lingering showers in the Metropolitan to the South Coast and into the adjacent Tablelands. Very wet for a time in the Central West Plains, with chance of localised flooding, possibly in the Coonamble district.	Showers		
5th	Mon	Showers around the Mid-North Coast to the North Coast, while mostly dry elsewhere over the next couple of days. Cool over the next couple of nights in the Central Tablelands.			
6th	Tue		Showers	Showers, Th	
7th	Wed	Showers in the southern parts of the State from the Lower Western and Riverina to the South Coast. Cold overnight in the Snowy Mountains with patches of frosts likely, particularly at Thredbo.			
8th	Thu	Mostly dry spell across the State over the next few days.			
9th	Fri				XhS
10th	Sat	Chance of isolated showers over the next couple of days in the North Coast.			
11th	Sun				3Q
12th	Mon	Some rain around the North Coast to the Northern Tablelands over the next couple of days. Hot spell over the next couple of days in the Lower Darling region, particularly in the Pooncarie district where maximum temperatures may reach into the low 40's.			
13th	Tue				
14th	Wed	The next 3-4 weeks ahead may be generally wetter and warmer than is typical for this time of year. Showers spread southward along the coast into the Hunter to the Greater Sydney region.	Showers		A
15th	Thur	Thunderstorms on the Southwest Slopes, with heavy rain which may cause some localised flash flooding, and occasional hail showers. Chance of storm activity, including large hail causing damaage to crops in the Batlow area. Heavy rain returns to the North Coast to the Hunter regions over the next couple of days, possibly also causing some outbreaks of flooding, particularly in the Lismore, Bellingen, Murmayumbah and Kyogle districts.	Showers		
16th	Fri	Showers spread southward along the Tablelands to the Riverina.			
17th	Sat	Mostly dry over the next couple of days, except for around the North Coast.			V
18th	Sun	Very heavy rain in the Northern Rivers around this time, which chance of localised flooding.			
19th	Mon	Thunderstorms with outbursts of heavy rain on the Northwest Slopes, particularly around Nundle. Thunderstorms with chance of flash flooding also on the Northern Tablelands and Mid-North Coast especially near Port Macquarie and Inverell.			N
20th	Tue				
21st	Wed				
22nd	Thu	Some showers possible around the Broken Hill area.		Showers, Th	
23rd	Fri	Showers and thunderstorms likely around the Hunter to the Metropolitan over the next couple of days. Very hot day in the Lower Darling with maximum temperatures into the mid-40's in some places.	Showers, Th		
24th	Sat	Strong winds with gusts in the Riverina with chance of causing damage in some parts, including in Deniliquin area. The unstable winds in this region may cause the formation of a tornado near Shepparton. Temperatures may also be unusually warm overnight in some parts of the Riverina and maximum temperatures into the low to mid 40's during the day.	Showers		XhN
25th	Sun	High temperatures generate thunderstorms with large hail and damaging winds in the Southern Tablelands.			
26th	Mon	Very strong and gusty winds in the Sydney region with chance of reaching gale force for a time. Very hot with maximum temperatures likely to reach into the upper 30's in a number of districts in the Metropolitan, Illawarra and South Coast regions.		Showers, Th	1Q
27th	Tue	Thunderstorms with large hail in the Central Tablelands, particularly in the Orange area, and thunderstorms with strong damaging wind gusts in the Metropolitan, including in the western suburbs.			
28th	Wed	Widespread outbreaks of rain and storms across the State over the next few days. Chance of flash flooding in a number of districts and along the river catchments during this wet spell. Temperatures may be very hot in the west of the State around this time, with daily maximas into the low to mid 40's likely in a number of districts.	Rain	Rain	

Date	Day	New South Wales / ACT - January 2026	Sydney	Canberra	Moon
29th	Thur	Thunderstorms with heavy rain in the Central Tablelands and Central West Slopes with chance of localised flooding in the Dubbo and Rylstone areas. Thunderstorms and hail with damaging wind gusts move into the Metropolitan, with particularly large hail possible in the Western suburbs.	Rain, Th	Rain	
30th	Fri		Showers, Th	Showers	P10 ^
31st	Sat	The widespread storm activity of the past few days eases around this time, clearing in the western half of the State, but with showers persisting in the Tablelands to the eastern coasts.	Rain		

February 2026

Date	Day	New South Wales / ACT - February 2026	Sydney	Canberra	Moon
1st	Sun	Patchy showers across the State.			
2nd	Mon	Mostly dry over the next couple of days.			F
3rd	Tue	Cold spell in the Snowy Mountains over the next couple of days including the likelihood of patches of frost in some areas, including in the Thredbo district.			
4th	Wed	Chance of showers around the North Coast and the Lower Western.			
5th	Thu	Showers in the northeast quarter of the State and in the Southwestern Plains to the Central West.			
6th	Fri				XhS
7th	Sat	Clearing to mostly dry throughout the State over the next couple of days.			
8th	Sun				
9th	Mon			Showers	
10th	Tue	A series of wet fronts and storms cross the State over the coming 8-10 days, with widespread precipitation in most districts on most days.	Rain, Th	Th	3Q
11th	Wed	Stormy in the Central Tablelands and Central West Slopes, with thunderstorms and large hail in some districts, including in Katoomba and Coonabarabran, where damaging wind gusts may occur.	Showers		A
12th	Thur	A trough embedded with heavy rain and thunderstorms cross the Upper Western over the next couple of days with chance of flash flooding in the Cobar area.	Showers		
13th	Fri	Scattered thunderstorms from the Hunter, down through the Metropolitan and into the Illawarra with heavy rain and large hail, and chance of local flash flooding in the Sydney Metropolitan area, and in some outer suburbs. Chance of flash flooding also in the Bellambi area. Chance of heavy rain and strong gusty winds accompanying scattered thunderstorm activity in the Northwest Slopes, particularly in the Tamworth area, with possible damage to property. The thunderstorms cells may generate very strong winds and cause some localised damage in the Metropolitan region.	Rain	Showers	V
14th	Sat	The next 3-4 weeks ahead may continue to be wetter than is typical for this time of year, but cooler than normal. Thunderstorms with heavy rain and bursts of flash flooding in the Mid-North Coast and Northwest Slopes.	Rain	Rain	
15th	Sun	Heavy rain and thunderstorms in the Upper Western, with possible minor flooding in the White Cliffs area. Very warm spell in the Southwest Slopes and through the Riverina regions, with maximum temperatures settling around the mid-30's in most districts.	Showers	Showers	
16th	Mon	Thunderstorms with heavy rain at times in the Metropolitan and Southwest Slopes regions. Heavy rain in the Northern Rivers region may produce localised flooding in some districts.	Showers, Th		
17th	Tue	Thunderstorms and spells of heavy rain in the Northwest Slopes region. Chance of localised flooding for a brief time in the Ogunbil area.	Rain, Th	Showers, Th	
18th	Wed	Rain activity contracts to the eastern parts of the State from the Tablelands to the coast over the next couple of days. Mostly dry in the west.	Rain		N
19th	Thu	A slow moving line of storms cross the east of the State bringing scattered thunderstorms across the Northern and Central Tablelands, through the Blue Mountains, Northern Rivers, Metropolitan and Illawarra regions. Heavy rainfall in many places with chance of flash flooding.	Rain		
20th	Fri	Further heavy rain and thunderstorms in the Metropolitan region, which chance of flash flooding in some Sydney suburbs. Thunderstorms in the Hunter may be accompanied by strong wind gusts.	Rain, Th		XhN
21st	Sat	Brief dry spell across the State.		Rain	
22nd	Sun	Further widespread rain throughout the State over the next couple of days.	Rain	Showers	
23rd	Mon	Cold overnight in the Northern Tablelands to the North Coast regions.			
24th	Tue	Rain clears to isolated showers over the next couple of days, confined mostly in the northeast.			
25th	Wed				1Q P14
26th	Thur	Brief dry spell across most of the State for the rest of the month, except for some shower activity returning in the North Coast to Northern Tablelands regions.			^
27th	Fri				
28th	Sat				

March 2026

Date	Day	New South Wales / ACT - March 2026	Sydney	Canberra	Moon
1st	Sun	Light showers in the northeast over the next few days, while mostly dry in the western half of the State over the coming week.			
2nd	Mon			Showers	
3rd	Tue	Brief showery spell around the South Coast over the next couple of days.		Rain, Th	F
4th	Wed				
5th	Thu	Mostly dry throughout the State over the next 3-4 days.			XhS
6th	Fri			Showers, Th	
7th	Sat	Very warm spell throughout the State, and particularly hot in the Far Northwest, with many districts maximum temperatures likely to rise into the mid-40's and overnight minimums only dropping into the late 20's to early 30's.			
8th	Sun	Thunderstorms, heavy rain and flash flooding in the Southern Tablelands over the next couple of days, with chance of damage to local property, including in Queanbeyan. Very strong and blustery winds a times in the Snowy Mountains, including around Thredbo. Very hot along the northern coastal regions to the Northern Tablelands with daily maximas in the mid to upper 30's in most districts.	Rain, Th		
9th	Mon	Showers turn to rain in the North Coast and Northern Tablelands, later spreading into the Hunter to the Metropolitan over the next couple of days.	Showers, Th		
10th	Tue	Brief incursion of rain activity into the Riverina.	Rain, Th	Showers	
11th	Wed	Thunderstorms wit heavy rain and possible flash flooding in the Metropolitan region. Scatttered thunderstorms also through the Central and Southern Tablelands and into the Illawarra, with chance of flash flooding in some places, and large hail showers at times in the Illawarra. Chance of damaging wind gusts in the Illawarra.	Rain, Th		3Q A
12th	Thur	Mostly dry over the coming week throughout the State with only a couple of brief exceptions.	Showers		V
13th	Fri	Scattered thunderstorms in the Hunter, with heavy rain and flash flooding in the Newcastle area, which may cause some damage to local property.		Fg	
14th	Sat	Some showers likely around the North Coast. Unusually cold in the Snowy Mountains to the Central Tablelands around this time with the chance of frosts overnight, particularly in the Thredbo and Katoomba areas.			
15th	Sun	The next 3-4 weeks may be warmer and drier than normal for this time of year. Temperatures may be sustained at heatwave levels over the next couple of weeks throughout the State, particularly in the western regions.			
16th	Mon				
17th	Tue			Th	
18th	Wed		Rain, Th	Showers, Th	
19th	Thu	Storms cross the eastern regions from the Hunter to the Metropolitan with thunderstorms, accompanying blustery winds, heavy rain and hail, with chance of flash flooding in some districts.			N XhN
20th	Fri				
21st	Sat	Brief showers likely in the North Coast. Very warm spell in the Central and Northwest Plains with daily maximas into the upper 30's in many districts.	Rain, Th		
22nd	Sun	Dry over most of the State over the remainder of the month, with just a couple of brief showery interludes, mostly in the North Coast region.		Showers, Th	P11
23rd	Mon			Fg	
24th	Tue	Frosts in the Snowy Mountains overnight, particularly in the Charlotte Pass area.			
25th	Wed				
26th	Thur				1Q ^
27th	Fri	Very hot in the Lower Darlings around this time with maximum temperatures into the low 40's particularly in the Pooncarie district. Warmer spell throughout most of the State over the next few days.			
28th	Sat	Light showers in the North Coast to the Northern Tablelands likely over the next couple of days. Unusually warm overnight in the Lower Darling region, including in Broken Hill, with minimum temperatures into the upper 20's likely. Hot spell in the Riverina, Southwest Slopes, Southern Tablelands and Goulburn regions over the next couple of days with daily maximas into the upper 30's to early 40's in many districts.			
29th	Sun				
30th	Mon				
31st	Tue	Chance of isolated showers in the North Coast to the Northern Tablelands.	Rain, Th		

April 2026

Date	Day	New South Wales / ACT - April 2026	Sydney	Canberra	Moon
1st	Wed	A showery week ahead, with showers starting in the southeast at first, spreading up the coast and adjacent Tablelands over the next couple of days, before becoming widespread throughout the State in the latter half of the coming week, then contracting back to the coast once more by early in the next week.	Showers		
2nd	Thu				F XhS

Date	Day	New South Wales / ACT - April 2026	Sydney	Canberra	Moon
3rd	Fri	Strong damaging wind gusts may accompany some thunderstorm activity in the Illawarra.	Rain		
4th	Sat	Unusually cold over the next couple of nights in the Southern Tablelands to the Illawarra regions, with maximim temperatures barely making it into double digits.	Showers, Fg	Showers, Fg	
5th	Sun	Widespread rain and scattered thunderstorms throughout the State over the next couple of days.	Rain, Th, Fg	Rain	
6th	Mon	A line of thunderstorms cross the Lower and Upper Western regions bringing heavy rain and possible flash flooding, particularly in the Broken Hill and Brewarrina areas. Strong gusty winds ahead of the storms may raise dust. Winds may be up to gale force at times across the Upper Darlings, particularly in the White Cliffs district.	Th		
7th	Tue	Rain contracts to the Riverina and eastern coastal fringes over the next couple of days.	Showers, Th		A
8th	Wed	Thunderstorms, heavy rain, and possible pockets of flash flooding in the Northern Rivers region, particularly around Ballina.	Showers, Th		
9th	Thur	Showers in the Northern Tablelands to the adjacent coast while mostly dry elsewhere.			V
10th	Fri	Strong winds through the Mid-North Coast region with chance of damage to property and infrastructure.			3Q
11th	Sat	Thunderstorms and hail showers in the Central Tablelands. Very strong winds with damaging gusts accompanying thunderstorm activity on the Mid-North Coast, particularly around Nambucca Heads. Generally very cold over the next few nights in many parts of the State.			
12th	Sun	Mostly dry over the next few days.		Fg	
13th	Mon	Unusually cold day in the Snowy Mountains, particularly at Thredbo, where the maximum temperature may only just rise above frost level. Chance of some patches of frost in sheltered parts of the Northern Tablelands to the North Coast regions.		Showers	
14th	Tue	Chance of isolated showers in a line from the Far Northwest to the South Coast.			
15th	Wed	The next 3-4 weeks may be cooler and drier than average for this time of year. Dry throughout the western half of the State over the remainder of the month.			
16th	Thu	Hot spell across the State over the next couple of days, particularly in the Far Northwest where maximum temperatures in the mid-30's are likely in a number of districts, including in the Tibooburra area, and maximum temperatures in the mid to upper 20's elsewhere. Very strong and unsettled winds in the Southern Tablelands around this time, with possible gales in the Snowy Mountains.			XhN
17th	Fri	Showers return to the North Coast region and into the adjacent Tablelands at first, spreading into the Hunter and Central Tablelands and later to the South Coast over the next couple of days.	Rain		N
18th	Sat	Thunderstorms and burst of heavy rain in the Mid-North Coast region over the next few days.	Rain		
19th	Sun		Rain		P7
20th	Mon		Showers	Fg	
21st	Tue	Scattered rain activity contracts to the northeast quarter of the State over the next couple of days, while remaining mostly dry elsewhere.		Fg	
22nd	Wed				^
23rd	Thur	Brief dry spell throughout the State over the next couple of days.		Showers, Th	
24th	Fri	Warmer spell across the State over the next few days.		Showers, Th	1Q
25th	Sat	Thunderstorms with large hail and heavy rain in the ACT region, particularly around Kingston around this time. Supercell activity in the Hunter to the Illawarra including in the Metropolitan and Mid-North Coast develops thunderstorms, large hail and heavy rain in the regions. Wind gusts accompanying the thunderstorms may cause some damage to property in the Metropolitan areas.	Rain, Th		
26th	Sun	Showers ease in the southeast to mostly dry, while reform in the northeast for a time. Unusually mild overnight in the Far Northwest.	Showers		
27th	Mon		Showers, Th	Rain, Th	
28th	Tue	Further outbreaks of showers and patches of rain over the eastern parts of the State from the North Coast to the South Coast over the next few days.	Showers		
29th	Wed		Showers		XhS
30th	Thur	Rain may be heavy for a time in the North Coast to the Northern Tablelands, and in parts of the Hunter.	Showers		

May 2026

Date	Day	New South Wales / ACT - May 2026	Sydney	Canberra	Moon
1st	Fri	Outbreaks of rain along the east of the State over the coming week, from the North Coast to the South Coast.	Rain	Fg	
2nd	Sat	Heavy rain at times in the Hunter to the adjacent Tablelands.	Rain		F
3rd	Sun		Rain		
4th	Mon	Pockets of heavy rain likely in the Hunter to the Mid-North Coast around this time and over the next couple of days, showers elsewhere east of the Tablelands.	Rain	Fg	
5th	Tue		Showers	Fg	A
6th	Wed		Showers	Fg	V
7th	Thu				

Date	Day	New South Wales / ACT - May 2026	Sydney	Canberra	Moon
8th	Fri	Mostly dry along the east of the State, while light showers move into the western districts.		Rain, Sn	
9th	Sat	An intense cold front crosses the State over the next couple of days. Heavy rain moves over the Upper Western to the Upper Darlings around this time, and down through the Central Plains. Strong winds accompany thunderstorm activity in the Central West Slopes and Plains, with chance of wind damage occuring in the Forbes and Condobolin areas. Snow in the NSW-ACT ski regions over the next couple of days, mostly in the higher reaches, but some light snow may also occur in the Central Tablelands including in the districts of Bathurst, Katoomba, Oberon, and Orange. Chance of very heavy rain around this time on the Mid-North Coast with possible pockets of flash flooding.			
10th	Sun	Mostly dry throughout the State over the next couple of weeks.			3Q
11th	Mon	Colder spell throughout the State over the next few days, including the arrival over patches of frost overnight in some districts, particularly in the Central Plateau to the Far Northwest, and in the Southern Tablelands to the Southwest Slopes and the South Coast, where some frosts may be severe.			
12th	Tue	Chance of isolated showers west of the Tablelands over the next couple of days. Very cold with severe frosts overnight in the Snowy Mountains, particularly at Charlotte Pass.			
13th	Wed				XhN
14th	Thur				
15th	Fri	The next 3-4 weeks may be drier throughout the State mostly as a result of persistent high pressure systems being the dominant feature, and while maximum temperatures may be warmer than average, minimum temperatures may be colder than the norm for this time of year. Generally, it won't be until late May into early June before shower embedded troughs cross the State bringing some light falls mostly in coastal regions, and only some in inland locations.			
16th	Sat				
17th	Sun				N
18th	Mon	Frosts over the next couple of nights in the Tablelands, including in the Northern Tablelands, as well as in the Central and Southwestern Slopes.		Fg	P3
19th	Tue			Fg	^
20th	Wed	Milder spell in the North Coast around this time.	Showers	Fg	
21st	Thu				
22nd	Fri	Milder spell in the southeast of the State over the next couple of days.			
23rd	Sat			Fg	1Q
24th	Sun	Chance of isolated showers in the Southern Tablelands and milder temperatures in the Central and Northern Tablelands.		Fg	
25th	Mon	Isolated showers likely along the coastal fringes of the State over the next couple of days. Hot spell over the next couple of days in the Upper Darling to the Northwest Plains region.	Fg	Fg	
26th	Tue		Showers	Showers	XhS
27th	Wed	Light showers throughout most of the State, although mostly dry in the northeast.		Showers	
28th	Thur			Showers	
29th	Fri	Widespread shower activity throughout the State. Chance of storm activity in the Hunter with possible large hail stones that may cause damage to property.		Fg	
30th	Sat	Mostly dry throughout the State over the next couple of days except for the chance of snow in the NSW-ACT ski fields.		Fg	
31st	Sun	Possible isolated showers in the Far Northwest. Very cold in the Southern Tablelands, with severe frosts in the Snowy Mountains around this time, particularly at Charlotte Pass. Winds may be particularly strong and gusty in the Snowy Mountains around this time, possibly to gale force at times.		Fg	F

June 2026

Date	Day	New South Wales / ACT - June 2026	Sydney	Canberra	Moon
1st	Mon	Mostly dry in the coming week, with a couple of exceptions.			A
2nd	Tue	Some showers in the Northern Tablelands to the North Coast regions over the next couple of days.			V
3rd	Wed				
4th	Thu			Fg	
5th	Fri	Chance of isolated showers over the next few days along the southern border districts from the Riverina to the South Coast.	Showers	Showers, Fg	
6th	Sat	A high pressure system near the south of Tasmania brings a very cold, dry airflow over the Alpine and Tableland regions over the next few days, with severe frosts likely. In contrast, the next couple of days may be unusually mild in the Far Northwest and the Darlings regions.		Rain, Fg	
7th	Sun			Fg	
8th	Mon		Showers, Th	Fg	3Q
9th	Tue	Rain moves into the North Coast to the Northern Tablelands over the next few days.	Showers	Fg	XhN
10th	Wed		Fg	Fg	
11th	Thur		Showers	Fg	
12th	Fri	Chance of storms with hail showers in the Metropolitan over the next couple of days, with large hail stones possible in some places.	Rain		

Date	Day	New South Wales / ACT - June 2026	Sydney	Canberra	Moon
13th	Sat	Widespread rain activity over the eastern half of the State over the next couple of days, with heavy falls around the North Coast to the Mid-North Coast, as well as precipitation spreading west from the South Coast and over the Southern Tablelands and into the Riverina.	Showers, Fg	Showers	
14th	Sun	Mild overnight temperatures in the North Coast region. In contrast, overnight minimum temperatures may be very low in the Riverina region, with chance of some patches of frost in sheltered areas.	Rain	Showers	
15th	Mon	Shower activity spreads throughout the western half of the State while persisting in the east with heavy falls likely in many districts from the North Coast to the Illawarra and up into the adjacent Tablelands. Unusually mild temperatures overnight on the North Coast.	Rain		N P2
16th	Tue	Rain activity clears in the west and into the North Coast regions, but continues with heavy falls around the Hunter to the Greater Sydney and Illawarra regions over the next couple of days.	Rain		^
17th	Wed	The next 3-4 weeks may be warmer over most of the State as well as wetter than average in the north and west, and drier than normal for this time of year in the south and east.			
18th	Thu	Another wet spell moves onto the State from the eastern coastal fringes around this time, and spreading eastward over the next couple of days. Very heavy rain likely in the Manning region, with chance of flash flooding.	Showers		
19th	Fri	Mild overnight temperatures in the Hunter.	Showers		
20th	Sat	Chance of heavy rain at times west of the Tablelands, particularly around the Broken Hill area. Patches of showers elsewhere over the next few days.	Showers, Fg	Showers	
21st	Sun	Very cold with severe frosts in the the Snowy Mountains around this time, particularly at Charlotte Pass.		Showers	
22nd	Mon		Showers	Showers	1Q XhS
23rd	Tue			Showers	
24th	Wed				
25th	Thur	Brief dry spell throughout the State over the next couple of days. Very strong winds in the Snowy Mountains, with winds to gale force likely at times, particularly in the Thredbo area.			
26th	Fri	Unusually milder day in the Southwest Plains.	Showers		
27th	Sat	Snow and frosts in the NSW-ACT ski regions. Light showers in the northeast quarter of the State over the next couple of days.	Showers		
28th	Sun			Fg	A
29th	Mon			Fg	V
30th	Tue	Widespread shower activity returns throughout most regions of the State around this time with some heavy falls at times in the Far Northwest, but remaining mostly dry along the South Coast to the Illawarra.	Showers	Showers	F

July 2026

Date	Day	New South Wales / ACT - July 2026	Sydney	Canberra	Moon
1st	Wed	Widespread rain throughout most of the State around this time, although mostly dry in the Broken Hill area. Rain may be heavy at times in the Far Northwest to the Central Plains and adjacent Tablelands, as well as in central parts of the Riverina.	Showers	Showers, Fg	
2nd	Thu	Rain swiftly clears to isolated showers, with dry conditions developing and prevailing throughout the State over the coming week to 10 days.			
3rd	Fri				
4th	Sat			Fg	
5th	Sun			Fg	
6th	Mon	Light snow in the NSW-ACT ski regions.			
7th	Tue				XhN
8th	Wed			Fg	3Q
9th	Thur				
10th	Fri	Light snow and frosts in the NSW-ACT ski regions. Cold spell throughout the State over the next 4-5 days, with scattered frosts in many districts.			
11th	Sat			Showers	
12th	Sun	Light showers in the Upper and Lower Western regions as well as around the South Coast, while remaining dry elsewhere.		Showers	
13th	Mon	Light showers may persist briefly around the South Coast. Dry elsewhere over the next few days. Milder maximum temperatures over the next couple of days west of the Tablelands.	Showers		P4 ^
14th	Tue	Snow in the NSW-ACT ski regions over the next few days.	Showers		N
15th	Wed	Snow may be heavy for a time in the Snowy Mountains and ACT ski fields over the next couple of days.	Rain		
16th	Thu	Showery down through the eastern coasts around this time. Winds may be very strong and gusty to gale force at times on the South Coast.		Fg	
17th	Fri	The next 3-4 weeks may bring near average rainfall in the east of the State and average to below average rainfall to the west, with temperatures slightly above normal for this time of year.		Showers	
18th	Sat	A very wet spell throughout the State over the next few days. Heavy rain at times in the Upper Western. Mostly dry briefly along the eastern coastal regions.		Rain	

Date	Day	New South Wales / ACT - July 2026	Sydney	Canberra	Moon
19th	Sun	Heavy rain at times in the Riverina to the South Coast, showers elsewhere.		Showers, Fg	
20th	Mon	Showers become isolated throughout the State.	Showers		XhS
21st	Tue	Light snow in the NSW-ACT ski fields.		Showers	1Q
22nd	Wed	Very cold in the Snowy Mountains with maximum temperatures not rising above frost levels in numerous ski regions, where snowfalls are also likely, particularly in Thredbo and Charlotte Pass fields. Isolated showers elsewhere.		Fg	
23rd	Thur	Mostly dry throughout the State over the next few days except for some snowflurries in the NSW-ACT ski fields.			
24th	Fri	Snow in the NSW-ACT ski fields over the next couple of days.			
25th	Sat	Chance of light snowfalls spreading briefly into the Central and Northern Tablelands around this time.	Fg		
26th	Sun	Spotty showers in the Upper and Lower Western region as well as in the Southern Tablelands.		Showers, Fg	A
27th	Mon	Patches of heavy rain in the Riverina, and the Upper and Lower Western.		Fg	V
28th	Tue	Mostly dry throughout the State for the remainder of this month. Mild maximum temperatures throughout much of the State around this time.		Fg	
29th	Wed	Chance of light snow in the NSW-ACT ski fields and isolated showers in the Upper Wester and Upper Darlings.		Showers	
30th	Thu				F
31st	Fri	Snowflurries in the NSW-ACT ski fields. Chance of isolated showers in the Lower Western.	Showers	Rain	

August 2026

Date	Day	New South Wales / ACT - August 2026	Sydney	Canberra	Moon
1st	Sat	Widespread rain and storm activity throughout the State. Snow in the NSW-ACT ski fields over the next 3-4 days.			
2nd	Sun	Mostly dry over the next couple of days except for the chance of some snow in the ski fields.			
3rd	Mon		Rain		XhN
4th	Tue	While remaining mostly dry in the west of the State over the next few days, rain returns to the northeast quarter of the State. Heavy rain at times in the North Coast to the Mid-North Coast and adjacent Tablelands over the next few days. Mild overnight in the North Coast region.	Rain		
5th	Wed	Frosts may be regular throughout the Central and Southern Tablelands and adjacent Western Slopes regions over the coming week.	Rain		
6th	Thur	Widespread scattered frosts through the Riverina region around this time.		Fg	3Q
7th	Fri	Brief dry spell throughout the State. Very cold overnight in the Snowy Mountains with severe frosts.	Showers, Th	Showers	
8th	Sat	Chance of isolated showers in most regions of the State.	Rain		
9th	Sun	Brief spell of rain activity in the Central Tablelands to the eastern coasts, and down from the Metropolitan to the South Coast. Dry elsewhere.	Showers	Fg	^
10th	Mon	Mostly dry throughout the State over the next few days. Chance of snow in the NSW-ACT ski regions.			P8
11th	Tue	Snowflurries in the Snowy Mountains.		Showers	
12th	Wed	Very heavy rain likely around the Mid-North Coast region.		Showers	
13th	Thu	Some showers in the Riverina to the South Coast, and isolated showers in the Northwest Plains and Slopes. Dry elsewhere.			N
14th	Fri	Light snow in the NSW-ACT ski fields. Dry elsewhere throughout the State over the next couple of days. In contrast, maximum temperatures may be unusually warm in the Far Northwest of the State, almost reaching into the 30's for a time in some locations.			
15th	Sat	While the next 3-4 weeks may be drier than is normal for this time of year due to a strong and persistent high pressure influence over the State, only the far southeast regions can expect above average rain, mostly as a result of a single low pressure system early in this lookahead period. Temperatures may also be much colder than normal except in the far northeast and far northern inland parts, which may be closer to the seasonal average. Snow almost daily over the next 2 weeks in the NSW-ACT ski fields.			
16th	Sun	Very strong winds in the South Coast region, possibly up to gale force at times.		Showers	XhS
17th	Mon	Some showers in the South Coast to the Southern Tablelands and isolated showers into the Riverina and up along the coasts to the North Coast.		Fg	
18th	Tue	Showers in most inland parts of the State over the next couple of days.	Showers	Showers	
19th	Wed	Showers in the coastal regions from the Metropolitan to the South Coast with chance of some heavy falls in some parts of the South Coast briefly.	Showers, Th		
20th	Thur	Brief dry spell throughout the State.			1Q
21st	Fri	Some light shower activity in the western and northern districts over the next couple of days. Snow likely in the Central and Southern Tablelands around this time, including in such areas as Orange, Batlow, and Oberon.	Rain	Showers	
22nd	Sat				A
23rd	Sun	Mostly dry throughout the State over the remainder of the month, except for the occasional snowfall events. Frosts in the Northern Tablelands around this time.			V
24th	Mon	Heavy snow in the Snowy Mountains, and ACT ski regions. Chance of snow in Canberra, Goulburn, and Yass, as well as up into the Central Tablelands.			

Date	Day	New South Wales / ACT - August 2026	Sydney	Canberra	Moon
25th	Tue	Daily snowfalls in the Snowy Mountains and ACT ski regions continue over the next week. Heavier snow likely in the Central Tablelands around this time, particularly in the Batlow and Oberon areas.			
26th	Wed			Showers	
27th	Thu				
28th	Fri	Severe frost patches likely in the North Tablelands to the Northwest Slopes over the next couple of days.			F
29th	Sat		Showers, Th	Fg	
30th	Sun				XhN
31st	Mon	Widespread frosts in the southeast of the State around this time, from the Southwest Slopes to the South Coast.			

September 2026

Date	Day	New South Wales / ACT - September 2026	Sydney	Canberra	Moon
1st	Tue	Light snow in the NSW-ACT ski fields.			
2nd	Wed	Some showers down through the Tablelands and into the Hunter, spreading by tomorrow along the eastern coastal regions, with heavier rain for a time in the Metropolitan to the Hunter.	Showers, Th		
3rd	Thur	Damaging winds accompany fast moving thunderstorms and hail showers through the Northern Rivers region around this time, particularly around Ballina.	Rain		
4th	Fri	Mostly dry and cold with scattered frosts throughout the State over the next 4-5 days.			3Q
5th	Sat	Snow in the NSW-ACT ski regions. Very low maximum temperatures in the Southwest Slopes and adjacent Tablelands.			
6th	Sun				^
7th	Mon	Very cold in the Snowy Mountains and Southern Tablelands with severe frosts, particularly in the Charlotte Pass area. Widespread frosts also likely in the Riverina.	Showers		P12
8th	Tue	Showers return to the North Coast over the next couple of days.			
9th	Wed	Mild maximum temperatures throughout most of the State over the next 4-5 days.			
10th	Thu	Rain threatening throughout the State, with some showers developing in the Lower Western and Mid-North Coast.	Rain	Rain	
11th	Fri	Widespread rain and storm activity spreads throughout the State over the next couple of days. Thunderstorms with hail in the Northern Tablelands, with possible very large hail in the Walcha area.	Showers, Th	Rain	N
12th	Sat				XhS
13th	Sun	Mostly dry throughout the State over the next couple of days. Unsually warm day in the Upper Darlings with maximum temperatures into the early 30's likely.			
14th	Mon	Unusually mild overnight in the North Coast to the Northwest Slopes regions.			
15th	Tue	A wet front crosses the State over the next few days. Light snow in the NSW-ACT ski fields over the next couple of days. Heavy rain may develop in the Hunter to the Mid-North Coast around this time, and spread into the North Coast and Northern Tablelands by tomorrow, then spread heavy rain down through the Tablelands to the Southern Tablelands and South Coast and into the Riverina regions in a couple of days time.	Showers		
16th	Wed	Light snow in the NSW-ACT ski fields. The next 3-4 weeks may be warmer and slighty wetter than average almost everywhere, although rainfall amounts may be fairly average for this time of year.	Showers		
17th	Thur	Heavy rain likely in the Metropolitan to the adjacent Central Tablelands as well as up around the Mid-North Coast over the next couple of days.	Rain	Fg	
18th	Fri	Some rain lingering around the Hunter to the Greater Sydney region and up into the adjacent Tablelands, while clearing to dry elsewhere.	Rain	Fg	
19th	Sat	Heavy rain around the Mid-North Coast at this time may cause some localised flooding.			1Q A V
20th	Sun	Dry throughout the State over the next 4-5 days.			
21st	Mon				
22nd	Tue				
23rd	Wed	Very cold with severe frosts in the the Snowy Mountains around this time, particularly at Charlotte Pass. Chance of heavy rain in some parts of the Central West.			
24th	Thu	A supercell thunderstorm system brings widespread hail showers and strong winds in the Northern Rivers region, with chance of very large hail stones particularly in Grafton and Swan Creek areas.	Showers	Rain	
25th	Fri	A squall line of thunderstorms cross the Riverina and Southern Tablelands with chance of small hail and damaging wind gusts, particularly around Narrandera and in the Thredbo areas. Light showers and passing thunderstorms elsewhere.	Showers, Th	Showers	
26th	Sat	Thunderstorm activity in the Southwest Slopes, and Southern and Central Tablelands may bring damaging wind gusts across the regions, including up through the Blue Mountains.		Showers	XhN
27th	Sun	Isolated showers in the South Coast. Mostly dry elsewhere throughout the State until the end of the month.			F
28th	Mon	Snow and very strong winds in the NSW-ACT ski fields over the next couple of days. Chance of gale force winds in the Thredbo area.		Fg	
29th	Tue				
30th	Wed	Very cold in the Snowy Mountains, particularly in the Thredbo area, with maximum temperatures possibly not rising out of the frost levels.			

October 2026

Date	Day	New South Wales / ACT - October 2026	Sydney	Canberra	Moon
1st	Thur	Mostly dry over the next couple of days throughout the State.			
2nd	Fri	Unusually hot day in the Far Northwest, particularly in the Tibooburra area with maximum temperatures into the mid-30's likely.			P13
3rd	Sat	A wet spell moves over the State over the next few days. Chance of thunderstorms in the Northern Rivers and Northwest Slopes regions which may develop large hail stones at times.	Showers	Rain, Th	
4th	Sun	Scattered thunderstorms across the Upper and Lower Western regions and into the Northern Tablelands, with chance of damaging wind gusts, particularly around Tenterfield, Broken Hill, and Fowlers Gap.	Showers		3Q ^
5th	Mon	Storm activity in the Upper and Lower Western, Riverina and into the Southern Tablelands, with the passage of a band of thunderstorms, bringing strong gusty and damaging winds and large hail showers, particularly in the Broken Hill, Tibooburra, Goulburn and Hay areas.		Fg	
6th	Tue	Mostly dry throughout the State over the coming 8-10 days.			
7th	Wed			Fg	
8th	Thu				
9th	Fri				
10th	Sat				XhS
11th	Sun	Chance of isolated showers along the coasts from the Hunter to the Illawarra. Unusually warm day in some southern parts of the State including in the Southern Tablelands, Southwest Slopes and throughout most of the Riverina.	Showers		N
12th	Mon	Unusually mild night in the Far Northwest, particularly in the Tibooburra district.			
13th	Tue				
14th	Wed	The next 3-4 weeks may be wetter and warmer than is typical for this time of year in most regions, although some districts in the southwest and in the far northeast may be drier than average.		Rain	
15th	Thur	Widespread rain activity throughout the State over the next few days. Thunderstorms, heavy rain, and strong wind gusts in the Upper Western and Central West Plains.	Rain, Th	Showers	
16th	Fri	Strong damaging winds accompanying passing thunderstorm activity in the Upper Western, with chance of property damage in the Bourke area.	Showers, Th		
17th	Sat	Pockets of heavy rain for a time in the Riverina. Showers elsewhere.	Showers	Rain, Th	A V
18th	Sun	Stormy air with damaging winds in the Metropolitan, particularly in the Cabramatta to Penrith areas. Chance of snow in the Snowy Mountains including the NSW-ACT ski fields. Showers elsewhere.	Showers		
19th	Mon	Mostly dry spell throughout the State over the next 3-4 days.			1Q
20th	Tue			Fg	
21st	Wed	Cold spell over the next few nights throughout most of the State. Severe frosts overnight in the Snowy Mountains, particularly at Thredbo.			
22nd	Thur		Th		
23rd	Fri	Chance of showers down through the Tablelands.			
24th	Sat	Isolated showers in the North Cosat to the Northern Tablelands.	Showers	Rain, Th	XhN
25th	Sun	Rain activity in the Upper Western to the Northern Tablelands and from the Riverina to the South Coast, while mostly dry through the central parts of the State.	Rain		
26th	Mon	Widespread rain through the eastern half of the State. Thunderstorms with damaging winds, heavy rain and hail in the Central Tablelands may bring some damage in places.		Fg	F
27th	Tue	Damaging wind gusts likely around the South Coast region, including across Monatgue Island. Heavy rain and scattered thunderstorms on the Mid-North Coast.			
28th	Wed	Mostly dry throughout the State for the remainder of the month with a couple of exceptions. Chance of isolated showers around the North Coast.		Fg	
29th	Thu				P9
30th	Fri	Chance of heavy rain for a brief time in the Northern Rivers region, particularly around Tweed Heads.			^
31st	Sat	Chance of isolated showers in the Southern Tablelands.	Showers, Th		

November 2026

Date	Day	New South Wales / ACT - November 2026	Sydney	Canberra	Moon
1st	Sun	Light showers in the Central Tablelands.	Rain	Showers	
2nd	Mon	Storms in the Central West Slopes and Plains to the Upper Western around this time may cause some localised damage to property, particularly in Tibooburra and Manildra.	Rain		3Q
3rd	Tue	Storm activity in the Northwest Slopes and Plains. Thunderstorms in the Central Tablelands may produce large hail at times, particularly in the Blue Mountains around the Mount Boyce area.	Showers		
4th	Wed	Supercell activity in the Mid-North Coast region brings thunderstorms and showers with large hail likely. Very strong winds with gusts at times in the Northern Rivers region over the next couple of days.			

Date	Day	New South Wales / ACT - November 2026	Sydney	Canberra	Moon
5th	Thur	Thunderstorms, gusty winds, and some hail showers in the Northern Rivers to Northern Tablelands regions. Very cold in the Snowy Mountains with maximum temperatures unlikely to rise above frost level in a number of districts, as well as the chance of light snow in the Snowy Mountains and NSW ski fields and flurries into the Central and Northern Tablelands.			
6th	Fri	Rain activity clears throughout the State around this time, and prevails for the next 3-4 days. Very cold in the Upper Darlings to the Northern Tablelands around this time, with maximum temperatures barely rising above frost level in a number of districts.			XhS
7th	Sat				
8th	Sun		Showers		
9th	Mon	Thunderstorms with heavy rain and some hail in the Northern Tablelands with possible flooding along local waterways, and strong gusty winds that may damage property. Showers elsewhere along the eastern coasts over the next couple of days.	Rain		N
10th	Tue		Showers	Showers	
11th	Wed	Mostly dry over the next couple of days.			
12th	Thu	Very warm spell over the next few days throughout most of the State with daily maximum temperatures ranging from the upper 20's to the low 40's.			
13th	Fri	Some showers in the Lower Western and around the Mid-North Coast. Mostly dry elsewhere. Unusually hot day in the Lower Darling with maximum temperatures into the low 40's likely in a number of districts, including in the Menindee district.	Showers	Rain	V
14th	Sat	Widespread rain activity throughout the State. Heavy falls at times in a line from the Far Northwest to the Southwest Riverina to the South Coast. Showers elsewhere. Strong blustery winds in the Northwest Slopes and Plains region, particularly around Narrabri. Very warm overnight in the Far Northwest.	Showers		A
15th	Sun	Rain activity eases to spotty showers mostly in the Riverina and Northern Tablelands. Dry elsewhere over the next couple of days.			
16th	Mon	The next 3-4 weeks may be warmer and wetter than is normal for this time of year.			
17th	Tue	Isolated showers return briefly into the North Coast region and adjacent Tablelands.			
18th	Wed	Some showers in the Lower Western.		Rain	1Q
19th	Thur	Widespread storm and rain activity throughout the State around this time with heavy falls in the Far Northwest to the Southwest Plains and in the Lower Western to the Riverina and Southern Tablelands.	Showers		
20th	Fri	Rain activity clears in the west of the State, but persists in the northeast regions for a couple more days.	Showers		XhN
21st	Sat				
22nd	Sun	Mostly dry over the next couple of days throughout the State, except for continuing isolated shower activity in the North Coast.			
23rd	Mon				
24th	Tue	Showers intensify to rain in the North Coast around this time before ebbing away by tomorrow. Mostly dry elsewhere.			
25th	Wed			Showers	F
26th	Thu	Showers around the central coastal regions from the Hunter to the South Coast and into the adjacent Tablelands over the next couple of days. Mostly dry elsewhere.	Showers, Th		P5
27th	Fri	Very warm overnight in the Southwest Plains with a warmer spell spreading throughout most of the State for the next couple of days.			^
28th	Sat	Gusty winds in the Central West Slopes and Plains and for a time in the Metropolitan regions. Outbreaks of heavy rain returns to the North Coast and Northern Tablelands regions. Chance of some showers in the Southwest Plains. Very hot in the Upper Darlings around this time, and unusually warm overnight.			
29th	Sun	Showers in the west from the Central Plains to the Southwest Riverina regions, and rain persisting in the northeast quarter of the State.	Showers	Showers	
30th	Mon	Widespreaad rain throughout the State, with heavy falls and thunderstorms in the North Coast to the Hunter and into the adjacent Tablelands, as well as in the Southwest Riverina to the Soutwest Slopes.	Rain, Th	Rain	

December 2026

Date	Day	New South Wales / ACT - December 2026	Sydney	Canberra	Moon
1st	Tue	Very cold with severe frosts overnight in the Snowy Mountains, particularly in the Charlotte Pass district.	Showers, Th	Showers	3Q
2nd	Wed	An east coast low bring widespread rain and storm activity throughout the State over the next couple of days. Thunderstorms in the Northern Rivers region with chance of hail. Strong blustery winds in the Upper Western, especially in the Wilcannia area.		Showers	
3rd	Thur	Strong winds with gusts in the Northwest Slopes and Plains region around this time. Lingering showers in most districts.	Showers	Rain	XhS
4th	Fri	Heavy rain around the South Coast easing to showers by tomorrow.	Showers	Showers	
5th	Sat			Rain	
6th	Sun	A cold snap develops throughout the State over the next few days with a strong southwesterly flow prevailing. Very cold with widespread frosts in the Snowy Mountains, as well as the chance of snow in the NSW-ACT ski regions over the next couple of days.			

Date	Day	New South Wales / ACT - December 2026	Sydney	Canberra	Moon
7th	Mon	Heavy snow possible in the Snowy Mountains and in the NSW-ACT ski regions and with the chance of light snow also occuring in the Katoomba, Oberon, and Orange districts.. Showers in the northeast of the State.			
8th	Tue	Rain in the northeast of the State with some heavy falls around the North Coast. Light showers in the Western districts over the next couple of days.	Showers	Rain, Th	
9th	Wed	A low pressure trough spreads rain throughout the State over the next few days.	Rain, Th	Rain, Th	N
10th	Thu	Severe hail storms likely in the Central West Slopes and Plains, particularly around Coonabarabran, with chance of damage to property. Heavy falls along the eastern coastal fringes from the South Coast to the Mid-North Coast and chance of heavy rain producing flooding for a time in the Tweed Heads area of the Northern Rivers region.	Rain, Th	Showers	V
11th	Fri	Hail storms and strong wind gusts likely in the Central Tablelands and onto the Northwest Slopes and Plains over the next couple of days, particularly around Moree, Tamworth and Bathhurst areas. Hail and wind gusts may cause some localised damage in some places. Gusty winds for a time in the Upper Western, particularly around the Tibooburra area and gales likely in the Northwest Plains, including at Moree.			A
12th	Sat	Chance of hail showers in the Hunter Valley. Strong damaging winds in the Northern Tablelands may produce tornado formations in the Walcha area bringing some localised wind damage, as well as brief localised flooding. Strong and unsettled winds in the Riverina and Central West Slopes and Plains, particularly around Griffith, Forbes and Parkes areas.			
13th	Sun	Hail showers likely in the Metropolitan region to the Northern Tablelands with thunderstorms likely and the chance of heavy rain in the Tamworth area producing localised flash flooding . Mostly dry elsewhere.	Hail		
14th	Mon	Some showers in the North Coast to the Northern Tablelands over the next couple of days. Mostly dry elsewhere.			
15th	Tue	The next 3-4 weeks may bring above average rainfall in the southwest of the State, and average rainfall elsewhere, along with seasonally average temperatures for this time of year.		Showers	
16th	Wed	Hail likely in the Northern Rivers region, mostly around the Grafton area.		Showers	
17th	Thur	Strong damaging wind gusts in the Northern Rivers region. Chance of pockets of rain in the Broken Hill area. Dry elsewhere.			1Q XhN
18th	Fri	Showers in the west to the northeast quarter of the State over the next couple of days. Cold with scattered frosts in the Snowy Mountains.	Showers		
19th	Sat	Chance of hail showers in the Hunter Valley.	Showers, Th	Fg	
20th	Sun	Scattered thunderstorms in the Hunter Valley and Northern Tablelands likely.			
21st	Mon	Light showers in the Upper and Lower Western regions. Mostly dry elsewhere.	Showers		
22nd	Tue	Showers in the eastern regions of the State, spreading and intensifying into heavy rain in most regions throughout the State over the next couple of days.	Showers	Showers	
23rd	Wed	Widespread rain as a low pressure system crosses the State over the next few days.	Rain	Rain	
24th	Thu	Very heavy rain and storms throughout the State. Chance of thunderstorms in the Mid-North Coast to the Northern Rivers regions, with showers of hail, some large, and strong gusty winds particularly in the Northern Rivers, which may cause some damage to property. Chance of flash flooding in a number of regions, and particularly in the river catchments.			P1 F ^
25th	Fri	Rain activity eases to light showers in most districts.			
26th	Sat	Damaging wind gusts in some parts of the Upper Western, including in the Brewarrina area. Some rain persists in the Upper Wester and Upper Darlings to the Central and Southwest Plains and Lower Western. Mostly dry along the eastern coastal regions.			
27th	Sun	Mostly dry throughout the State over the next couple of days.			
28th	Mon	Very cold with low maximum temperatures in the Snowy Mountains. Very strong winds around the South Coast with chance of gale force gusts at times.		Showers, Th	
29th	Tue	Showers along the border regions of the west and south of the State.		Showers, Th	
30th	Wed	Spotty showers in most districts, although mostly dry in the Lower Western.			XhS
31st	Thur	Mostly dry throughout the State around this time.			3Q

NEW SOUTH WALES / ACT GRAPHS AND TABLES 2026

Seasonal Breakdown – State Overall Rainfall Timing and Estimated mm Trend Graphs
State Overall Maximum Temperature Trend Graphs
Regional Rainfall Expectations Tables - mm and Trends
Regional Maximum and Minimum Temperature Trends Tables

Monthly Breakdown - **Rainfall:**
Regional Expectations by mm
Regional Expectations by mm Variance to Average
Regional Variance Trend by mm
Regional Estimated Expectations by % of Variance to Average
Regional Estimatedl Variance Trend by %

Maximum & Minimum Temperature Trends:
Regional Estimated Average Temperature
Regional Estimated Degree of Variation from the Average
Regional Estimated Temperature Trend

Seasonal Breakdowns

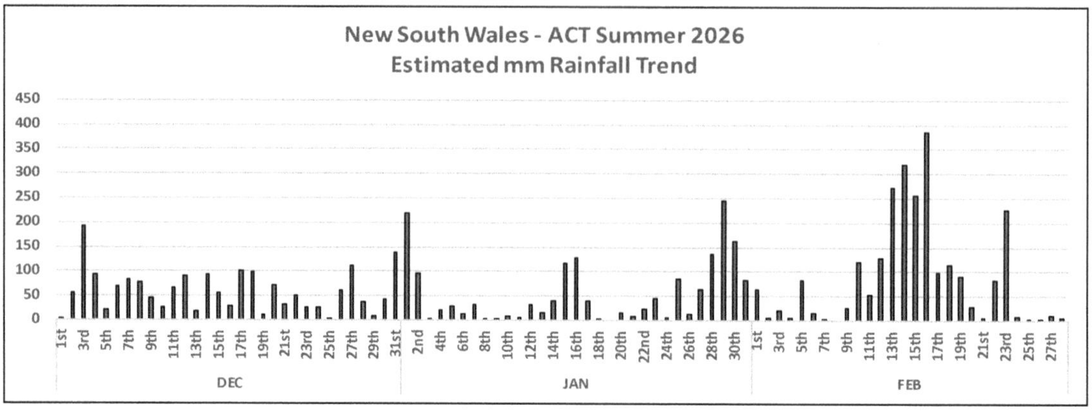

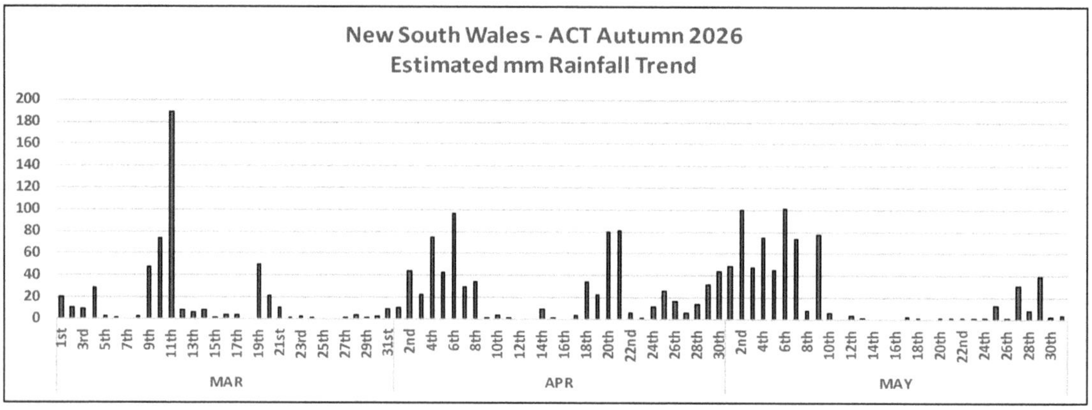

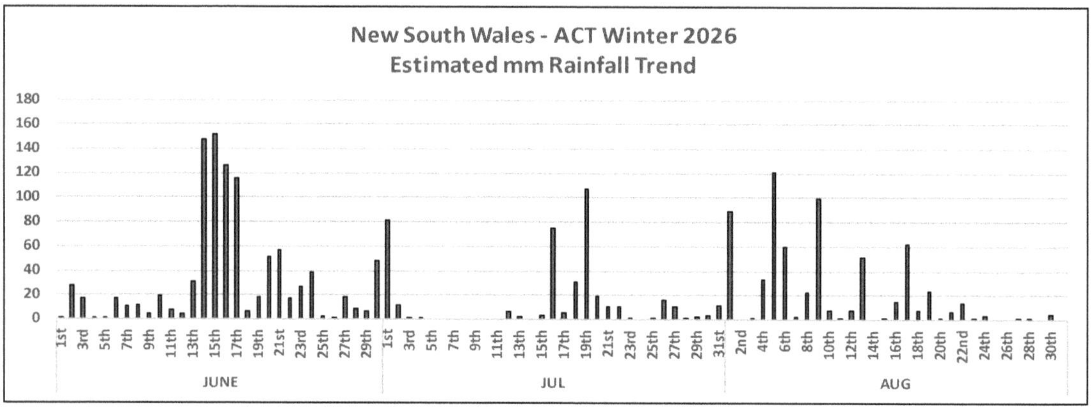

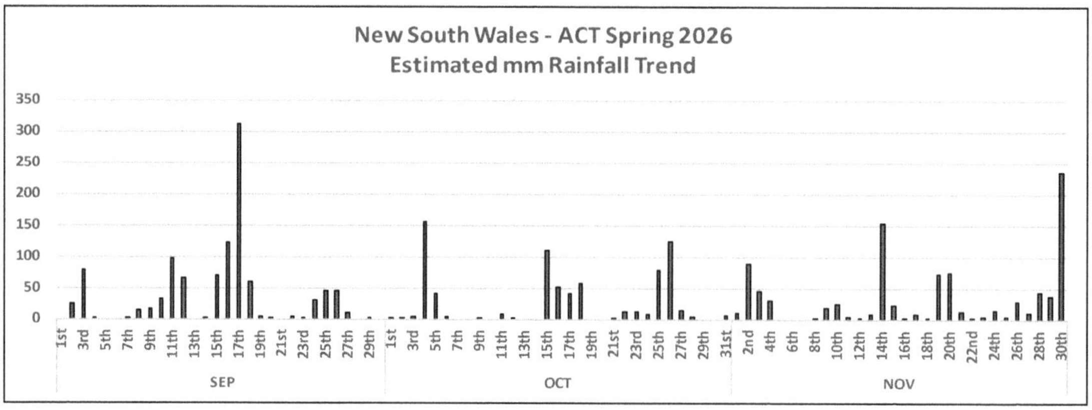

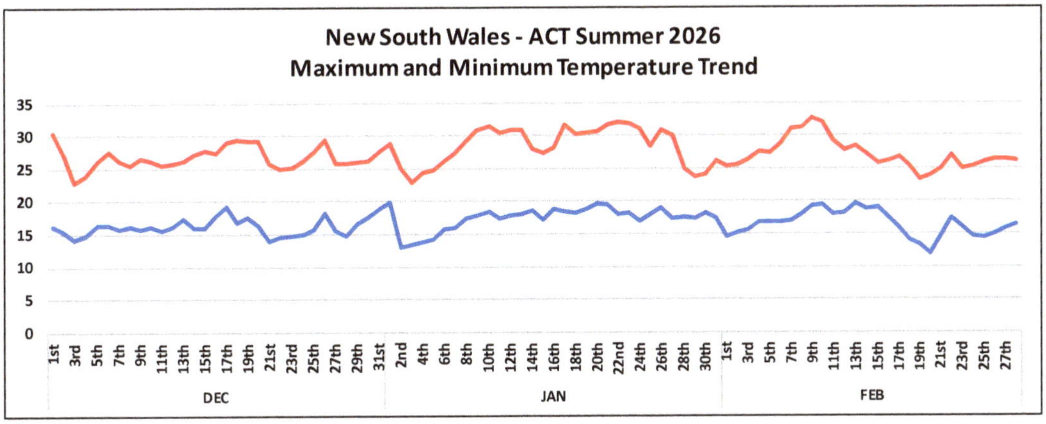
New South Wales - ACT Summer 2026
Maximum and Minimum Temperature Trend
DEC
JAN
FEB

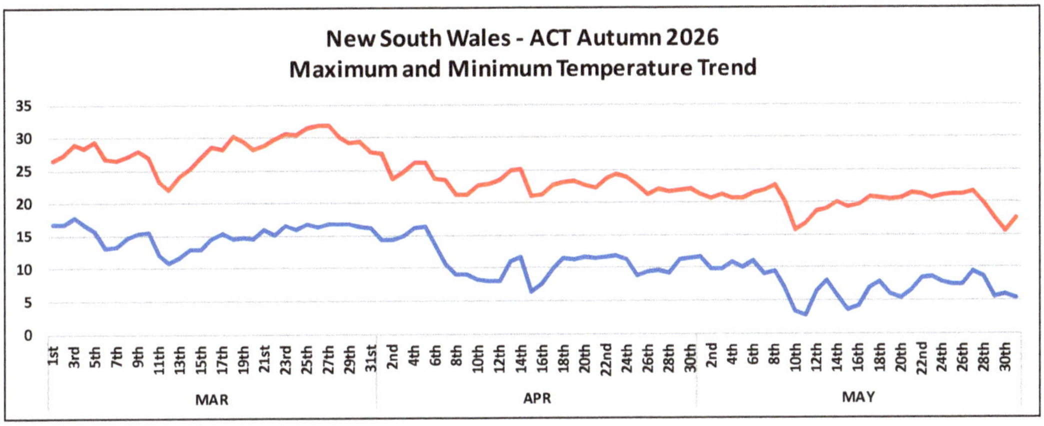
New South Wales - ACT Autumn 2026
Maximum and Minimum Temperature Trend
MAR
APR
MAY

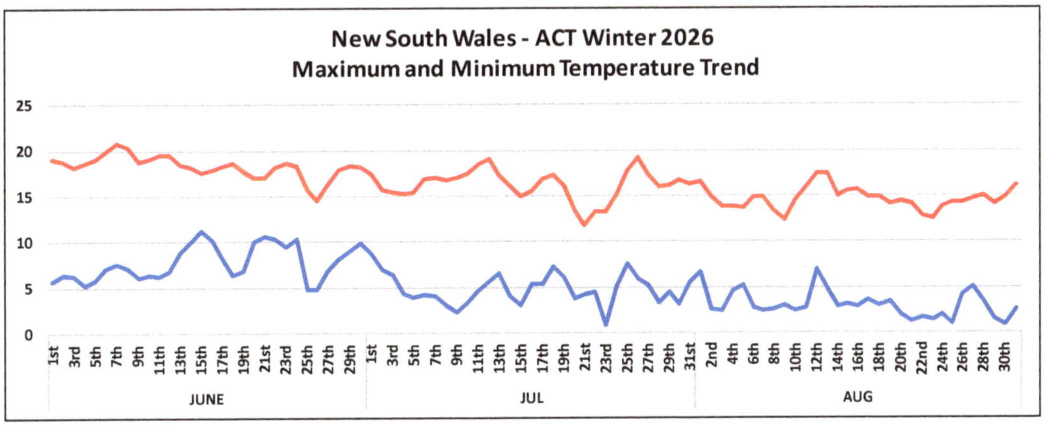
New South Wales - ACT Winter 2026
Maximum and Minimum Temperature Trend
JUNE
JUL
AUG

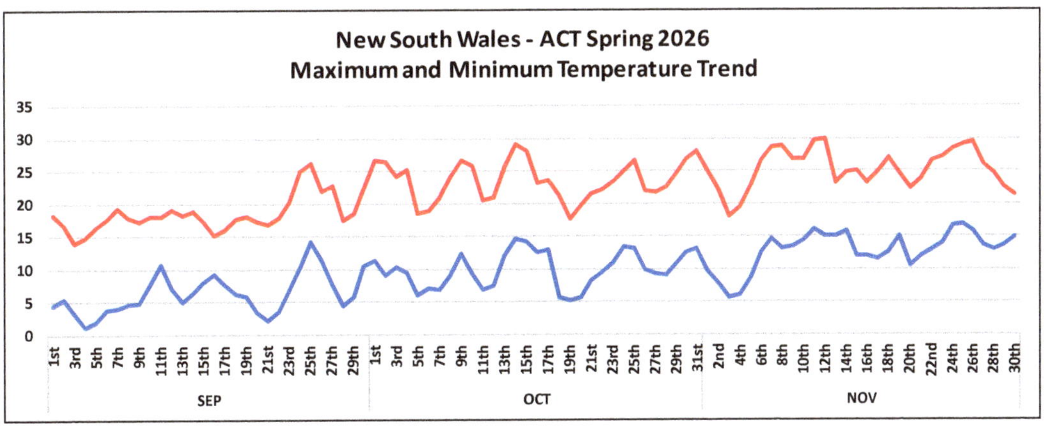
New South Wales - ACT Spring 2026
Maximum and Minimum Temperature Trend
SEP
OCT
NOV

2026 — NSW / ACT - Estimated Seasonal Rainfall Expectations (mm)

REGION	Upper West	Lower West	Riverina	NW Slopes & Plains	Northern Tablelds	CW Slopes & Plains	Central Tablelds	SW Slopes & Plains	Southern Tablelds	Northern Rivers	Mid-North Coast	Hunter	Metropolitan	Illawarra	South Coast	ACT	NSW / ACT
Town Sample	Wilcannia	Broken Hill	Hay	Moree	Armidale	Dubbo	Orange	Wagga Wagga	Goulburn	Grafton	Coffs Harbour	Newcastle	Sydney	Wollongong	Bega	Canberra	Overall
Summer	143	52	196	415	390	335	270	260	286	681	849	366	452	666	349	223	5935
Autumn	26	35	40	23	91	57	57	68	102	127	372	325	271	217	42	91	1945
Winter	37	102	36	72	109	99	99	113	136	135	248	213	225	213	105	90	2031
Spring	29	56	57	209	238	202	202	97	162	259	326	303	222	222	57	100	2742
Overall	268	287	281	619	748	669	604	512	692	1199	1851	1156	1095	1179	563	516	12237

2026 — NSW / ACT - Estimated Seasonal Rainfall Variation from the Average (mm)

REGION	Upper West	Lower West	Riverina	NW Slopes & Plains	Northern Tablelds	CW Slopes & Plains	Central Tablelds	SW Slopes & Plains	Southern Tablelds	Northern Rivers	Mid-North Coast	Hunter	Metropolitan	Illawarra	South Coast	ACT	NSW / ACT
Town Sample	Wilcannia	Broken Hill	Hay	Moree	Armidale	Dubbo	Orange	Wagga Wagga	Goulburn	Grafton	Coffs Harbour	Newcastle	Sydney	Wollongong	Bega	Canberra	Overall
Summer	67	-7	113	201	98	171	31	129	98	251	327	100	180	313	128	60	2261
Autumn	-48	-23	-53	-91	-51	-71	-117	-59	-35	-191	-197	-36	-42	-153	-163	-33	-1361
Winter	-23	48	-63	-25	-20	-42	-179	-49	-10	-26	4	-36	-22	-45	-43	-48	-578
Spring	-47	-5	-35	48	52	56	-32	-50	-7	88	-0	86	12	-73	-136	-71	-114
Overall	-17	55	-85	33	-1	89	-322	-54	53	119	191	64	53	-98	-204	-82	-207

2026 — NSW / ACT - Estimated Seasonal Rainfall Trend Variation by 10mm

REGION	Upper West	Lower West	Riverina	NW Slopes & Plains	Northern Tablelds	CW Slopes & Plains	Central Tablelds	SW Slopes & Plains	Southern Tablelds	Northern Rivers	Mid-North Coast	Hunter	Metropolitan	Illawarra	South Coast	ACT	NSW / ACT
Town Sample	Wilcannia	Broken Hill	Hay	Moree	Armidale	Dubbo	Orange	Wagga Wagga	Goulburn	Grafton	Coffs Harbour	Newcastle	Sydney	Wollongong	Bega	Canberra	Overall
Summer	wtr	av	wtr	wtr	wtr	wtr	wtr	wtr	wtr	wtr	wtr	wtr	wtr	wtr	wtr	wtr	wtr
Autumn	drr	av	drr	drr	drr	drr	drr	drr	drr	drr	drr	drr	drr	drr	drr	drr	drr
Winter	av	wtr	drr	av	av	drr	drr	drr	av	av	av	drr	av	drr	drr	drr	drr
Spring	drr	av	drr	wtr	wtr	wtr	drr	drr	av	wtr	av	wtr	av	drr	drr	drr	drr
Overall	av	av	av	av	av	av	drr	av	av	wtr	wtr	av	av	av	drr	av	drr

2026 — NSW / ACT - Estimated Seasonal Rainfall Expectation by %

REGION	Upper West	Lower West	Riverina	NW Slopes & Plains	Northern Tablelds	CW Slopes & Plains	Central Tablelds	SW Slopes & Plains	Southern Tablelds	Northern Rivers	Mid-North Coast	Hunter	Metropolitan	Illawarra	South Coast	ACT	NSW / ACT
Town Sample	Wilcannia	Broken Hill	Hay	Moree	Armidale	Dubbo	Orange	Wagga Wagga	Goulburn	Grafton	Coffs Harbour	Newcastle	Sydney	Wollongong	Bega	Canberra	Overall
Summer	88%	-12%	134%	94%	33%	103%	13%	99%	52%	58%	63%	38%	66%	89%	58%	36%	62%
Autumn	-65%	-40%	-57%	-80%	-36%	-55%	-67%	-46%	-25%	-60%	-35%	-10%	-13%	-41%	-80%	-26%	-41%
Winter	-39%	88%	-63%	-26%	-16%	-30%	-64%	-30%	-7%	-16%	2%	-14%	-9%	-18%	-29%	-35%	-22%
Spring	-62%	-8%	-38%	30%	28%	38%	-14%	-34%	-4%	52%	0%	40%	6%	-25%	-70%	-42%	-4%
Overall	-6%	24%	-23%	6%	0%	15%	-35%	-10%	8%	11%	12%	6%	5%	-8%	-27%	-14%	-2%

2026 — NSW / ACT - Estimated Seasonal Rainfall Trend Variation by 25%

REGION	Upper West	Lower West	Riverina	NW Slopes & Plains	Northern Tablelds	CW Slopes & Plains	Central Tablelds	SW Slopes & Plains	Southern Tablelds	Northern Rivers	Mid-North Coast	Hunter	Metropolitan	Illawarra	South Coast	ACT	NSW / ACT
Town Sample	Wilcannia	Broken Hill	Hay	Moree	Armidale	Dubbo	Orange	Wagga Wagga	Goulburn	Grafton	Coffs Harbour	Newcastle	Sydney	Wollongong	Bega	Canberra	Overall
Summer	wtr	av	wtr	wtr	wtr	wtr	wtr	av	wtr	wtr	wtr	wtr	wtr	wtr	wtr	wtr	wtr
Autumn	drr	drr	drr	drr	drr	drr	drr	drr	av	drr	drr	av	av	drr	drr	drr	drr
Winter	drr	wtr	drr	drr	av	drr	drr	drr	av	av	av	av	av	av	drr	drr	av
Spring	drr	av	drr	wtr	wtr	wtr	drr	av	av	wtr	av	wtr	av	av	drr	drr	av
Overall	av	av	av	av	av	av	av	drr	av	av	av	av	av	av	drr	av	av

2026 NSW/ACT - Estimated Seasonal Maximum Temperature Average and Trend

Region	Upper West	Lower West	Riverina	NW Slopes & Plains	Northern Tablelds	CW Slopes & Plains	Central Tablelds	SW Slopes & Plains	Southern Tablelds	Northern Rivers	Mid-North Coast	Hunter	Metropolitan	Illawarra	South Coast	ACT	NSW / ACT
Town Sample	Wilcannia	Broken Hill	Hay	Moree	Armidale	Dubbo	Orange	Wagga Wagga	Goulburn	Grafton	Coffs Harbour	Newcastle	Sydney	Wollongong	Bega	Canberra	Overall
Summer Est	34.1	32.6	32.0	30.3	23.3	29.5	23.9	30.2	24.5	27.8	25.9	26.4	25.0	24.2	23.7	26.4	27.5
Trend	clr	av	clr	clr	clr	clr	clr	a-c	clr	clr	av	wmr	a-c	av	clr	clr	clr
Autumn Est	28.3	26.2	26.2	27.2	20.3	25.8	19.1	24.6	20.8	26.2	24.1	23.6	23.1	22.2	21.8	22.0	23.8
Trend	wmr	wmr	wmr	av	a-w	wmr	av	wmr	a-w	av	av	a-w	a-w	av	av	a-w	a-w
Winter Est	18.5	16.0	16.5	19.3	13.5	16.8	10.4	14.7	12.4	21.4	19.7	18.0	18.0	17.4	16.8	13.3	16.4
Trend	av	av	av	av	av	av	av	a-w	av	av	av	av	a-w	av	av	av	av
Spring Est	26.9	24.3	24.3	26.3	19.8	24.1	17.5	22.8	19.5	25.8	23.0	22.9	22.0	20.6	19.4	20.6	22.5
Trend	clr	clr	clr	clr	a-c	clr	a-c	wmr	av	a-c	a-c	a-w	av	a-c	clr	clr	a-c

2026 NSW/ACT - Estimated Seasonal Minimum Temperature Average and Trend

Region	Upper West	Lower West	Riverina	NW Slopes & Plains	Northern Tablelds	CW Slopes & Plains	Central Tablelds	SW Slopes & Plains	Southern Tablelds	Northern Rivers	Mid-North Coast	Hunter	Metropolitan	Illawarra	South Coast	ACT	NSW / ACT
Town Sample	Wilcannia	Broken Hill	Hay	Moree	Armidale	Dubbo	Orange	Wagga Wagga	Goulburn	Grafton	Coffs Harbour	Newcastle	Sydney	Wollongong	Bega	Canberra	Overall
Summer Est	18.5	20.0	17.6	18.5	12.4	16.8	12.3	17.0	11.6	19.3	19.0	19.4	19.1	17.2	14.6	14.0	16.7
Trend	av	av	av	clr	av	a-c	av	wmr	av	a-w	av	av	a-w	clr	av	a-w	av
Autumn Est	14.2	12.4	11.7	12.0	5.4	9.7	6.2	9.6	8.9	14.1	14.5	15.4	15.1	13.5	9.0	7.5	11.2
Trend	wmr	av	wmr	a-c	clr	a-c	av	av	wmr	a-w	av	av	av	clr	av	a-w	av
Winter Est	6.7	5.0	5.4	4.8	0.0	4.0	1.7	4.1	2.3	7.4	8.2	9.8	9.7	10.2	3.3	1.8	5.3
Trend	wmr	av	wmr	av	clr	av	a-w	a-w	wmr	wmr	av	a-w	a-w	av	av	wmr	av
Spring Est	11.6	11.6	9.7	12.0	5.5	8.9	5.8	7.0	2.0	13.5	13.8	14.0	13.6	13.1	7.1	6.0	9.7
Trend	av	av	av	a-c	clr	a-c	av	a-c	clr	a-w	av	av	av	a-c	clr	av	a-c

Monthly Breakdowns

2026	NEW SOUTH WALES / ACT - Estimated Rainfall Expectations (mm)																
REGION	Upper West	Lower West	Riverina	NW Slopes & Plains	Northern Tablelds	CW Slopes & Plains	Central Tablelds	SW Slopes & Plains	Southern Tablelds	Northern Rivers	Mid-North Coast	Hunter	Metropolitan	Illawarra	South Coast	ACT	NSW / ACT
Town Sample	Wilcannia	Broken Hill	Hay	Moree	Armidale	Dubbo	Orange	Wagga Wagga	Goulburn	Grafton	Coffs Harbour	Newcastle	Sydney	Wollongong	Bega	Canberra	Overall
January	83	25	79	128	56	145	79	79	109	210	232	109	71	100	106	76	1686
February	13	2	8	134	168	97	97	94	60	339	524	153	207	327	167	36	2427
March	2	0	0	9	20	20	20	1	31	60	63	88	83	65	16	37	515
April	12	31	6	2	16	27	27	27	47	31	154	87	118	96	22	42	745
May	12	4	33	12	56	11	11	40	23	36	155	149	70	56	5	12	686
June	22	80	11	37	59	34	34	32	33	104	103	136	123	111	48	20	987
July	4	17	11	20	27	39	39	48	41	9	53	8	23	23	14	30	409
August	11	5	14	14	23	26	26	32	61	23	92	68	78	78	43	39	635
September	7	19	14	58	53	61	61	19	43	107	144	144	131	131	19	28	1042
October	8	20	23	41	74	87	87	36	100	26	55	57	30	30	19	47	739
November	14	16	20	111	111	54	54	43	18	127	126	102	61	61	19	25	961
December	81	67	62	53	86	68	68	60	124	127	151	53	99	99	86	122	1407
Overall	268	287	281	619	748	669	604	512	692	1199	1851	1156	1095	1179	563	516	12237

2026	NEW SOUTH WALES / ACT - Estimated Rainfall Variation from the Average (mm)																
REGION	Upper West	Lower West	Riverina	NW Slopes & Plains	Northern Tablelds	CW Slopes & Plains	Central Tablelds	SW Slopes & Plains	Southern Tablelds	Northern Rivers	Mid-North Coast	Hunter	Metropolitan	Illawarra	South Coast	ACT	NSW / ACT
Town Sample	Wilcannia	Broken Hill	Hay	Moree	Armidale	Dubbo	Orange	Wagga Wagga	Goulburn	Grafton	Coffs Harbour	Newcastle	Sydney	Wollongong	Bega	Canberra	Overall
January	55	6	52	48	-56	82	3	40	45	63	71	38	-9	-4	46	22	503
February	-17	-17	-22	66	85	52	24	54	2	176	303	28	88	177	74	-17	1057
March	-30	-25	-30	-53	-21	-19	-38	-41	-20	-78	-164	-26	-5	-55	-65	-8	-679
April	-9	18	-21	-20	-36	-16	-28	-10	4	-67	-26	-33	-1	-42	-53	1	-339
May	-9	-16	-1	-18	7	-36	-51	-8	-19	-46	-7	23	-36	-56	-46	-26	-344
June	-5	63	-24	-0	23	-10	-42	-20	-17	59	-3	19	23	8	-5	-24	44
July	-14	-4	-21	-13	-27	-12	-61	-11	-4	-53	-29	-66	-49	-44	-46	-18	-471
August	-4	-11	-18	-12	-15	-20	-76	-18	11	-31	36	11	5	-9	8	-7	-150
September	-13	2	-17	21	5	21	-15	-30	-7	61	84	87	70	55	-34	-24	265
October	-23	-3	-12	-11	9	39	9	-17	46	-33	-48	-14	-34	-68	-41	-4	-206
November	-11	-4	-6	38	38	-3	-26	-2	-46	61	-36	13	-25	-60	-61	-44	-173
December	62	46	35	-13	-12	12	-20	9	57	8	10	-16	25	1	17	66	286
Overall	-17	55	-85	33	-1	89	-322	-54	53	119	191	64	53	-98	-204	-82	-207

2026	NEW SOUTH WALES / ACT - Estimated Rainfall Trend Variation by 10mm															
REGION	Upper West	Lower West	Riverina	NW Slopes & Plains	Northern Tablelds	CW Slopes & Plains	Central Tablelds	SW Slopes & Plains	Southern Tablelds	Northern Rivers	Mid-North Coast	Hunter	Metropolitan	Illawarra	South Coast	ACT	NSW / ACT
Town Sample	Wilcannia	Broken Hill	Hay	Moree	Armidale	Dubbo	Orange	Wagga Wagga	Goulburn	Grafton	Coffs Harbour	Newcastle	Sydney	Wollongong	Bega	Canberra	Overall
January	wtr	av	wtr	wtr	drr	wtr	av	wtr	wtr	wtr	wtr	wtr	av	av	wtr	wtr	wtr
February	drr	drr	drr	wtr	wtr	wtr	wtr	wtr	av	wtr	wtr	wtr	wtr	wtr	wtr	drr	wtr
March	drr	drr	drr	drr	drr	drr	drr	drr	drr	drr	drr	drr	av	drr	drr	av	drr
April	av	av	drr	drr	drr	drr	drr	av	av	drr	drr	drr	av	drr	drr	av	drr
May	av	drr	av	drr	av	drr	drr	av	drr	drr	av	wtr	drr	drr	drr	drr	drr
June	av	wtr	drr	av	wtr	av	drr	drr	drr	wtr	av	wtr	wtr	av	av	drr	wtr
July	drr	av	drr	drr	drr	drr	drr	drr	av	drr	drr	drr	drr	drr	drr	drr	drr
August	av	drr	drr	drr	drr	drr	drr	drr	wtr	drr	wtr	wtr	av	av	av	av	drr
September	drr	av	drr	wtr	av	wtr	drr	drr	av	wtr	wtr	wtr	wtr	wtr	drr	drr	wtr
October	drr	av	drr	drr	av	wtr	av	drr	wtr	drr	drr	drr	drr	drr	drr	av	drr
November	drr	av	av	wtr	wtr	av	drr	av	drr	wtr	drr	wtr	drr	drr	drr	drr	drr
December	wtr	wtr	wtr	drr	drr	wtr	drr	av	wtr	av	av	drr	wtr	av	wtr	wtr	wtr
Overall	av	av	av	av	av	av	drr	av	av	wtr	wtr	av	av	av	drr	av	drr

2026	NEW SOUTH WALES / ACT - Estimated Rainfall Expectation by %															
REGION	Upper West	Lower West	Riverina	NW Slopes & Plains	Northern Tablelds	CW Slopes & Plains	Central Tablelds	SW Slopes & Plains	Southern Tablelds	Northern Rivers	Mid-North Coast	Hunter	Metropolitan	Illawarra	South Coast	ACT	NSW / ACT
Town Sample	Wilcannia	Broken Hill	Hay	Moree	Armidale	Dubbo	Orange	Wagga Wagga	Goulburn	Grafton	Coffs Harbour	Newcastle	Sydney	Wollongong	Bega	Canberra	Overall
January	203%	30%	189%	61%	-50%	130%	4%	105%	72%	43%	44%	53%	-11%	-4%	78%	40%	42%
February	-56%	-90%	-73%	97%	102%	114%	33%	133%	4%	108%	137%	22%	74%	118%	81%	-31%	77%
March	-93%	-99%	-100%	-85%	-51%	-49%	-66%	-99%	-39%	-56%	-72%	-23%	-6%	-45%	-81%	-18%	-57%
April	-43%	131%	-77%	-92%	-70%	-38%	-51%	-27%	10%	-68%	-14%	-27%	-1%	-31%	-71%	3%	-31%
May	-43%	-79%	-4%	-59%	13%	-76%	-82%	-16%	-44%	-56%	-4%	18%	-34%	-50%	-90%	-68%	-33%
June	-19%	391%	-68%	-1%	63%	-23%	-55%	-38%	-34%	129%	-3%	16%	23%	8%	-10%	-55%	5%
July	-77%	-20%	-66%	-39%	-50%	-24%	-61%	-18%	-8%	-86%	-35%	-89%	-68%	-65%	-77%	-37%	-54%
August	-29%	-67%	-55%	-46%	-40%	-43%	-75%	-36%	22%	-58%	65%	20%	6%	-11%	24%	-14%	-19%
September	-65%	10%	-56%	58%	10%	52%	-20%	-62%	-14%	136%	139%	149%	113%	72%	-63%	-45%	34%
October	-74%	-12%	-34%	-21%	14%	79%	12%	-33%	84%	-57%	-47%	-20%	-52%	-69%	-69%	-9%	-22%
November	-44%	-20%	-22%	52%	53%	-6%	-33%	-5%	-71%	91%	-22%	15%	-29%	-50%	-76%	-63%	-15%
December	328%	223%	130%	-20%	-12%	21%	-23%	17%	85%	6%	7%	-22%	34%	1%	25%	117%	26%
Overall	-6%	24%	-23%	6%	0%	15%	-35%	-10%	8%	11%	12%	6%	5%	-8%	-27%	-14%	-2%

2026 NEW SOUTH WALES / ACT - Estimated Rainfall Trend Variation by 25%

REGION	Upper West	Lower West	Riverina	NW Slopes & Plains	Northern Tablelds	CW Slopes & Plains	Central Tablelds	SW Slopes & Plains	Southern Tablelds	Northern Rivers	Mid-North Coast	Hunter	Metropolitan	Illawarra	South Coast	ACT	NSW / ACT
Town Sample	Wilcannia	Broken Hill	Hay	Moree	Armidale	Dubbo	Orange	Wagga Wagga	Goulburn	Grafton	Coffs Harbour	Newcastle	Sydney	Wollongong	Bega	Canberra	Overall
January	wtr	wtr	wtr	wtr	drr	wtr	av	wtr	wtr	wtr	wtr	wtr	av	av	wtr	wtr	wtr
February	drr	drr	drr	wtr	wtr	wtr	wtr	wtr	av	wtr	wtr	av	wtr	wtr	wtr	drr	wtr
March	drr	drr	drr	drr	drr	drr	drr	drr	drr	drr	drr	av	av	drr	drr	av	drr
April	drr	wtr	drr	drr	drr	drr	drr	drr	av	drr	av	drr	av	drr	drr	av	drr
May	drr	drr	av	drr	av	drr	drr	av	drr	drr	av	av	drr	drr	drr	drr	drr
June	av	wtr	drr	av	wtr	av	drr	drr	drr	wtr	av	av	av	av	av	drr	av
July	drr	av	drr	drr	drr	av	drr	av	av	drr	drr	drr	drr	drr	drr	drr	drr
August	drr	drr	drr	drr	drr	drr	drr	drr	av	drr	wtr	av	av	av	av	av	av
September	drr	av	drr	wtr	av	wtr	av	drr	av	wtr	wtr	wtr	wtr	wtr	drr	drr	wtr
October	drr	av	drr	av	av	wtr	av	drr	wtr	drr	drr	av	drr	drr	drr	av	av
November	drr	av	av	wtr	wtr	av	drr	av	drr	wtr	av	av	drr	drr	drr	av	av
December	wtr	wtr	wtr	av	av	av	av	av	wtr	av	av	av	wtr	av	av	wtr	wtr
Overall	av	av	av	av	av	av	av	drr	av	av	av	av	av	av	drr	av	av

2026 NEW SOUTH WALES / ACT - Estimated Average Maximum Temperature

Region	Upper West	Lower West	Riverina	NW Slopes & Plains	Northern Tablelds	CW Slopes & Plains	Central Tablelds	SW Slopes & Plains	Southern Tablelds	Northern Rivers	Mid-North Coast	Hunter	Metropolitan	Illawarra	South Coast	ACT	NSW / ACT
Town Sample	Wilcannia	Broken Hill	Hay	Moree	Armidale	Dubbo	Orange	Wagga Wagga	Goulburn	Grafton	Coffs Harbour	Newcastle	Sydney	Wollongong	Bega	Canberra	Overall
January	36	34	34	30	24	30	25	33	27	28	26	27	26	25	25	29	**29**
February	33	33	32	30	23	29	24	29	24	28	26	26	25	24	24	26	**27**
March	35	33	34	31	23	31	25	31	26	28	26	26	25	25	24	28	**28**
April	27	25	25	28	21	25	18	24	20	26	24	24	23	22	21	20	**23**
May	23	21	20	23	17	22	14	19	17	24	22	21	21	20	20	18	**20**
June	21	18	19	22	16	20	14	18	15	23	21	19	19	18	18	16	**18**
July	18	16	16	19	14	16	10	14	12	21	20	18	18	18	17	13	**16**
August	17	15	15	17	11	14	7	13	11	20	18	17	17	16	15	12	**15**
September	22	20	20	22	16	19	13	18	15	23	21	19	19	17	17	16	**19**
October	29	25	25	28	21	24	18	24	21	27	24	25	23	22	20	22	**24**
November	30	28	28	30	22	29	21	27	23	27	24	25	23	22	21	25	**25**
December	30	28	27	31	23	28	21	26	23	29	27	26	25	23	20	24	**26**
Overall	**27**	**24**	**24**	**26**	**19**	**24**	**18**	**23**	**19**	**25**	**23**	**23**	**22**	**21**	**20**	**20**	**22**

2026 NEW SOUTH WALES / ACT - Estimated Average Minimum Temperature

Region	Upper West	Lower West	Riverina	NW Slopes & Plains	Northern Tablelds	CW Slopes & Plains	Central Tablelds	SW Slopes & Plains	Southern Tablelds	Northern Rivers	Mid-North Coast	Hunter	Metropolitan	Illawarra	South Coast	ACT	NSW / ACT
Town Sample	Wilcannia	Broken Hill	Hay	Moree	Armidale	Dubbo	Orange	Wagga Wagga	Goulburn	Grafton	Coffs Harbour	Newcastle	Sydney	Wollongong	Bega	Canberra	Overall
January	19	21	18	19	13	17	13	18	12	20	20	20	20	18	15	15	**17**
February	18	20	17	18	12	16	12	17	13	19	19	19	19	17	14	13	**16**
March	20	18	17	16	8	14	10	14	12	17	17	19	18	16	13	12	**15**
April	12	12	11	12	6	9	6	9	9	15	15	16	15	13	9	7	**11**
May	10	8	7	8	2	5	3	6	6	10	11	12	12	11	6	3	**7**
June	9	8	7	9	4	8	4	6	3	11	12	12	12	12	5	4	**8**
July	7	4	6	4	-1	3	1	4	4	7	7	9	9	10	2	1	**5**
August	5	3	4	2	-2	1	0	2	1	4	5	8	8	8	2	0	**3**
September	8	6	6	8	2	5	2	3	-1	11	11	11	10	11	4	2	**6**
October	11	12	10	12	6	9	7	6	2	14	14	15	15	14	7	6	**10**
November	15	16	13	15	9	12	8	12	5	16	16	16	16	15	10	10	**13**
December	15	17	13	17	10	13	9	12	8	18	18	18	17	15	13	11	**14**
Overall	13	12	11	12	6	10	6	9	6	13	14	15	14	13	8	7	**11**

2026 NEW SOUTH WALES / ACT - Estimated Degree of Variation from the Maximum Average

Region	Upper West	Lower West	Riverina	NW Slopes & Plains	Northern Tablelds	CW Slopes & Plains	Central Tablelds	SW Slopes & Plains	Southern Tablelds	Northern Rivers	Mid-North Coast	Hunter	Metropolitan	Illawarra	South Coast	ACT	NSW / ACT
Town Sample	Wilcannia	Broken Hill	Hay	Moree	Armidale	Dubbo	Orange	Wagga Wagga	Goulburn	Grafton	Coffs Harbour	Newcastle	Sydney	Wollongong	Bega	Canberra	Overall
January	-1.6	-0.2	-1.7	-4.0	-2.9	-3.6	-2.2	0.7	-1.6	-2.7	-0.9	1.6	-0.5	0.2	-2.4	-2.0	-1.5
February	-2.5	0.1	-1.5	-3.4	-2.2	-3.3	-2.6	-2.1	-2.7	-2.2	-0.8	0.5	-0.8	-0.7	-2.6	-2.5	-1.8
March	3.2	4.3	3.8	-0.7	-0.2	1.5	1.8	3.8	2.5	-0.4	0.2	1.3	0.7	0.5	-0.7	2.5	1.5
April	-0.4	0.0	-0.1	0.2	1.1	0.2	-0.7	0.9	-0.6	0.6	0.0	1.1	0.5	-0.4	-0.9	-0.8	0.0
May	1.5	1.6	0.2	0.8	1.1	1.5	0.3	1.4	0.6	1.0	0.7	0.8	1.3	0.1	0.5	0.9	0.9
June	2.8	1.7	2.7	2.8	2.9	3.4	3.1	3.7	2.2	2.0	1.4	1.3	1.8	0.7	1.0	2.4	2.2
July	0.4	-0.1	0.1	0.6	1.4	0.6	-0.2	1.1	0.3	0.6	1.2	1.5	1.9	0.7	0.7	-0.3	0.7
August	-3.2	-3.2	-2.7	-3.4	-2.8	-3.2	-3.8	-1.9	-2.8	-2.1	-1.5	-1.3	-1.0	-2.1	-2.5	-2.5	-2.5
September	-2.6	-2.2	-2.1	-3.1	-1.6	-2.3	-1.7	0.0	-1.9	-1.5	-1.0	-0.9	-1.1	-2.7	-2.9	-2.4	-1.9
October	0.0	-0.2	-1.0	-0.7	0.5	-1.0	-0.2	2.2	0.7	0.1	0.2	2.7	1.2	0.8	-1.9	0.0	0.2
November	-1.6	-0.8	-1.9	-1.0	-0.7	0.1	-0.7	1.1	0.1	-1.3	-0.9	1.0	-0.3	-0.4	-2.7	-0.8	-0.7
December	-4.5	-3.8	-5.6	-2.3	-1.8	-3.5	-4.0	-3.2	-3.1	0.2	0.6	1.5	-0.1	-0.9	-4.6	-4.2	-2.5
Overall	-0.7	-0.2	-0.8	-1.2	-0.4	-0.8	-0.9	0.7	-0.5	-0.5	-0.1	0.9	0.3	-0.3	-1.6	-0.8	-0.4

2026 — NEW SOUTH WALES / ACT - Estimated Degree of Variation from the Minimum Average

Region	Upper West	Lower West	Riverina	NW Slopes & Plains	Northern Tablelds	CW Slopes & Plains	Central Tablelds	SW Slopes & Plains	Southern Tablelds	Northern Rivers	Mid-North Coast	Hunter	Metropolitan	Illawarra	South Coast	ACT	NSW / ACT
Town Sample	Wilcannia	Broken Hill	Hay	Moree	Armidale	Dubbo	Orange	Wagga Wagga	Goulburn	Grafton	Coffs Harbour	Newcastle	Sydney	Wollongong	Bega	Canberra	Overall
January	0.1	-0.2	-0.2	-1.5	-0.3	-1.3	0.0	1.6	-0.4	1.1	0.3	0.7	1.1	-1.2	0.1	0.8	0.0
February	-0.2	-0.5	0.1	-1.7	-1.6	-1.7	-1.0	0.1	0.4	0.3	-0.5	-0.2	-0.1	-2.2	-0.5	-0.1	-0.6
March	4.8	1.5	2.0	-1.2	-3.0	-0.4	-0.1	0.9	2.1	-0.2	-0.8	0.2	0.7	-1.8	-0.2	0.6	0.3
April	0.6	-0.4	0.8	-1.1	-2.0	-1.0	0.2	-0.2	3.2	1.4	0.3	0.2	0.2	-2.5	-0.7	0.4	0.0
May	2.2	0.2	0.7	-0.2	-2.6	-1.2	0.5	-0.4	3.1	0.6	-0.9	0.2	0.7	-2.1	0.0	1.0	0.1
June	3.0	2.7	2.7	3.0	1.1	3.4	2.7	2.1	1.3	3.7	2.9	2.6	3.1	0.6	1.8	3.0	2.5
July	1.9	-0.1	2.2	-0.5	-2.6	-0.1	0.8	1.7	3.3	1.5	-0.3	0.8	1.1	0.3	0.2	1.5	0.7
August	-0.7	-1.4	-0.2	-3.6	-4.0	-1.9	-0.9	-1.5	0.1	-1.5	-2.9	-1.6	-1.4	-2.1	-1.0	-1.1	-1.6
September	0.0	-1.8	-0.4	-0.7	-2.5	-0.7	-0.9	-2.2	-3.9	1.0	0.2	-0.5	-0.8	-1.8	-1.8	-1.1	-1.1
October	-0.3	0.2	0.2	-0.5	-1.7	-0.4	1.0	-1.4	-3.4	0.9	0.3	0.7	0.9	-0.6	-1.2	0.5	-0.3
November	0.2	0.5	-0.4	-0.9	-1.5	-1.2	0.1	0.9	-3.1	0.0	0.0	0.0	0.1	-0.5	-1.4	-0.3	-0.5
December	-2.4	-2.0	-2.9	-1.9	-2.1	-2.7	-1.6	-2.1	-3.0	0.4	-0.6	-0.5	-0.2	-2.2	-0.9	-1.3	-1.6
Overall	0.8	-0.1	0.4	-0.9	-1.9	-0.8	0.1	0.0	0.0	0.8	-0.2	0.2	0.4	-1.3	-0.5	0.3	-0.2

2026 — NEW SOUTH WALES / ACT - Estimated Maximum Temperature Trend

Region	Upper West	Lower West	Riverina	NW Slopes & Plains	Northern Tablelds	CW Slopes & Plains	Central Tablelds	SW Slopes & Plains	Southern Tablelds	Northern Rivers	Mid-North Coast	Hunter	Metropolitan	Illawarra	South Coast	ACT	NSW / ACT
Town Sample	Wilcannia	Broken Hill	Hay	Moree	Armidale	Dubbo	Orange	Wagga Wagga	Goulburn	Grafton	Coffs Harbour	Newcastle	Sydney	Wollongong	Bega	Canberra	Overall
January	clr	av	clr	clr	clr	clr	clr	a-w	clr	clr	a-c	wmr	av	av	clr	clr	clr
February	clr	av	clr	clr	clr	clr	clr	clr	clr	clr	a-c	av	a-c	a-c	clr	clr	clr
March	wmr	wmr	wmr	a-c	av	wmr	wmr	wmr	wmr	av	av	wmr	a-w	av	a-c	wmr	wmr
April	av	av	av	av	wmr	av	a-c	a-w	a-c	a-w	av	wmr	av	av	a-c	a-c	av
May	wmr	wmr	av	a-w	wmr	wmr	av	wmr	a-w	a-w	a-w	a-w	wmr	av	av	a-w	a-w
June	wmr	wmr	wmr	wmr	wmr	wmr	wmr	wmr	wmr	wmr	wmr	wmr	wmr	a-w	a-w	wmr	wmr
July	av	av	av	a-w	wmr	a-w	av	wmr	av	a-w	wmr	wmr	wmr	a-w	a-w	av	a-w
August	clr	clr	clr	clr	clr	clr	clr	clr	clr	clr	clr	clr	a-c	clr	clr	clr	clr
September	clr	clr	clr	clr	clr	clr	clr	av	clr	clr	a-c	a-c	clr	clr	clr	clr	clr
October	av	av	a-c	a-c	av	a-c	av	wmr	a-w	av	av	wmr	wmr	a-w	clr	av	av
November	clr	a-c	clr	a-c	a-c	av	a-c	wmr	av	clr	a-c	a-w	av	av	clr	a-c	a-c
December	clr	clr	clr	clr	clr	clr	clr	clr	clr	av	a-w	wmr	av	a-c	clr	clr	clr
Overall	a-c	av	a-c	clr	av	a-c	a-c	a-w	av	av	av	a-w	av	av	clr	a-c	av

2026 NEW SOUTH WALES / ACT - Estimated Minimum Temperature Trend

Region	Upper West	Lower West	Riverina	NW Slopes & Plains	Northern Tablelds	CW Slopes & Plains	Central Tablelds	SW Slopes & Plains	Southern Tablelds	Northern Rivers	Mid-North Coast	Hunter	Metropolitan	Illawarra	South Coast	ACT	NSW / ACT
Town Sample	Wilcannia	Broken Hill	Hay	Moree	Armidale	Dubbo	Orange	Wagga Wagga	Goulburn	Grafton	Coffs Harbour	Newcastle	Sydney	Wollongong	Bega	Canberra	Overall
January	av	av	av	clr	av	clr	av	wmr	av	wmr	av	a-w	wmr	clr	av	a-w	av
February	av	av	av	clr	clr	clr	a-c	av	av	av	av	av	av	clr	av	av	a-c
March	wmr	wmr	wmr	clr	clr	av	av	a-w	wmr	av	a-c	av	a-w	clr	av	a-w	av
April	a-w	av	a-w	clr	clr	a-c	av	av	wmr	wmr	av	av	av	clr	a-c	av	av
May	wmr	av	a-w	av	clr	clr	av	av	wmr	a-w	a-c	av	a-w	clr	av	a-w	av
June	wmr	wmr	wmr	wmr	wmr	wmr	wmr	wmr	wmr	wmr	wmr	wmr	wmr	a-w	wmr	wmr	wmr
July	wmr	av	wmr	av	clr	av	a-w	wmr	wmr	wmr	av	a-w	wmr	av	av	wmr	a-w
August	a-c	clr	av	clr	clr	clr	a-c	clr	av	clr	clr	clr	clr	clr	a-c	clr	clr
September	av	clr	av	a-c	clr	a-c	a-c	clr	clr	a-w	av	av	a-c	clr	clr	clr	clr
October	av	av	av	av	clr	av	a-w	clr	clr	a-w	av	a-w	a-w	a-c	clr	av	av
November	av	av	av	av	clr	clr	av	a-w	clr	av	av	av	av	av	clr	av	av
December	clr	clr	clr	clr	clr	clr	clr	clr	clr	av	a-c	av	av	clr	a-c	clr	clr
Overall	a-w	av	av	a-c	clr	a-c	av	av	av	a-w	av	av	av	clr	av	av	av

VICTORIA 2026

GENERAL TRENDS

Year Overall:
Drier by around 10-15% below the average
Slightly warmer but within 0.5°C of the average

Summer:
Wetter by up to 25-30% more than the average
Wettest –Over the New Year from end of December to early January
Driest – Second to fourth weeks of January
Slightly cooler but within 0.5°C of the average
Hottest – Second to fourth weeks of January
Coolest – First weeks of December and January

Autumn:
Drier by around 30-35% below the average
Wettest – First half of April
Driest – Second half of March
Warmer by up to 1.5-2°C above the average
Hottest – Second half of March
Coolest – Second week of May to the end of the month

Winter:
Drier by around 20-25% below the average
Wettest – Third week of July and second half of August
Driest – First half of June
Slightly warmer but within 0.5°C of the average
Hottest – First half of June
Coolest – First and third weeks of August

Spring:
Drier by around 45-50% below the average
Wettest –Second week of September and mid-November
Driest – Second week of October and last 10 days of October
Average temperatures overall
Hottest – Last half of November
Coolest – First half of September

Major Weather Events

January 2026

Date	Day	Victoria - January 2026	Melbourne	Moon
1st	Thur	Very heavy rain in the Hume to the Alpine regions, particularly in the Mansfield area and at Mt Hotham. Very strong and unsettled winds with gale gusts likely across the Gippslands.	Rain	
2nd	Fri	Unusually cold in the Alpine region over the next few days with many areas likely to see maximum temperatures only reaching into the single digits, particularly in the Mt Baw Baw vicinity. Rain clearing to isolated showers.		P6
3rd	Sat	Mostly dry over the next couple of weeks.		F ^
4th	Sun	Very cold overnight on Mt Hotham, with frosts.		
5th	Mon			
6th	Tue	Very hot spell over the next 2-3 weeks in the Lodden-Mallee region, particularly in the Bendigo district.		
7th	Wed			
8th	Thu	Chance of heat wave conditions becoming established around this time throughout the State.		
9th	Fri			XhS
10th	Sat	Very hot temperatures across the State with particularly high maximum temperatures in the Wimmera region, where a number of districts may see highs into the mid-40's. Unusually warm overnight in the Central Highlands-Grampians.		
11th	Sun	The next 3-4 weeks may be warmer than is typical for this time of year, with northern parts of the State wetter than average, and southern parts of the State drier than average.		3Q
12th	Mon	Very hot spell in the Wimmera to Central regions with maximum temperatures into the mid-40's in a number of districts and overnight minimums into the late 20's.		
13th	Tue			
14th	Wed			A
15th	Thur	Light showers across the State.		
16th	Fri	Mostly dry over the next week to 10 days.		
17th	Sat	Cool overnight in the Wimmera with temperatures possibly dropping into the single digits in a number of areas.		V
18th	Sun			
19th	Mon			N
20th	Tue	Hot spell in the West Coast to Central regions, with maximum temperatures into the lower 40's in a number of districts.		
21st	Wed	Unsettled conditions in the Northern Country region accompany passing thunderstorms where microburst winds may develop that may cause some damage to local property. Unusually warm overnight in the Wimmera with minimum temperatures possibly into the early 30's and hot during the day in the Gippsland to Central regions with maximum temperatures into the early 40's for a time in some districts.		
22nd	Thu	Cold overnight in the West Coast region, with minimum temperatures dropping into the single digits in a number of districts.		
23rd	Fri			
24th	Sat	Unusually cold overnight in some parts of the Alpine region, where chance of some isolated patches of frost may occur, particularly in the Mt Baw Baw area. Also very cool in the Wimmera, Central, Western Plains, and Gippsland regions overnight with minimum temperatures dropping into the single digits.		XhN
25th	Sun			
26th	Mon	Severe thunderstorms with heavy rain in the Wimmera region and Western District, with chance of flash flooding in the Ararat area.		1Q
27th	Tue	Cold overnight in the Gippslands with single digit minimum temperatures likely in many districts.		
28th	Wed	Wet spell over the next few days with showers turning to widespread heavy rain before easing back to showers again by the end of the month.	Showers	
29th	Thur	Very heavy rain throughout the State with chance of flash flooding in some places, particularly in river catchments.	Rain	
30th	Fri	Very wet spell in the Alpine region, particularly in the Mt Buffalo area.		P10 ^
31st	Sat	Unusually cool day in the Alpine region, particularly in the Mt Baw Baw area where maximum temperatures may not reach into the double digits. Also very cool in the Wimmera overnight with minimum temperatures dropping into the single digits. Mostly dry conditions returns across the State.		

February 2026

Date	Day	Victoria - February 2026	Melbourne	Moon
1st	Sun	Cold over the next couple of nights in the Gippslands with single digit minimum temperatures likely in many districts. Mostly dry the next few days.		
2nd	Mon		Showers	F
3rd	Tue			
4th	Wed	Brief showers likely in most districts.	Rain, Th	
5th	Thu	Prolonged dry spell ahead with mostly dry conditions prevailing for the remainder of this month.		
6th	Fri			XhS
7th	Sat			
8th	Sun			
9th	Mon		Showers	
10th	Tue		Rain	3Q
11th	Wed	The next 3-4 weeks may be cooler than average, with close to normal amounts of rainfall for this time of year overall, slightly wetter in the east, and slightly drier than average in the northwest.		A
12th	Thur			
13th	Fri			V
14th	Sat	High maximum temperatures into the upper 30's in a number of districts in the North Mallee over the next couple of days.	Showers, Th	
15th	Sun			
16th	Mon		Showers	
17th	Tue	Very wet in the Alpine region, particularly in the Mt Baw Baw district.		
18th	Wed		Showers	N
19th	Thu		Showers	
20th	Fri	Cold over the next couple of nights in the West Coast region.		XhN
21st	Sat			
22nd	Sun			
23rd	Mon		Showers	
24th	Tue		Showers	
25th	Wed			1Q P14
26th	Thur	Very warm day in the central parts of the State and in the Grampians over the next couple of days.		^
27th	Fri			
28th	Sat			

March 2026

Date	Day	Victoria - March 2026	Melbourne	Moon
1st	Sun	Very hot day across the State with maximum temperatures into the upper 30's in many districts. Very little precipitation is expected during this coming month, with mostly dry conditions prevailing throughout.	Showers	
2nd	Mon	Very warm overnight in the Lodden-Mallee region.	Showers	
3rd	Tue			F
4th	Wed		Showers	
5th	Thu			XhS
6th	Fri	Very strong windy spells with occasional gusts across the Gippslands region, particularly around the coasts.		
7th	Sat			
8th	Sun			
9th	Mon	Cold spell across the State over the next few days.		
10th	Tue	Very cold spell in in the Alpine region over the next couple of days, particularly in the Mt Buller and Mt Hotham districts, where maximum temperatures may struggle to rise above frost levels and overnight minimas may dip into frosts.	Showers	
11th	Wed	Many regions across the State may see overnight minimas drop into single digits around this time. Scattered frosts likely in the Alpine region overnight.		3Q A
12th	Thur	The next 3-4 weeks ahead may be unusually hotter and drier than is normal for this time of year, with most expected rainfall coming towards the end of this lookahead in early April.		V
13th	Fri	Overnight frosts in the Alpine region, particularly at Mt Hotham and Mt Baw Baw as well as very cold overnight with chance of some patches of frost in some Central districts.		
14th	Sat			

Date	Day	Victoria - March 2026	Melbourne	Moon
15th	Sun			
16th	Mon			
17th	Tue			
18th	Wed	Dry conditions continue throughout the State.	Showers	
19th	Thu			**N XhN**
20th	Fri			
21st	Sat			
22nd	Sun			**P11**
23rd	Mon			
24th	Tue	Very hot spell across the State over the next 3-4 days with many districts seeing maximum temperatures into the upper 30's to lower 40's.		
25th	Wed			
26th	Thur			**1Q ^**
27th	Fri	Very hot in the Barwon region where maximum temperatures may reach into the lower 40's in some areas, including in the Geelong district.		
28th	Sat	Mild overnight temperatures in the Alpine region, including around Mt Buller.		
29th	Sun	Very warm night in the Melbourne area with minimum temperatures likely in the mid-to-upper 20's.		
30th	Mon			
31st	Tue			

April 2026

Date	Day	Victoria - April 2026	Melbourne	Moon
1st	Wed	Dry conditions continue throughout the State over the next few days.		
2nd	Thu			**F XhS**
3rd	Fri	Heavy rain accompanying thunderstorm activity in the Wimmera region and Western District, with chance of flash flooding in the Halls Gap area.		
4th	Sat	Heavy rain and hail showers in the North Central, Mallee and Wimmera regions. Chance of flash flooding in the Mildura area	Rain, Th	
5th	Sun	Heavy rain for a time in the Northeast to the Wimmera regions, including in the Tarrawingee and Starwell districts.		
6th	Mon	Cold spell over the State during the next few days. Cool day in the Barwon.	Showers	
7th	Tue	Overnight frosts in the Alpine region, particularly at Mt Hotham.	Showers	**A**
8th	Wed	Very cold overnight in the Wimmera with chance of patches of frost in sheltered areas.		
9th	Thur	Cold in the Alpine region, when maximum temperatures may sit close to frost levels in some areas, including at Mt Hotham.	Showers	**V**
10th	Fri	Light showers across the State.		**3Q**
11th	Sat	Very strong and blustery winds with occasional gale gusts in the Gippslands, particularly around the coasts. Another prolonged dry spell may begin around this time and continue to the end of the month in most parts of the State.		
12th	Sun	The next 3-4 weeks may be drier than normal for this time of year, as well as slightly cooler, with a very severe cold outbreak around the second week of May resulting in some significant snowfalls in the Alpine region.		
13th	Mon			
14th	Tue	Heavy rain in the Alpine region, particularly in the Falls Creek area. Very strong winds at times in the Gippslands, particularly around the coasts.		
15th	Wed			
16th	Thu			**XhN**
17th	Fri			**N**
18th	Sat			
19th	Sun			**P7**
20th	Mon			
21st	Tue	Warm spell throughout the State over the next few days with maximum temperatures into the late 20's to early 30's in many districts.		
22nd	Wed	Unusually warm in the Lodden-Mallee to the Barwon regions and very mild overnight in the Gippslands.		**^**
23rd	Thur		Showers	
24th	Fri			**1Q**
25th	Sat		Showers	
26th	Sun			
27th	Mon			
28th	Tue			
29th	Wed			**XhS**
30th	Thur			

May 2026

Date	Day	Victoria - May 2026	Melbourne	Moon
1st	Fri	Mostly dry throughout the State over the coming week except for some brief showers possible in the next few days in the far northeast of the Lakes region.	Fg	
2nd	Sat	Frosts likely in the Western Plains.	Fg	F
3rd	Sun	Milder spell in the Gippslands to the Alpine regions over the next couple of days.		
4th	Mon			
5th	Tue			A
6th	Wed		Showers	V
7th	Thu	Cold spell throughout the State over the next few days, with many districts daily maximas struggling to reach into the low teens, and closer to frost levels remaining throughout the day in the Alpine region. Scattered overnight frosts may frequent many areas over the next few days.	Rain	
8th	Fri	Light snow in the Victorian Alps over the next couple of days, and light showers elsewhere across the State.		
9th	Sat	Cold with frosts in the Alpine region, with particularly severe frosts in the Mt Hotham district.		
10th	Sun	Very cold overnight in the Lodden-Mallee and Wimmera regions with widespread frosts likely. Mostly dry over the next couple of days.		3Q
11th	Mon		Showers	
12th	Tue	The next few weeks ahead may be predominately dry in many districts, except in the Western to Central regions, where closer to normal rainfall can be expected. It may be generally cooler than normal overnight in the 3-4 weeks ahead, with maximum daytime temperatures closer to what is typical for this time of year. In a couple of weeks time, a spell of severe frosts arrives and lingers for several days.	Showers	
13th	Wed	Heavy rain with chance of flash flooding and small hail in the Central region, particularly in the Doncaster district. Light showers elsewhere.	Showers	XhN
14th	Thur	Mostly dry over the next 3-4 days across the State.		
15th	Fri			
16th	Sat			
17th	Sun		Showers	N
18th	Mon	A band of showers cross the State.		P3
19th	Tue	Mostly dry in all districts over the next week.		^
20th	Wed	Mild spell over the next couple of days in the Northeast of the State and in the Grampians.		
21st	Thu	Mild night on Gabo Island around this time.		
22nd	Fri			
23rd	Sat			1Q
24th	Sun			
25th	Mon	Unusually warm day in the Lodden-Mallee to the West Coast regions, particularly in the Ouyen district.		
26th	Tue	Mild spell over the next couple of days in the Gippsland to central parts of the State.		XhS
27th	Wed	A wet front moves over the State over the next few days. Widespread showers prevail with an outbreak of heavy precipitation for a time.	Rain	
28th	Thur	Very wet day in the Wimmera. Very cold with frosts, strong winds, and snow in the Victorian Alps over the next couple of days.	Rain	
29th	Fri	Very cold with overnight frosts in the Northeast region and in some parts of the Grampians.		
30th	Sat	Dry conditions return across the State.		
31st	Sun			F

June 2026

Date	Day	Victoria - June 2026	Melbourne	Moon
1st	Mon	Chance of isolated showers in the State.		A
2nd	Tue	Mostly dry over the next 4-5 days.		V
3rd	Wed			
4th	Thu	Very cold with severe frosts overnight in the Alpine region, particularly in the Mt Hotham area and cold with frosts in the Central region, including in some sheltered parts of Melbourne.		
5th	Fri			
6th	Sat		Showers, Fg	
7th	Sun	Chance of some brief showers in the northern Mallee region and along the western border fringes.	Showers	

Date	Day	Victoria - June 2026	Melbourne	Moon
8th	Mon	Dry throughout the State over the next week. Very low overnight temperatures across the State over the next few days with scattered frosts in many districts.	Fg	3Q
9th	Tue			XhN
10th	Wed			
11th	Thur		Fg	
12th	Fri	Overall, the next 3-4 weeks may be unusually warmer and drier than is typical for this time of year, with may regions possibly only seeing around half their normal amount of rainfall during this spell.	Fg	
13th	Sat	Widespread frosts over the next couple of nights.	Fg	
14th	Sun		Showers	
15th	Mon	Outbreaks of heavy rain possible in the Lakes region.	Fg	N P2
16th	Tue	Mostly dry throughout the State once more over the next few days.		^
17th	Wed		Showers	
18th	Thu	Frosts in the Hume region.	Fg	
19th	Fri			
20th	Sat	A wet front moves onto the State around this time bringing some precipitation into most districts over the next several days.		
21st	Sun	Wet day in the Grampians with more frequent showers elsewhere.		
22nd	Mon	Mild overnight in the Melbourne region. Shower activity becoming isolated around this time.		1Q XhS
23rd	Tue		Showers	
24th	Wed	Cold with frosts and snowflurries in the Victorian Alpine region, particularly at Mt Hotham. Chance of isolated showers elsewhere.	Showers	
25th	Thur	Mostly dry in the State over the remainder of this month. Very cold over the next few nights throughout the State with scattered frosts in many districts likely.		
26th	Fri		Fg	
27th	Sat	Severe frosts over the next couple of nights in some parts of the Gippsland to Central regions.		
28th	Sun			A
29th	Mon			V
30th	Tue		Showers	F

July 2026

Date	Day	Victoria - July 2026	Melbourne	Moon
1st	Wed	A mostly dry spell prevails over the next 7-10 days.	Showers	
2nd	Thu		Showers	
3rd	Fri			
4th	Sat			
5th	Sun			
6th	Mon			
7th	Tue	Frosts and snowflurries in the Victorian Alpine region, and particularly frosty on Mt Hotham over the next couple of days. Strong, blustery winds in the Gippsland, particularly around the coasts.		XhN
8th	Wed	Severe frosts in the Alpine region.		3Q
9th	Thur			
10th	Fri			
11th	Sat	Snowflurries in the Victorian Alpine region. Unusually mild day in the Gippsland.	Showers	
12th	Sun	The next 4 weeks may bring nearer to average rainfall across the State, and average to slightly below normal temperatures for this time period. Heavy snow in the Victorian Alpine region and ski fields. A weak front crosses the State over the nex couple of days bringing some brief precipitation into most districts.		
13th	Mon	Snowflurries in Falls Creek to Mt Hotham.	Showers	P4 ^
14th	Tue	Mostly dry over the next 3-4 days.	Showers, Fg	N
15th	Wed			
16th	Thu			
17th	Fri	Unusually mild day in the Gippsland, particularly in the Lakes Entrance and Orbost districts.		
18th	Sat	Mild overnight temperatures in the Gippsland region, particularly around the coasts. Wet spell across the State over the next couple of days.	Showers	
19th	Sun	Very cold day with sub-zero maximum temperatures and some light snow in the Victorian Alps. Heavy rain for a time in the Barwon - Western District.	Showers	
20th	Mon	Mostly dry interlude.	Showers	XhS
21st	Tue	Snow in the Victorian Alps and ski fields and light showers elsewhere over the next couple of days.	Showers	1Q

Date	Day	Victoria - July 2026	Melbourne	Moon
22nd	Wed			
23rd	Thur	Dry spell over the next couple of days.		
24th	Fri	Milder daytime temperatures over much of the State during the next few days.		
25th	Sat	Light showers in some parts of the State.		
26th	Sun			A
27th	Mon	Widespread heavy precipitation throughout the State with chance of flash flooding in some districts, and heavy snow for a time in the Alps.		V
28th	Tue		Showers	
29th	Wed	Snowflurries in the Victorian Alps.	Showers	
30th	Thu	Snow in the Victorian Alpine regions and ski fields over the next few days.	Showers	F
31st	Fri	Light showers over most of the State.	Rain	

August 2026

Date	Day	Victoria - August 2026	Melbourne	Moon
1st	Sat	Very strong winds to gale force at times in the Gippsland, particularly around the coasts. Chance of brief outbreaks of rain in the Lakes region. Mostly dry elsewhere over the next 5-6 days.	Showers	
2nd	Sun	Snowflurries in the Victorian Alps.	Showers	
3rd	Mon	Widespread frosts across much of the State over the next few nights.	Fg	XhN
4th	Tue	Very cold overnight with severe frosts in the Alpine region, particularly at Mt Hotham.		
5th	Wed			
6th	Thur	Snowflurries in Falls Creek to Mt Hotham.	Rain	3Q
7th	Fri	Light snow in the Victorian Alpine region and ski fields over the next couple of days and light showers elsewhere. Widespread overnight frosts across the State over the next few days.	Showers, Fg	
8th	Sat			
9th	Sun	Snowflurries in Falls Creek to Mt Hotham. Very cold day in the Barwon region. Dry elsewhere over the next few days.		^
10th	Mon		Showers	P8
11th	Tue	Snowfurries on some of the Victorian Alpine peaks. Milder maximum temperatures in the Mallee to the Wimmera, including in the Grampians, around this time.		
12th	Wed	Over the next 3-4 weeks, temperatures may be cooler than normal for this time of year, while rainfall may be nearer to average overall, although some heavy rain in the last couple of days during the fourth week may cause some minor flooding in the Northeast region. Light snow in the Victorian Alps. Some light showers in the State over the next couple of days.	Showers	
13th	Thu	Snowflurries in some of the Victorian Alpine ski fields.		N
14th	Fri			
15th	Sat	Snowflurries in Falls Creek to Mt Hotham.		
16th	Sun	Light showers over most of the State.		XhS
17th	Mon	Scattered frosts in parts of the Gippsland around this time.		
18th	Tue	Snowflurries in Falls Creek to Mt Hotham with light showers elsewhere over the next 3-4 days.	Showers	
19th	Wed	Light snow in the Victorian Alps and widespread frosts in Central to Western Plains districts.	Showers	
20th	Thur	Snow in the Victorian Alps on most days over the next week ahead.	Showers	1Q
21st	Fri	Hail showers likely in the Central region. Frosts in the Pyrenees district.	Rain	
22nd	Sat	Isolated showers in the State over the next few days.		A
23rd	Sun		Showers	V
24th	Mon			
25th	Tue	Mostly dry throughout the State over the remainder of the month.		
26th	Wed	Snowflurries in Falls Creek to Mt Hotham over the next couple of days.	Rain	
27th	Thu		Showers	
28th	Fri		Showers	F
29th	Sat	Light snow in the Victorian Alpine region and ski fields.		
30th	Sun			XhN
31st	Mon			

September 2026

Date	Day	Victoria - September 2026	Melbourne	Moon
1st	Tue	Light snow in the Victorian Alps and ski fields over the next couple of days. Chance of brief showers in the western border fringe districts.	Rain	
2nd	Wed			
3rd	Thur	Dry and very cold over the next 3-5 nights in the State with scattered and some severe frosts in many districts, particularly in the Wimmera, and Northeast regions.		
4th	Fri			3Q
5th	Sat			
6th	Sun			^
7th	Mon			P12
8th	Tue	Dry conditions continue for the next few days.	Fg	
9th	Wed	Mild temperatures across the State over the next couple of days with maximum temperatures fairly consistently in the mid to upper teens.		
10th	Thu		Showers	
11th	Fri	A wet front crosses the State over the next couple of days bringing widespread showers throughout and some snowflurries in Falls Creek to Mt Hotham.		N
12th	Sat	Persistent high pressure systems lingering around the southeast of the country over the coming 3-4 weeks may result in this coming look-ahead period being drier with above average maximum temperatures and below average minimum temperatures than what is typical for this time of year.		XhS
13th	Sun	Chance of brief showers in the west of the Wimmera.	Showers	
14th	Mon	Dry spell over the next couple of days with some patches of frost overnight in the West Coast to the Western Plains and Central regions.		
15th	Tue			
16th	Wed	Brief showery spelly across the State.		
17th	Thur	Mostly dry throughout the State over the next 7-10 days.		
18th	Fri		Showers	
19th	Sat			1Q A V
20th	Sun	Severe overnight frosts in the Alpine region, particularly at Mt Hotham, with scattered frosts throughout the State over the next couple of days.		
21st	Mon			
22nd	Tue			
23rd	Wed			
24th	Thu	Warm spell briefly in the West Coast region.		
25th	Fri	Unusually mild overnight in the Gippsland region.	Showers	
26th	Sat	A weak front crosses the State over the next couple of days bringing light showers into most districts and some snow in the Victorian Alps and ski fields.	Showers	XhN
27th	Sun	Very cold in the Alpine region with maximum temperatures not lifting out of frost levels in a number of districts, particularly in the Mt Hotham area.		F
28th	Mon	Mostly dry throughout the State over the next couple of days.		
29th	Tue		Showers	
30th	Wed	Severe thundersorms with possible large hail in the northern Dandenong area in the Central region. Mild maximum temperatures into the mid 20's in the Gippsland districts.	Th	

October 2026

Date	Day	Victoria - October 2026	Melbourne	Moon
1st	Thur	Mostly dry throughout the State over the next couple of weeks, with just a couple of brief exceptions.		
2nd	Fri			P13
3rd	Sat	Very strong and blustery winds with gales gusts possible at times in the Alpine region, particularly in the Mt Hotham area.	Showers	
4th	Sun	Heavy rain in the Alpine region, particularly in the Falls Creek area.		3Q ^
5th	Mon	Some patches of frost in the Mallee and Wimmera regions overnight.		
6th	Tue			
7th	Wed	Warm spell across the State over the next couple of days with many districts seeing maximum temperatures reach into the mid-to-upper 20's to early 30's at times.	Fg	
8th	Thu	Unusually hot day in the Loddon-Mallee region, particularly in the Mildura to Swan Hill areas, where maximum temperatures may rise into the low 30's for a time.		
9th	Fri		Showers	
10th	Sat	Chance of isolated showers in the Wimmera.		XhS

Date	Day	Victoria - October 2026	Melbourne	Moon
11th	Sun			N
12th	Mon	The next 3-4 weeks may be drier and warmer than what is typical for this time of year.		
13th	Tue	Very strong winds gusts in the Alpine region, particularly at Mt Hotham.		
14th	Wed		Showers	
15th	Thur	Wet spell across the State over the next few days. Very wet in the Alpine to the Gippsland regions, particularly in the Snowy River catchment. Widespread showers elsewhere.	Showers	
16th	Fri	Cold day in Melbourne with the maximum temperature possibly not reaching into double digits.	Showers	
17th	Sat	Cold spell across the State over the next few days. Very cold day with snowflurries in the Victoria Alpine region and ski fields. Maximum temperatures may remain at frost levels throughout the day, particularly on Mt Baw Baw. Chance of hail showers in the Melbourne area.	Showers, Sn	A V
18th	Sun	Very cold temperatures persist in the Alpine region with severe overnight frosts likely, particularly at Mt Hotham. Chance of isolated showers elsewhere.		
19th	Mon	Mostly dry throughout the State over the remainder of the month.		1Q
20th	Tue			
21st	Wed			
22nd	Thur			
23rd	Fri			
24th	Sat	Mild spell overnight in many parts of the State.		XhN
25th	Sun	Chance of brief shower activity over the next couple of days in the Gippsland to the Northeast regions.		
26th	Mon			F
27th	Tue			
28th	Wed			
29th	Thu	Unusually warm spell across the State with maximum temperatures possibly into the low 30's in a number of places including in the Melbourne region.		P9
30th	Fri	Chance of scattered showers of large hail in some parts of the State.		^
31st	Sat			

November 2026

Date	Day	Victoria - November 2026	Melbourne	Moon
1st	Sun	Cold spell over the next few days across the State with some patches of frosts in sheltered and higher areas. Mostly dry spell throughout the State over the next couple of weeks.	Showers	
2nd	Mon	Snowflurries in the Victorian Alpine region. Showery spell in the Central to Western regions of the State.		3Q
3rd	Tue	Snowflurries in the Falls Creek to Mt Hotham region. Some pockets of frost in the Gippsland region, including in the Noojee area.		
4th	Wed			
5th	Thur			
6th	Fri	Unusually mild night across the State, particularly in the Ouyen district of the Loddon-Mallee region.		XhS
7th	Sat	Strong winds to gale force at times in the Alpine to the Gippsland regions, particularly along the coastal fringes. Showery spell in the Wimmera, Western Plains, Central, and West Coast regions over the next few days.	Showers	
8th	Sun			
9th	Mon			N
10th	Tue	Unusually hot day in the Loddon-Mallee region, particularly in the Mildura district where maximum temperatures may rise into the upper 30's for a time. Very wind in the Melbourne region around this time.	Showers	
11th	Wed		Showers	
12th	Thu	A series of cold fronts may cross the State over the next 3-4 weeks, bringing average to above average rainfall overall for this lookahead period, as well as average to slightly above average temperatures.	Fg	
13th	Fri		Rain	V
14th	Sat	A weak front may cross the State bringing light showers into most districts.		A
15th	Sun	Cool over the next couple of nights in the Northeast to the South Mallee, Central, and Gippsland regions with minimum temperatures persisting in the lower single digits.		
16th	Mon			
17th	Tue			
18th	Wed	A cold front may cross the State over the next couple of days bringing widespread precipitation with its passage. Light showers throughout.	Showers	1Q

Date	Day	Victoria - November 2026	Melbourne	Moon
19th	Thur	Widespread heavy rain throughout the State with chance of pockets of localised flooding possible.		
20th	Fri	Rain activity swiftly clears to mostly dry, with dry conditions returning for the next 4-5 days.		XhN
21st	Sat			
22nd	Sun			
23rd	Mon	Very warm spell across the State over the next couple of days with many districts seeing maximum temperatures into the mid to upper 30's.		
24th	Tue		Rain, Th	
25th	Wed	Light showers over most of the State over the next couple of days.		F
26th	Thu			P5
27th	Fri	Cold over the next couple of nights in the Wimmera and Gippsland regions. Dry conditions return for the next few days.		^
28th	Sat	Some frosts overnight in the Barwon region.		
29th	Sun			
30th	Mon	A wet front nears the State spreading isolated showers.	Showers	

December 2026

Date	Day	Victoria - December 2026	Melbourne	Moon
1st	Tue	Scattered showers throughout the State.		3Q
2nd	Wed		Showers	
3rd	Thur	Cold spell overnight across much of the State with many districts seeing minimum temperatures in the single digits to frost levels in sheltered parts. Widespread light shower activity throughout with snow in the Victorian Alpine region and ski fields over the next couple of days.	Rain, Th	XhS
4th	Fri	Heavy snow Falls Creek to Mt Hotham.		
5th	Sat	Mostly dry over the next few days.		
6th	Sun			
7th	Mon			
8th	Tue	Storms cross the North Central, Central, and West Gippsland regions bringing heavy rain, large hail at times, and damaging winds in some areas over the next few days.	Showers, Th	
9th	Wed		Showers	N
10th	Thu			V
11th	Fri	Shower activity becomes more isolated over the next couple of days.	Showers	A
12th	Sat	The remainder of this year and into early next year over the upcoming 3-4 weeks may be cooler and wetter than is normal for this lookahead period.		
13th	Sun	Cold and dry over the next few nights in many parts of the State , with minimum temperatures remaining in single digits.		
14th	Mon			
15th	Tue	Cold overnight in the Barwon region as well as on Phillip Island.	Showers	
16th	Wed	Chance of some showers around the western border districts.		
17th	Thur	Mostly dry over the next 3-4 days.		1Q XhN
18th	Fri	Cold in the Wimmera over the next couple of nights.		
19th	Sat			
20th	Sun		Showers	
21st	Mon	A wet front swiftly crosses the State around this time bringing some patchy shower activity into most districts.		
22nd	Tue			
23rd	Wed	An intense front crosses the State over the nex few days. Widespread showers prevail throughout.	Rain	
24th	Thu	Cold day in the Barwon region. Widespread heavy rain throughout, with chance of localised flooding in some areas, particularly in the river catchments.	Rain	P1 F ^
25th	Fri	Very wet in the Alpine and cold in the region with rain activity clearing to lingering light showers elsewhere. Maximum temperatures may linger close to frost levels in a number of districts, particularly at Mt Hotham.	Showers	
26th	Sat	Mostly dry over the next couple of days.		
27th	Sun		Showers	
28th	Mon	Passing shower activity over the next couple of days in most districts.		
29th	Tue	Gusty winds and hail showers in the Northeast and Northern Country regions that may cause some damage to local property.	Showers, Th	
30th	Wed	Strong winds with gusts possibly to gale force at times in the Gippsland region, particularly around the coasts.		XhS
31st	Thur	Very cold overnight with frosts in the Alpine region, particularly at Mt Hotham. Mostly dry throughout the State.	Fg	3Q

VICTORIA GRAPHS AND TABLES 2026

Seasonal Breakdown – State Overall Rainfall Timing and Estimated mm Trend Graphs
State Overall Maximum Temperature Trend Graphs
Regional Rainfall Expectations Tables - mm and Trends
Regional Maximum and Minimum Temperature Trends Tables

Monthly Breakdown - **Rainfall:**
Regional Expectations by mm
Regional Expectations by mm Variance to Average
Regional Variance Trend by mm
Regional Estimated Expectations by % of Variance to Average
Regional Estimatedl Variance Trend by %

Maximum & Minimum Temperature Trends:
Regional Estimated Average Temperature
Regional Estimated Degree of Variation from the Average
Regional Estimated Temperature Trend

Seasonal Breakdowns

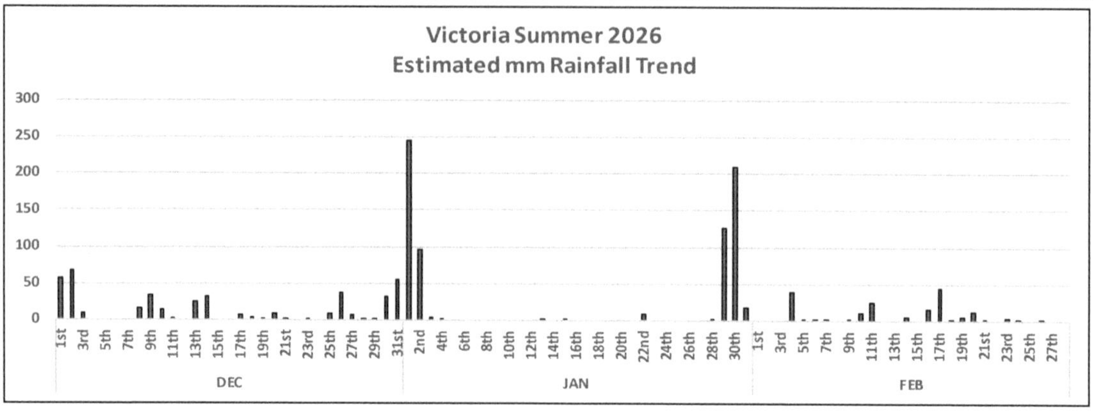

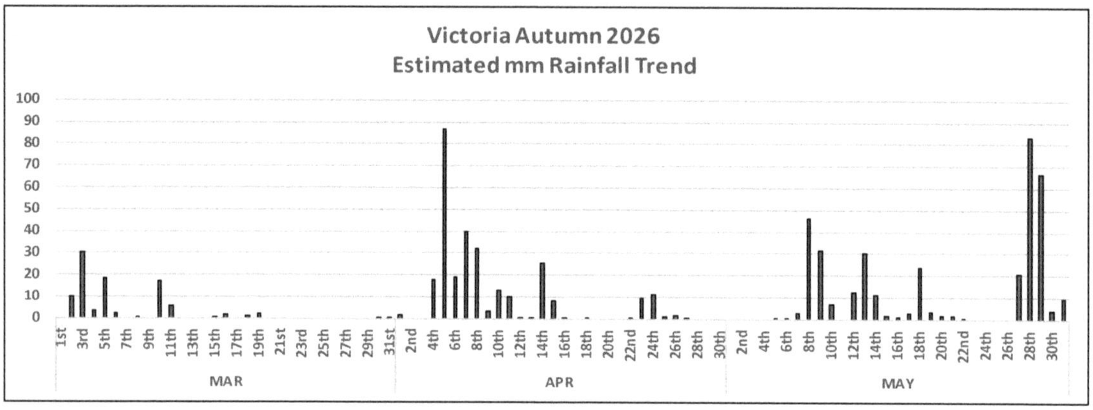

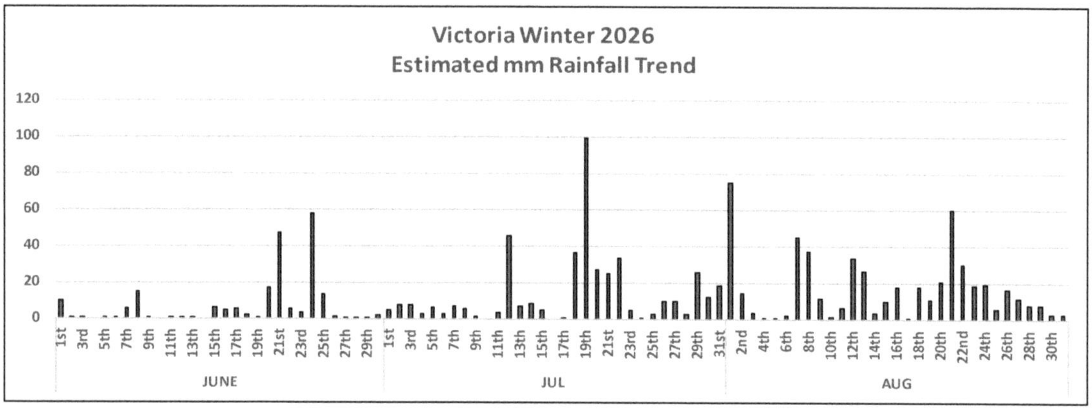

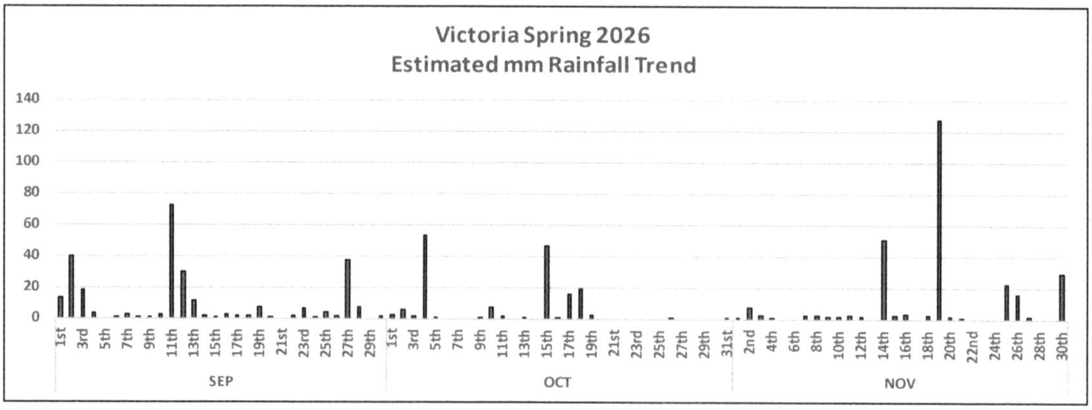

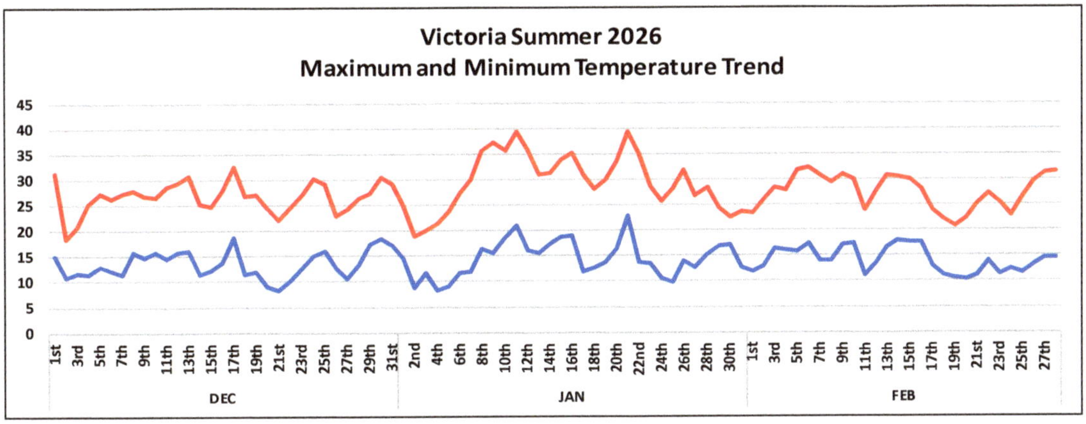
Victoria Summer 2026
Maximum and Minimum Temperature Trend
45
40
35
30
25
20
15
10
5
0
DEC
JAN
FEB

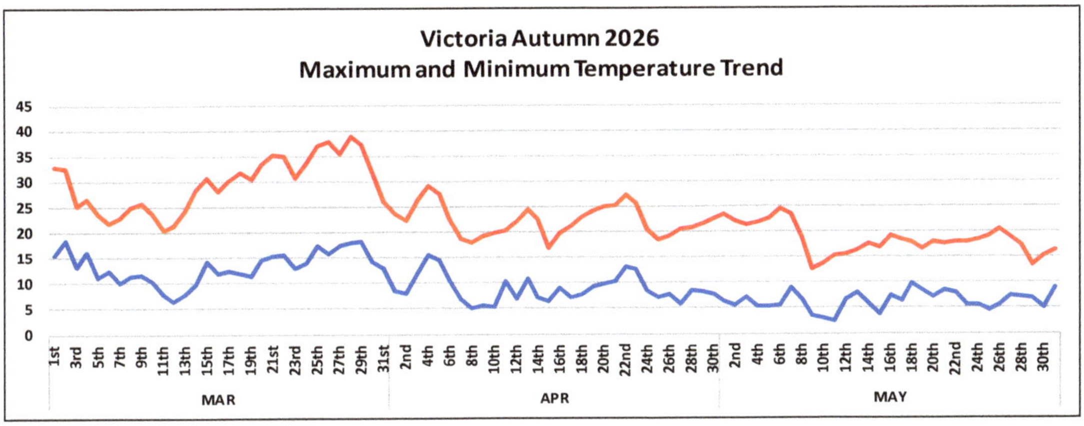
Victoria Autumn 2026
Maximum and Minimum Temperature Trend
45
40
35
30
25
20
15
10
5
0
MAR
APR
MAY

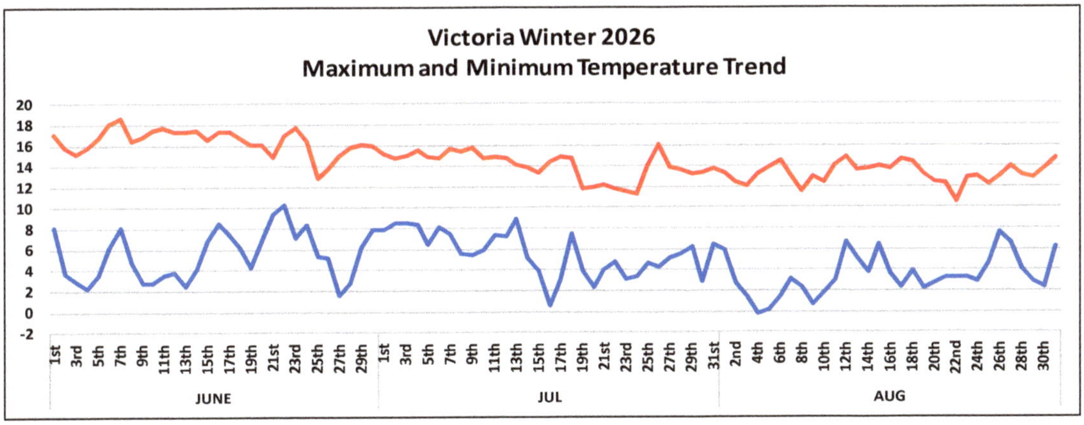
Victoria Winter 2026
Maximum and Minimum Temperature Trend
20
18
16
14
12
10
8
6
4
2
0
-2
JUNE
JUL
AUG

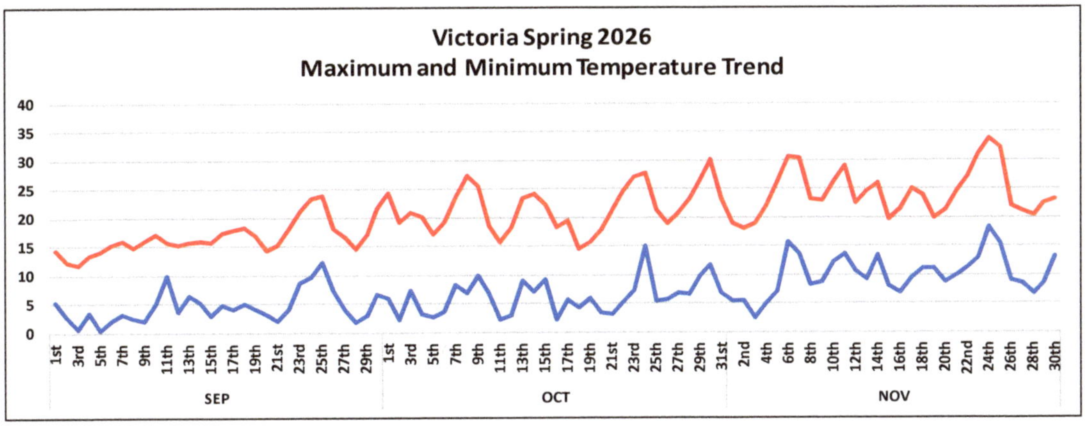
Victoria Spring 2026
Maximum and Minimum Temperature Trend
40
35
30
25
20
15
10
5
0
SEP
OCT
NOV

2026	VICTORIA - Estimated Seasonal Rainfall Expectations (mm)									
REGION	Mallee	Wimmera	Western District	Northern Country	North Central	Central	Northeast	W & S Gippsland	East Gippsland	VIC
Town Sample	Mildura	Horsham	Warrnambool	Echuca	Bendigo	Melbourne	Benalla	Traralgon	Bairnsdale	Overall
Summer	59	107	113	171	125	145	186	166	216	1288
Autumn	32	55	183	33	58	138	96	82	64	742
Winter	56	110	319	69	136	92	136	161	72	1150
Spring	48	56	142	60	82	65	76	107	75	710
Overall	215	354	826	384	437	479	497	580	550	4324

2026	VIC - Estimated Seasonal Rainfall Variation from the Average (mm)									
REGION	Mallee	Wimmera	Western District	Northern Country	North Central	Central	Northeast	W & S Gippsland	East Gippsland	VIC
Town Sample	Mildura	Horsham	Warrnambool	Echuca	Bendigo	Melbourne	Benalla	Traralgon	Bairnsdale	Overall
Summer	-8	31	-4	92	29	-4	66	13	51	266
Autumn	-17	-30	24	-52	-56	1	-45	-80	-75	-330
Winter	-20	-44	57	-56	-43	-50	-83	-19	-80	-338
Spring	-32	-65	-56	-52	-58	-110	-100	-111	-121	-704
Overall	-57	-81	90	-16	-91	-123	-158	-133	-101	-670

2026	VIC - Estimated Seasonal Rainfall Trend Variation by 10mm									
REGION	Mallee	Wimmera	Western District	Northern Country	North Central	Central	Northeast	W & S Gippsland	East Gippsland	VIC
Town Sample	Mildura	Horsham	Warrnambool	Echuca	Bendigo	Melbourne	Benalla	Traralgon	Bairnsdale	Overall
Summer	av	wtr	av	wtr	av	av	wtr	av	wtr	wtr
Autumn	av	drr	av	drr	drr	av	drr	drr	drr	drr
Winter	av	drr	wtr	drr	drr	drr	drr	av	drr	drr
Spring	drr	drr	drr	drr	drr	drr	drr	drr	drr	drr
Overall	av	av	av	av	av	drr	drr	drr	drr	drr

2026	VIC - Estimated Seasonal Rainfall Expectation by %									
REGION	Mallee	Wimmera	Western District	Northern Country	North Central	Central	Northeast	W & S Gippsland	East Gippsland	VIC
Town Sample	Mildura	Horsham	Warrnambool	Echuca	Bendigo	Melbourne	Benalla	Traralgon	Bairnsdale	Overall
Summer	-13%	41%	-4%	117%	31%	-3%	55%	9%	31%	26%
Autumn	-35%	-35%	15%	-61%	-49%	1%	-32%	-49%	-54%	-31%
Winter	-27%	-29%	22%	-45%	-24%	-35%	-38%	-11%	-53%	-23%
Spring	-41%	-54%	-28%	-46%	-41%	-63%	-57%	-51%	-62%	-50%
Overall	-21%	-19%	12%	-4%	-17%	-20%	-24%	-19%	-16%	-13%

2026	VIC - Estimated Seasonal Rainfall Trend Variation by 25%									
REGION	Mallee	Wimmera	Western District	Northern Country	North Central	Central	Northeast	W & S Gippsland	East Gippsland	VIC
Town Sample	Mildura	Horsham	Warrnambool	Echuca	Bendigo	Melbourne	Benalla	Traralgon	Bairnsdale	Overall
Summer	av	wtr	av	wtr	wtr	av	wtr	av	wtr	wtr
Autumn	drr	drr	av	drr	drr	av	drr	drr	drr	drr
Winter	drr	drr	av	drr	av	drr	drr	av	drr	av
Spring	drr	drr	drr	drr	drr	drr	drr	drr	drr	drr
Overall	wtr	wtr	wtr	wtr	wtr	wtr	wtr	wtr	wtr	wtr

2026	VIC - Estimated Seasonal Maximum Temperature Average and Trend									
Region	Mallee	Wimmera	Western District	Northern Country	North Central	Central	Northeast	West and South Gippsland	East Gippsland	VIC
Town Sample	Mildura	Horsham	Warrnambool	Echuca	Bendigo	Melbourne	Benalla	Traralgon	Bairnsdale	Overall
Summer Est	32.1	30.0	23.2	30.9	29.0	26.1	30.4	26.3	24.4	28.0
Trend	av	av	a-c	a-w	av	av	av	av	a-c	av
Autumn Est	26.2	24.4	21.6	24.7	23.3	22.1	24.0	22.6	22.1	23.4
Trend	wmr	wmr	wmr	wmr	wmr	wmr	wmr	wmr	wmr	wmr
Winter Est	16.4	14.3	13.8	15.1	13.4	13.9	14.5	14.4	15.6	14.6
Trend	av	av	av	a-w	av	av	a-w	av	av	av
Spring Est	23.9	21.2	17.7	22.8	20.6	20.0	21.6	19.9	20.5	20.9
Trend	av	a-c	av	a-w	av	av	av	av	a-w	av

2026	VIC - Estimated Seasonal Minimum Temperature Average and Trend									
Region	Mallee	Wimmera	Western District	Northern Country	North Central	Central	Northeast	West and South Gippsland	East Gippsland	VIC
Town Sample	Mildura	Horsham	Warrnambool	Echuca	Bendigo	Melbourne	Benalla	Traralgon	Bairnsdale	Overall
Summer Est	16.6	12.0	11.1	15.0	14.3	16.4	15.6	12.8	12.9	14.0
Trend	av	av	av	av	a-w	wmr	wmr	av	av	a-w
Autumn Est	11.2	7.7	9.2	9.4	8.8	12.6	8.6	9.1	9.1	9.5
Trend	a-w	av	av	av	av	wmr	wmr	av	av	a-w
Winter Est	5.0	3.6	5.4	4.1	3.9	7.9	4.2	4.7	4.7	5.0
Trend	av	av	av	av	a-w	wmr	wmr	a-w	av	a-w
Spring Est	9.3	4.9	6.9	6.9	5.7	10.7	5.5	6.1	6.7	7.0
Trend	a-c	clr	a-c	clr	clr	wmr	clr	clr	a-c	a-c

Monthly Breakdowns

2026	VICTORIA - Estimated Rainfall Expectations (mm)									
REGION	Mallee	Wimmera	Western District	Northern Country	North Central	Central	Northeast	W& S Gippsland	East Gippsland	VIC
Town Sample	Mildura	Horsham	Warrnambool	Echuca	Bendigo	Melbourne	Benalla	Traralgon	Bairnsdale	Overall
January	32	59	56	119	79	62	111	81	108	708
February	0	4	23	6	12	15	9	35	60	164
March	0	2	23	0	5	13	3	20	29	95
April	13	15	67	7	16	76	41	26	21	282
May	19	39	93	26	37	50	52	36	14	365
June	14	30	28	14	31	12	26	26	24	205
July	19	46	156	24	47	24	47	55	9	426
August	22	34	136	31	57	56	64	79	39	518
September	14	32	78	16	25	20	23	40	23	271
October	4	14	37	8	10	17	11	36	21	158
November	30	10	27	36	46	28	42	31	30	281
December	47	70	104	97	72	106	70	114	171	851
Overall	215	354	826	384	437	479	497	580	550	4324

2026	VICTORIA - Estimated Rainfall Variation from the Average (mm)									
REGION	Mallee	Wimmera	Western District	Northern Country	North Central	Central	Northeast	W& S Gippsland	East Gippsland	VIC
Town Sample	Mildura	Horsham	Warrnambool	Echuca	Bendigo	Melbourne	Benalla	Traralgon	Bairnsdale	Overall
January	10	34	17	91	48	17	68	31	58	374
February	-14	-16	-5	-14	-10	-25	-26	-5	7	-106
March	-14	-17	-22	-21	-24	-28	-34	-31	-13	-202
April	-0	-8	16	-20	-18	26	-2	-28	-35	-69
May	-3	-5	30	-12	-15	3	-9	-21	-27	-58
June	-8	-20	-56	-25	-28	-34	-43	-30	-37	-281
July	-8	-7	70	-21	-13	-21	-31	-7	-45	-82
August	-4	-18	44	-10	-3	6	-9	18	2	25
September	-14	-16	-3	-21	-25	-33	-43	-36	-33	-223
October	-22	-27	-27	-31	-37	-42	-49	-35	-39	-308
November	3	-22	-26	0	4	-35	-9	-40	-49	-172
December	16	39	53	66	29	42	28	52	109	433
Overall	-57	-81	90	-16	-91	-123	-158	-133	-101	-670

2026	VICTORIA - Estimated Rainfall Trend Variation by 10mm									
REGION	Mallee	Wimmera	Western District	Northern Country	North Central	Central	Northeast	W& S Gippsland	East Gippsland	VIC
Town Sample	Mildura	Horsham	Warrnambool	Echuca	Bendigo	Melbourne	Benalla	Traralgon	Bairnsdale	Overall
January	av	wtr	wtr	wtr	wtr	wtr	wtr	wtr	wtr	wtr
February	drr	drr	av	drr	av	drr	drr	av	av	drr
March	drr	drr	drr	drr	drr	drr	drr	drr	drr	drr
April	av	av	wtr	drr	drr	wtr	av	drr	drr	drr
May	av	av	wtr	drr	drr	av	av	drr	drr	drr
June	av	drr	drr	drr	drr	drr	drr	drr	drr	drr
July	av	av	wtr	drr	drr	drr	drr	av	drr	drr
August	av	drr	wtr	av	av	av	av	wtr	av	av
September	drr	drr	av	drr	drr	drr	drr	drr	drr	drr
October	drr	drr	drr	drr	drr	drr	drr	drr	drr	drr
November	av	drr	drr	av	av	drr	av	drr	drr	drr
December	wtr	wtr	wtr	wtr	wtr	wtr	wtr	wtr	wtr	wtr
Overall	av	av	av	av	av	drr	drr	drr	drr	drr

2026	VICTORIA - Estimated Rainfall Expectation by %									
REGION	Mallee	Wimmera	Western District	Northern Country	North Central	Central	Northeast	W& S Gippsland	East Gippsland	VIC
Town Sample	Mildura	Horsham	Warrnambool	Echuca	Bendigo	Melbourne	Benalla	Traralgon	Bairnsdale	Overall
January	45%	137%	45%	331%	151%	38%	158%	61%	116%	112%
February	-97%	-79%	-19%	-68%	-46%	-61%	-75%	-13%	13%	-39%
March	-100%	-89%	-48%	-100%	-84%	-69%	-92%	-61%	-31%	-68%
April	-2%	-36%	32%	-73%	-52%	51%	-5%	-52%	-62%	-20%
May	-14%	-11%	47%	-31%	-29%	7%	-15%	-36%	-66%	-14%
June	-35%	-40%	-67%	-64%	-47%	-74%	-63%	-53%	-61%	-58%
July	-29%	-13%	81%	-47%	-21%	-47%	-40%	-11%	-83%	-16%
August	-17%	-35%	47%	-24%	-5%	12%	-12%	29%	6%	5%
September	-50%	-33%	-4%	-57%	-49%	-62%	-65%	-47%	-59%	-45%
October	-85%	-66%	-42%	-80%	-79%	-72%	-81%	-49%	-65%	-66%
November	13%	-68%	-49%	0%	9%	-55%	-17%	-56%	-62%	-38%
December	53%	124%	105%	210%	67%	66%	66%	82%	176%	104%
Overall	-21%	-19%	12%	-4%	-17%	-20%	-24%	-19%	-16%	-13%

2026	VICTORIA - Estimated Rainfall Trend Variation by 25%									
REGION	Mallee	Wimmera	Western District	Northern Country	North Central	Central	Northeast	W& S Gippsland	East Gippsland	VIC
Town Sample	Mildura	Horsham	Warrnambool	Echuca	Bendigo	Melbourne	Benalla	Traralgon	Bairnsdale	Overall
January	wtr	wtr	wtr	wtr	wtr	wtr	wtr	wtr	wtr	wtr
February	drr	drr	av	drr	drr	drr	drr	av	av	drr
March	drr	drr	drr	drr	drr	drr	drr	drr	drr	drr
April	av	drr	wtr	drr	drr	wtr	av	drr	drr	av
May	av	av	wtr	drr	drr	av	av	drr	drr	av
June	drr	drr	drr	drr	drr	drr	drr	drr	drr	drr
July	drr	av	wtr	drr	av	drr	drr	av	drr	av
August	av	drr	wtr	av	av	av	av	wtr	av	av
September	drr	drr	av	drr	drr	drr	drr	drr	drr	drr
October	drr	drr	drr	drr	drr	drr	drr	drr	drr	drr
November	av	drr	drr	av	av	drr	av	drr	drr	drr
December	wtr	wtr	wtr	wtr	wtr	wtr	wtr	wtr	wtr	wtr
Overall	wtr	wtr	wtr	wtr	wtr	wtr	wtr	wtr	wtr	wtr

2026	VICTORIA - Estimated Average Maximum Temperatures									
Region	Mallee	Wimmera	Western District	Northern Country	North Central	Central	Northeast	West and South Gippsland	East Gippsland	VIC
Town Sample	Mildura	Horsham	Warrnambool	Echuca	Bendigo	Melbourne	Benalla	Traralgon	Bairnsdale	Overall
January	33	31	25	33	31	28	32	28	26	**30**
February	32	29	22	30	28	26	30	26	24	**28**
March	33	32	27	32	31	28	31	28	26	**30**
April	24	22	20	23	22	21	23	22	22	**22**
May	21	19	18	19	17	17	18	18	19	**18**
June	18	16	15	17	16	15	17	16	17	**16**
July	16	13	13	15	13	13	14	14	16	**14**
August	15	13	13	14	12	13	13	13	15	**13**
September	19	16	15	18	15	16	17	16	17	**17**
October	25	22	17	24	22	21	23	20	22	**22**
November	28	26	21	26	25	23	26	23	23	**24**
December	27	24	19	25	23	21	24	22	20	**23**
Overall	**24**	**22**	**19**	**23**	**21**	**20**	**22**	**21**	**20**	**21**

2026 VICTORIA - Estimated Average Minimum Temperatures

Region	Mallee	Wimmera	Western District	Northern Country	North Central	Central	Northeast	West and South Gippsland	East Gippsland	VIC
Town Sample	Mildura	Horsham	Warrnambool	Echuca	Bendigo	Melbourne	Benalla	Traralgon	Bairnsdale	Overall
January	17	12	11	15	15	17	16	13	13	**15**
February	16	12	11	15	15	17	16	13	13	**14**
March	17	12	12	14	13	16	13	11	12	**13**
April	10	7	9	9	9	12	8	8	9	**9**
May	7	5	7	5	5	10	4	8	7	**6**
June	6	4	6	5	5	9	5	5	5	**5**
July	6	5	6	5	5	9	5	5	5	**6**
August	4	2	4	2	2	7	3	4	4	**3**
September	6	3	6	4	3	9	3	4	4	**5**
October	9	4	7	7	5	10	4	6	7	**6**
November	13	9	8	11	10	13	10	8	9	**10**
December	13	9	9	11	10	13	11	10	11	**11**
Overall	10	7	8	9	8	12	8	8	8	**9**

2026 VICTORIA - Estimated Degree of Variation from the Maximum Average

Region	Mallee	Wimmera	Western District	Northern Country	North Central	Central	Northeast	West and South Gippsland	East Gippsland	VIC
Town Sample	Mildura	Horsham	Warrnambool	Echuca	Bendigo	Melbourne	Benalla	Traralgon	Bairnsdale	Overall
January	1.0	0.2	0.1	1.8	0.4	1.0	0.2	1.5	0.4	0.7
February	-0.1	-1.0	-2.5	-0.2	-1.2	-0.9	-1.1	-0.3	-1.7	-1.0
March	4.9	4.6	3.5	4.8	4.7	3.4	3.3	3.3	1.6	3.8
April	0.7	0.1	0.3	1.1	0.1	1.1	0.4	1.2	1.2	0.7
May	1.6	1.4	1.2	1.1	0.7	0.5	0.5	1.2	1.2	1.0
June	2.5	2.0	1.2	2.9	2.3	1.6	3.1	2.0	1.7	2.1
July	0.3	-0.2	0.0	1.1	0.1	0.3	0.7	0.1	1.1	0.4
August	-2.4	-2.1	-1.6	-1.8	-2.5	-1.4	-1.5	-1.6	-1.2	-1.8
September	-1.6	-1.9	-1.0	-0.1	-1.6	-0.7	-0.9	-0.6	-0.5	-1.0
October	1.0	0.1	-0.7	1.6	0.9	1.3	0.9	0.6	1.5	0.8
November	-0.2	-0.5	0.2	0.3	-0.2	1.1	-0.2	0.9	0.8	0.3
December	-3.7	-5.3	-3.5	-4.5	-4.7	-3.3	-5.1	-2.9	-3.6	-4.1
Overall	0.4	-0.2	-0.2	0.7	-0.1	0.3	0.1	0.5	0.2	0.2

2026 VICTORIA - Estimated Degree of Variation from the Minimum Average

Region	Mallee	Wimmera	Western District	Northern Country	North Central	Central	Northeast	West and South Gippsland	East Gippsland	VIC
Town Sample	Mildura	Horsham	Warrnambool	Echuca	Bendigo	Melbourne	Benalla	Traralgon	Bairnsdale	Overall
January	0.4	-0.2	-0.3	0.1	0.3	3.0	1.7	0.5	0.5	0.7
February	-0.1	-0.1	-1.1	-0.2	0.2	2.4	1.9	-0.3	0.2	0.3
March	2.7	1.3	0.5	1.5	1.5	3.1	2.5	-0.2	0.2	1.4
April	-0.5	-0.4	-0.1	-0.6	0.6	1.8	1.4	-0.4	0.0	0.2
May	-0.1	-0.7	-0.3	-1.7	-0.7	1.6	-0.2	1.2	0.4	-0.1
June	0.8	0.7	-0.2	0.2	1.8	2.3	1.9	0.4	0.4	0.9
July	1.2	1.8	0.7	1.5	2.0	3.1	2.4	1.1	0.9	1.6
August	-1.7	-1.6	-1.7	-2.5	-1.2	0.7	0.0	0.1	-0.5	-0.9
September	-1.3	-1.8	-0.9	-2.7	-1.9	1.9	-1.5	-1.4	-1.5	-1.2
October	-0.9	-2.1	-0.9	-2.1	-1.9	1.9	-2.5	-1.9	-1.0	-1.3
November	0.4	0.2	-0.6	-0.7	0.0	2.5	0.2	-1.3	-0.2	0.0
December	-2.0	-1.9	-0.9	-2.2	-2.1	0.4	-1.4	-1.2	-0.7	-1.3
Overall	-0.1	-0.4	-0.5	-0.8	-0.1	2.1	0.5	-0.3	-0.1	0.0

2026 VICTORIA - Estimated Maximum Temperature Trend

Region	Mallee	Wimmera	Western District	Northern Country	North Central	Central	Northeast	West and South Gippsland	East Gippsland	VIC
Town Sample	Mildura	Horsham	Warrnambool	Echuca	Bendigo	Melbourne	Benalla	Traralgon	Bairnsdale	Overall
January	a-w	av	av	wmr	av	a-w	av	wmr	av	a-w
February	av	a-c	clr	av	clr	a-c	clr	av	clr	a-c
March	wmr	wmr	wmr	wmr	wmr	wmr	wmr	wmr	wmr	wmr
April	a-w	av	av	wmr	av	wmr	av	wmr	wmr	a-w
May	wmr	wmr	wmr	wmr	a-w	av	av	wmr	wmr	a-w
June	wmr	wmr	wmr	wmr	wmr	wmr	wmr	wmr	wmr	wmr
July	av	av	av	wmr	av	av	a-w	av	wmr	av
August	clr	clr	clr	clr	clr	clr	clr	clr	clr	clr
September	clr	clr	a-c	av	clr	a-c	a-c	a-c	av	a-c
October	a-w	av	a-c	wmr	a-w	wmr	a-w	a-w	wmr	a-w
November	av	av	av	av	av	wmr	av	a-w	a-w	av
December	clr	clr	clr	clr	clr	clr	clr	clr	clr	clr
Overall	av	av	av	a-w	av	av	av	av	av	av

2026 VICTORIA - Estimated Minimum Temperature Trend

Region	Mallee	Wimmera	Western District	Northern Country	North Central	Central	Northeast	West and South Gippsland	East Gippsland	VIC
Town Sample	Mildura	Horsham	Warrnambool	Echuca	Bendigo	Melbourne	Benalla	Traralgon	Bairnsdale	Overall
January	av	av	av	av	av	wmr	wmr	av	av	a-w
February	av	av	clr	av	av	wmr	wmr	av	av	av
March	wmr	wmr	av	wmr	wmr	wmr	wmr	av	av	wmr
April	av	av	av	a-c	a-w	wmr	wmr	av	av	av
May	av	a-c	av	clr	a-c	wmr	av	wmr	av	av
June	a-w	a-w	av	av	wmr	wmr	wmr	av	av	a-w
July	wmr	wmr	a-w	wmr	wmr	wmr	wmr	wmr	a-w	wmr
August	clr	clr	clr	clr	clr	a-w	av	av	av	a-c
September	clr	clr	a-c	clr	clr	wmr	clr	clr	clr	clr
October	a-c	clr	a-c	clr	clr	wmr	clr	clr	a-c	clr
November	av	av	a-c	a-c	av	wmr	av	clr	av	av
December	clr	clr	a-c	clr	clr	av	clr	clr	a-c	clr
Overall	av	av	av	a-c	av	wmr	av	av	av	av

TASMANIA 2026

GENERAL TRENDS

Year Overall: Drier by around 10-15% below the average
Slightly warmer but within 0.5°C of the average

Summer: Drier by around 20-25% below the average
Wettest – New Year to the end of the first week of January
Driest – Second and third weeks of January
Warmer by around 0.5-1°C above the average
Hottest – Second and third weeks of January
Coolest – First week of January and mid-February

Autumn: Drier by around 10-15% below the average
Wettest – First half of April
Driest – Second half of March
Warmer by up to 1.5°C above the average
Hottest – Second half of March
Coolest – Second week of May

Winter: Drier by around 15-20% below the average
Wettest – First half of July
Driest – Mid-June
Slightly cooler but within 0.5°C of the average
Hottest – Second to third weeks of both June and July
Coolest – First weeks of June and August

Spring: Drier by around 20-25% below the average
Wettest –First half of October
Driest – First half of November
Slightly cooler but within 0.5°C of the average
Hottest – First and fourth weeks of November
Coolest – First half of September

Major Weather Events

January 2026

Date	Day	Tasmania - January 2026	Hobart	Moon
1st	Thur	Heavy rain likely in the Northwest region, particularly in the Erriba vicinity.	Show ers	
2nd	Fri		Show ers	P6
3rd	Sat	Colder than of recent in the south and west of the State.	Show ers	F ^
4th	Sun			
5th	Mon		Show ers	
6th	Tue			
7th	Wed	Very warm spell across the State over the next couple of days, with many areas seeing maximum temperatures into the 30's.		
8th	Thu			
9th	Fri			XhS
10th	Sat	Unusually high maximas into the high 30's in the Central Highlands to some parts of the West Coast likely.		
11th	Sun	Very hot over the next couple of weeks with daily maximas into the 30's in many parts of the State.		3Q
12th	Mon			
13th	Tue			
14th	Wed			A
15th	Thur			
16th	Fri	Hot spell across the State, particularly in the Southeast region.		
17th	Sat			V
18th	Sun			
19th	Mon			N
20th	Tue	Very hot on the Central Plateau with chance of maximum temperatures in the upper 30's possible.		
21st	Wed	High maximum temperatures persist along the northern coasts. Strong unsettled winds around Mt Wellington, with chance of reaching gale force gusts at times.		
22nd	Thu			
23rd	Fri	Showers across the State.		
24th	Sat	Cooler spell becomes established around this time and persists for the next week to 10 days.		XhN
25th	Sun			
26th	Mon			1Q
27th	Tue			
28th	Wed			
29th	Thur	Wet spell across most of the State over the next couple of days with some heavy falls in the northern parts of the East Coast region.	Show ers	
30th	Fri	Heavy rain in the Northeast region.	Show ers	P10 ^
31st	Sat	Unusually colder than normal in the Central Plateau to the Southeast, particularly on Mt Wellington, and chance of frosts around Lake St Clair.		

February 2026

Date	Day	Tasmania - February 2026	Hobart	Moon
1st	Sun			
2nd	Mon	Warmer change.		F
3rd	Tue			
4th	Wed			
5th	Thu		Show ers	
6th	Fri	Very hot on Flinders Island. Rainy spell across the State over the next 3-4 days.		XhS
7th	Sat			
8th	Sun			
9th	Mon			
10th	Tue	Cool change.		3Q
11th	Wed	Chance of frosts overnight on the Central Plateau and in the Southeast regions.		A
12th	Thur			
13th	Fri			V
14th	Sat	Mild overnight temperatures.		

Date	Day	Tasmania - February 2026	Hobart	Moon
15th	Sun	Very wet spell across the State, with very heavy falls and possible localised flooding in the Northeast region. Unusually warm overnight on Flinders Island.	Show ers	
16th	Mon	Heavy rain in the Central North to the Midlands regions.	Rain	
17th	Tue	Unusually cool spell over the State in the coming week. Heavy rain in the Northern region.	Show ers	
18th	Wed		Show ers	N
19th	Thu			
20th	Fri			XhN
21st	Sat			
22nd	Sun		Show ers	
23rd	Mon	Wet spell across the State. Chance of frost or snow flurries around Lake Leake.		
24th	Tue			
25th	Wed			1Q P14
26th	Thur			^
27th	Fri	Very warm spell across the State over the next few days, with temperatures in the 30°C range in many areas.		
28th	Sat	Unusually mild overnight temperatures.		

March 2026

Date	Day	Tasmania - March 2026	Hobart	Moon
1st	Sun	Very hot wave across the State, particularly in the Derwent Valley where maximum temperatures may reach into the upper 30's.	Rain	
2nd	Mon			
3rd	Tue	Unusually cool spell over the State in the coming week.		F
4th	Wed		Rain	
5th	Thu	A cold front crosses the State over the next couple of days, bringing cooler temperatures throughout. Chance of snow falling below 900 metres, particularly on the Central Plateau. Very strong winds with gale gusts likely across Hogan Island.	Rain	XhS
6th	Fri			
7th	Sat		Show ers	
8th	Sun	Heavy rain in the Northeast region with the chance of localised flooding.	Show ers	
9th	Mon	Unusually warm in the Derwent Valley over the next couple of days.	Show ers	
10th	Tue	Colder than normal across much of the State with many areas not seeing maximum temperatures into the 20's.	Show ers	
11th	Wed			3Q A
12th	Thur			V
13th	Fri			
14th	Sat	Cold front crosses the State bringing widespread scattered outbreaks of rain.	Show ers	
15th	Sun			
16th	Mon	Chance of light snow in the Central Highlands over the next couple of days.		
17th	Tue	A warmer than average spell across the State over the next two weeks.		
18th	Wed			
19th	Thu			N XhN
20th	Fri			
21st	Sat	Chance of snowflurries in the Central Highlands. In contrast, unusually warmer than normal elsewhere with many areas maximum temperatures possibly reaching into the mid-30's.		
22nd	Sun			P11
23rd	Mon			
24th	Tue			
25th	Wed	Very hot day across the State with many places possibly seeing maximum temperatures in the upper 30's. Very strong winds across Maatsuyker Island. Very hot in Hobart, with maximum temperatures possibly reaching to almost 40°C.		
26th	Thur			1Q ^
27th	Fri	Warmer spell over the next couple of days, with many areas maximum temperatures possibly reaching into the mid-30's		
28th	Sat	Very hot on Maatsuyker Island.		
29th	Sun	Wet spell across the State. Unusually mild temperatures on Hogan Island.	Show ers	
30th	Mon	Cold change, particularly over the next 3-4 days. Frosts and possible light snow on the Central Highlands.		
31st	Tue			

April 2026

Date	Day	Tasmania - April 2026	Hobart	Moon
1st	Wed			
2nd	Thu			F XhS
3rd	Fri	Bands of thunderstorms cross the State over the next 3-4 days.		
4th	Sat	Brief respite from cooler change with the next couple of days warmer, with daily maximas possibly back up into the high 20's.	Showers	
5th	Sun		Showers, Th	
6th	Mon	Cooler temperatures return.	Rain	
7th	Tue	Unusually cold day on Mt Wellington.	Showers	A
8th	Wed	A cool, southwesterly change prevails over the State during the next 3-4 days, with outbreaks of moderate to heavy rain at times, particularly in the west and south. Chance of overnight frost level temperatures around Lake Leake in the East Coast region.		
9th	Thur			V
10th	Fri		Showers	3Q
11th	Sat	Chance of snowflurries in the Central Highlands and frosts on Mt Wellington and on the higher reaches of the West Coast ranges including Mt Read area. Heavy rain on the West Coast, with particularly heavy falls in the Mount Read area.		
12th	Sun	A mostly settled spell over the next 3-4 weeks as high pressure systems dominate, with just a couple of low pressure interruptions.		
13th	Mon	An intense depression near Tasmania brings strong and damaging winds across the State and through the Bass Strait over the next couple of days, particularly in the Southeast.	Rain	
14th	Tue	Widespread rain across the State with heavy falls at times in the north and west.	Showers	
15th	Wed	Wet westerlies cross the State over the next few days. Very wet in the West Coast, particularly around Mt Read.		
16th	Thu			XhN
17th	Fri	High pressure and mostly settled conditions over the next 3-4 days with mostly light breezes and mostly dry and cool flow.		N
18th	Sat	Chance of snowflurries in the Central Highlands.		
19th	Sun			P7
20th	Mon	A series of cold fronts cross the State over the next 3-4 days with some rain mostly in the west. Chance of light snow in the Central Highlands. In contrast, it may be unusually warm in many places seeing maximum temperatures for a time in the low to mid- 20's.	Showers	
21st	Tue			
22nd	Wed	Chance of light snow in the Central Highlands. In contrast, very warm on King Island.	Showers	^
23rd	Thur			
24th	Fri			1Q
25th	Sat	High pressure and mostly settled conditions over the next 7-10 days with mostly light breezes and mostly dry and cool flow. Chance of heavy snow in the Central Highlands and snowflurries into the Mt Field National Park region including on Mt Mawson.		
26th	Sun			
27th	Mon			
28th	Tue			
29th	Wed			XhS
30th	Thur			

May 2026

Date	Day	Tasmania - May 2026	Hobart	Moon
1st	Fri			
2nd	Sat			F
3rd	Sun		Fg	
4th	Mon			
5th	Tue			A
6th	Wed	Southerly winds bring a cold and unsettled change with spells of rain over the next three days, and particularly wet in the Western Highlands. Mostly dry in the eastern half of the State. Unusually warm on King Island.	Showers, Fg	V
7th	Thu			
8th	Fri			
9th	Sat	Frosts on Mt Wellington.	Showers	
10th	Sun			3Q

Date	Day	Tasmania - May 2026	Hobart	Moon
11th	Mon	Scattered frosts across the State.		
12th	Tue	Cold fronts may regularly cross the State over the coming 3-4 weeks, but with most precipitation mainly confined to the west. Overall, daily maximum temperatures may be above average on most days in the next 3 weeks.		
13th	Wed			XhN
14th	Thur			
15th	Fri		Show ers	
16th	Sat			
17th	Sun		Show ers	N
18th	Mon		Show ers	P3
19th	Tue			^
20th	Wed	Chance of light snow in the Central Highlands. In contrast, warm daily maximas are likely over the next couple of days, particularly in the north and east.	Fg	
21st	Thu	Unusually warm overnight on Flinders Island.		
22nd	Fri			
23rd	Sat			1Q
24th	Sun			
25th	Mon			
26th	Tue			XhS
27th	Wed			
28th	Thur	Unusually mild temperatures, particularly in the south and west.		
29th	Fri	Chance of light snow in the Central Highlands.		
30th	Sat	Very strong winds with severe gusts in the Southeast, particularly over Mt Wellington.		
31st	Sun		Show ers	F

June 2026

Date	Day	Tasmania - June 2026	Hobart	Moon
1st	Mon	Cooler spell across the State over the coming week with chance of frequent frosts in many districts over the next 3-4 days.		A
2nd	Tue	A high pressure system follows a depression and brings unusually cold temperatures across the State with many areas possibly seeing overnight frosts and very low minimum temperatures. Severe frosts at Lake Leake.		V
3rd	Wed		Fg	
4th	Thu		Fg	
5th	Fri			
6th	Sat	A slow moving trough crosses the Northwest to the King Island regions over the next couple of days, producing thunderstorms, heavy rain, damaging winds and scattered hail showers in some areas. Very heavy rain in the Elliot district and chance of flash flooding in the Wynyard and Burnie areas.	Show ers	
7th	Sun	Heavy rain moves through the Northern coastal region.		
8th	Mon		Show ers	3Q
9th	Tue		Show ers	XhN
10th	Wed	Unusually warm maximum temperatures in most districts.		
11th	Thur	Snowflurries likely in the Central Highlands over the next couple of days.	Fg	
12th	Fri	Generally a warmer than normal spell over the coming week as a high pressure system in the Tasman Sea directs a milder, northerly flow.		
13th	Sat			
14th	Sun		Fg	
15th	Mon	Unusually milder than average temperatures overnight.		N P2
16th	Tue			^
17th	Wed	Unusually warmer daytime temperatures but colder than average overnight in most districts over the next 2-3 days.	Show ers	
18th	Thu	A cold front crossing from the west brings colder temperatures across the State. A westerly flow may dominate much of the next 3 weeks.		
19th	Fri			
20th	Sat			
21st	Sun	A series of cold fronts cross the State over the next 3-4 days bringing widespread rain activity to most areas, and some heavy rain at times in the Northeast. Unusually milder than average temperatures overnight likely over the next 2-3 days.	Show ers	

Date	Day	Tasmania - June 2026	Hobart	Moon
22nd	Mon			1Q XhS
23rd	Tue			
24th	Wed	Downpours likely in the Northeast region. Colder in most parts of the State.	Show ers	
25th	Thur	Very cold overnight in many districts over the next 3-4 days.		
26th	Fri			
27th	Sat	Very cold on the Central Plateau with severe frosts.	Fg	
28th	Sun	Very cold in the Derwent Valley with thick fog prevailing.		A
29th	Mon	Milder northwestery flow prevails over the State ahead of a front.		V
30th	Tue	Unusually milder than average temperatures overnight.	Fg	F

July 2026

Date	Day	Tasmania - July 2026	Hobart	Moon
1st	Wed	Very strong westerly winds prevail over the coming week, occasionally reaching storm-force at times, and gales likely in many areas.		
2nd	Thu	A cold front crossing from the west brings colder temperatures across the State. Severe frosts in the southeast likely, particularly on Mt Wellington.		
3rd	Fri	Very cold overnight in many districts over the next 3-4 days with scattered local frosts in many districts, but particularly in the east and southeast.		
4th	Sat			
5th	Sun	Light snow in the Central Highlands.		
6th	Mon	Snowflurries in the Central Highlands.	Show ers	
7th	Tue	A cold front crossing from the west brings colder temperatures across the State and very strong winds through the Bass Strait with severe gale gusts over Hogan Island. Wet spell across the State with heavy falls at times in the Southeast. Chance of snow reaching levels below 500 metres in some places.	Rain	XhN
8th	Wed	Very heavy rain likely in west of the State, particularly in the Northwest and Cradle Valley district. Unusually mild maximum temperatures in the Northeast region.		3Q
9th	Thur			
10th	Fri	Northwesterly winds ahead of a front brings milder temperatures across the State over the next couple of days.		
11th	Sat		Show ers	
12th	Sun	Very west across the State today. Some rain activity can be expected somewhere in the State almost daily over the next four weeks as the passage of several low pressure systems may slowly pass across and widespread scattered frosts may be frequent. Strong and unsettled winds over the next few days.	Rain	
13th	Mon	Cold spell. Light snow in the Central Highlands. Strong gusty winds to gale force at times on Maatsuyker Island.	Show ers	P4 ^
14th	Tue	Snowflurries in the Central Highlands.	Show ers	N
15th	Wed	Very cold across the State over the next few days. Severe frosts in the Central Plateau.		
16th	Thu	Chance of snow in Hobart and severe frost temperatures in the Central Plateau.		
17th	Fri			
18th	Sat	Heavy snow in the Central Highlands, light snow on Mt Mawson, and light snow on Mt Wellington. Milder overnight temperatures on Hogan Island.	Rain, Fg	
19th	Sun	Snow in the Central Highlands. Milder temperatures on the East Coast.	Fg	
20th	Mon	Milder temperatures around this time.	Show ers	XhS
21st	Tue	Cold southerly winds prevail over the next 2-3 days.	Show ers	1Q
22nd	Wed	Light snow in the Central Highlands. Heavy rain in the West Coast.		
23rd	Thur	Heavy snow in the Central Highlands over the next couple of days, and light snow on Mt Mawson.		
24th	Fri	Snow on Mt Mawson.		
25th	Sat	Light snow in the Central Highlands. Widespread patches of fog likely over the next 3 days.	Show ers	
26th	Sun		Show ers, Fg	A
27th	Mon		Fg	V
28th	Tue		Fg	
29th	Wed	Low pressure system crosses the State bringing widespread outbreaks of rain.		
30th	Thu	A cold southerly air prevails over the State. Very cold temperatures over the next 8-10 days.		F
31st	Fri		Show ers	

August 2026

Date	Day	Tasmania - August 2026	Hobart	Moon
1st	Sat	A wet front crosses the State bringing strong and blustyer winds and widespread outbreaks of precipitation and very cold temperatures. Light snow in the Central Highlands and snowflurries down to 2000 meter level in the south of the State as well as snowflurries in the Mt Field National Park including on Mt Mawson.	Showers, Sn	
2nd	Sun	Heavy snow in the Central Highlands, light snow on Mt Welllington and in the Mt Field National Park region including Mt Mawson, and light snow at Ben Lomond.		
3rd	Mon	Snowflurries in the Central Highlands over the next couple of days.		XhN
4th	Tue			
5th	Wed			
6th	Thur		Showers, Fg	3Q
7th	Fri		Showers	
8th	Sat	Very cold across the State over the next 3-4 days.	Showers	
9th	Sun	Light snow in the Central Highlands over the next few days.		^
10th	Mon	Very cold with severe frosts on the Central Plateau.		P8
11th	Tue		Showers	
12th	Wed	A cold south to southwest flow prevails over the State. Snow in the Central Highlands and also on Mt Mawson and the Mt Field National Park over the next couple of days. Snowflurries possible on Mt Wellington.		
13th	Thu	Widespread scattered frosts throughout the coming 3-4 weeks.		N
14th	Fri			
15th	Sat			
16th	Sun			XhS
17th	Mon			
18th	Tue			
19th	Wed	A cold airmass lingers over the State with south to southwest winds prevailing for the next 8-10 days with snow falling in some part of the State on each day.	Showers	
20th	Thur			1Q
21st	Fri	A cold south to southwest flow prevails over the State during the coming week. Very cold spell over the next few days.		
22nd	Sat	Light snow in the Central Highlands, and snow flurries at Ben Lomond.	Showers	A
23rd	Sun	Very cold in the south of the State over the next 3-4 days. A cold front crossing the State brings snowflurries in the Central Highlands.	Showers	V
24th	Mon	Light snow in the Central Highlands. Severe frosts on Mt Wellington.	Rain	
25th	Tue	Heavy rain in the West Coast and Central Plateau regions, particularly in the Warra district.	Showers	
26th	Wed	Very cold in the Southeast particularly on Mt Wellington.	Showers	
27th	Thu		Showers	
28th	Fri		Showers	F
29th	Sat	Unusually cold day.		
30th	Sun	Light snow in the Central Highlands over the next couple of days. Very strong and gusty winds in the Southeast, particularly around Mt Wellington.	Showers	XhN
31st	Mon			

September 2026

Date	Day	Tasmania - September 2026	Hobart	Moon
1st	Tue	A cold south to southwest flow develops over the State. Chance of significant snow events around this time.	Showers	
2nd	Wed	Very cold spell in the State over the next 3-4 days. Snowflurries and severe frost temperatures in the Central Highlands.		
3rd	Thur	Snow in the Mt Field National Park including on Mt Mawson and heavy snow in the Central Highlands over the next few days, and light snow on Mt Wellington over the next couple of days.		
4th	Fri			3Q
5th	Sat	Snow in the Central Highlands daily over the coming week.		
6th	Sun			^
7th	Mon	Light snow likely in the Mt Field National Park, including on Mt Mawson.		P12
8th	Tue			
9th	Wed			
10th	Thu	Light snow or snowflurries in the Mt Field National Park including on Mt Mawson over the next few days.	Showers	

Date	Day	Tasmania - September 2026	Hobart	Moon
11th	Fri	Milder spell through the Southeast and East Coast regions as well as on Flinders Island.		N
12th	Sat	Wet spell across the State. Light snow or snowflurries in the Mt Field National Park including on Mt Mawson.	Showers	XhS
13th	Sun	Very cold across the State for the next couple of days.	Sn	
14th	Mon	Very cold in the Midlands to the eastern coasts over the next couple of days, with widespread frosts in the Northern, Eastern and Midlands regions.		
15th	Tue	Some patches of severe frosts around the East Coast region.		
16th	Wed			
17th	Thur	Light snow in the Central Highlands over the next couple of days.		
18th	Fri		Showers	
19th	Sat	Very cold in the State.	Showers	1Q A V
20th	Sun			
21st	Mon		Showers	
22nd	Tue	Brief spell of milder temperatures over the next few days.	Showers	
23rd	Wed	The passage of an active cold front brings strong winds and rain across the State over the next few days. Snow in the Central Highlands over the next couple of days.	Showers	
24th	Thu	Chance of snowflurries on Mt Mawson.		
25th	Fri		Showers	
26th	Sat	Wet spell across the State.	Showers	XhN
27th	Sun	Very cold across the State with very strong winds in the southwest of the State. Very heavy rain likely in the West Coast to adjacent Highlands.		F
28th	Mon			
29th	Tue	Light snow in the Central Highlands over the next couple of days.		
30th	Wed	Chance of snowflurries on Mt Mawson. Brief spell of milder temperatures over the next few days.		

October 2026

Date	Day	Tasmania - October 2026	Hobart	Moon
1st	Thur	A series of wet fronts cross the State over the next few days bringing blustery winds and widespread scattered rain.	Rain	
2nd	Fri			P13
3rd	Sat	An active cold front brings scattered precipitation across the State. Chance of snowflurries in the Central Highlands.	Showers	
4th	Sun			3Q ^
5th	Mon			
6th	Tue	Light snow in the Central Highlands over the next couple of days.		
7th	Wed	Unseasonal mild temperatures likely around this time for a couple of days. Strong winds may cause some damage particularly in the southeast of the State.	Showers	
8th	Thu	Snow in the Central Highlands, and light snow on Mt Mawson.		
9th	Fri		Showers	
10th	Sat		Showers	XhS
11th	Sun			N
12th	Mon	Snow in the Central Highlands over the next 3-4 days, and light snow on Mt Mawson. Chance of snowflurries at Ben Lomond.	Showers	
13th	Tue			
14th	Wed		Showers	
15th	Thur	A cooler change over the coming week.		
16th	Fri			
17th	Sat	A wet front crosses the State bringing some moderate rain mostly over the western half of the State. Heavy rain in the Strathgordon area.	Showers	A V
18th	Sun	A cold front crosses the State with light snow down to 700 metres level in the South and West including in the Central Highlands and on Mt Mawson. Strong winds to gale force at times likely in the southeast of the State, particularly around Mt Wellington.	Showers	
19th	Mon			1Q
20th	Tue	Snow in the Central Highlands over the next few days and light snow on Mt Mawson over the next couple of days.		
21st	Wed	Warmer change with temperatures steadily building over the coming week.		
22nd	Thur		Fg	
23rd	Fri	A wet front crosses the State.	Showers	
24th	Sat			XhN

Date	Day	Tasmania - October 2026	Hobart	Moon
25th	Sun	Light snow in the Central Highlands over the next couple of days.		
26th	Mon			F
27th	Tue			
28th	Wed	Snow in the Central Highlands over the next few days.	Showers	
29th	Thu	Chance of snowflurries on Mt Mawson. Milder temperatures over the next couple of days.		P9
30th	Fri			^
31st	Sat			

November 2026

Date	Day	Tasmania - November 2026	Hobart	Moon
1st	Sun	A cold change with the passage of a wet front and brings widespread outbreaks of rain across the State. Cooler particularly in the south and west.	Rain	
2nd	Mon			3Q
3rd	Tue	Chance of severe overnight frosts in some parts of the State.		
4th	Wed			
5th	Thur			
6th	Fri	Milder temperatures over the next couple of days.		XhS
7th	Sat	High pressure directs a warmer airflow over the State with chance of maximum temperatures briefly reaching into the low 30's in some areas. Unusually warmer on Flinders Island.	Th	
8th	Sun			
9th	Mon			N
10th	Tue			
11th	Wed		Showers	
12th	Thu	Light snow in the Central Highlands over the next couple of days. Chance of snowflurries on Mt Mawson.		
13th	Fri			V
14th	Sat	Cold southerly winds prevail over the next 2-3 days.	Showers	A
15th	Sun			
16th	Mon	Very cold overnight on the Central Plateau with severe frosts likely.		
17th	Tue		Showers, Th, Fg	
18th	Wed	A wet front crosses the State over the next couple of days with heavy rain at times in the Northern region, and West Coast and adjacent Highlands.	Rain	1Q
19th	Thur	Very heavy rain in the West Coast, particularly in the Preolenna area. Very strong and blustery winds around Mt Wellington.		
20th	Fri	Average temperatures prevail over the next 4-5 days.		XhN
21st	Sat			
22nd	Sun		Showers	
23rd	Mon			
24th	Tue	Unusually hot day in the southeast of the State, including in Hobart, where maximum temperatures may reach into the low 30's.		
25th	Wed	Chance of snowflurries in the Central Highlands.		F
26th	Thu	Cold southerly winds prevail over the next 2-3 days.	Showers	P5
27th	Fri			^
28th	Sat			
29th	Sun			
30th	Mon		Showers	

December 2026

Date	Day	Tasmania - December 2026	Hobart	Moon
1st	Tue	A series of low pressure systems pass near the State over the next few days, bringing outbreaks of rain and scattered thunderstorms, with heavy rain at times in the central and eastern districts.	Showers	3Q
2nd	Wed		Rain	
3rd	Thur	Cold southerly winds cross the State over the next 3-4 days.	Showers	XhS
4th	Fri		Showers	
5th	Sat	Very cold on Mt Wellington.		
6th	Sun			

Date	Day	Tasmania - December 2026	Hobart	Moon
7th	Mon	Snow in the Central Highlands and light snow in the Mt Field National Park including on Mt Mawson over the next couple of days.		
8th	Tue		Showers, Th	
9th	Wed	A wet front brings scattered thunderstorms across the State, with particularly heavy rain in the north.	Showers	N
10th	Thu	Cold winds in a southerly flow prevail.		V
11th	Fri		Showers	A
12th	Sat	Cooler than average for this time of year over the next three weeks with many areas maximum temperatures only reaching into the late teens to early 20's. A cold front crosses the State over the next 2-3 days.	Showers	
13th	Sun	Light snow in the Central Highlands over the next couple of days.	Showers	
14th	Mon		Showers	
15th	Tue	A clear airmass and light winds over the coming week brings unusually cold overnight temperatures. Chance of frosts in sheltered areas in the Cressey to Lake Leake areas as well as in the Southeast and the highlands of the West Coast.	Showers	
16th	Wed		Showers	
17th	Thur			1Q XhN
18th	Fri	Cold day particularly in the West Coast region.		
19th	Sat			
20th	Sun	A band of thunderstorms cross the State.	Showers	
21st	Mon	Cold front crosses the State. Chance of frosts in the Central Plateau.		
22nd	Tue			
23rd	Wed	A low pressure system passes near the northwest of the State and over Flinders Island during the next couple of days. Storm activity with gusty winds and outbreaks of rain in many districts, particularly in the northeast. Chance of snowflurries in the Central Highlands over the next couple of days.	Showers	
24th	Thu	The passage of a low pressure system brings widespread rain across the State with chance of localised flooding in the Northwest to the Northeast regions.	Rain	P1 F ^
25th	Fri	Chance of light snow in the Central Highlands over the next couple of days. Heavy rain in the past few days may have caused some moderate flooding on the South Eask River catchment by this time as well as in some pockets on Flinders Island. Heavy rain in the East Coast, particularly around in the St Mary area.		
26th	Sat			
27th	Sun	Thunderstorms cross the State.	Showers	
28th	Mon			
29th	Tue	Chance of snowflurries in the Central Highlands and scattered thunderstorms across the State.	Rain, Th	
30th	Wed			XhS
31st	Thur	A storm in the Central North and Midlands regions brings heavy rain. Scattered thunderstorms in most districts.		3Q

TASMANIA GRAPHS AND TABLES 2026

Seasonal Breakdown – State Overall Rainfall Timing and Estimated mm Trend Graphs
State Overall Maximum Temperature Trend Graphs
Regional Rainfall Expectations Tables - mm and Trends
Regional Maximum and Minimum Temperature Trends Tables

Monthly Breakdown - **Rainfall:**
Regional Expectations by mm
Regional Expectations by mm Variance to Average
Regional Variance Trend by mm
Regional Estimated Expectations by % of Variance to Average
Regional Estimatedl Variance Trend by %

Maximum & Minimum Temperature Trends:
Regional Estimated Average Temperature
Regional Estimated Degree of Variation from the Average
Regional Estimated Temperature Trend

Seasonal Breakdowns

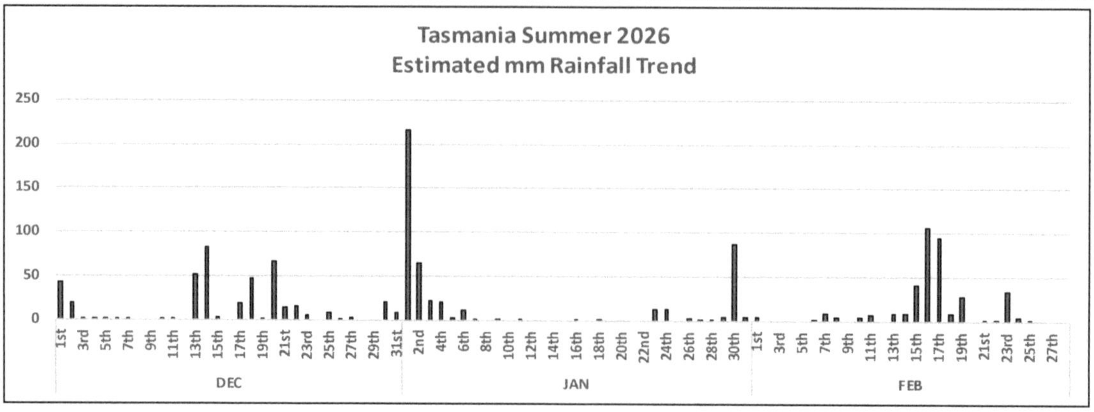

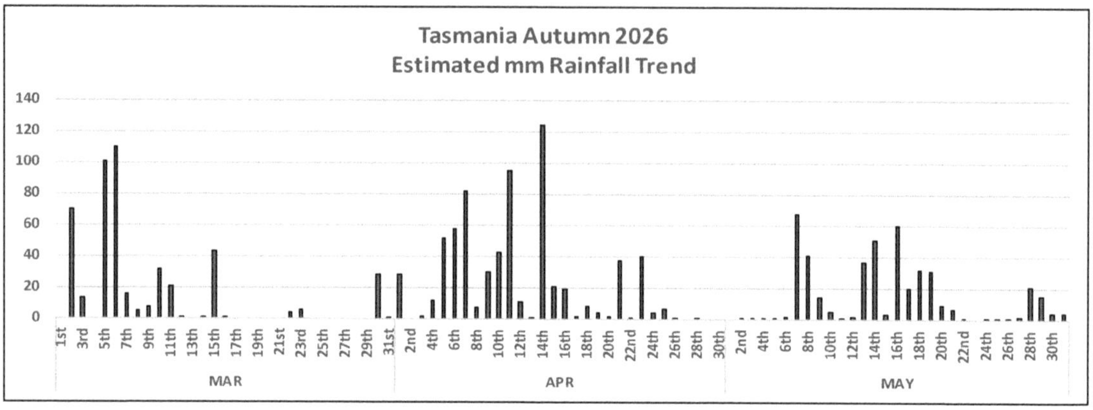

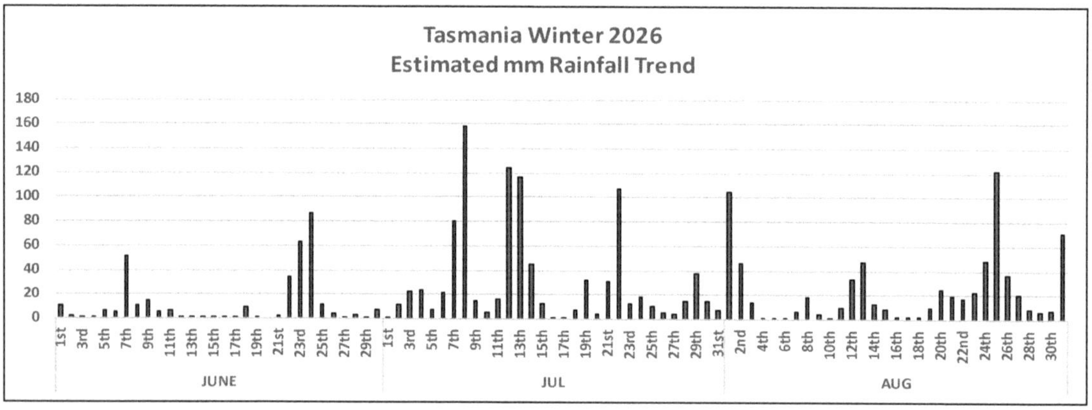

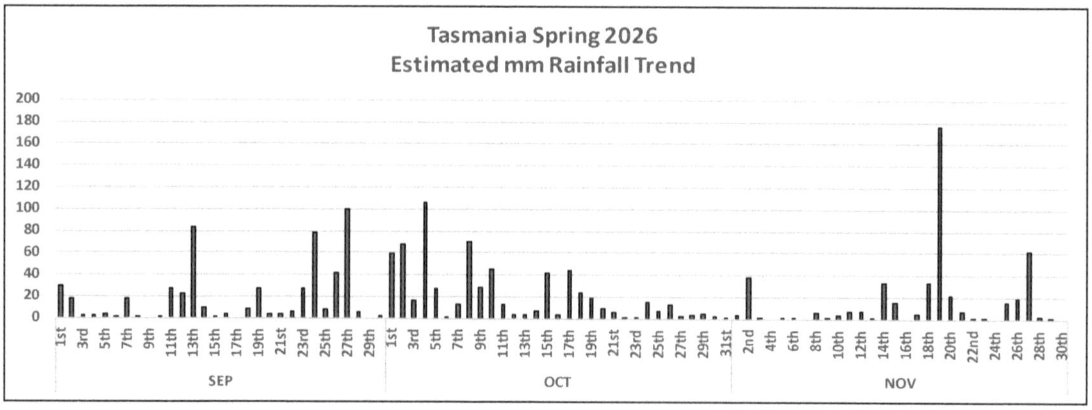

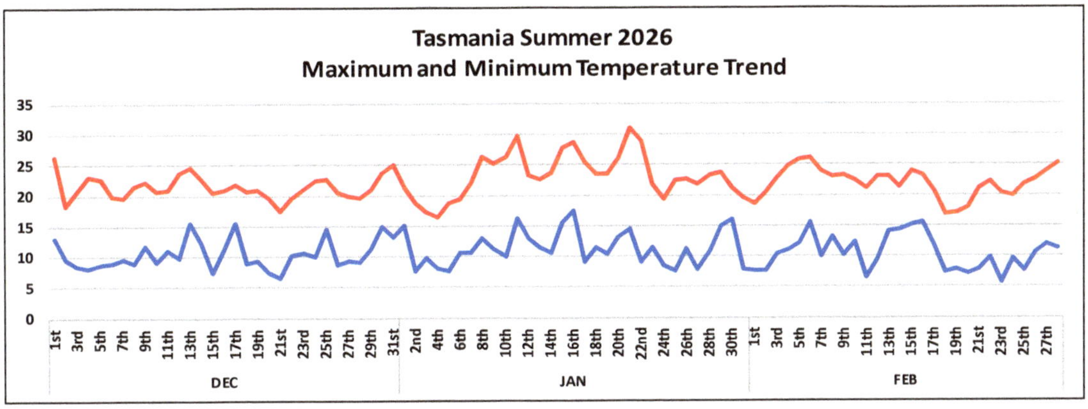
Tasmania Summer 2026
Maximum and Minimum Temperature Trend
DEC
JAN
FEB

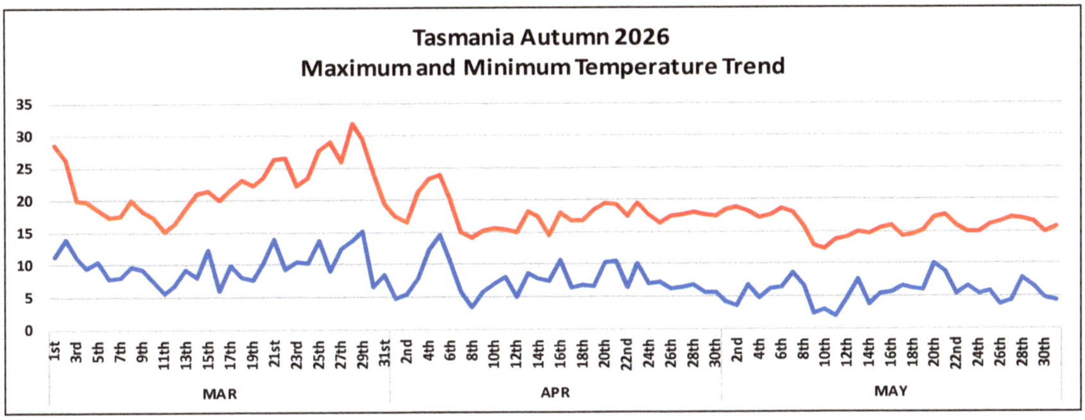
Tasmania Autumn 2026
Maximum and Minimum Temperature Trend
MAR
APR
MAY

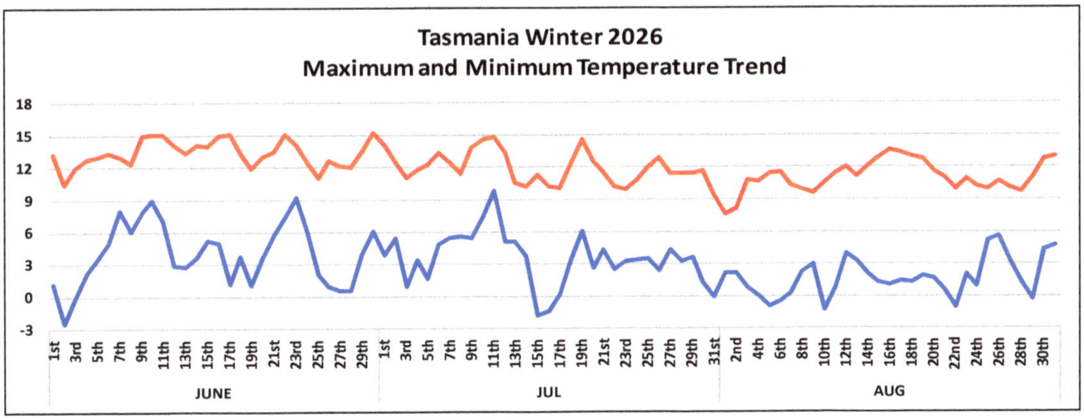
Tasmania Winter 2026
Maximum and Minimum Temperature Trend
JUNE
JUL
AUG

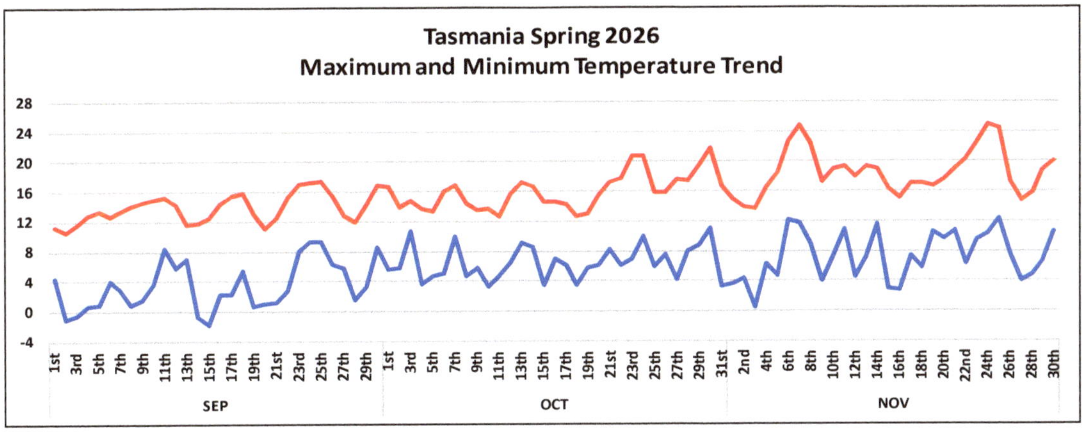
Tasmania Spring 2026
Maximum and Minimum Temperature Trend
SEP
OCT
NOV

2026 TAS - Estimated Seasonal Rainfall Expectations (mm)

REGION	NW Coast & King Is	W & S Coasts & Highlands	Central Plateau	Upper Derwent	South East	Central North	Lower Derwent	Midlands Alpine	East Coast	NE Coast & Flinders Is	TAS
Town Sample	Devonport	Strathan	Lake St Clair	Bushy Park	Hobart	Launceston	Huonville	Fingal	Swansea	Killiecrankie	Overall
Summer	124	186	172	83	110	149	100	139	96	80	1239
Autumn	125	445	371	69	97	100	114	86	55	110	1571
Winter	209	372	532	134	121	165	168	84	98	150	2033
Spring	130	361	496	95	79	139	125	85	56	88	1653
Overall	599	1417	1649	402	434	608	555	478	424	606	7173

2026 TAS - Estimated Seasonal Rainfall Variation from the Average (mm)

REGION	NW Coast & King Is	W & S Coasts & Highlands	Central Plateau	Upper Derwent	South East	Central North	Lower Derwent	Midlands Alpine	East Coast	NE Coast & Flinders Is	TAS
Town Sample	Devonport	Strathan	Lake St Clair	Bushy Park	Hobart	Launceston	Huonville	Fingal	Swansea	Killiecrankie	Overall
Summer	0	-60	-130	-42	-15	29	-55	-9	-60	-54	-395
Autumn	-43	63	-10	-62	-4	-32	-31	-43	-65	-48	-275
Winter	-43	-143	-16	-17	3	-36	-32	-76	-20	-60	-439
Spring	-62	-4	-13	-68	-48	-19	-79	-61	-76	-77	-507
Overall	-136	-90	-91	-167	-35	-4	-148	-105	-102	-61	-939

2026 TAS - Estimated Seasonal Rainfall Trend Variation by 10mm

REGION	NW Coast & King Is	W & S Coasts & Highlands	Central Plateau	Upper Derwent	South East	Central North	Lower Derwent	Midlands Alpine	East Coast	NE Coast & Flinders Is	TAS
Town Sample	Devonport	Strathan	Lake St Clair	Bushy Park	Hobart	Launceston	Huonville	Fingal	Swansea	Killiecrankie	Overall
Summer	av	drr	drr	drr	av	av	drr	av	drr	drr	drr
Autumn	drr	wtr	av	drr	av	drr	drr	drr	drr	drr	drr
Winter	drr	drr	av	av	av	drr	drr	drr	av	drr	drr
Spring	drr	av	av	drr	drr	av	drr	drr	drr	drr	drr
Overall	drr	av	av	drr	av	av	drr	drr	drr	av	drr

2026 TAS - Estimated Seasonal Rainfall Expectations by %

REGION	NW Coast & King Is	W & S Coasts & Highlands	Central Plateau	Upper Derwent	South East	Central North	Lower Derwent	Midlands Alpine	East Coast	NE Coast & Flinders Is	TAS
Town Sample	Devonport	Strathan	Lake St Clair	Bushy Park	Hobart	Launceston	Huonville	Fingal	Swansea	Killiecrankie	Overall
Summer	0%	-24%	-43%	-34%	-12%	24%	-36%	-6%	-38%	-40%	-24%
Autumn	-26%	17%	-3%	-47%	-4%	-24%	-21%	-33%	-55%	-31%	-15%
Winter	-17%	-28%	-3%	-11%	3%	-18%	-16%	-48%	-17%	-29%	-18%
Spring	-32%	-1%	-3%	-42%	-38%	-12%	-39%	-42%	-57%	-47%	-23%
Overall	-18%	-6%	-5%	-29%	-7%	-1%	-21%	-18%	-19%	-9%	-12%

2026 TAS - Estimated Seasonal Rainfall Trend Variation by 25%

REGION	NW Coast & King Is	W & S Coasts & Highlands	Central Plateau	Upper Derwent	South East	Central North	Lower Derwent	Midlands Alpine	East Coast	NE Coast & Flinders Is	TAS
Town Sample	Devonport	Strathan	Lake St Clair	Bushy Park	Hobart	Launceston	Huonville	Fingal	Swansea	Killiecrankie	Overall
Summer	av	av	drr	drr	av	av	drr	av	drr	drr	av
Autumn	drr	av	av	drr	av	av	av	drr	drr	drr	av
Winter	av	drr	av	av	av	av	av	drr	av	drr	av
Spring	drr	av	av	drr	drr	av	drr	drr	drr	drr	av
Overall	av	wtr	av	av	wtr	wtr	wtr	wtr	wtr	wtr	wtr

2026 TAS - Estimated Seasonal Maximum Temperature Average and Trend

Region	NW Coast & King Is	W & S Coasts & Highlands	Central Plateau	Upper Derwent	South East	Central North	Lower Derwent	Midlands Alpine	East Coast	NE Coast & Flinders Is	TAS
Town Sample	Devonport	Strathan	Lake St Clair	Bushy Park	Hobart	Launceston	Huonville	Fingal	Swansea	Killiecrankie	Overall
Summer Est	21.5	20.9	19.8	25.1	24.2	23.2	22.9	22.9	19.8	22.9	22.3
Trend	av	av	a-w	wmr	av	av	wmr	av	a-w	wmr	a-w
Autumn Est	19.0	17.8	14.8	20.2	20.5	19.4	19.6	19.6	16.1	20.1	18.7
Trend	wmr	a-w	wmr	wmr	wmr	a-w	wmr	wmr	wmr	wmr	wmr
Winter Est	13.3	12.8	7.8	12.2	13.5	12.6	13.3	12.7	9.0	14.4	12.2
Trend	av	av	av	av	wmr	av	a-w	av	av	a-w	av
Spring Est	15.5	15.4	11.8	18.1	17.5	17.8	17.8	16.8	13.6	17.2	16.2
Trend	av	a-c	clr	av	av	av	a-w	a-c	av	av	av

2026 TAS - Estimated Seasonal Minimum Temperature Average and Trend

Region	NW Coast & King Is	W & S Coasts & Highlands	Central Plateau	Upper Derwent	South East	Central North	Lower Derwent	Midlands Alpine	NE Coast & Flinders Is	TAS
Town Sample	Devonport	Strathan	Lake St Clair	Bushy Park	Hobart	Launceston	Huonville	Fingal	Killiecrankie	Overall
Summer Est	12.2	10.2	5.3	9.4	12.3	12.7	9.4	10.2	14.1	10.6
Trend	av	av	av	av	a-w	wmr	av	av	a-w	av
Autumn Est	9.7	8.6	3.1	5.8	9.0	9.0	5.9	6.0	11.2	7.6
Trend	a-w	av	av	av	av	wmr	av	av	av	av
Winter Est	5.4	5.2	-0.4	1.2	4.3	3.2	1.3	1.0	5.6	3.0
Trend	av	av	a-c	a-c	a-c	a-w	a-c	av	a-c	av
Spring Est	6.8	7.1	2.0	4.4	7.3	6.0	4.8	3.4	8.3	5.6
Trend	a-c	av	av	clr	av	av	a-c	clr	av	a-c

Monthly Breakdowns

2026 TASMANIA - Estimated Rainfall Expectations (mm)

REGION	NW Coast & King Is	W & S Coasts & Highlands	Central Plateau	Upper Derwent	South East	Central North	Lower Derwent	Midlands Alpine	East Coast	NE Coast & Flinders Is	TAS
Town Sample	Devonport	Strathan	Lake St Clair	Bushy Park	Hobart	Launceston	Huonville	Fingal	Swansea	Killiecrankie	Overall
January	48	76	86	33	30	42	50	47	26	23	462
February	23	31	28	13	22	80	14	68	60	26	366
March	30	93	98	33	48	15	61	33	34	11	457
April	59	183	165	26	43	54	46	37	17	58	687
May	35	170	108	10	6	31	7	16	4	41	427
June	65	81	61	11	7	63	9	20	12	12	340
July	88	179	264	75	71	64	83	31	40	74	968
August	56	111	207	49	43	39	76	32	46	64	724
September	40	121	184	21	15	52	30	30	14	27	533
October	38	168	214	36	26	46	52	29	23	26	658
November	52	72	98	38	38	41	42	26	19	35	461
December	64	131	137	58	86	82	84	109	129	209	1089
Overall	599	1417	1649	402	434	608	555	478	424	606	7173

2026 TASMANIA - Estimated Rainfall Variation from the Average (mm)

REGION	NW Coast & King Is	W & S Coasts & Highlands	Central Plateau	Upper Derwent	South East	Central North	Lower Derwent	Midlands Alpine	East Coast	NE Coast & Flinders Is	TAS
Town Sample	Devonport	Strathan	Lake St Clair	Bushy Park	Hobart	Launceston	Huonville	Fingal	Swansea	Killiecrankie	Overall
January	7	-5	-17	-7	-11	-7	1	-7	-31	-22	-100
February	-6	-39	-46	-21	-11	54	-31	32	24	-3	-46
March	-8	-13	-1	-6	12	-17	13	-6	-12	-31	-69
April	-3	67	33	-20	5	8	-5	-13	-28	5	50
May	-32	9	-43	-36	-21	-23	-38	-25	-26	-22	-256
June	-14	-74	-93	-37	-28	2	-46	-27	-25	-55	-398
July	-1	-0	84	26	34	-0	18	-18	-1	-1	140
August	-28	-69	-7	-6	-3	-37	-4	-31	6	-4	-181
September	-35	-27	-16	-32	-29	-10	-47	-22	-22	-36	-276
October	-24	47	36	-22	-17	-4	-17	-19	-28	-26	-75
November	-3	-24	-33	-14	-1	-5	-15	-20	-26	-14	-157
December	11	37	12	8	35	36	24	50	66	150	429
Overall	-136	-90	-91	-167	-35	-4	-148	-105	-102	-61	-939

2026	TASMANIA - Estimated Rainfall Trend Variation by 10mm										
REGION	NW Coast & King Is	W & S Coasts & Highlands	Central Plateau	Upper Derwent	South East	Central North	Lower Derwent	Midlands Alpine	East Coast	NE Coast & Flinders Is	TAS
Town Sample	Devonport	Strathan	Lake St Clair	Bushy Park	Hobart	Launceston	Huonville	Fingal	Swansea	Killiecrankie	Overall
January	av	av	drr	av	drr	av	av	av	drr	drr	drr
February	av	drr	drr	drr	drr	wtr	drr	wtr	wtr	av	drr
March	av	drr	av	av	wtr	drr	wtr	av	drr	drr	drr
April	av	wtr	wtr	drr	av	av	av	drr	drr	av	wtr
May	drr	av	drr	drr	drr	drr	drr	drr	drr	drr	drr
June	drr	drr	drr	drr	drr	av	drr	drr	drr	drr	drr
July	av	av	wtr	wtr	wtr	av	wtr	drr	av	av	wtr
August	drr	drr	av	av	av	drr	av	drr	av	av	drr
September	drr	drr	drr	drr	drr	av	drr	drr	drr	drr	drr
October	drr	wtr	wtr	drr	drr	av	drr	drr	drr	drr	drr
November	av	drr	drr	drr	av	av	drr	drr	drr	drr	drr
December	wtr	wtr	wtr	av	wtr	wtr	wtr	wtr	wtr	wtr	wtr
Overall	drr	av	av	drr	av	av	drr	drr	drr	av	drr

2026	TASMANIA - Estimated Rainfall Expectations by %										
REGION	NW Coast & King Is	W & S Coasts & Highlands	Central Plateau	Upper Derwent	South East	Central North	Lower Derwent	Midlands Alpine	East Coast	NE Coast & Flinders Is	TAS
Town Sample	Devonport	Strathan	Lake St Clair	Bushy Park	Hobart	Launceston	Huonville	Fingal	Swansea	Killiecrankie	Overall
January	17%	-6%	-16%	-18%	-27%	-14%	1%	-13%	-54%	-49%	-18%
February	-20%	-55%	-62%	-61%	-34%	209%	-68%	92%	68%	-12%	-11%
March	-22%	-12%	-1%	-16%	34%	-52%	27%	-15%	-26%	-73%	-13%
April	-5%	58%	25%	-43%	14%	17%	-10%	-25%	-62%	9%	8%
May	-48%	6%	-28%	-78%	-77%	-43%	-85%	-61%	-87%	-35%	-38%
June	-18%	-48%	-60%	-78%	-81%	3%	-83%	-57%	-68%	-83%	-54%
July	-1%	0%	46%	52%	93%	-1%	27%	-37%	-2%	-1%	17%
August	-33%	-38%	-3%	-10%	-6%	-49%	-4%	-49%	16%	-6%	-20%
September	-47%	-18%	-8%	-60%	-66%	-16%	-61%	-43%	-61%	-57%	-34%
October	-39%	38%	20%	-38%	-40%	-8%	-24%	-39%	-55%	-50%	-10%
November	-6%	-25%	-25%	-27%	-3%	-12%	-26%	-44%	-58%	-29%	-25%
December	20%	39%	10%	15%	69%	78%	41%	84%	105%	256%	65%
Overall	-18%	-6%	-5%	-29%	-7%	-1%	-21%	-18%	-19%	-9%	-12%

2026	TASMANIA - Estimated Rainfall Trend Variation by 25%										
REGION	NW Coast & King Is	W & S Coasts & Highlands	Central Plateau	Upper Derwent	South East	Central North	Lower Derwent	Midlands Alpine	East Coast	NE Coast & Flinders Is	TAS
Town Sample	Devonport	Strathan	Lake St Clair	Bushy Park	Hobart	Launceston	Huonville	Fingal	Swansea	Killiecrankie	Overall
January	av	av	av	av	drr	av	av	av	drr	drr	av
February	av	drr	drr	drr	drr	wtr	drr	wtr	wtr	av	av
March	av	av	av	av	wtr	drr	wtr	av	drr	drr	av
April	av	wtr	av	drr	av	av	av	av	drr	av	av
May	drr	av	drr	drr	drr	drr	drr	drr	drr	drr	drr
June	av	drr	drr	drr	drr	av	drr	drr	drr	drr	drr
July	av	av	wtr	wtr	wtr	av	wtr	drr	av	av	av
August	drr	drr	av	av	av	drr	av	drr	av	av	av
September	drr	av	av	drr	drr	av	drr	drr	drr	drr	drr
October	drr	wtr	av	drr	drr	av	av	drr	drr	drr	av
November	av	av	av	drr	av	av	drr	drr	drr	drr	av
December	av	wtr	av	av	wtr	wtr	wtr	wtr	wtr	wtr	wtr
Overall	av	wtr	av	av	wtr	wtr	wtr	wtr	wtr	wtr	wtr

2026	TASMANIA - Estimated Average Maximum Temperature										
Region	NW Coast & King Is	W & S Coasts & Highlands	Central Plateau	Upper Derwent	South East	Central North	Lower Derwent	Midlands Alpine	East Coast	NE Coast & Flinders Is	TAS
Town Sample	Devonport	Strathan	Lake St Clair	Bushy Park	Hobart	Launceston	Huonville	Fingal	Swansea	Killiecrankie	Overall
January	22	21	21	25	24	25	24	25	21	25	**23**
February	22	21	20	26	22	24	22	23	19	22	**22**
March	22	21	20	25	23	24	23	23	20	24	**22**
April	19	17	13	19	19	20	18	18	15	19	**18**
May	16	15	12	17	17	17	17	17	13	17	**16**
June	15	14	10	13	14	15	14	14	10	16	**13**
July	13	12	8	12	13	13	13	13	9	14	**12**
August	12	12	6	12	12	13	12	11	8	13	**11**
September	14	14	9	16	15	15	15	15	11	15	**14**
October	15	15	11	18	19	17	16	18	13	18	**16**
November	18	17	15	21	19	20	19	21	17	19	**19**
December	18	17	14	20	18	20	17	18	14	18	**17**
Overall	**17**	**16**	**13**	**19**	**18**	**19**	**18**	**18**	**14**	**18**	**17**

2026	TASMANIA - Estimated Average Minimum Temperature									
Region	NW Coast & King Is	W & S Coasts & Highlands	Central Plateau	Upper Derwent	South East	Central North	Lower Derwent	Midlands Alpine	NE Coast & Flinders Is	TAS
Town Sample	Devonport	Strathan	Lake St Clair	Bushy Park	Hobart	Launceston	Huonville	Fingal	Killiecrankie	Overall
January	12	11	6	10	13	13	10	11	15	**11**
February	13	9	4	9	12	13	9	10	14	**10**
March	12	9	4	8	11	12	8	9	13	**10**
April	10	8	3	5	9	9	6	6	11	**7**
May	8	8	2	3	7	6	4	2	9	**6**
June	7	6	0	2	5	5	2	2	6	**4**
July	5	5	0	2	5	3	2	1	6	**3**
August	4	4	-2	0	3	1	0	0	4	**2**
September	5	6	0	1	5	3	3	0	6	**3**
October	7	8	3	6	8	7	6	4	9	**6**
November	8	8	3	6	9	8	6	6	10	**7**
December	9	8	4	8	9	9	7	8	11	**8**
Overall	8	8	2	5	8	7	5	5	10	**6**

2026	TASMANIA - Estimated Degree of Variation from the Maximum Average										
Region	NW Coast & King Is	W & S Coasts & Highlands	Central Plateau	Upper Derwent	South East	Central North	Lower Derwent	Midlands Alpine	East Coast	NE Coast & Flinders Is	TAS
Town Sample	Devonport	Strathan	Lake St Clair	Bushy Park	Hobart	Launceston	Huonville	Fingal	Swansea	Killiecrankie	Overall
January	0.4	0.6	0.9	1.3	2.6	-0.3	0.6	1.0	1.5	2.4	1.1
February	0.3	-0.5	0.4	1.6	-0.1	-0.1	-0.5	-0.4	0.4	-0.2	0.1
March	1.3	1.2	2.7	2.6	2.5	1.8	2.1	1.3	2.3	2.4	2.0
April	0.8	0.5	0.1	0.7	1.3	1.6	0.0	0.0	0.6	0.5	0.6
May	1.1	0.8	1.8	3.1	2.8	2.7	2.4	1.5	1.9	0.8	1.9
June	1.0	1.2	1.7	1.3	2.4	2.5	1.2	1.2	1.3	1.9	1.6
July	0.5	0.0	0.5	0.6	1.5	1.4	0.4	0.3	0.2	0.9	0.6
August	-0.9	-0.8	-1.8	-1.0	-1.0	-0.4	-1.7	-1.8	-1.6	-0.9	-1.2
September	-0.6	-0.4	-1.2	0.2	0.4	-0.3	-0.5	-0.5	-0.7	0.0	-0.3
October	-0.7	-1.5	-1.8	0.0	1.6	-0.6	-1.0	0.2	-0.6	0.6	-0.4
November	-0.3	-0.6	-0.5	0.8	0.6	0.2	-0.3	1.0	0.7	0.3	0.2
December	-1.7	-2.8	-3.4	-2.5	-2.6	-3.0	-3.9	-3.8	-3.7	-2.5	-3.0
Overall	0.1	-0.2	0.0	0.7	1.0	0.4	-0.1	0.0	0.2	0.5	0.3

2026 TASMANIA - Estimated Degree of Variation from the Minimum Average

Region	NW Coast & King Is	W & S Coasts & Highlands	Central Plateau	Upper Derwent	South East	Central North	Lower Derwent	Midlands Alpine	NE Coast & Flinders Is	TAS
Town Sample	Devonport	Strathan	Lake St Clair	Bushy Park	Hobart	Launceston	Huonville	Fingal	Killiecrankie	Overall
January	-0.1	0.0	-0.3	-0.3	1.3	2.1	0.1	0.4	1.4	0.5
February	-0.2	-1.9	-1.9	-0.8	-0.1	2.0	-0.1	0.1	0.5	-0.3
March	0.7	-0.5	-0.6	-0.1	0.3	2.0	0.2	1.4	0.5	0.4
April	1.0	0.0	0.3	-0.6	-0.2	2.6	-0.3	0.8	0.2	0.4
May	0.8	0.4	0.2	-0.6	-0.2	2.0	-0.6	-0.5	0.4	0.2
June	1.9	0.8	-0.2	-0.2	-0.2	2.7	-0.4	0.6	-0.8	0.5
July	0.7	0.2	0.2	0.3	0.0	1.1	0.2	0.6	0.0	0.4
August	-1.2	-2.0	-1.9	-2.3	-1.9	-1.4	-2.7	-1.0	-2.4	-1.9
September	-1.0	-0.5	-0.8	-2.5	-1.2	-0.5	-1.8	-2.9	-1.5	-1.4
October	-0.2	0.5	0.6	-0.2	0.4	1.3	0.5	-1.4	0.3	0.2
November	-1.0	-0.8	-0.6	-1.4	-0.7	0.3	-1.5	-0.7	-0.4	-0.7
December	-2.0	-1.4	-1.4	-1.3	-1.5	-0.3	-1.9	-1.2	-1.2	-1.4
Overall	0.0	-0.4	-0.5	-0.9	-0.3	1.2	-0.7	-0.3	-0.2	-0.3

2026 TASMANIA - Estimated Maximum Temperature Trend

Region	NW Coast & King Is	W & S Coasts & Highlands	Central Plateau	Upper Derwent	South East	Central North	Lower Derwent	Midlands Alpine	East Coast	NE Coast & Flinders Is	TAS
Town Sample	Devonport	Strathan	Lake St Clair	Bushy Park	Hobart	Launceston	Huonville	Fingal	Swansea	Killiecrankie	Overall
January	av	a-w	a-w	wmr	wmr	av	a-w	a-w	wmr	wmr	wmr
February	av	av	av	wmr	av	av	av	av	av	av	av
March	wmr	wmr	wmr	wmr	wmr	wmr	wmr	wmr	wmr	wmr	wmr
April	a-w	av	av	a-w	wmr	wmr	av	av	a-w	av	a-w
May	wmr	a-w	wmr	wmr	wmr	wmr	wmr	wmr	wmr	a-w	wmr
June	a-w	wmr	wmr	wmr	wmr	wmr	wmr	wmr	wmr	wmr	wmr
July	av	av	av	a-w	wmr	wmr	av	av	av	a-w	a-w
August	a-c	a-c	clr	a-c	a-c	av	clr	clr	clr	a-c	clr
September	a-c	av	clr	av	av	av	av	av	a-c	av	av
October	a-c	clr	clr	av	wmr	a-c	a-c	av	a-c	a-w	av
November	av	a-c	av	a-w	a-w	av	av	a-w	a-w	av	av
December	clr	clr	clr	clr	clr	clr	clr	clr	clr	clr	clr
Overall	av	av	av	a-w	a-w	av	av	av	av	av	av

2026 TASMANIA - Estimated Minimum Temperature Trend

Region	NW Coast & King Is	W & S Coasts & Highlands	Central Plateau	Upper Derwent	South East	Central North	Lower Derwent	Midlands Alpine	NE Coast & Flinders Is	TAS
Town Sample	Devonport	Strathan	Lake St Clair	Bushy Park	Hobart	Launceston	Huonville	Fingal	Killiecrankie	Overall
January	av	av	av	av	wmr	wmr	av	av	wmr	av
February	av	clr	clr	a-c	av	wmr	av	av	av	av
March	a-w	av	a-c	av	av	wmr	av	wmr	av	av
April	a-w	av	av	a-c	av	wmr	av	a-w	av	av
May	a-w	av	av	a-c	av	wmr	a-c	av	av	av
June	wmr	a-w	av	av	av	wmr	av	a-w	a-c	av
July	a-w	av	av	av	av	wmr	av	a-w	av	av
August	clr	clr	clr	clr	clr	clr	clr	a-c	clr	clr
September	a-c	av	a-c	clr	clr	av	clr	clr	clr	clr
October	av	av	a-w	av	av	wmr	av	clr	av	av
November	a-c	a-c	a-c	clr	a-c	av	clr	a-c	av	a-c
December	clr	clr	clr	clr	clr	av	clr	clr	clr	clr
Overall	av	av	av	a-c	av	wmr	a-c	av	av	av

SOUTH AUSTRALIA 2026

GENERAL TRENDS

Year Overall:
Drier by around 15-20% below the average
Slightly warmer but within 0.5°C of the average

Summer:
Drier by around 25-30% below the average
Wettest – Last week of December to first few days of January
Driest – First three weeks of December and most of January
Slightly warmer but within 0.5°C of the average
Hottest – Second week of January and last few days of February
Coolest – First week of January

Autumn:
Drier by around 30-35% below the average
Wettest – Second week and last week of May
Driest – Throughout March
Warmer by up to 2-2.5°C above the average
Hottest – Second half of March
Coolest – Second week and last few days of May

Winter:
Drier by around 15-20% below the average
Wettest – Second to fourth weeks of August
Driest – Second week of June
Average temperatures overall
Hottest – First half of June
Coolest – First and fourth weeks of August

Spring:
Drier by around 35-40% below the average
Wettest –Mid-November to end of the month
Driest – Second week of October and fourth week to the end of month
Slightly cooler but within 0.5°C of the average
Hottest – Second half of November
Coolest – First half of September

Major Weather Events

January 2026

Date	Day	South Australia - January 2026	Adelaide	Moon
1st	Thur	Some breakouts of rain likely in the Western Agricultural region. Unusually cold day in the Mt Lofty Ranges.	Showers	
2nd	Fri			P6
3rd	Sat			F ^
4th	Sun	Scattered rain activity through the coastal regions of the Bight. Cool overnight in the Northwest Pastoral to the Eyre regions, particularly in the Nonning area, and down into some of the Southeast districts.		
5th	Mon			
6th	Tue	Heavy rain for a time in the Western Agricultural region to the Northwest.		
7th	Wed	Rain spreads throughout the western half of the State around this time before easing to isolated showers by tomorrow.		
8th	Thu			
9th	Fri	Widespread rain builds throughout the State over the next couple of days.		XhS
10th	Sat	Heavy rain in the western half of the State to the coastal Eyre Peninsula regions, spreading throughout the State by tomorrow. Very hot in the State around this time, particularly in the Far North to the Western Agricultural regions and in some parts of the Southeast region when daily maximas into the mid 40's can be expected.		
11th	Sun	Rain eases to mostly dry over the next 3-4 weeks, with just a couple of interruptions, mostly in the Far North, or as a result of accompanying thunderstorm activity. The next 3-4 weeks may also bring heat-wave temperatures at times, with the overall period much warmer than is typical for this time of year, especially during the daytime, while overnight temperatures may bring some relief by being closer to the average.		3Q
12th	Mon			
13th	Tue			
14th	Wed			A
15th	Thur	Chance of isolated showers in the Far North region over the next couple of days.		
16th	Fri			
17th	Sat			V
18th	Sun			
19th	Mon	The current heat-wave temperatures may reach their peak over the next couple of days in many parts of the State.		N
20th	Tue	Very hot spell across the State continues particularly over the Eyre Peninsula around this time, with daily maximum temperatures likely to be in the mid-40's, including in the Port Augusta and Wudunna areas.		
21st	Wed	Unusually hot overnight in the Flinders Ranges, particularly in the Arkaroola district.		
22nd	Thu	Unusually cool over the next few nights in the Western Agricultural region, with a number of districts mininmum temperatures dropping into single digits for a time.	Showers	
23rd	Fri	Unusually cool overnight in the Southeast region, with a number of districts minimum temperatures dropping into the lower single digits, including in the Naracoorte area.		
24th	Sat			XhN
25th	Sun	Chance of isolated showers in the Far North region over the next couple of days.		
26th	Mon			1Q
27th	Tue			
28th	Wed	Thunderstorms cross the Adelaide region and over the Mt Lofty Ranges, bringing outbursts of heavy rain, flash flooding and wind damage in some Adelaide suburbs, and in pockets in the surrounding ranges. Accompanying this thunderstorm activity may also be lower temperatures in many parts of the State, particulary in the Far North and Pastoral regions, including an unusually cool day in the Marree district.		
29th	Thur	Chance of rain activity in the Far North region regularly over the next week.		
30th	Fri	Showers in the Far North may be heavy over the next couple of days.		P10 ^
31st	Sat	Cool temperatures across Kangaroo and Fleurieu Islands around this time.		

February 2026

Date	Day	South Australia - February 2026	Adelaide	Moon
1st	Sun			
2nd	Mon		Rain	F
3rd	Tue	Some outbreaks of heavy rain over the next few days in the Far North region.		
4th	Wed			
5th	Thu	Strong winds with severe gusts accompanying thunderstorm activity in the Northeast Pastoral region, particularly around the Marree area.		
6th	Fri	Unusually cool over the next few nights in the Western Agricultural region, with a number of districts mininmum temperatures dropping into single digits for a time.		XhS
7th	Sat			
8th	Sun	Chance of rain along the eastern border fringes of the State over the next couple of days.		
9th	Mon			
10th	Tue			3Q
11th	Wed	Mostly dry over the coming week except for some isolated shower activity mostly in the northern half of the State in the Far North to the Northwest Pastoral regions.		A
12th	Thur	Rainfall amounts may continue to be below normal over much of the State during the coming 3-4 weeks, while the heat wave temperatures of late may break under the arrival of a cooler southerly airflow and increased cloud bands, resulting in the next 3-4 weeks being generally cooler than is normal for this time of year.		
13th	Fri			V
14th	Sat			
15th	Sun			
16th	Mon	Cool overnight on the Yorke Peninsula.		
17th	Tue		Showers	
18th	Wed	Mostly dry conditions continue over the southern half of the State over the next 4-5 days. Cooler overnight temperatures over the next few days in the southeast quarter of the State.		N
19th	Thu	The next few nights may see cooler temperatures in the Northwest, Western Agricultural and Far North regions, bringing some respite from the heat.		
20th	Fri			XhN
21st	Sat			
22nd	Sun			
23rd	Mon	Chance of rain in the Western Agricultural to the Eyre Peninsula around this time.		
24th	Tue	Some showers in the North East region.		
25th	Wed	Isolated shoewrs along the southern coastal regions of the State over the next couple of days, and briefly expanding into the North East to the Far North regions.		1Q P14
26th	Thur			^
27th	Fri	A hot spell prevails over most of the State during the next few days.		
28th	Sat	Heavy rain threatens into the western border districts for a time.		

March 2026

Date	Day	South Australia - March 2026	Adelaide	Moon
1st	Sun	Mostly dry over the coming 8-10 days throughout the State with just a couple of interruptions mostly in the Far North. Hot day in the Murray Bridge area with maximum temperatures likely to reach into the low 40's, and very warm overnight in the Western Agricultural region.	Showers	
2nd	Mon	Unusually cool in the Mt Lofty Ranges, particularly on Mt Crawford.		
3rd	Tue	Showers in the Far North over the next several days.		F
4th	Wed	Very hot in the Far North region around this time, particularly in the Moomba area, where maximum temperatures may return to the mid-40's. In contrast, the passage of a front may bring unusually cooler than normal temperatures across the Eyre Peninsula. Blustery with some wind gusts for a time in the Murray-Mallee region, particularly in the Lameroo district.	Showers	
5th	Thu			XhS
6th	Fri			
7th	Sat			
8th	Sun			
9th	Mon			
10th	Tue			
11th	Wed	Cold overnight in some of the Southeast districts, particularly in the Naracoorte area, where minimum temperatures may not reach into the double digits, and the Keith area may not be far away from frost level temperatures.		3Q A

Date	Day	South Australia - March 2026	Adelaide	Moon
12th	Thur	The next few weeks may see a prevailing northerly air flow dominating over the country bringing very high temperatures across the State and especially in the southeast quarter, with possible heat-wave temperatures being reached and sustained in the upper 30's to early 40's for a number of days, particularly in a couple of weeks time.		V
13th	Fri	Showers over the southern half of the State around this time.		
14th	Sat	Mostly dry over the next 3-4 weeks throughout the State, with a number of districts being much drier than average, and some districts may not see any rainfall at all.		
15th	Sun			
16th	Mon			
17th	Tue	Chance of isolated showers around the Flinders Ranges.		
18th	Wed			
19th	Thu	Very hot in the Northwest, Western Agricultural, and Far North regions over the next couple of days with many districts maximum temperatures reaching into the low 40's.		N XhN
20th	Fri			
21st	Sat			
22nd	Sun	Isolated showers possible in the Far North.		P11
23rd	Mon	Thunderstorms with heavy rain, large hail and severe wind gusts over the next couple of days in the Mid-North and Lower North regions with chance of damage to property and local infrastructure, particularly around the Riverton and Hoyleton areas.		
24th	Tue			
25th	Wed	Very hot overnight in the Adelaide area, with chance of some minimum temperatures into the upper 20's to early 30's likely.		
26th	Thur			1Q ^
27th	Fri	Very warm overnight in the Far North, particularly in the Oodnadatta district.		
28th	Sat	Very strong winds across the Western Agricultural region, particularly in the Nullarbor district. Very hot in the Southeast, particularly around Mt Gambier and in the Victor Harbor area.		
29th	Sun			
30th	Mon			
31st	Tue			

April 2026

Date	Day	South Australia - April 2026	Adelaide	Moon
1st	Wed			
2nd	Thu			F XhS
3rd	Fri			
4th	Sat		Showers	
5th	Sun		Showers	
6th	Mon		Showers	
7th	Tue	A cooler change sweeps across the State around this time, bringing below average temperatures for this time of year over much of the State. Unusually cool in the Yunta area of the Far North and in the Mt Lofty Ranges with maximum temperatures unlikely to reach into the 20's and minimum temperatures close to zero in some districts.		A
8th	Wed	Cold overnight in the southeast quarter of the State from the Eyre Peninsula to the Southeast regions over the next couple of days. Chance of patches of frost overnight in the Southeast, particularly in the Keith area.	Showers	
9th	Thur			V
10th	Fri		Showers	3Q
11th	Sat	The next 3-4 weeks ahead may be fairly typical for this time of year overall for both rainfall amounts and temperatures, with only the far northern parts of the States slightly drier than average.		
12th	Sun	Storm activity across the southeast quarter of the State over the next couple of days including in the Yorke Peninsula to the Lower South East and over Kangaroo Island, as well as in the Adelaide region and into the Mt Lofty Ranges where funnel formation winds or microbursts may develop, with wind damage to property and infrastructure including power outages possible in many parts of the southeast. Strong winds may also cause sandstorms in the Riverlands and Murraylands regions over the next couple of days.	Showers	
13th	Mon	Strong stormy winds continue across the southeast regions of the State, with further chance of the development of mini-twisters and wind gusts causing damage to property and infrastructure, and sandstorms in the Riverlands and Murraylands. The strong winds around this time may generate dust storms in some inland areas. Very strong gusty winds across Kangaroo Island.	Showers, Sn	
14th	Tue	Cold over the next couple of nights in the Western Agricultural region.	Showers	

Date	Day	South Australia - April 2026	Adelaide	Moon
15th	Wed			
16th	Thu			XhN
17th	Fri	Occasional shower activity over the next couple of days in the southern half of the State.		N
18th	Sat	Very hot day in the Western Agricultural region, particularly in the Nullabor area, with maximum temperatures into the upper 30's likely in a number of districts.		
19th	Sun	Widespread warm spell across the State over the next few days with daily maximum temperatures into the low 30's likely in many districts.		P7
20th	Mon	Warm in the Pastoral regions, particularly in the Oodnadatta and Marla districts over the next couple of days, while there is also a chance of showers across the Eyre Peninsula and into the Southeast during this same period.		
21st	Tue		Showers	
22nd	Wed	Warm overnight in the Far North, particularly in the Marla area.	Showers	^
23rd	Thur		Showers	
24th	Fri	Isolated showers in the Far North.		1Q
25th	Sat	Showers in the Eyre Peninsula to the Southeast region.		
26th	Sun	Mostly dry throughout the State over the next week ahead.		
27th	Mon			
28th	Tue	Very cold overnight in the Clare Valley and Mt Lofty Ranges, with chance of isolated pockets of frost.		
29th	Wed			XhS
30th	Thur			

May 2026

Date	Day	South Australia - May 2026	Adelaide	Moon
1st	Fri			
2nd	Sat			F
3rd	Sun	Isolated showers in the Western Agricultural to the Northwest Pastoral regions.		
4th	Mon	Light showers clear out of the west and move through the southeast quarter of the State.		
5th	Tue	Mostly dry throughout the State for the next week ahead.		A
6th	Wed		Showers	V
7th	Thu		Rain, Th	
8th	Fri	Unusually cool spell in the southeast quarter of the State over the next couple of days, including in the Murray-Mallee region to the Mt Lofty Ranges.	Showers	
9th	Sat	Unusually low minimum temperatures overnight spread into the Far North over the next couple of days.	Showers	
10th	Sun	Chance of some spill over showers from Western Australia into the border areas of South Australia. Cold overnight with patches of frost likely in the Far North, including in the Yunta district.	Showers	3Q
11th	Mon	The next 3-4 weeks may be fairly typical for this time of year, with maximum temperatures slightly above normal overall, while minimum temperatures and rainfall expectations may be fair average in most districts, although the far northern parts may be drier than is normal.	Showers	
12th	Tue	Dry throughout the State over the next 5-6 days.	Showers	
13th	Wed	Warm overnight in the Far North, particularly in the Moomba district.	Showers	XhN
14th	Thur			
15th	Fri			
16th	Sat	Warmer in the Far North and Northwest regions around this time, particularly in the Marla district.		
17th	Sun		Showers	N
18th	Mon	Wet spell over the southern half of the State over the next few days.		P3
19th	Tue	Rain in the Western Agricultural region to the Eyre Peninsula and into the Spencer Gulf and Southeast regions.		^
20th	Wed			
21st	Thu			
22nd	Fri	Isolated shower activity through the coastal regions.		
23rd	Sat	Widespread light shower activity spreads throughout the State around this time.		1Q
24th	Sun			
25th	Mon	Thunderstorms with possible microburst damaging winds in the Mid-North region, particularly in the Burra area. It may be an unusually hot spell in the Western Agricultural to the Spencer Gulf around this time, particularly in the Nullabor area, where maximum temperatures may reach into the lower 30's, and mild overnight minimum temperatures, particularly on Neptune Island.		

Date	Day	South Australia - May 2026	Adelaide	Moon
26th	Tue			XhS
27th	Wed	An intense cold front brings severe wind gusts across the Lower Eyre Peninsula and across the Spencer Gulf, particularly across Neptune Island. Chance of light snow in the Mid-North and Flinders Ranges regions.	Rain	
28th	Thur	Cold in the Far North region to the Mt Lofty Ranges, particularly in the Yunta district and on Mt Crawford.	Showers	
29th	Fri	Light showers in the coastal regions over the next couple of days.	Showers	
30th	Sat		Showers	
31st	Sun	Heavy rain crosses the Lower Eyre Peninsula, with chance of some local flash flooding in the Port Lincoln area.	Showers	F

June 2026

Date	Day	South Australia - June 2026	Adelaide	Moon
1st	Mon	Clearing to mostly dry throughout the State over the next 5-6 days.		A
2nd	Tue			V
3rd	Wed	Cool overnight in the southeast quarter of the State over the next couple of days with chance of some patches of frost here and there in sheltered parts.		
4th	Thu		Showers, Th	
5th	Fri		Fg	
6th	Sat		Rain, Th	
7th	Sun	Heavy rain likely around the Eyre Peninsula over the next couple of days.		
8th	Mon	Rain activity spreads throughout the southern half of the State with heavy falls in particularly in the Eyre Peninsula to the Southeast regions. Cold minimum temperatures in the Lower Southeast region likely, with chance of frost patches or wintry showers, including in the Coonawarra area.		3Q
9th	Tue	Rain eases to widespread showers throughout most of the State although mostly dry in the Far North. Very cool overnight in the Far North with patches of frost likely in a number of areas, including in the Yunta district.		XhN
10th	Wed	Isolated showers linger in the southeast regions over the next couple of days.		
11th	Thur			
12th	Fri	Brief showers in the southeast of the State. Mild temperatures prevail over the next few days in the southeastern quarter of the State, with a number of districts seeing daily maximas into the early to mid 20's.		
13th	Sat	Generally, the next 3-4 weeks may be drier than average for this time of year, while slightly warmer overall.		
14th	Sun	Warm overnight in the Spencer Gulf region, including on Neptune Island.		
15th	Mon			N P2
16th	Tue			^
17th	Wed			
18th	Thu	A band of showers cross the southern half of the State.		
19th	Fri	Mostly dry throughout the State over the next week.	Showers	
20th	Sat			
21st	Sun	A passing front may bring gale force winds through the Mt Lofty Ranges and over Kangaroo and Fleurieu Islands.	Showers	
22nd	Mon			1Q XhS
23rd	Tue	Thunderstorm with strong downburst winds may cause some damage as it crosses the Lower South East region.	Rain, Th	
24th	Wed			
25th	Thur	Cold day in the Mt Lofty Ranges, particularly in the Mt Crawford area where maximum temperatures may not reach into the double digits in some areas.		
26th	Fri			
27th	Sat	Some showers in the southeast of the State over the next couple of days. Very cold overnight with possible patches of frost in the Mt Lofty Ranges, through the Murray Valley region and into some parts of the Southeast districts.		
28th	Sun	Heavier showers for a time in the Flinders Ranges.		A
29th	Mon	Showers clear in the south but build in the northern half of the State.	Showers, Fg	V
30th	Tue	Widespread rain activity across the State over the next couple of days.		F

July 2026

Date	Day	South Australia - July 2026	Adelaide	Moon
1st	Wed	Heavy rain at times through the southern coastal districts from the Western Agricultural to the Southeast as well a across the northern Eyre Peninsula.	Showers	
2nd	Thu	Light showers linger mostly through the southeast quarter of the State over the next few days.	Showers, Fg	
3rd	Fri	Showers briefly extend into the Far North.		
4th	Sat	Cold over the next few nights in the State, particularly in the Western Agricultural districts with some patches of frost possible here and there.		
5th	Sun	Some heavy rain for a time on the Eyre Peninsula.		
6th	Mon	An intense cold front directs very strong winds, possibly to gale force, up through the Mt Lofty Ranges and across Kangaroo Island. Chance of frosts overnight in the Far North Nonning district.	Showers	
7th	Tue			XhN
8th	Wed	Very cold overnight in the Upper North region with scattered frosts likely in the Yongala district of the Flinders Ranges.		3Q
9th	Thur		Showers	
10th	Fri	Cold fronts cross the Bight directing squally storms with strong wind gusts over the Yorke Peninsula. Warmer in the Far North, particularly in the Marla area.	Showers	
11th	Sat	Cold fronts cross the southeast of the State bringing strong wind gusts and possible rumbling thunder through the Lower South East region and in the Mt Lofty Ranges. Winds may be particularly strong and gusty through the Spencer Gulf area.	Showers	
12th	Sun	Generally, the next 3-4 weeks may be fairly average compared to what is typical for this time of year, with a mix of some cold and frosty spells and milder spells. Rainfall may also be fairly average overall, despite the predominately westerly airflow bringing above average rain into the Mt Lofty Ranges, other parts of the State may see average rainfall, and the central and eastern pastoral areas may be drier than normal.		
13th	Mon	Isolated showers in the southeast over the next couple of days. Warm overnight on the Eyre and Yorke Peninsulas, particularly in the Streaky Bay and Maitland areas.		P4 ^
14th	Tue			N
15th	Wed			
16th	Thu			
17th	Fri		Showers	
18th	Sat	Isolated showers around the Eyre Peninsula. Cooler in many parts of the State, particularly in the Mt Lofty Ranges, where maximum temperatures may sit in the single digit ranges at this time.	Rain	
19th	Sun		Showers	
20th	Mon		Showers	XhS
21st	Tue			1Q
22nd	Wed	Spell of light shower activity over the southeastern regions of the State over the next few days from the Eyre Peninsula to the Southeast.		
23rd	Thur	An intense depression crossing the Bight brings very strong wind gusts across Kangaroo Island. Chance of some heavy rain in the inland Eyre Peninsula for a time.		
24th	Fri	A small, intense low pressure system crosses the Bright during the next couple of days extending strong, blustery winds, up to gale force at times, over the Yorke Peninsula and across Kangaroo Island.	Showers	
25th	Sat	Gale gusts likely around Cape mayoughby.	Showers	
26th	Sun			A
27th	Mon	Isolated showers in the western half of the State.		V
28th	Tue	Showers clear in the west, but linger over the Western Agricultural region.	Showers	
29th	Wed	Rain across the southern parts of the State over the next couple of days from the Western Agricultural to the Southeast, with some heavy falls at times.	Showers	
30th	Thu	Heavy rain in the Southeast region.	Showers	F
31st	Fri		Showers	

August 2026

Date	Day	South Australia - August 2026	Adelaide	Moon
1st	Sat	Isolated showers in the southeast quarter of the State over the next couple of days.		
2nd	Sun			
3rd	Mon			XhN

Date	Day	South Australia - August 2026	Adelaide	Moon
4th	Tue	Very cold minimum temperatures over the next couple of nights, with overnight frosts in many parts of the State, including frosts in the Far North and Northeast Pastoral regions. Frosts may be severe in the Yunta district around this time.		
5th	Wed	Light showers in the west of the State at first, shifting into the southeast over the following day.	Showers	
6th	Thur		Showers	**3Q**
7th	Fri	Heavy rain in Western Australia may drift for a time in to the western border areas of South Australia, while heavy rain in Victoria may briefly spread into the southeast regions of South Australia.		
8th	Sat	Very cold with scattered frosts in the Adelaide region to the Mt Lofty Ranges as well as in some parts of the Northwest, Western Agricultural, Southeast, and Far North regions over the next couple of days.		
9th	Sun	Strong winds to gale force at times across the West Coast region, including in Wudinna. Light shower activity in the southeast.	Showers	**^**
10th	Mon	Showers in the western border regions associated with storm activity over the southern half of Western Australia, with isolated showers into the Far North of South Australia. Slightly warmer in the Western Agricultural region over the next couple of days.	Showers	**P8**
11th	Tue	Light showers over much of the State. Warmer over the State around this time, particularly in the Far North where some areas may seeing maximum temperatures up to the early 30's, including in the Oodnadatta area.	Showers	
12th	Wed	Storms with severe wind gusts across the Murraylands to the Adelaide region and widespread scattered rain activity elsewhere throughout the State.	Showers, Th	
13th	Thu	The next 3-4 weeks may be cooler than normal for this time of year, while rainfall may be near average overall, with the Far North wetter than average, and the southern Pastoral regions drier than average.	Showers	**N**
14th	Fri	Some showers over the next couple of days mostly in the Far North of the State. Some frosts in the Flinders Ranges.	Showers	
15th	Sat		Showers, Fg	
16th	Sun		Showers	**XhS**
17th	Mon	Isolated showers in the southeast of the State over the next few days.	Rain	
18th	Tue	Cold day in the Mt Lofty Ranges, particularly in the Mt Crawford area where single digit maximum temperatures are likely.	Showers	
19th	Wed	Scattered frosts in northern parts of the State from the Northwest to the Far North, as well as down through the Flinders Ranges.	Showers	
20th	Thur		Rain, Sn	**1Q**
21st	Fri		Showers	
22nd	Sat	Frosts in the Far North, Northwest and Flinders Ranges over the next couple of nights, including in the Oodnadatta and Hawker areas.	Showers	**A**
23rd	Sun	Scattered rain activity through the southern half of the State.	Showers, Fg	**V**
24th	Mon	Lingering showers likely in the southern half of the State.	Fg	
25th	Tue		Showers	
26th	Wed	Further shower activity over the next week in most regions in the southern half of the State.	Showers	
27th	Thu		Showers	
28th	Fri			**F**
29th	Sat	Rain may be heavy for a time in the Eyre Peninsula to the Southeast regions.		
30th	Sun			**XhN**
31st	Mon		Showers	

September 2026

Date	Day	South Australia - September 2026	Adelaide	Moon
1st	Tue	Heavy rain at times in the southeast quarter of the State, particularly over the Eyre Peninsula and in the Southeast. Chance of snow in the Mt Lofty Ranges.	Showers	
2nd	Wed	Light showers in the southeast of the State over the next few days. Very cold with frosts in the Mallee region.		
3rd	Thur	Very strong blustery winds in the southeast, particularly in the Murraylands region in the Renmark area. Scattered patches of frost in the Western Agricultural region.		
4th	Fri			**3Q**
5th	Sat			
6th	Sun			**^**
7th	Mon	Brief showers in the southern half of the State over the next few days.		**P12**

Date	Day	South Australia - September 2026	Adelaide	Moon
8th	Tue	Milder daytime temperatures in the State over the next few days, while overnight frosts likely in many southeastern parts of the State.		
9th	Wed	Warm day in the Far North, particularly in the Marla area where maximum temperatures may rise into the upper 20's and the Oodnadatta district may reach up to 30°C.	Showers	
10th	Thu	Unusually warm overnight in the Flinders Ranges and warmer daytime temperatures in many districts in the south of the State.	Showers	
11th	Fri			N
12th	Sat	Scattered rain activity over the next couple of days through the southern half of the State.	Showers	XhS
13th	Sun	The next 3-4 weeks may be drier than normal overall, with fairly average maximum temperatures and below average minimum temperatures.		
14th	Mon	Chance of some patches of frost in the Western Agricultural region, particularly in the Kyancutta to Wudinna districts.		
15th	Tue			
16th	Wed			
17th	Thur			
18th	Fri	Storm activity over the Yorke Peninsula with heavy rain and chance of flash flooding in some places, including in the Minlaton area.	Showers	
19th	Sat	Cold overnight with frosts in the Mt Lofty Ranges.		1Q A V
20th	Sun	Widespread scattered rain activity across the State over the next few days. Chance of frost patches in the Lower Southeast and Murray Mallee around this time.		
21st	Mon		Showers	
22nd	Tue		Fg	
23rd	Wed			
24th	Thu	Warm day in the Western Agricultural region to the Yorke Peninsula.	Showers, Th	
25th	Fri		Showers	
26th	Sat	Mostly dry over the next week. Cold in the Mt Lofty Ranges with maximum temperatures only in the single digits in many parts of the Ranges. Blustery with very strong wind gusts in the Spencer Gulf region around this time, particularly across Neptune Island.	Showers	XhN
27th	Sun			F
28th	Mon			
29th	Tue			
30th	Wed		Showers	

October 2026

Date	Day	South Australia - October 2026	Adelaide	Moon
1st	Thur			
2nd	Fri	A vigorous front crosses the North West Pastoral region, particularly in the Coober Pedy district, with a line of thunderstorms and damaging wind gusts possible.		P13
3rd	Sat	Unusually warm overnight in the Far North, particularly in the Oodnadatta area.	Showers	
4th	Sun			3Q ^
5th	Mon	Chance of frost patches overnight in the Yorke Peninsula to the Flinders Ranges.		
6th	Tue	Light shower activity in the southeastern quarter of the State over the next few days.		
7th	Wed			
8th	Thu	Very warm day with maximum temperatures into the upper 30's in the Northwest and Far North regions, particularly in the Tarcoola district. Also unseasonally warmer than average in the Eyre Peninsula and some Western Agricultural districts, in the Mid-North region, Murraylands and other districts in the southeast of the State.		
9th	Fri		Showers	
10th	Sat			XhS
11th	Sun			N
12th	Mon	The next 3-4 weeks may be warmer and drier than is typical for this time of year. Blustery in the southeast, particularly in the Clare Valley.		
13th	Tue		Showers	
14th	Wed	Some strong downburst winds in the North East Pastoral region, including in the Moobma area.	Showers	
15th	Thur	Some showers in the southeast of the State around this time.		
16th	Fri			
17th	Sat	Mostly dry over the coming week throughout the State.		A V
18th	Sun	Cold day over the State, particularly in the Mt Lofty and Flinders Ranges, with chance of some overnight frost patches in the Upper North region in the Yongala area.	Showers	

Date	Day	South Australia - October 2026	Adelaide	Moon
19th	Mon	Minimum temperatures in the single digits may be common over the next few days in many parts of the State, particularly in the northern regions but also in some parts of the Western Agricultural districts.		1Q
20th	Tue			
21st	Wed			
22nd	Thur			
23rd	Fri		Showers	
24th	Sat			XhN
25th	Sun			
26th	Mon	Showers in the southeast of the State.		F
27th	Tue			
28th	Wed	Some showers around Eyre Peninsula over the next couple of days. Isolated showers in the Western Agricultural and Northwest Pastoral regions.		
29th	Thu			P9
30th	Fri			^
31st	Sat	The dry spell in the State continues over most of the next 2-3 weeks. Cold night in the Flinders Ranges to the Southeast over the next few nights with chance of some pockets of frost.		

November 2026

Date	Day	South Australia - November 2026	Adelaide	Moon
1st	Sun			
2nd	Mon			3Q
3rd	Tue			
4th	Wed			
5th	Thur	Warmer spell over the next few days across much of the State. Unusually warm overnight in the Far North, with minimum temperatures into the high 20's, particularly in the Arkaroola district.		
6th	Fri	Very warm overnight in the Murraylands.	Showers	XhS
7th	Sat			
8th	Sun			
9th	Mon			N
10th	Tue	Very hot in the State, particularly in the Far North around this time, with maximum temperatures possibly into the early 40's, particularly in the Marree area. Maximum temperatures may also reach into the upper 30's in the Murraylands to the York Peninsula.		
11th	Wed			
12th	Thu	Rain returns over the next 3-4 weeks, with this period being wetter than is typical for this time of year, and subsequently, temepratures may be cooler than average. Several low pressure troughs can be expected to bring moist northwest airflows and scattered thunderstorms during these coming weeks, particularly in the last week of November and in early December.	Showers, Th	
13th	Fri	Chance of tornado formations in the Mid-North region around the Spalding area, as well as accompanying large hail likely. Winds may reach storm force for a time over the Eyre		V
14th	Sat			A
15th	Sun		Showers	
16th	Mon			
17th	Tue	Strong winds across the North West Pastoral and the Flinders Ranges for a time, possibly causing some local damage in the Port Augusta and Mount Barry areas.	Showers, Th	
18th	Wed	Strong winds across the Lower Southeast, particularly around Mt Gambier. Chance of some drizzly showers in the southeast around this time.	Showers	1Q
19th	Thur			
20th	Fri			XhN
21st	Sat			
22nd	Sun			
23rd	Mon	Strong winds and showers through the North West Pastoral to East Eyre Peninsula over the next couple of days, particularly in the Tarcoola and Whyalla districts. Very warm spell across the State with many districts seeing maximum temperatures into the mid to upper 30's, and near to the low 40's in the Far North and Northwest regions.		
24th	Tue	Showers in the Eyre Peninsula to the Southeast region over the next couple of days. Very warm overnight in the Southeast region, including in the Keith district.		
25th	Wed	Warm overnight in the Far North region, particularly in the Moomba district where minimum temperatures into the upper 20's is expected.		F

Date	Day	South Australia - November 2026	Adelaide	Moon
26th	Thu			P5
27th	Fri	Widespread light shower activity spreads throughout the State around this time.		^
28th	Sat	Very cold overnight in the Southeast region, particularly in the Naracoorte district, which may see some patches of frost.	Showers	
29th	Sun	A passing band of thunderstorms bring severe wind gusts across the North East Pastoral region. Scattered shower activity in most districts over the next couple of days.		
30th	Mon			

December 2026

Date	Day	South Australia - December 2026	Adelaide	Moon
1st	Tue	Mostly dry in the State over the next few days.		3Q
2nd	Wed		Showers	
3rd	Thur	Unusually cool day over much of the State, including in the Spencer Gulf region over the Fleurieu and Kangaroo Islands.		XhS
4th	Fri	Cool overnight in the southeast quarter of the State with minimum temperatures only in the single digits in many areas, and some areas possibly developing some patches of frost.		
5th	Sat			
6th	Sun	Brief showers in the Flinders Ranges to the North East region.		
7th	Mon			
8th	Tue	Thunderstorm activity in the Flinders region brings hail showers. Thunderstorms also over the North East Pastoral region, particularly around Marree, Woomera, and Coober Pedy, with associated gusty winds possibly causing some localised damage. Very hot in the Far North with maximum temperatures reaching into the early 40's, particularly in the Oodnadatta area.	Showers	
9th	Wed	Outbursts of heavy rain in the Northwest Pastoral region.		N
10th	Thu	Rain activity spreads throughout the State over the next couple of days.		V
11th	Fri	Heavy rain in the eastern half of the State north of the Flinders Ranges.		A
12th	Sat	Rain activity clears in the west but lingers in the eastern regions along the border areas of the State.		
13th	Sun	The next 3-4 weeks may be wetter and warmer than is normal for this time of year, with the wettest spell expected over the Christmas period.		
14th	Mon	Spotty shower activity in most regions over the next few days. Cooler minimum temperatures briefly around this time in the Western Agricultural region.		
15th	Tue		Showers	
16th	Wed	Very strong and unsettled winds in the Far North to the Flinders Ranges, particularly in the Coober Pedy area.		
17th	Thur			1Q XhN
18th	Fri			
19th	Sat	Very cold overnight in the Keith area of the Southeast region, with the chance of frost patches in sheltered areas.	Showers	
20th	Sun	A thunderstorm embedded trough with severe wind gusts crosses through the North East Pastoral region. Light showers elsewhere.		
21st	Mon	Outbreaks of rain in the eastern half of the State.		
22nd	Tue	Possible thunderstorm activity crosses the Western district around the Penong and Ceduna areas bringing some heavy rain for a time.	Showers	
23rd	Wed	Mostly dry over the next few days.	Rain	
24th	Thu	Cooler spell overnight over much of the State over the next few days. Cold day in the far southeast of the State in the Mt Gambier area.	Showers	P1 F ^
25th	Fri	Cold minimum temperatures over the next couple of days in the Northeast region to the Southeast region, as well as in a number of districts on the Yorke Peninsula.		
26th	Sat	Cold overnight in the Mallee region, where minimum temperatures may fail to reach double digits.		
27th	Sun	Some rain activity in most regions across the State.		
28th	Mon	Showers in the Western Agricultural and the Northwest Pastoral regions.		
29th	Tue	Chance of rain for a short time in the northern Eyre region.		
30th	Wed			XhS
31st	Thur	Showers throughout the State over the next few days.		3Q

SOUTH AUSTRALIA AND TABLES 2026

Seasonal Breakdown – State Overall Rainfall Timing and Estimated mm Trend Graphs
State Overall Maximum Temperature Trend Graphs
Regional Rainfall Expectations Tables - mm and Trends
Regional Maximum and Minimum Temperature Trends Tables

Monthly Breakdown - **Rainfall:**
Regional Expectations by mm
Regional Expectations by mm Variance to Average
Regional Variance Trend by mm
Regional Estimated Expectations by % of Variance to Average
Regional Estimatedl Variance Trend by %

Maximum & Minimum Temperature Trends:
Regional Estimated Average Temperature
Regional Estimated Degree of Variation from the Average
Regional Estimated Temperature Trend

Seasonal Breakdowns

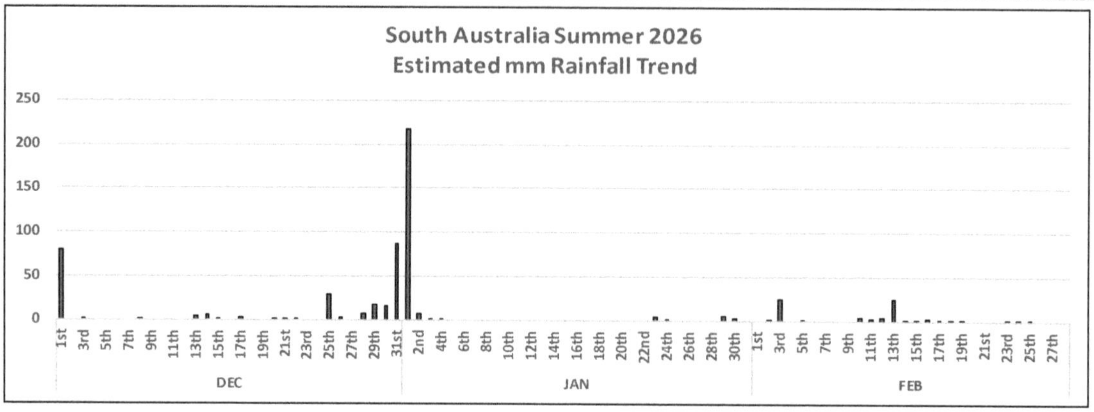

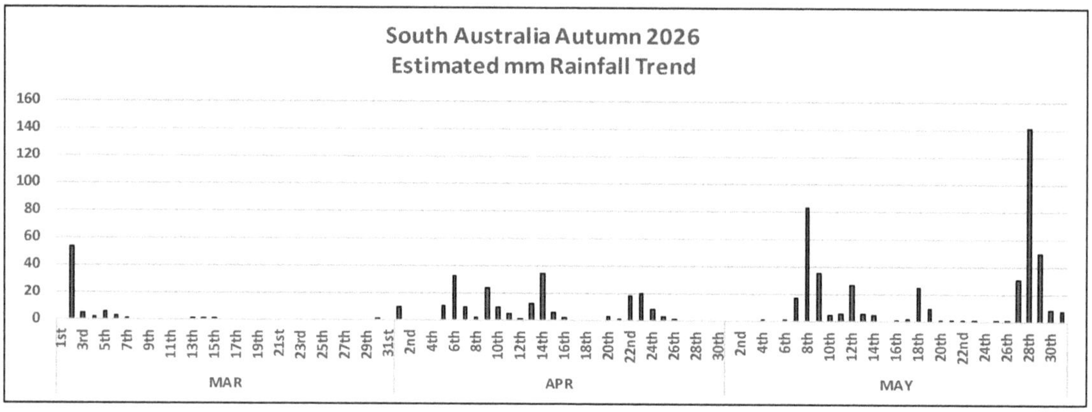

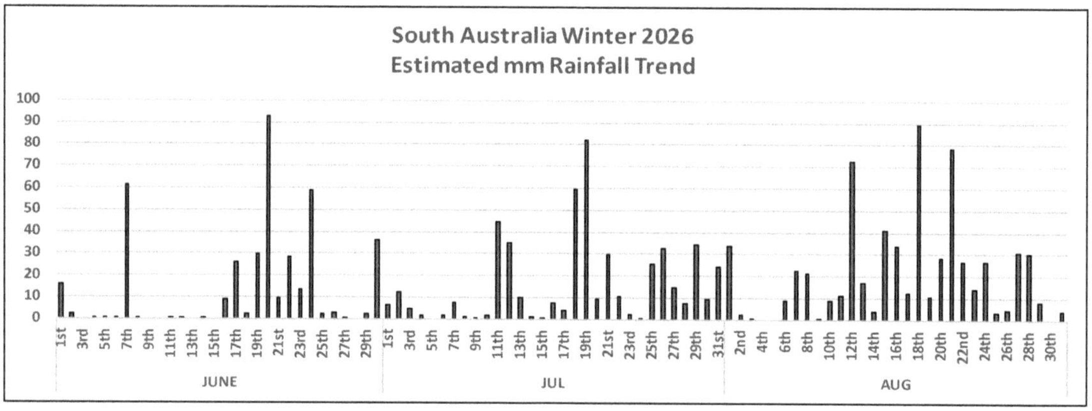

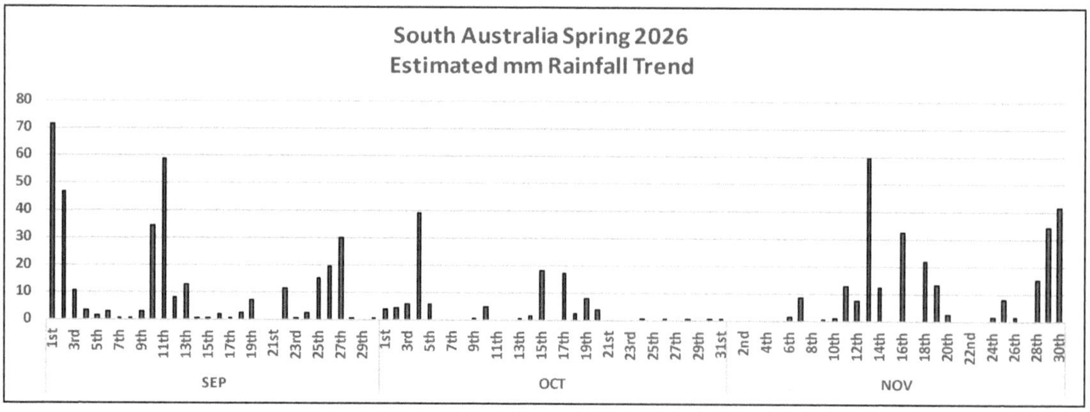

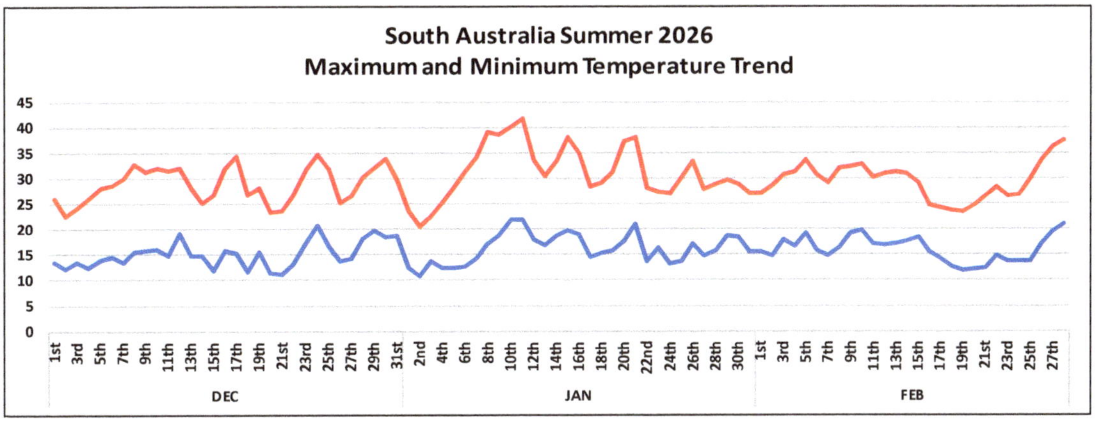
South Australia Summer 2026
Maximum and Minimum Temperature Trend
DEC
JAN
FEB

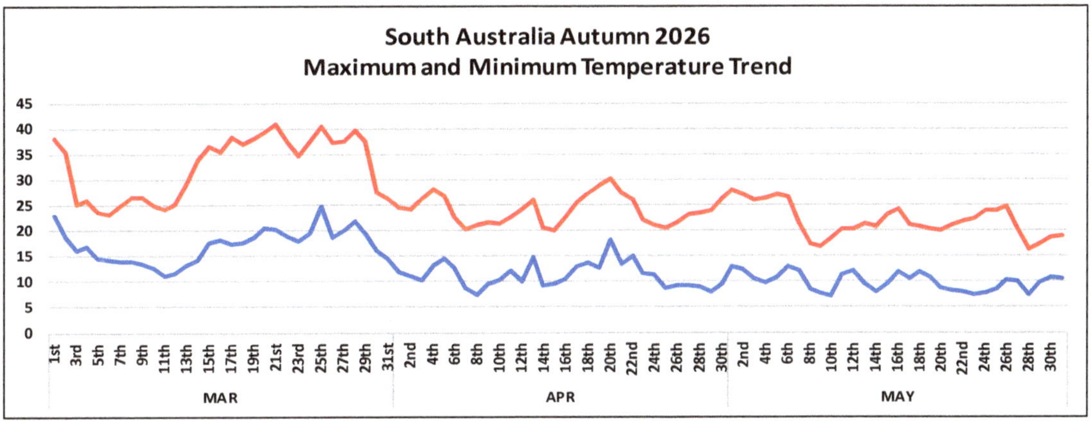
South Australia Autumn 2026
Maximum and Minimum Temperature Trend
MAR
APR
MAY

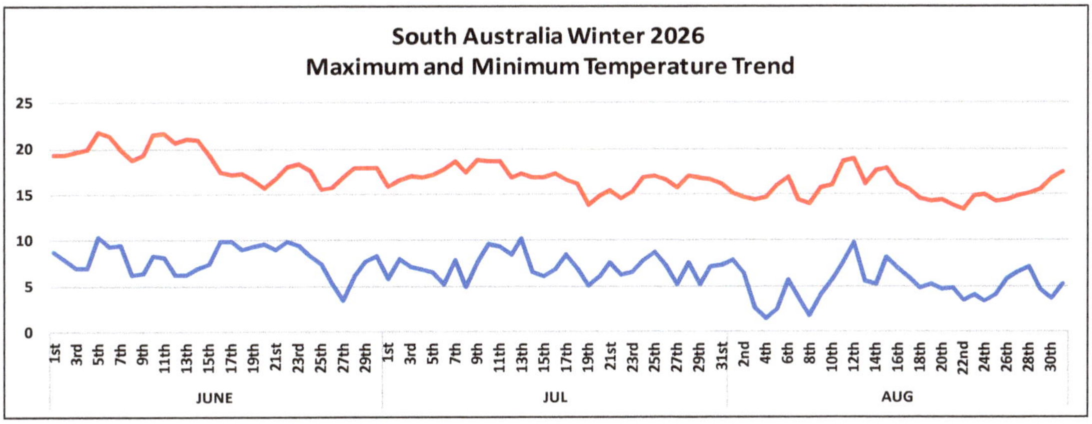
South Australia Winter 2026
Maximum and Minimum Temperature Trend
JUNE
JUL
AUG

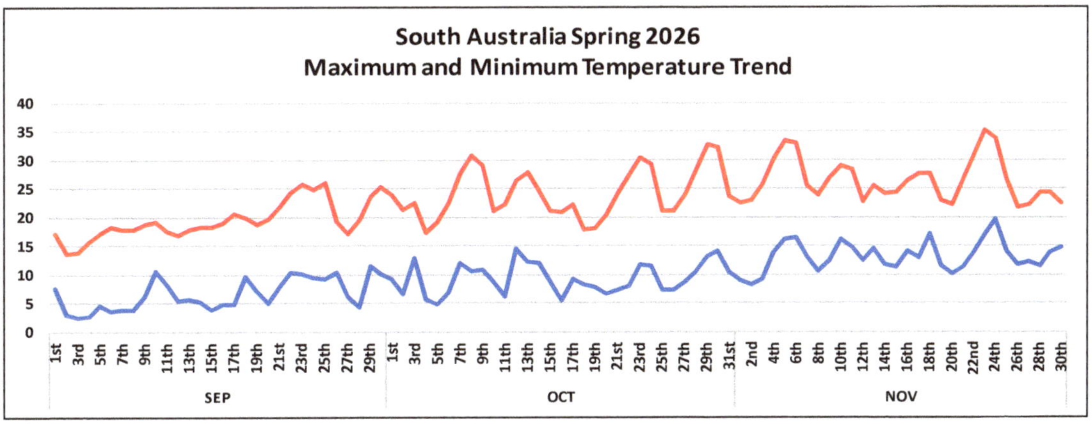
South Australia Spring 2026
Maximum and Minimum Temperature Trend
SEP
OCT
NOV

2026 SA - Estimated Seasonal Rainfall Expectations (mm)

REGION	NW Pastoral	West Coast	Inland West Coast	Far NE	Lower Eyre Peninsula	Eastern Eyre Peninsula	Spencers Gulf	Greater Adelaide	Mid-North	Murraylands	Yorke Peninsula	Upper Southeast	Lower Southeast	SA
Town Sample	Coober Pedy	Ceduna	Wudinna	Marree	Pt Lincoln	Cleve	Pt Augusta	Adelaide	Port Pirie	Murrays Bridge	Victor Harbour	Keith	Mt Gambier	Overall
Summer	10	38	65	35	55	52	44	49	40	42	51	43	40	564
Autumn	0	42	47	2	73	55	14	73	39	57	132	77	129	740
Winter	15	59	73	25	187	147	50	135	93	117	244	158	227	1529
Spring	35	39	47	39	85	58	50	35	37	45	92	50	126	735
Overall	91	209	246	156	407	330	229	317	278	290	554	365	639	4111

2026 SA - Estimated Seasonal Rainfall Variation from the Average (mm)

REGION	NW Pastoral	West Coast	Inland West Coast	Far NE	Lower Eyre Peninsula	Eastern Eyre Peninsula	Spencers Gulf	Greater Adelaide	Mid North	Murraylands	Yorke Peninsula	Upper SE	Lower SE	SA
Town Sample	Coober Pedy	Ceduna	Wudinna	Marree	Pt Lincoln	Cleve	Pt Augusta	Adelaide	Port Pirie	Murrays Bridge	Victor Harbour	Keith	Mt Gambier	Overall
Summer	-53	-6	19	-23	5	-13	-11	-6	-20	-17	-9	-23	-48	-205
Autumn	-34	-13	-9	-30	-40	-36	-38	-29	-35	-22	13	-29	-23	-325
Winter	-15	-45	-32	-8	-53	6	-18	-37	-18	-10	17	-9	-64	-285
Spring	-2	-28	-22	-2	-23	-48	-20	-72	-60	-54	-34	-73	-53	-491
Overall	-72	-61	-30	-7	-103	-73	-15	-121	-64	-74	23	-95	-69	-761

2026 SA - Estimated Seasonal Rainfall Trend Variation by 10mm

REGION	NW Pastoral	West Coast	Inland West Coast	Far NE	Lower Eyre Peninsula	Eastern Eyre Peninsula	Spencers Gulf	Greater Adelaide	Mid North	Murraylands	Yorke Peninsula	Upper SE	Lower SE	SA
Town Sample	Coober Pedy	Ceduna	Wudinna	Marree	Pt Lincoln	Cleve	Pt Augusta	Adelaide	Port Pirie	Murrays Bridge	Victor Harbour	Keith	Mt Gambier	Overall
Summer	drr	av	av	av	av	av	av	av	av	av	av	av	drr	drr
Autumn	drr	av	av	drr	drr	drr	drr	drr	av	av	av	av	av	drr
Winter	av	drr	drr	drr	av	av	av	av	drr	av	av	av	drr	drr
Spring	av	av	av	av	drr	av	av	drr	drr	drr	drr	drr	drr	drr
Overall	av	av	av	drr	av	av	av	av	av	av	av	av	av	drr

2026 SA - Estimated Seasonal Rainfall Expectation by %

REGION	NW Pastoral	West Coast	Inland West Coast	Far NE	Lower Eyre Peninsula	Eastern Eyre Peninsula	Spencers Gulf	Greater Adelaide	Mid North	Murraylands	Yorke Peninsula	Upper SE	Lower SE	SA
Town Sample	Coober Pedy	Ceduna	Wudinna	Marree	Pt Lincoln	Cleve	Pt Augusta	Adelaide	Port Pirie	Murrays Bridge	Victor Harbour	Keith	Mt Gambier	Overall
Summer	-84%	-14%	40%	-40%	11%	-20%	-19%	-12%	-33%	-29%	-15%	-35%	-54%	-27%
Autumn	-100%	-24%	-16%	-94%	-36%	-39%	-73%	-28%	-47%	-28%	11%	-27%	-15%	-31%
Winter	-51%	-43%	-30%	-24%	-22%	4%	-26%	-21%	-16%	-8%	7%	-5%	-22%	-16%
Spring	-5%	-42%	-32%	-4%	-21%	-45%	-29%	-67%	-62%	-55%	-27%	-60%	-30%	-40%
Overall	-44%	-23%	-11%	-4%	-20%	-18%	-6%	-28%	-19%	-20%	4%	-21%	-10%	-16%

2026 SA - Estimated Seasonal Rainfall Trend Variation by 25%

REGION	NW Pastoral	West Coast	Inland West Coast	Far NE	Lower Eyre Peninsula	Eastern Eyre Peninsula	Spencers Gulf	Greater Adelaide	Mid North	Murraylands	Yorke Peninsula	Upper SE	Lower SE	SA
Town Sample	Coober Pedy	Ceduna	Wudinna	Marree	Pt Lincoln	Cleve	Pt Augusta	Adelaide	Port Pirie	Murrays Bridge	Victor Harbour	Keith	Mt Gambier	Overall
Summer	drr	av	wtr	drr	av	av	av	av	drr	drr	av	drr	drr	drr
Autumn	drr	av	av	drr	drr	drr	drr	drr	drr	drr	av	drr	av	drr
Winter	drr	drr	drr	av	av	av	drr	av	av	av	av	av	av	av
Spring	av	drr	drr	av	av	drr	drr	drr	drr	drr	drr	drr	drr	drr
Overall	wtr	wtr	wtr	wtr	wtr	wtr	wtr	wtr	wtr	wtr	wtr	wtr	wtr	wtr

2026 SA - Estimated Seasonal Maximum Temperature Average and Trend

Region	NW Pastoral	West Coast	Inland West Coast	Far NE	Lower Eyre Peninsula	Eastern Eyre Peninsula	Spencers Gulf	Greater Adelaide	Mid North	Murraylands	Yorke Peninsula	Upper SE	Lower SE	SA
Town Sample	Coober Pedy	Ceduna	Wudinna	Marree	Pt Lincoln	Cleve	Pt Augusta	Adelaide	Port Pirie	Murrays Bridge	Victor Harbour	Keith	Mt Gambier	Overall
Summer Est	35.6	28.5	32.7	25.5	27.9	37.7	33.4	31.9	28.1	23.2	29.9	29.9	25.3	30.0
Trend	av	av	av	av	av	av	av	a-w	a-w	clr	av	a-w	a-w	av
Autumn Est	28.6	27.0	27.8	24.0	25.3	30.8	28.3	27.2	24.5	23.0	26.5	24.8	22.1	26.1
Trend	wmr	wmr	wmr	wmr	wmr	wmr	wmr	wmr	wmr	wmr	wmr	wmr	wmr	wmr
Winter Est	18.9	18.8	17.9	16.4	16.0	20.1	18.2	17.4	15.8	15.8	16.8	15.4	13.8	17.0
Trend	av	a-w	av	av	av	av	av	av	av	av	av	av	av	av
Spring Est	27.2	24.8	25.3	20.5	22.7	29.0	26.1	24.6	21.3	20.1	23.5	21.5	18.2	23.4
Trend	clr	a-w	a-c	av	av	clr	clr	av	av	av	av	av	av	av

2026 SA - Estimated Seasonal Minimum Temperature Average and Trend

Region	NW Pastoral	West Coast	Inland West Coast	Far NE	Lower Eyre Peninsula	Eastern Eyre Peninsula	Spencers Gulf	Greater Adelaide	Mid North	Murraylands	Yorke Peninsula	Upper SE	Lower SE	SA
Town Sample	Coober Pedy	Ceduna	Wudinna	Marree	Pt Lincoln	Cleve	Pt Augusta	Adelaide	Port Pirie	Murrays Bridge	Victor Harbour	Keith	Mt Gambier	Overall
Summer Est	20.9	14.6	14.1	21.7	15.4	15.4	18.7	15.6	17.9	13.9	13.7	12.8	11.5	15.9
Trend	av	av	a-c	av	av	av	av	av	a-w	av	av	av	av	av
Autumn Est	15.4	12.3	10.8	15.3	14.5	13.6	13.8	13.4	15.4	10.7	10.8	10.6	10.0	12.8
Trend	wmr	wmr	av	wmr	wmr	wmr	a-w	wmr	wmr	a-w	av	wmr	wmr	wmr
Winter Est	6.5	6.2	4.6	6.4	7.9	7.8	6.2	7.8	8.9	6.2	7.3	5.7	6.1	6.7
Trend	av	av	clr	a-w	av	av	a-w	av	a-w	wmr	av	av	a-w	av
Spring Est	13.0	9.9	7.8	13.1	9.0	9.6	11.3	10.2	12.1	8.2	8.6	7.5	7.0	9.8
Trend	clr	av	clr	a-c	a-c	a-c	a-c	a-c	av	av	a-c	a-c	av	a-c

Monthly Breakdowns

2026 SOUTH AUSTRALIA - Estimated Rainfall Expectations (mm)

REGION	NW Pastoral	West Coast	Inland West Coast	Far NE	Lower Eyre Peninsula	Eastern Eyre Peninsula	Spencers Gulf	Greater Adelaide	Mid-North	Murraylands	Yorke Peninsula	Upper Southeast	Lower Southeast	SA
Town Sample	Coober Pedy	Ceduna	Wudinna	Marree	Pt Lincoln	Cleve	Pt Augusta	Adelaide	Port Pirie	Murrays Bridge	Victor Harbour	Keith	Mt Gambier	Overall
January	0	3	10	9	18	25	18	26	27	24	28	32	19	239
February	0	3	4	22	7	0	6	10	0	3	5	1	8	70
March	0	8	6	0	10	6	1	2	2	2	16	2	16	72
April	0	6	7	2	17	15	0	24	14	11	42	16	56	209
May	0	28	34	0	46	34	13	47	23	44	74	59	57	459
June	15	17	9	19	44	45	30	29	27	28	52	32	49	397
July	0	22	33	2	55	38	6	38	22	28	89	65	85	483
August	0	20	31	4	88	64	15	68	44	60	103	61	93	649
September	4	14	18	5	38	33	8	18	20	23	57	29	78	344
October	0	5	3	0	3	10	1	9	6	14	21	14	30	117
November	31	19	26	34	44	15	41	8	10	8	14	7	18	274
December	41	63	65	59	38	45	91	36	82	44	55	48	131	798
Overall	91	209	246	156	407	330	229	317	278	290	554	365	639	4111

2026 SOUTH AUSTRALIA - Estimated Rainfall Variation from the Average (mm)

REGION	NW Pastoral	West Coast	Inland West Coast	Far NE	Lower Eyre Peninsula	Eastern Eyre Peninsula	Spencers Gulf	Greater Adelaide	Mid North	Murraylands	Yorke Peninsula	Upper SE	Lower SE	SA
Town Sample	Coober Pedy	Ceduna	Wudinna	Marree	Pt Lincoln	Cleve	Pt Augusta	Adelaide	Port Pirie	Murrays Bridge	Victor Harbour	Keith	Mt Gambier	Overall
January	-20	-9	-5	-12	5	8	2	10	7	8	9	12	-9	6
February	-18	-5	-11	3	-6	-22	-11	-3	-13	-9	-9	-19	-11	-135
March	-18	-8	-10	-13	-16	-15	-16	-21	-19	-18	-10	-19	-23	-206
April	-6	-11	-10	-4	-17	-14	-13	-7	-8	-16	3	-16	7	-111
May	-10	5	11	-13	-8	-6	-9	-2	-9	12	20	7	-7	-8
June	1	-15	-27	8	-35	-2	4	-32	-13	-16	-23	-20	-42	-211
July	-9	-18	0	-11	-27	-8	-15	-20	-15	-11	12	9	-13	-124
August	-8	-12	-5	-5	9	16	-6	15	10	17	27	2	-9	51
September	-5	-12	-7	-5	-13	-9	-17	-30	-18	-16	1	-21	2	-150
October	-15	-20	-18	-16	-33	-26	-21	-27	-25	-19	-21	-27	-28	-295
November	18	4	3	19	23	-13	18	-16	-17	-20	-14	-25	-26	-46
December	16	40	49	42	13	19	70	11	55	14	27	22	90	468
Overall	-72	-61	-30	-7	-103	-73	-15	-121	-64	-74	23	-95	-69	-761

2026 SOUTH AUSTRALIA - Estimated Rainfall Trend Variation by 10mm

REGION	NW Pastoral	West Coast	Inland West Coast	Far NE	Lower Eyre Peninsula	Eastern Eyre Peninsula	Spencers Gulf	Greater Adelaide	Mid North	Murraylands	Yorke Peninsula	Upper SE	Lower SE	SA
Town Sample	Coober Pedy	Ceduna	Wudinna	Marree	Pt Lincoln	Cleve	Pt Augusta	Adelaide	Port Pirie	Murrays Bridge	Victor Harbour	Keith	Mt Gambier	Overall
January	drr	av	av	av	av	drr	av	av	av	av	av	wtr	av	av
February	drr	av	drr	av	drr	av	drr	drr	av	av	av	drr	drr	drr
March	drr	av	av	drr	drr	drr	drr	drr	drr	av	drr	drr	drr	drr
April	av	drr	av	drr	drr	av	drr	av	av	av	drr	drr	av	drr
May	av	av	wtr	av	av	drr	av	av	av	wtr	wtr	av	av	av
June	av	drr	drr	drr	av	av	av	drr	drr	drr	drr	drr	drr	drr
July	av	drr	av	drr	av	drr	drr	drr	drr	wtr	drr	av	drr	drr
August	av	drr	av	av	wtr	av	av	av	wtr	wtr	wtr	av	av	wtr
September	av	drr	av	drr	av	av	drr	drr	drr	av	drr	drr	av	drr
October	drr	drr	drr	drr	drr	drr	drr	drr	drr	drr	drr	drr	drr	drr
November	wtr	av	av	wtr	drr	wtr	wtr	drr	drr	drr	drr	drr	drr	drr
December	wtr	wtr	wtr	wtr	wtr	wtr	wtr	wtr	wtr	wtr	wtr	wtr	wtr	wtr
Overall	av	av	av	drr	av	av	av	av	av	av	av	av	av	drr

2026 SOUTH AUSTRALIA - Estimated Rainfall Expectation by %

REGION	NW Pastoral	West Coast	Inland West Coast	Far NE	Lower Eyre Peninsula	Eastern Eyre Peninsula	Spencers Gulf	Greater Adelaide	Mid North	Murraylands	Yorke Peninsula	Upper SE	Lower SE	SA
Town Sample	Coober Pedy	Ceduna	Wudinna	Marree	Pt Lincoln	Cleve	Pt Augusta	Adelaide	Port Pirie	Murrays Bridge	Victor Harbour	Keith	Mt Gambier	Overall
January	-100%	-74%	-33%	-56%	37%	50%	12%	56%	37%	50%	51%	65%	-33%	3%
February	-100%	-61%	-73%	17%	-47%	-99%	-66%	-22%	-100%	-74%	-65%	-95%	-57%	-66%
March	-100%	-51%	-63%	-100%	-61%	-71%	-93%	-90%	-89%	-89%	-39%	-92%	-58%	-74%
April	-100%	-63%	-56%	-70%	-50%	-49%	-100%	-22%	-36%	-60%	8%	-50%	15%	-35%
May	-100%	24%	45%	-100%	-14%	-15%	-41%	-3%	-27%	36%	37%	13%	-11%	-2%
June	9%	-46%	-75%	73%	-44%	-5%	15%	-52%	-33%	-36%	-31%	-38%	-46%	-35%
July	-100%	-45%	1%	-83%	-33%	-17%	-71%	-34%	-40%	-29%	16%	16%	-13%	-20%
August	-100%	-38%	-14%	-57%	12%	33%	-31%	29%	30%	39%	36%	4%	-9%	9%
September	-57%	-46%	-28%	-53%	-25%	-22%	-69%	-62%	-47%	-40%	2%	-42%	2%	-30%
October	-100%	-79%	-85%	-100%	-92%	-71%	-95%	-74%	-80%	-57%	-50%	-66%	-49%	-72%
November	139%	28%	13%	130%	107%	-45%	75%	-66%	-63%	-72%	-51%	-79%	-59%	-14%
December	66%	170%	301%	241%	54%	74%	334%	43%	207%	47%	97%	83%	222%	142%
Overall	-44%	-23%	-11%	-4%	-20%	-18%	-6%	-28%	-19%	-20%	4%	-21%	-10%	-16%

2026 SOUTH AUSTRALIA - Estimated Rainfall Trend Variation by 25%

REGION	NW Pastoral	West Coast	Inland West Coast	Far NE	Lower Eyre Peninsula	Eastern Eyre Peninsula	Spencers Gulf	Greater Adelaide	Mid North	Murraylands	Yorke Peninsula	Upper SE	Lower SE	SA
Town Sample	Coober Pedy	Ceduna	Wudinna	Marree	Pt Lincoln	Cleve	Pt Augusta	Adelaide	Port Pirie	Murrays Bridge	Victor Harbour	Keith	Mt Gambier	Overall
January	drr	drr	drr	drr	wtr	wtr	av	wtr	wtr	wtr	wtr	wtr	drr	av
February	drr	drr	drr	av	drr	drr	drr	av	drr	drr	drr	drr	drr	drr
March	drr	drr	drr	drr	drr	drr	drr	drr	drr	drr	drr	drr	drr	drr
April	drr	drr	drr	drr	drr	drr	drr	av	drr	drr	av	drr	av	drr
May	drr	av	wtr	drr	av	av	drr	av	drr	wtr	av	av	av	av
June	av	drr	drr	wtr	drr	av	av	drr	drr	drr	drr	drr	drr	drr
July	drr	drr	av	drr	drr	av	drr	drr	drr	drr	av	av	av	av
August	drr	drr	av	drr	av	wtr	drr	wtr	wtr	wtr	wtr	av	av	av
September	drr	drr	drr	drr	av	av	drr	drr	drr	drr	av	drr	av	drr
October	drr	drr	drr	drr	drr	drr	drr	drr	drr	drr	drr	drr	drr	drr
November	wtr	wtr	av	wtr	wtr	drr	wtr	drr	drr	drr	drr	drr	drr	av
December	wtr	wtr	wtr	wtr	wtr	wtr	wtr	wtr	wtr	wtr	wtr	wtr	wtr	wtr
Overall	wtr	wtr	wtr	wtr	wtr	wtr	wtr	wtr	wtr	wtr	wtr	wtr	wtr	wtr

2026 SOUTH AUSTRALIA - Estimated Average Maximum Temperature

Region	NW Pastoral	West Coast	Inland West Coast	Far NE	Lower Eyre Peninsula	Eastern Eyre Peninsula	Spencers Gulf	Greater Adelaide	Mid-North	Murraylands	Yorke Peninsula	Upper Southeast	Lower Southeast	SA
Town Sample	Coober Pedy	Ceduna	Wudinna	Marree	Pt Lincoln	Cleve	Pt Augusta	Adelaide	Port Pirie	Murrays Bridge	Victor Harbour	Keith	Mt Gambier	Overall
January	37	29	34	39	27	29	35	29	33	31	25	31	27	**31**
February	36	28	33	37	25	28	33	28	32	30	22	30	25	**30**
March	35	33	35	37	29	32	35	31	34	33	28	32	28	**33**
April	27	24	25	29	22	23	26	22	25	24	21	23	20	**24**
May	24	23	23	26	21	21	24	20	23	22	20	20	18	**22**
June	20	21	20	22	18	18	20	18	20	18	17	17	15	**19**
July	19	19	18	19	16	16	18	15	17	16	16	15	13	**17**
August	18	17	16	19	15	14	17	14	16	16	14	14	13	**16**
September	24	21	21	24	17	18	22	18	21	19	17	17	15	**20**
October	29	26	26	31	22	24	27	22	25	24	21	22	18	**24**
November	29	28	29	32	23	26	30	24	28	27	23	25	22	**27**
December	32	25	27	32	22	25	29	23	26	25	21	23	19	**25**
Overall	**27**	**24**	**26**	**29**	**21**	**23**	**26**	**22**	**25**	**24**	**20**	**22**	**19**	**24**

2026 SOUTH AUSTRALIA - Estimated Average Minimum Temperature

Region	NW Pastoral	West Coast	Inland West Coast	Far NE	Lower Eyre Peninsula	Eastern Eyre Peninsula	Spencers Gulf	Greater Adelaide	Mid-North	Murraylands	Yorke Peninsula	Upper Southeast	Lower Southeast	SA
Town Sample	Coober Pedy	Ceduna	Wudinna	Marree	Pt Lincoln	Cleve	Pt Augusta	Adelaide	Port Pirie	Murrays Bridge	Victor Harbour	Keith	Mt Gambier	Overall
January	22	15	15	22	16	16	19	16	18	14	14	13	12	**16**
February	22	15	14	21	17	16	19	16	18	14	13	13	12	**16**
March	21	16	16	21	17	18	19	18	20	15	14	15	13	**17**
April	14	10	8	15	14	12	13	12	13	9	10	9	9	**11**
May	11	10	9	10	13	11	9	10	12	8	8	8	8	**10**
June	8	7	6	9	9	10	9	9	11	7	8	6	7	**8**
July	6	7	5	6	7	8	6	9	9	7	8	7	7	**7**
August	5	4	3	5	7	6	4	6	7	5	6	4	4	**5**
September	9	6	4	9	7	7	7	7	9	5	6	5	5	**7**
October	13	10	7	13	8	9	11	10	12	7	8	7	7	**9**
November	17	14	12	17	12	12	16	14	16	12	12	11	9	**13**
December	18	14	12	19	14	12	16	14	15	12	12	11	10	**14**
Overall	14	11	9	14	12	11	12	12	13	10	10	9	9	**11**

2026 SOUTH AUSTRALIA - Estimated Degree of Variation from the Maximum Average

Region Town Sample	NW Pastoral Coober Pedy	West Coast Ceduna	Inland West Coast Wudinna	Far NE Marree	Lower Eyre Peninsula Pt Lincoln	Eastern Eyre Peninsula Cleve	Spencers Gulf Pt Augusta	Greater Adelaide Adelaide	Mid North Port Pirie	Murraylands Murrays Bridge	Yorke Peninsula Victor Harbour	Upper SE Keith	Lower SE Mt Gambier	SA Overall
January	0.2	0.3	0.3	0.3	0.5	0.0	0.4	1.3	1.3	0.5	-0.2	1.2	1.4	0.6
February	0.0	-0.3	0.0	-0.1	-1.1	-0.9	-0.1	0.0	0.2	-0.2	-2.6	-0.1	-0.5	-0.4
March	3.2	6.4	5.6	3.6	4.3	5.5	4.4	5.3	5.1	5.5	4.5	5.2	5.0	4.9
April	-0.4	0.2	-0.1	0.1	-0.4	-0.4	-0.7	0.1	-0.2	0.8	-0.6	0.0	0.6	-0.1
May	1.9	2.5	2.1	2.4	1.6	1.8	2.0	1.7	2.1	2.7	1.1	1.1	1.7	1.9
June	1.5	2.4	2.5	2.7	1.1	1.9	2.0	1.8	2.6	2.3	1.2	1.7	1.5	1.9
July	0.4	1.4	0.8	0.0	0.2	0.1	0.0	0.4	0.5	0.7	0.1	-0.2	0.1	0.3
August	-3.3	-1.8	-2.5	-3.0	-1.9	-2.6	-3.3	-1.8	-2.4	-1.5	-1.9	-2.1	-1.4	-2.3
September	-2.2	-0.2	-1.4	-2.5	-1.8	-1.7	-1.8	-0.4	-0.8	-1.2	-1.6	-1.5	-1.0	-1.4
October	-0.5	1.6	0.1	0.4	1.1	0.6	-0.7	0.6	0.6	-0.1	0.2	0.3	-0.1	0.3
November	-2.7	1.4	-0.9	-1.5	-0.6	-0.1	-0.8	0.4	0.5	0.1	0.0	0.4	1.2	-0.2
December	-2.9	-2.2	-4.1	-4.7	-2.9	-2.6	-3.8	-2.5	-3.5	-3.9	-3.4	-4.6	-3.8	-3.4
Overall	-0.4	1.0	0.2	-0.2	0.1	0.1	-0.2	0.5	0.4	0.5	-0.3	0.1	0.4	0.2

2026 SOUTH AUSTRALIA - Estimated Degree of Variation from the Minimum Average

Region Town Sample	NW Pastoral Coober Pedy	West Coast Ceduna	Inland West Coast Wudinna	Far NE Marree	Lower Eyre Peninsula Pt Lincoln	Eastern Eyre Peninsula Cleve	Spencers Gulf Pt Augusta	Greater Adelaide Adelaide	Mid North Port Pirie	Murraylands Murrays Bridge	Yorke Peninsula Victor Harbour	Upper SE Keith	Lower SE Mt Gambier	SA Overall
January	-0.6	-0.3	-1.0	-1.6	0.2	0.2	-0.7	-0.2	0.4	-0.1	-0.6	0.1	0.4	-0.3
February	-0.1	-0.3	-1.0	-0.8	0.4	-0.6	0.6	-0.1	0.4	-0.5	-1.2	-0.4	-0.2	-0.3
March	2.5	2.9	2.6	2.2	2.0	3.3	2.8	3.3	4.3	2.3	1.6	3.1	2.2	2.7
April	-0.5	-0.6	-2.0	0.5	1.5	-0.5	-0.6	0.1	0.3	-0.4	-0.4	-0.5	0.1	-0.2
May	1.3	1.9	-0.1	0.7	2.6	0.7	0.5	0.6	1.7	0.7	-0.7	0.6	1.0	0.9
June	1.3	0.9	-0.4	2.9	0.7	1.6	2.6	0.8	2.1	1.5	0.6	-0.2	1.0	1.2
July	-0.2	1.1	-0.9	1.0	-0.4	0.3	1.5	1.6	1.6	2.4	1.4	1.4	1.9	1.0
August	-2.5	-1.8	-3.1	-1.9	0.2	-1.4	-1.3	-1.5	-1.4	0.6	-0.6	-1.6	-1.0	-1.3
September	-2.1	-1.6	-2.7	-1.8	-1.1	-1.6	-0.9	-2.1	-0.7	-0.1	-1.4	-2.0	-1.2	-1.5
October	-1.3	-0.4	-1.5	-0.7	-1.2	-0.5	-0.9	-0.9	-0.3	-0.1	-1.0	-1.2	-0.5	-0.8
November	-0.8	1.7	0.1	-0.6	-0.1	0.1	0.2	1.0	1.2	1.4	0.4	0.7	0.7	0.5
December	-2.0	-0.4	-2.1	-2.0	0.2	-2.2	-1.6	-0.7	-1.3	-0.7	-0.7	-1.4	-0.1	-1.2
Overall	-0.4	0.3	-1.0	-0.2	0.4	-0.1	0.2	0.2	0.7	0.6	-0.2	-0.1	0.3	0.0

2026 SOUTH AUSTRALIA - Estimated Maximum Temperature Trend

Region Town Sample	NW Pastoral Coober Pedy	West Coast Ceduna	Inland West Coast Wudinna	Far NE Marree	Lower Eyre Peninsula Pt Lincoln	Eastern Eyre Peninsula Cleve	Spencers Gulf Pt Augusta	Greater Adelaide Adelaide	Mid North Port Pirie	Murraylands Murrays Bridge	Yorke Peninsula Victor Harbour	Upper SE Keith	Lower SE Mt Gambier	SA Overall
January	av	av	av	av	av	av	av	wmr	wmr	av	av	wmr	wmr	a-w
February	av	av	av	av	clr	a-c	av	av	av	av	clr	av	av	av
March	wmr	wmr	wmr	wmr	wmr	wmr	wmr	wmr	wmr	wmr	wmr	wmr	wmr	wmr
April	av	av	av	av	av	av	a-c	av	av	a-w	a-c	av	a-w	av
May	wmr	wmr	wmr	wmr	wmr	wmr	wmr	wmr	wmr	wmr	wmr	wmr	wmr	wmr
June	wmr	wmr	wmr	wmr	wmr	wmr	wmr	wmr	wmr	wmr	wmr	wmr	wmr	wmr
July	av	wmr	a-w	av	av	av	av	av	av	a-w	av	av	av	av
August	clr	clr	clr	clr	clr	clr	clr	clr	clr	clr	clr	clr	clr	clr
September	clr	av	clr	clr	clr	clr	clr	av	a-c	clr	clr	clr	a-c	clr
October	av	wmr	av	av	wmr	a-w	a-c	a-w	a-w	av	av	av	av	av
November	clr	wmr	a-c	clr	a-c	av	a-c	av	av	av	av	av	wmr	av
December	clr	clr	clr	clr	clr	clr	clr	clr	clr	clr	clr	clr	clr	clr
Overall	av	a-w	av	av	av	av	av	av	av	av	av	av	av	av

2026 SOUTH AUSTRALIA - Estimated Minimum Temperature Trend

Region Town Sample	NW Pastoral Coober Pedy	West Coast Ceduna	Inland West Coast Wudinna	Far NE Marree	Lower Eyre Peninsula Pt Lincoln	Eastern Eyre Peninsula Cleve	Spencers Gulf Pt Augusta	Greater Adelaide Adelaide	Mid North Port Pirie	Murraylands Murrays Bridge	Yorke Peninsula Victor Harbour	Upper SE Keith	Lower SE Mt Gambier	SA Overall
January	a-c	av	clr	av	av	av	a-c	av	av	av	a-c	av	av	av
February	av	av	a-c	a-c	av	a-c	a-w	av	av	av	clr	av	av	av
March	wmr	wmr	wmr	wmr	wmr	wmr	wmr	wmr	wmr	wmr	wmr	wmr	wmr	wmr
April	av	a-c	clr	av	wmr	av	a-c	av	av	av	av	av	av	av
May	wmr	wmr	av	a-w	wmr	a-w	av	a-w	wmr	a-w	a-c	a-w	a-w	a-w
June	wmr	a-w	av	wmr	a-w	wmr	wmr	a-w	wmr	wmr	a-w	av	a-w	wmr
July	av	wmr	a-c	a-w	av	av	wmr	wmr	wmr	wmr	wmr	wmr	wmr	a-w
August	clr	clr	clr	clr	av	clr	clr	clr	clr	a-w	a-c	clr	a-c	clr
September	clr	clr	clr	clr	clr	clr	a-c	clr	a-c	av	clr	clr	clr	clr
October	clr	av	clr	a-c	clr	av	a-c	a-c	av	av	a-c	clr	av	a-c
November	a-c	wmr	av	a-c	av	av	av	a-w	wmr	wmr	av	a-w	a-w	av
December	clr	av	clr	clr	av	clr	clr	a-c	clr	a-c	a-c	clr	av	clr
Overall	av	av	a-c	av	av	av	av	av	a-w	a-w	av	av	av	av

WESTERN AUSTRALIA 2026

GENERAL TRENDS

Year Overall: Slightly drier but within 5% of the average
Cooler by around 2-2.5°C below the average

Summer: Drier by around 10-15% below the average
Wettest – Second half of February
Driest – First half of December and second half of January
Cooler by around 4.5-5°C below the average
Hottest – First week of January
Coolest – Second half of December

Autumn: Slightly wetter but within 5% of the average
Wettest – First week of March and second week of April
Driest – Second to fourth weeks of March
Slightly warmer but within 0.5°C of the average
Hottest – Second and third weeks of March
Coolest – Second half of May

Winter: Drier by around 10-15% below the average
Wettest – Second half of June and first half of August
Driest – First week of July
Warmer by up to 0.5-1°C above the average
Hottest – First half of June
Coolest – Last week of July and mid-August

Spring: Wetter by up to 20-25% more than the average
Wettest – First three weeks of November
Driest – Second half of October
Cooler by around 4.5-5°C below the average
Hottest – First week of November
Coolest – First half of September

Major Weather Events

January 2026

Date	Day	Western Australia - January 2026	Perth	Moon
1st	Thur	Mostly dry over the first couple of weeks this month with just a couple of brief showery interruptions, mostly outside of the populous areas.		
2nd	Fri	Some showers around the Nullarbor Plains.		P6
3rd	Sat			F ^
4th	Sun			
5th	Mon	Very hot in the southwest of the State over the next couple of days from the Central Coast, through the Central Wheat Belt and Great Southern regions to the South Coast and Esperance regions.		
6th	Tue	A tropical low may develop in the Timor Sea around this time and drift off the northern coasts of the Pilbara for about a week, but is not expected to make landfall. Very hot in the North Coast region, with maximum temperatures reching into the mid-to-upper 40's in some places.		
7th	Wed			
8th	Thu	Chance of rain in the Interior. Monsoonal showers over the North Kimberley over the next few days.		
9th	Fri			XhS
10th	Sat			
11th	Sun	The next few weeks may be wetter and warmer than is typical for this time of year.		3Q
12th	Mon	Very hot in the Pilbara, Gascoyne, Murchison, and Southwest Land Division over the next couple of days with maximum temperatures into the upper 40's in many districts.		
13th	Tue	Very hot in the Central Wheat Belt region, with maximas into the mid-to-upper 40's.		
14th	Wed	Strong winds and thunderstorms in the Pilbara. Isolated shower activity through the central regions from the Murchison to the Interior, on the Nullarbor Plains, and around the North Kimberley regions. Chance of heavy rain for a time along the Kimberley coastal fringes with downpours possibly around the Faraway Bay district. Very hot in the Gascoyne - Murchison regions over the next couple of days with maximas into the upper 40's likely.		A
15th	Thur			
16th	Fri	Chance of showers in the Nullabor Plains and Eucla.		
17th	Sat		Th	V
18th	Sun	Severe thunderstorms cross the Southern Coastal region, particularly in the Mount Barker area, with large hail showers and strong gusty winds at times, that may cause some damage to local property. Severe thunderstorms with hail in the Gascoyne region, including around Mt Magnet and gusts around Lake Grace.		
19th	Mon			N
20th	Tue	Very hot in the Pilbara over the next couple of days with temperatures in the high 40's likely.		
21st	Wed			
22nd	Thu	A squall line crosses the Kimberleys bringing strong gusty winds and some heavy rain at times, particularly in the Kununurra to Fitzroy Crossing areas over the next couple of days. Chance of some brief rain in the Interior, through the Eucla and across the Nullarbor Plains to the southern coasts over the next couple of days.		
23rd	Fri	Very strong winds with gusts around the South Coast region.		
24th	Sat	Cold overnight in the Eucla, particularly in the Eyre district.		XhN
25th	Sun			
26th	Mon	A line of thunderstorms cross the Pilbara, with heavy rain, some hail showers, and possible wind damage in their path.		1Q
27th	Tue			
28th	Wed	Some light showers around the western coastal regions over the next few days.		
29th	Thur			
30th	Fri			P10 ^
31st	Sat	Thunderstorms and strong gusty winds in the Kimberley around the Fitzroy Crossing area. Very warm overnight in the Pilbara with minimum temperatures still likely to be in the mid-30's in a number of districts.		

February 2026

Date	Day	Western Australia - February 2026	Perth	Moon
1st	Sun	Heavy monsoonal rains in the North Kimberley.		
2nd	Mon	Outbreaks of showers in the eastern half of the State.		F
3rd	Tue	Thunderstorms and strong winds with gusts in the Goldfields, particularly around Leonora.		

Date	Day	Western Australia - February 2026	Perth	Moon
4th	Wed	Isolated showers mostly in the South Coast and Central Wheatbelt districts over the next couple of days.		
5th	Thu	Isolated showers in the southeast of the State.		
6th	Fri			XhS
7th	Sat			
8th	Sun	Thunderstorms with wind gusts in the Eucla area, around the Forrest district. Monsoonal shower activity in the Kimberleys on most days for the remainder of the month.		
9th	Mon	Cool spell around the South Coast region.		
10th	Tue			3Q
11th	Wed	The following 3-4 weeks may be cooler and wetter than is typical for this time of year. A wide cloud band brings outbreaks of rain in the eastern half of the State at first, before spreading throughout most regions over the week to come. Rain may be heavy at times.		A
12th	Thur	Thunderstorms cross the Kimberley, Pilbara and Great Southern regions which may bring large hail and strong gusty winds, including in the Newman area.		
13th	Fri			V
14th	Sat			
15th	Sun			
16th	Mon	Outbreaks of heavy rain over the next 3 days in the western half of the State from the Hamersley Ranges southward.	Rain	
17th	Tue		Rain, Th	
18th	Wed	Widespread heavy rain throughout the State, especially from the Kimberley through the Interior and down through the Goldfields to the southern coasts and Central Wheat Belt districts.		N
19th	Thu			
20th	Fri	Rain activity contracts to the Kimberleys around this time.		XhN
21st	Sat	A depression may form off the Kimberley coast and swiftly build to cyclonic strength as it drifts around offshore for around a week or more before eventually making landfall near the Fortescue region nearer the end of the month.		
22nd	Sun			
23rd	Mon	Very heavy rain in the Kimberley, particularly in the Kalumburu district.		
24th	Tue	Some scattered rain activity in the eastern half of the State over the remaining days of this month.		
25th	Wed			1Q P14
26th	Thur	Very warm overnight in the Pilbara.		^
27th	Fri	Very hot in the Eucla with maximum temperatures around the mid-40's particularly in the Eyre district.		
28th	Sat	A cyclonic system off the Pilbara coast directs very strong winds into the region, particularly around the Onslow, Learmonth, and Thevenard Island districts.		

March 2026

Date	Day	Western Australia - March 2026	Perth	Moon
1st	Sun	Cyclonic system may make landfall in the Fortescue region around this time before swiftly decaying, then spreading widespread rain throughout the State in the next 2-3 days. This system may bring unusually cooler than normal temperatures across the Goldfields over the next few days. Very strong winds likely in the Gascoyne region.		
2nd	Mon	Cold overnight around the South Coast, particularly in the Jarrahwood area.		
3rd	Tue	Cool day in some parts of the Gascoyne, particularly around the Geraldton district.		F
4th	Wed	Rain activity in the Kimberley over the next few days. Unusually cold in some parts of the Murchison, particularly in the Kalbarri district.		
5th	Thu	A tropical low near the Pilbara may bring strong gusty winds over Varanus Island.		XhS
6th	Fri			
7th	Sat	A strong burst of heavy rain likely in the North Kimberley.		
8th	Sun			
9th	Mon	Some isolated showers in the Nullarbor Plains around this time.		
10th	Tue			
11th	Wed			3Q A
12th	Thur	The next 3-4 weeks may see near average rainfall and temperatures throughout the State, although it may be warmer than average for overnight minimas, and there is likely to be at least two cyclonic systems affecting the State. A low pressure center that may have formed over the NT may have reached cyclonic intensity by around this time, as it crosses the Kimberleys before heading offshore in a couple of days. This system is likely to drift in and southwesterly track parallel to the Kimberley and Pilbara coastlines, before degrading without making landfall in the De Grey region within a week.		V

Date	Day	Western Australia - March 2026	Perth	Moon
13th	Fri	The next couple of weeks ahead may see heatwave temperature conditions across the Nullarbor and Eucla regions, with many days possibly seeing maximum temperatures into the late 30's to early 40's.		
14th	Sat	Rain activity becomes scarce over the southern half of the State from around this point until the end of the month, while some spotty shower activity occasionally breaks out in the Kimberley.		
15th	Sun	Thunderstorms with gusts in the Learmonth area of the Pilbara. Very hot in the Eucla over the next 4-5 days.		
16th	Mon			
17th	Tue		Showers, Th	
18th	Wed	A decaying cyclonic system of the past week may distribute some heavy rain over the Pilbara, Gascoyne, and Southwest Land Division regions about this time. Very hot overnight in the Murchison regin, particularly around the Leonora district.	Showers	
19th	Thu	Thunderstorm with gusts in the Roebourne area of the Pilbara with the chance of wind damage occuring.		**N XhN**
20th	Fri	Thunderstorms and gusty winds in the Pilbara near Thevenard Island.		
21st	Sat			
22nd	Sun			**P11**
23rd	Mon			
24th	Tue	A cold front crosses the South West bringing strong gusty winds around Cape Leeuwin. Very hot in the Eucla.		
25th	Wed			
26th	Thur			**1Q ^**
27th	Fri	Heavy rain in the North Kimberley over the next few days.		
28th	Sat			
29th	Sun	A supercell system embedded with thunderstorms cross the Central West, with heavy rain, hail showers, and damaging wind gusts in some places. Chance of funnel clouds in the Dartmoor to Geraldton areas.		
30th	Mon	Light shower activity in the northern half of the State and in the Eucla to the Nullarbor Plains.		
31st	Tue			

April 2026

Date	Day	Western Australia - April 2026	Perth	Moon
1st	Wed	Mostly dry across the State over the next few days.		
2nd	Thu	Another cyclonic system may form in the Indian Ocean during this week and track towards northern WA, but dissipate by the end of the week as it nears the Fortescue coast.		**F XhS**
3rd	Fri	Thunderstorms and strong winds in the Paraburdoo area of the Pilbara.		
4th	Sat			
5th	Sun			
6th	Mon	Heavy rainfall associated with an approaching cyclonic system likely to affect the west Pilbara and Gascoyne regions over the next couple of days.		
7th	Tue	Very cold in the Eucla overnight, particularly in the Eyre district. Very heavy rain in the Gascoyne region, particularly in the Exmouth district where pockets of localised flooding may occur.		**A**
8th	Wed	Shower activity in the northern coastal regions over the next few days.		
9th	Thur	The cyclonic system threatening the State over the past few days may have degraded to a tropical low around this time, but still brings widespread rain in the Pilbara, Gascoyne, and Southwest Land Division over the next few days.		**V**
10th	Fri	Heavy rainfall associated with a decaying cyclonic system in Perth with unusually cooler temperatures in the city over the next couple of days.	Rain	**3Q**
11th	Sat	The remnants of a tropical cyclone may bring heavy rain and possible localised flooding over the next couple of days in the Central Wheat Belt particulary in the Northam area. Strong gusty winds across Rottnest and Garden Islands in the Lower West. Chance of thunderstorms in the Southwest Land Division, with strong winds that may raise dust storms.	Rain, Th	
12th	Sun	The next 3-4 weeks may bring around average rainfall overall for this time of year, wih parts of the Kimberley, Interior, Goldfieds and southern coasts, much drier than normal, while a series of tropical lows may bring above average rainfall along the western coasts and into the Pilbar. Temperatures may also average out to close to normal for the State as a whole, despite the southwest quarter generally being colder during the next 3-4 weeks compared to normal for this time of year, elsewhere other regions may be much warmer than normal.		

Date	Day	Western Australia - April 2026	Perth	Moon
13th	Mon			
14th	Tue	Some rain in the Kimberleys.		
15th	Wed	A slow moving band of heavy rain crosses the Lower West over the next couple of days bringing some shower activity mostly into Central Wheat Belt region.	Rain	
16th	Thu		Rain, Fg	XhN
17th	Fri			N
18th	Sat	Very warm over the next 4-5 nights in the Bedout Island region.	Showers	
19th	Sun	Chance of showers in the west of the State from the Hamersley Ranges to the South Coast, with isolated showers also possible to briefly sprinkle into the Wes Kimberleys. Very hot around the South Coast region at this time with maximum temperatures possibly reaching into the low 40's in some districts.		P7
20th	Mon	Mostly dry over the coming week throughout the State.	Showers	
21st	Tue	Severe thunderstorms with strong wind gusts cross the Nullarbor and Eucla regions, with severe winds likely for a time in the Forrest area.		
22nd	Wed			^
23rd	Thur	Very warm overnight in the Kimberley region, particularly in the Broome district.		
24th	Fri	Cold overnight in the Eucla with chance of frosts in some areas, particularly in the Eyre district.		1Q
25th	Sat			
26th	Sun			
27th	Mon	Storms cross the South West region over the next couple of days bringing heavy rain, instensive quantity of hail, some large stones, and severe wind gusts, including the chance of the formation of tornadic activity, with some damage to property and local infrastructure, particularly in the Margaret River district.	Rain, Th	
28th	Tue		Rain	
29th	Wed	Widespread rain throughout most of the State except in the North Kimberleys.	Showers	XhS
30th	Thur	Cold spell around the South Coast.		

May 2026

Date	Day	Western Australia - May 2026	Perth	Moon
1st	Fri	Shower activity in the Pilbara over the next 3-4 days.	Showers, Fg	
2nd	Sat	Isolated showers briefly extend into the Gibson Desert region.	Showers, Fg	F
3rd	Sun	Light shower activity drifts southward through the southwest of the State over the next couple of days.	Showers	
4th	Mon		Showers, Fg	
5th	Tue			A
6th	Wed			V
7th	Thu			
8th	Fri	Rain moves into the Pilbara, turning to showers by tomorrow.	Showers, Th	
9th	Sat		Showers	
10th	Sun	Widespread rain activity throughout the State with patches of heavy falls, partricularly in the De Grey to the East Kimberleys. Hail showers in the Central Wheat Belt, with chance of unusually large stones in the Mukinbudin area. Heavy rain likely in the Margaret River area.	Showers	3Q
11th	Mon	Rain eases to showers and contracts to the Pilbara around this time.		
12th	Tue	The next 3-4 weeks may be unusually drier and warmer than is normal for this time of year. Unusually warm overnight in the Pilbara.		
13th	Wed	Mostly dry throughout the State over the coming week ahead.		XhN
14th	Thur	Very hot in the Kimberley with maximum temperatures possibly up into the early 40's in some districts.		
15th	Fri			
16th	Sat			
17th	Sun			N
18th	Mon			P3
19th	Tue		Fg	^
20th	Wed		Showers, Fg	
21st	Thu		Showers	
22nd	Fri	Chance of outbreaks of rain in the Gascoyne and Murchison regins.	Fg	
23rd	Sat		Rain, Fg	1Q
24th	Sun	Showers develop in the Gascoyne region and then spreads to widespread rain activity over the next couple of days throughout the southern half of the State in most regions south of the Kimberleys.	Showers, Fg	

Date	Day	Western Australia - May 2026	Perth	Moon
25th	Mon		Showers	
26th	Tue		Showers, Th	XhS
27th	Wed			
28th	Thur			
29th	Fri			
30th	Sat			
31st	Sun			F

June 2026

Date	Day	Western Australia - June 2026	Perth	Moon
1st	Mon		Showers, Fg	A
2nd	Tue	Wet spell over the southern half of the State over the next 3-4 days with passing showers in most districts south of the Gascoyne. Unusually cool around the South Coast. Very strong gusty winds at times around Cape Leeuwin.	Showers	V
3rd	Wed		Showers	
4th	Thu		Rain, Th, Fg	
5th	Fri		Showers, Fg	
6th	Sat	Very cold overnight in the Eucla with chance of frost patches in some places, particularly in the Eyre district.		
7th	Sun			
8th	Mon	Rain in the Gascoyne and moving into the Murchison over the next couple of days.	Showers	3Q
9th	Tue			XhN
10th	Wed	Thunderstorms with strong winds in the South West region, which may cause power outages and localised flooding, including in the Perth and Rockingham areas. Isolated showers in the Nullarbor Plains, Goldfields, and up into the Ecula.		
11th	Thur	Chance of thunderstorms and tornado activity crossing the Southwest region, including possibly in the Busselton and Perth districts. Very heavy rain likely in the South Coast region, particularly around the Jarrahwood district.	Rain, Th, Fg	
12th	Fri	Overall, the next 3-4 weeks may be warmer than normal for this time of year with close to average rainfall. Generally, it may be drier than average in the south and west, especially in the Goldfields and adjacent areas, while wetter than average in the Kimberley and Pilbara.		
13th	Sat	Isolated showers in the De Grey to the Fortescue regions. Very heavy rain on the South Coast around this time, particularly in the Dwellingup district. Warm spell in the Eucla to the Central Wheat Belt districts over the next couple of days, with maximum temperatures in the upper 20's in a number of areas.		
14th	Sun			
15th	Mon			N P2
16th	Tue			^
17th	Wed	Warm spell in the Kimberley, particularly overnight on Troughton Island.		
18th	Thu	Brief outbreak of rain in the Nullarbor Plains. Some heavy rain in parts of the Kimberley, particularly in the Halls Creek area.		
19th	Fri		Showers	
20th	Sat	Widespread storm activty over the southern half of the State from the Fortescue to the South Coast with showers into the Interior over the next couple of days. Chance of a tornado forming in the Rockingham area of the Lower West district causing some localised damage.	Rain, Fg	
21st	Sun	Showers spreading into the Kimberleys.		
22nd	Mon	Showers clearing throughout the State with just some isolated showers around the Pilbara and West Kimberley coasts over the next few days. Chance of lingering showers in the Murchison and Gascoyne region for a brief time.	Showers, Fg	1Q XhS
23rd	Tue	Very warm in the Kimberleys with maximum temperatures in the mid-30's in a number of districts.	Fg	
24th	Wed			
25th	Thur		Rain	
26th	Fri	A deep low pressure system off the southwest coast directs strong to gale force winds and pockets of heavy rain onto coastal areas of the Southwest region, including in the Walpole and the Albany to Geraldton districts. Wind gusts may cause some damage to property and power outages.	Rain	
27th	Sat	Widepread rain activity spreads throughout the State except in the North Kimberleys, with storms and heavy rain through the Gascoyne and Murchison over the next couple of days, clearing to mostly dry by the end of the month. Cool spell around the South Coast region.	Rain, Sn, Fg	
28th	Sun		Showers	A
29th	Mon			V
30th	Tue			F

July 2026

Date	Day	Western Australia - July 2026	Perth	Moon
1st	Wed	Mostly dry over the coming week throughout the State.		
2nd	Thu		Fg	
3rd	Fri			
4th	Sat			
5th	Sun		Showers	
6th	Mon	Brief spell of isolated showers in the Gascoyne region and around the South Coast.		
7th	Tue	Unusually cool in the Central Coast to the South Coast regions and east into the adjacent regions. Very cold overnight in the Goldfields - Esperance regions, with chance of frost patches in some areas, including in the Norseman district.	Showers	**XhN**
8th	Wed	Widespread rain activity across the State over the next few days except for in the North Kimberleys. Storms cross the Lower West, with heavy rain, possible hail, and severe winds and gusts, which may turn tornadic in some areas, possibly in the Champion Lakes district. Some damage to property and infrastructure likely with the passage of the storm.	Rain, Th	**3Q**
9th	Thur	Storms and very heavy rain spreads into the Murchison to the Gascoyne as well as along the southern coasts.	Showers	
10th	Fri	Rain eases to isolated showers in most districts south of the North Kimberley over the next couple of days.		
11th	Sat			
12th	Sun	Isolated showers linger around the Fortescue region for the next couple of days.		
13th	Mon	The next 3-4 weeks may see close to normal rainfall and minimum temperatures, while maximum temperatures may be above normal for this time of year.	Showers	**P4 ^**
14th	Tue	Isolated showers return over the southern half of the State over the next few days.	Showers, Fg	**N**
15th	Wed		Showers, Fg	
16th	Thu		Showers	
17th	Fri			
18th	Sat	Cold overnight in the Murchinson with chance of frosts in some districts, including in the Yeelirrie district.		
19th	Sun	Warm spell in the Kimberley, particularly overnight on Troughton Island.	Showers, Fg	
20th	Mon	Showers move into the Gascoyne and Murchison at first, spreading to throughout the State over the next couple of days.	Showers, Fg	**XhS**
21st	Tue	A cut-off low may form near the lower western coasts of the State and direct damaging wind gusts over the next couple of days around the South Coastal region, including in the Albany to Geraldton areas, Walpole, Esperance, Munglinup and Cape Leeuwin areas.	Showers	**1Q**
22nd	Wed	Unusually cool in the Kimberley around this time, particularly in the Wyndham district.	Showers	
23rd	Thur			
24th	Fri			
25th	Sat			
26th	Sun	Rain in the Gascoyne and Murchison at first, spreading and intensifying by tomorrow. Cold overnight in the Murchinson with chance of frosts in some districts, including in the Yeelirrie district.	Rain	**A**
27th	Mon	Widespread and heavy rain and storm activity throughout all regions south of the North Kimberley, easing to showers over the next couple of days before another burst of heavy rain in many districts near the end of the month.	Showers	**V**
28th	Tue		Rain	
29th	Wed	Cold fronts crossing the Central West to the South and Lower West regions brings severe damaging winds across the regions and adjacent coastal areas, particularly in the Geraldton area but also in the Perth and Freemantle areas, where storm surges may cause some coastal damage. Chance of funnel formation winds briefly developing with the passage of the severe squalls and storms. Power outages and damage to infrastructure in some districts likely.	Rain, Th	
30th	Thu	Very heavy rain in many districts throughout the State with only the far northern coasts of the Kimberley likely to remain dry.		**F**
31st	Fri			

August 2026

Date	Day	Western Australia - August 2026	Perth	Moon
1st	Sat	Isolated lingering showers remain over the northern parts of the State briefly.		
2nd	Sun			
3rd	Mon	Another widespread wet spell develops around this time, starting in the Gascoyne and spreading swiftly by tomorrow into most regions of the State except the North Kimberleys once more.	Rain, Fg	**XhN**

Date	Day	Western Australia - August 2026	Perth	Moon
4th	Tue	Heavy rain in many regions over the southern half of the State at first, easing by tomorrow for a short time, and then returning further widespread showers and outbreaks of rain before the end of the week in most districts except in the North Kimberley's once more.	Showers	
5th	Wed		Rain, Fg	
6th	Thur		Showers, Fg	3Q
7th	Fri		Showers, Fg	
8th	Sat	Widespread rain activity continues on most days throughout the State over the coming week ahead. Heavy rain in some Central Wheat Belt districts, including in the Paynes Find area. Very warm day in the Murchison region.	Rain	
9th	Sun	Storms across the Lower West, with heavy rain, severe wind gusts, possibly to gale force at times, and chance of storm surges around the coastal fringes.	Rain, Th	^
10th	Mon	Strong winds with sustained gales across the Southwest region with chance of storm surges along the coasts. Very heavy rain likely in the Perth region, with the chance of flash flooding in some districts, while unusually cool across the South Coast region and in contrast, unusually hot in the Eucla-Nullarbor region, particularly in the Giles district.	Showers	P8
11th	Tue	Very warm in the Kimberleys, with maximum temperatures in the mid-30's in a number of districts.	Showers	
12th	Wed	The next 3-4 weeks may be wetter than normal in the Kimberleys and parts of the Pilbara and Interior, while very much drier than normal in large areas in the west and south of the State, and temperatures overall, may be below average. Cool day in the Central Wheat Belt region. Very strong and blustery winds around Cape Leeuwin.	Rain	
13th	Thu	Rain activity clears over the northern half of the State around this time, while still persisting but easing to isolated for another day or so in the southern half. Chance of heavy rain for a time in the Medina area of the Perth region.		N
14th	Fri			
15th	Sat		Fg	
16th	Sun	Mostly dry over the next few days.		XhS
17th	Mon			
18th	Tue			
19th	Wed			
20th	Thur			1Q
21st	Fri	Chance of a brief rainy spell in the Gascoyne.		
22nd	Sat	Mostly dry throughout the State for the remainder of the month. Cold overnight in the South Coast with chance of some patches of frost in sheltered districts.		A
23rd	Sun			V
24th	Mon			
25th	Tue			
26th	Wed			
27th	Thu			
28th	Fri	Very cold spell in the Eucla, with frosts in numerous areas, particularly in the Eyre district. Chance of overnight frosts also spilling over into some Central Wheat Belt districts, including in the Northam area.		F
29th	Sat			
30th	Sun	Chance of a brief showery spell in the Gascoyne.	Showers	XhN
31st	Mon			

September 2026

Date	Day	Western Australia - September 2026	Perth	Moon
1st	Tue	Dry throughout the State over most of the coming week ahead.		
2nd	Wed			
3rd	Thur			
4th	Fri			3Q
5th	Sat	Chance of isolated showers in the Gascoyne coastal areas over the next couple of days.		
6th	Sun			^
7th	Mon	Some isolated showers in the Nullarbor Plains around this time, with shower activity increasing and becoming more widespread across the State over the next few days. Hot in the Kimberleys over the next couple of days with maximas into the upper 30's in many districts.		P12
8th	Tue		Showers, Fg	
9th	Wed		Showers, Fg	
10th	Thu	Cold day around the South Coast.		

Date	Day	Western Australia - September 2026	Perth	Moon
11th	Fri		Showers	N
12th	Sat	The next 3-4 weeks may be warmer than is normal for this time of year, while rainfall may be a mix of wetter and drier than normal areas, with the Kimberley, northern Interior, Pilbara, Gascoyne and some of the Southwest Land Division to the Goldfields possibly wetter than average, while elsewhere drier than average. Spells of heavy rain for a time in the Gascoyne to the Murchison regions.	Fg	XhS
13th	Sun	Mostly dry throughout the State over the coming week ahead.		
14th	Mon	Storm activity over the next few days in the Kimberley region with heavy rainfall in a number of districts.		
15th	Tue		Showers	
16th	Wed			
17th	Thur			
18th	Fri			
19th	Sat			1Q A V
20th	Sun			
21st	Mon		Showers	
22nd	Tue	Isolated showers in the Gascoyne, spreading into the Interior over the next couple of days.	Showers	
23rd	Wed		Showers	
24th	Thu	Heavier showers in the Gascoyne coastal areas, with light showers elsewhere over the southern half of the State.	Showers	
25th	Fri			
26th	Sat			XhN
27th	Sun		Showers, Fg	F
28th	Mon	Light showers in the Fortescue to the Murchison regions.	Showers, Fg	
29th	Tue	Showers spread throughout the State over the next couple of days, although the North Kimberleys may remain mostly dry.	Showers	
30th	Wed			

October 2026

Date	Day	Western Australia - October 2026	Perth	Moon
1st	Thur	Showers in the Gascoyne over the next couple of days.	Showers	
2nd	Fri	A cold front crosses the Lower West, bringing damaging storms into the region, including severe wind gusts and patches of heavy falls at times. Strong winds, mostly southerly, may cause some damage in some parts, including in the Norseman, Kalgoorlie, and Swanbourne districts and across Rottnest Island. Very hot temperatures move over the Kimberley region around this time, with maximum temperatures possibly reaching into the low 40's in a number of districts, including in the Curtin district. In contrast, it may be a cold day in the Great Southern region, particularly in the Katanning district.	Showers	P13
3rd	Sat	Widespread rain activity and warm temperatures throughout most of the State except in the North Kimberley. Widespread frost over the southern half of the State likely over the next couple of days, particularly in the Great Southern districts, including in parts of the Central Wheat Belt including in the Wandering district.		
4th	Sun			3Q ^
5th	Mon		Rain	
6th	Tue	Heavy rain moves onto the west of the State and spreads into all regions except the North Kimberleys over the next couple of days. Mild temperatures overnight in the Central Wheat Belt districts.	Rain	
7th	Wed	Very wet over the next couple of days across the Central Wheat Belt to the Goldfields regions with heavy rain in many districts.		
8th	Thu	Pockets of heavy rain likely in the Goldfield-Esperance region, particularly in the Norseman district.		
9th	Fri			
10th	Sat		Showers	XhS
11th	Sun	Strong winds across the Eucla, with chance of microbursts at times, possibly in the Forrest area. Light to isolated showers in the southern half of the State over the next few days.	Showers	N
12th	Mon	In a similar pattern to the previous 3-4 weeks, the next few weeks ahead may also be warmer than is normal for this time of year, and with a mix of wetter and drier than normal regions compared to what is typical for this time of year. Overall, there may be above average rain over the coming few weeks in the Kimberley, Pilbara, Gascoyne and Goldfields, while below average rainfall elsewhere.	Showers	

Date	Day	Western Australia - October 2026	Perth	Moon
13th	Tue			
14th	Wed		Showers	
15th	Thur	Mostly dry in the State over the remainder of the month, with just a couple of brief exceptions. Unusually cold overnight in the South Coast to the Eucla regions, including the chance of patches of frost, particularly in the Eyre district.	Showers	
16th	Fri	Some showers in the coastal parts of the Fortescue and Gascoyne around this time.		
17th	Sat			A V
18th	Sun			
19th	Mon	Chance of outbreaks of rain in the coastal North Kimberley as well as in the Gibson Desert.		1Q
20th	Tue	Rain activity moves into the Pilbara over the next couple of days with chance of some heavy falls at times.		
21st	Wed			
22nd	Thur			
23rd	Fri			
24th	Sat			XhN
25th	Sun			
26th	Mon			F
27th	Tue			
28th	Wed			
29th	Thu	Chance of some isolated showers around the Gascoyne.		P9
30th	Fri	Very warm spell over the coming few days in the Southwest Land Division to the Perth regions, with many areas seeing maximum temperatures into the early 30's on most days. Very warm over the next couple of nights in the Murchison.		^
31st	Sat	Unusually hot day around Cape Leeuwin.		

November 2026

Date	Day	Western Australia - November 2026	Perth	Moon
1st	Sun	Severe thunderstorm activity across the Central Wheatbelt over the next couple of days with bursts of heavy rain and strong wind gusts, including in the Cunderdin district.		
2nd	Mon	Widespread light shower activity in most districts over the next few days except the North Kimberley.		3Q
3rd	Tue		Showers	
4th	Wed	A cold front crosses from the Central Coast into the Goldfields region, as well as in the Lower West, bringing thunderstorms and strong winds with gusts and the chance of some damage to property, particularly in the Kalgoorlie area and across Rottnest Island. Very hot with maximum temperatures possibly into the early 40's in the Eucla around this time, particularly in the Eyre area.	Rain, Th	
5th	Thur	Widespread rain throughout most of the State except in the North Kimberleys.		
6th	Fri			XhS
7th	Sat			
8th	Sun	A deep surface trough brings widespread storm activity across the Central Wheat Belt to the SouthWest and Southeast Coastal regions, including the chance of severe thunderstorms, heavy rain and hail and strong gusty winds. Chance of power outages, flash flooding in a number of areas and some damage to property, including in the York, Cunderdin, Albany, Newdegate, Cape Leeuwin, Hopetoun, and mayiams districts. Large hail may damage crops.	Showers, Th	
9th	Mon	Widespread shower activity across the State over the next few days with heavy falls likely in the coastal Fortescue region at first and later along the Gascoyne coasts to the North and Central Coast regions, as well as moving into the Central Wheat Belt districts for a time.		N
10th	Tue			
11th	Wed	Very warm overnight in the Kimberleys, particularly in the Halls Creek district. Heavy rain likely around the South Coast region to the Central Wheat Belt, particularly in the Jacup and Karlgarin districts.		
12th	Thu	Generally, the next 3-4 weeks may be wetter and cooler than normal for this time of year. Mostly dry over the next few days.		
13th	Fri	Cool overnight in the Central Wheat Belt region.		V
14th	Sat		Rain	A
15th	Sun	Some shower activity in the southwest of the State over the next 3-4 days.	Showers	
16th	Mon		Rain	
17th	Tue	Cool day around the South Coast region, particularly in the Shannon district. Very unsettled and blustery winds around Cape Leeuwin.	Showers	
18th	Wed			1Q
19th	Thur	Mostly dry across the State over the coming week ahead.		

Date	Day	Western Australia - November 2026	Perth	Moon
20th	Fri			XhN
21st	Sat			
22nd	Sun	Cool day around the South Coast region, particularly around Mt Barker.		
23rd	Mon	Strong winds across the Interior, possibly accompanying thunderstorm activity, particularly in the Warburton area.		
24th	Tue			
25th	Wed	Hailstorms likely in the Interior region, including in the Warburton area. Cooler spell in the Murchison.		F
26th	Thu	Isolated showers around the South Coast over the next few days.		P5
27th	Fri	Showers over the Gibson Desert to the Kimberleys, with chance of some heavy falls briefly in the De Grey and into the Murchison over the next couple of days.		^
28th	Sat	Heavy rain around the South Coast to the Central Wheat Belt districts.		
29th	Sun	Chance of some rain along the coasts of the Nullarbor Plains over the next couple of days.		
30th	Mon		Showers, Th, Fg	

December 2026

Date	Day	Western Australia - December 2026	Perth	Moon
1st	Tue	A surface trough crosses the South Coastal region over the next couple of days bringng heavy rain into the region, and the chance of strong gusty winds at times that may turn tornadic in the Ongerup to Jerramungup areas. Heavy rain and strong winds also in the King River catchment and in the Albany districts with the passage of this trough with possible localised flooding developing.	Showers	3Q
2nd	Wed	Some rain activity in the western half of the State on most days throughout the month ahead. Some heavy rain around the Gascoyne region over the next couple of days with showers spreading into the De Grey for a time. Very wet in some parts of the South Coast to the Great Southern regions with chance of localised flooding, particularly in the Peaceful Bay and Denmark areas, and chance of flooding along the King River catchment.		
3rd	Thur	Very hot around the South Coast region, with some districts seeing maximum temperatures into the mid-40's. In contrast, it may be unusually cool in the Pilbara.	Rain, Fg	XhS
4th	Fri	The cloud band remnants of a ex-tropical cyclone crosses the southern Pilbara to the north Gascoyne and east into the Interior regions over the next few days, bringing very heavy rain as it passes. Widespread rain activity throughout the State, with heavy falls at imes around the Central Coast to the North Central regions. Unusually warm overnight in the Kimberley.	Showers	
5th	Sat	Rain contracts to heavy falls along the coasts from the Fortescue to the Central Coast.	Rain	
6th	Sun	Light shower activity throughout the southern half of the State over the next few days.		
7th	Mon	Thunderstorms with hail and strong wind gusts in the Southeast Coastal region with possible large size hail at times, possibly in the Salmon Gums area, and later crossing into the Goldfields bringing gusty winds into the Norseman area. In the north, thunderstorms in the Kimberleys around the Wyndham area may see very strong wind gusts at times.	Showers	
8th	Tue			
9th	Wed	Mostly dry in the State over the next few days.		N
10th	Thu	Strong winds accompany thunderstorm activity is in the Kimberley Fitzroy Crossing area.		V
11th	Fri			A
12th	Sat	Overall, the next 3-4 weeks may see average to below average temperatures across the State, while it may be much wetter than normal in the Kimberleys, Pilbara, Interior, Eucla, and some parts of the Southwest Land Division to the southern Goldfields, while much drier than average for this time of year elsewhere, including in some parts of the Gascoyne.		
13th	Sun	Chance of rain in the Interior as well as in the North Kimberley over the next couple of days.	Showers, Th	
14th	Mon	Stormy wind activity across the Goldfields, including the chance of tornado activity or strong downbursts of wind, with chance of raised dust storms, property damage, and including power outages, particularly in the Laverton area.	Showers	
15th	Tue	Dry over the northern half of the State over the coming week.		
16th	Wed	Some sowers in the Gascoyne and around the South Coast in the next couple of days.		
17th	Thur			1Q XhN
18th	Fri	Cold overnight in the Eucla, particularly in the Eyre district.		
19th	Sat	Blustery winds across the Eucla. Dry throughout the State over the next couple of days.		
20th	Sun		Showers	
21st	Mon	Thunderstorms with heavy rain, hail, and severe winds cross the southeast coastal region, with chance of damage to property, crops, and infrastructure, including in the Fitzgerald area. Storms also cross the Great Southern region, with heavy rain and hail in the Hyden area. Chance of flash flooding with the passage of the storms. Cooler temperatures around the South Coast.	Showers, Th	

Date	Day	Western Australia - December 2026	Perth	Moon
22nd	Tue	Rain activity spreads throughout the State into almost all regions. Wind gusts accompany thunderstorm activity in the Giles area of the Goldfields region.		
23rd	Wed	Rain activity contracts to the coastal fringes of the southern half of the State, including in the Nullarbor Plains to the South Coast and north to the Gascoyne coasts.		
24th	Thu	Mostly dry throughout the State for the remainder of the month.		**P1 F ^**
25th	Fri			
26th	Sat			
27th	Sun	Chance of isolated showers around the Central Coast region.		
28th	Mon			
29th	Tue			
30th	Wed	A tropical low that formed near Darwin and crossed the Josephine Bonaparte Gulf over the last few days, may cross into the Kimberleys around this time as a fully developed low grade cyclonic system bringing storm winds, heavy rain and flooding likely in many parts of the Kimberleys over the next few days. Very hot in the Pilbara over the next couple of days with maximum temperatures into the mid-to-upper 40's likely in some districts, particularly in the Marble Bay area.		**XhS**
31st	Thur	Thunderstorms with very strong wind gusts in the Munglinup area of the Southeast Coastal region. Chance of shower activity in the Gibson Desert. Chance of flooding in some parts of the north Kimberley over the next few days.	Showers	**3Q**

WESTERN AUSTRALIA GRAPHS AND TABLES 2026

Seasonal Breakdown – State Overall Rainfall Timing and Estimated mm Trend Graphs
State Overall Maximum Temperature Trend Graphs
Regional Rainfall Expectations Tables - mm and Trends
Regional Maximum and Minimum Temperature Trends Tables

Monthly Breakdown - **Rainfall:**
Regional Expectations by mm
Regional Expectations by mm Variance to Average
Regional Variance Trend by mm
Regional Estimated Expectations by % of Variance to Average
Regional Estimatedl Variance Trend by %

Maximum & Minimum Temperature Trends:
Regional Estimated Average Temperature
Regional Estimated Degree of Variation from the Average
Regional Estimated Temperature Trend

Seasonal Breakdowns

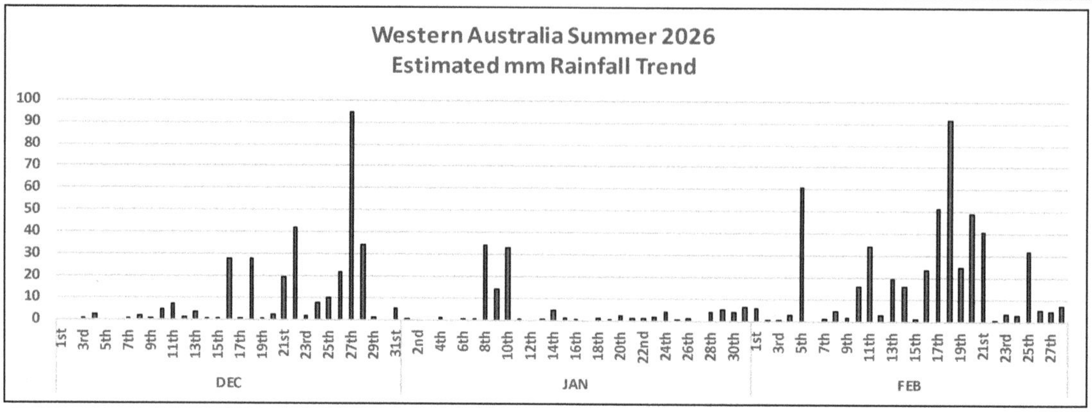

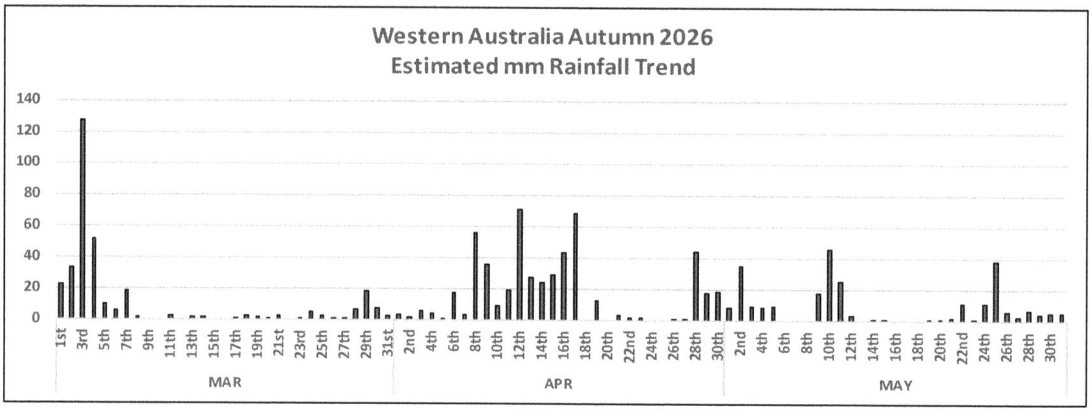

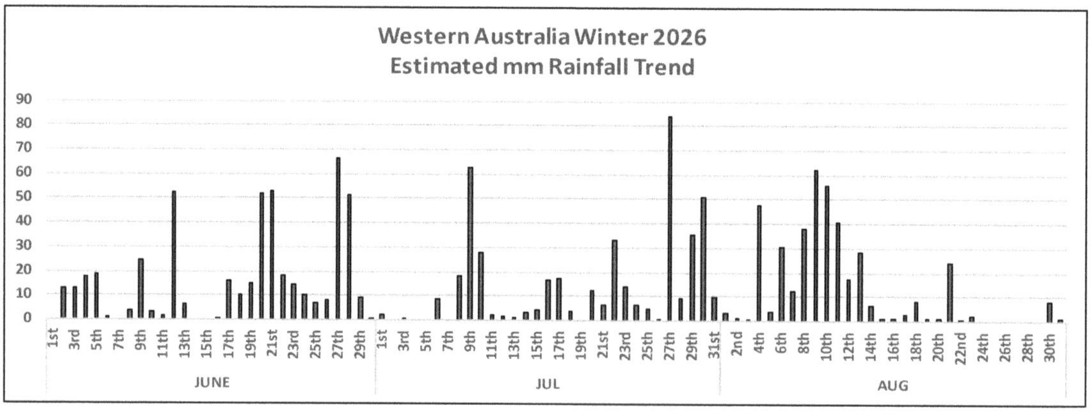

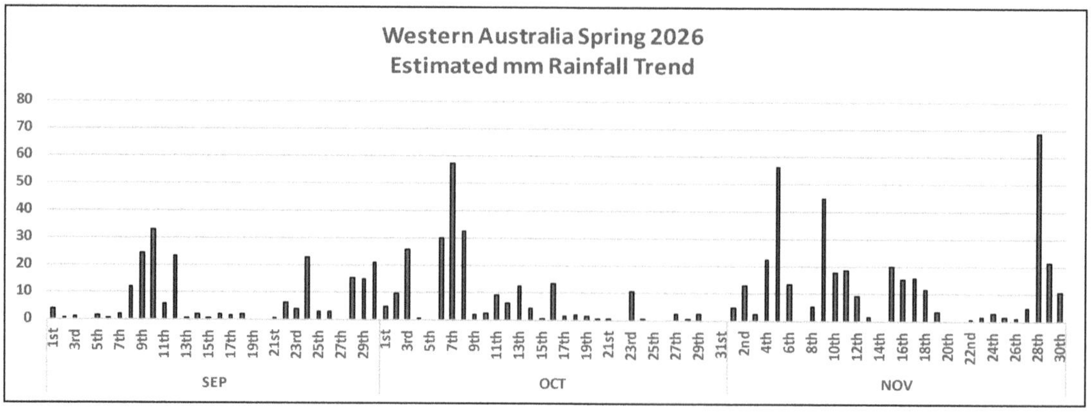

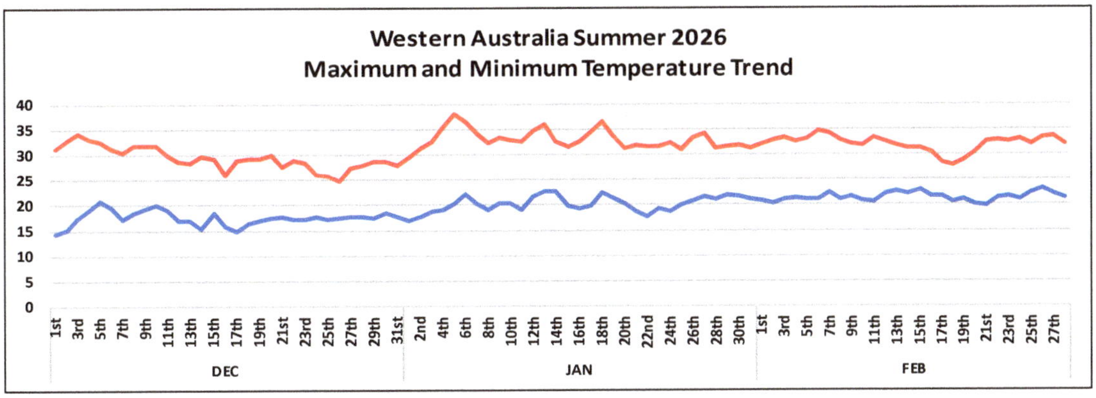
Western Australia Summer 2026
Maximum and Minimum Temperature Trend
DEC
JAN
FEB

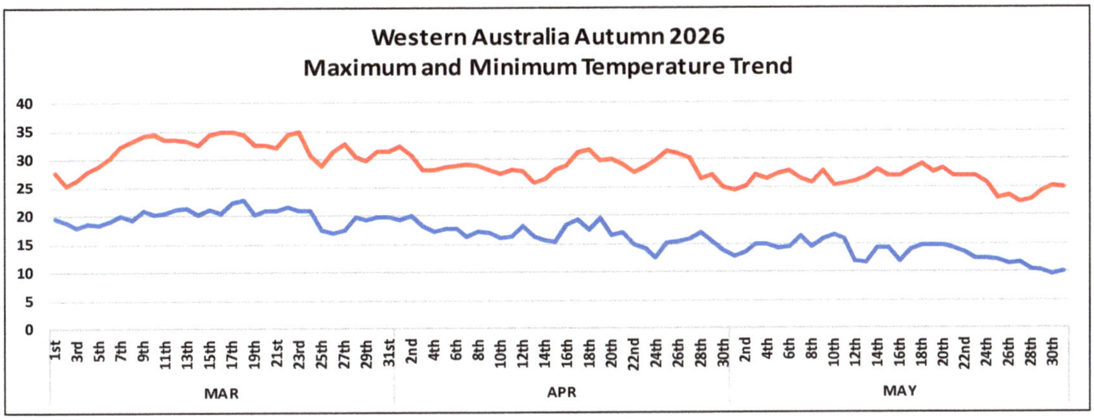
Western Australia Autumn 2026
Maximum and Minimum Temperature Trend
MAR
APR
MAY

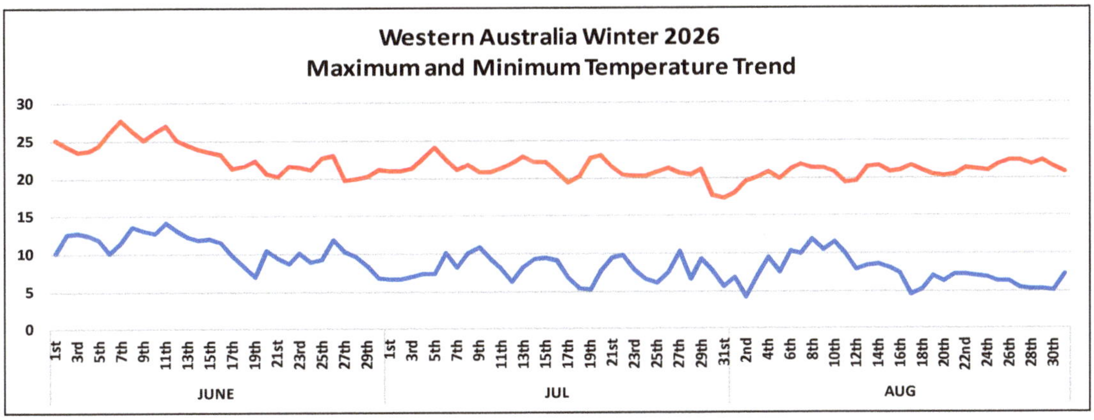
Western Australia Winter 2026
Maximum and Minimum Temperature Trend
JUNE
JUL
AUG

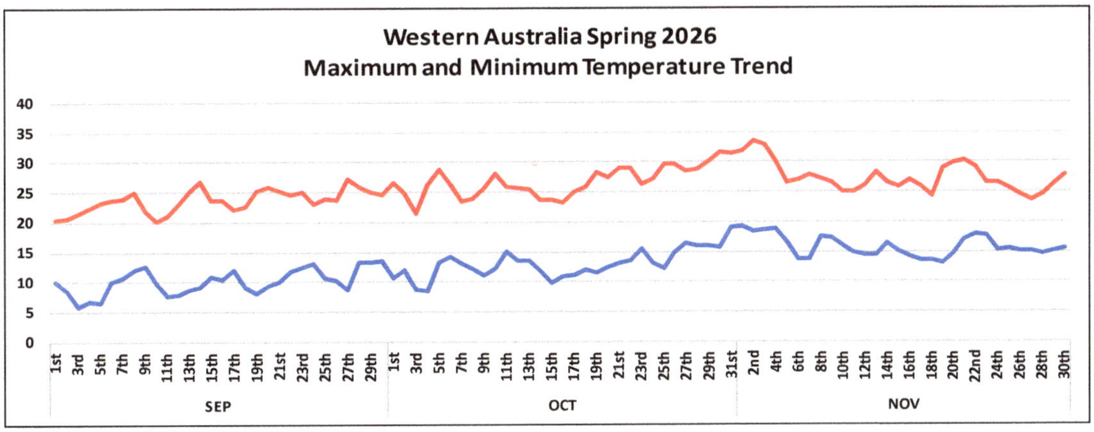
Western Australia Spring 2026
Maximum and Minimum Temperature Trend
SEP
OCT
NOV

2026 WA - Estimated Seasonal Rainfall Expectations (mm)

REGION	Kimberleys	Pilbara	Gascoyne	Central West	Lower & South West	Great Southern	Central Wheat Belt	Goldfields	Eucla		WA
Town Sample	Broome	Port Hedland	Carnarvon	Geraldton	Perth	Albany	Merredin	Kalgoorlie	Eucla	Giles	Overall
Summer	349	54	2	38	69	107	71	54	60	143	949
Autumn	132	56	229	80	192	195	37	80	88	9	1099
Winter	2	35	82	180	383	389	107	48	91	1	1318
Spring	3	0	6	59	145	230	130	67	83	96	819
Overall	478	146	329	347	806	1080	317	263	312	221	4299

2026 WA - Estimated Seasonal Rainfall Variation from the Average (mm)

REGION	Kimberleys	Pilbara	Gascoyne	Central West	Lower & South West	Great Southern	Central Wheat Belt	Goldfields	Eucla		WA
Town Sample	Broome	Port Hedland	Carnarvon	Geraldton	Perth	Albany	Merredin	Kalgoorlie	Eucla	Giles	Overall
Summer	-119	-113	-34	17	31	34	27	-25	7	15	-161
Autumn	-21	-42	169	-23	50	-4	-48	3	8	-74	18
Winter	-23	-7	-27	-53	-20	8	-31	-23	9	-35	-203
Spring	-9	-3	-11	-2	4	13	71	17	23	37	140
Overall	-181	-164	108	-72	81	210	-10	-14	37	-86	-91

2026 WA - Estimated Seasonal Rainfall Trend Variation by 10mm

REGION	Kimberleys	Pilbara	Gascoyne	Central West	Lower & South West	Great Southern	Central Wheat Belt	Goldfields	Eucla		WA
Town Sample	Broome	Port Hedland	Carnarvon	Geraldton	Perth	Albany	Merredin	Kalgoorlie	Eucla	Giles	Overall
Summer	drr	drr	drr	av	wtr	wtr	av	av	av	av	drr
Autumn	av	drr	wtr	av	wtr	av	drr	av	av	drr	av
Winter	av	av	av	drr	av	av	drr	av	av	drr	drr
Spring	av	av	av	av	av	av	wtr	av	av	wtr	wtr
Overall	drr	drr	wtr	av	av	wtr	av	av	av	av	av

2026 WA - Estimated Seasonal Rainfall Expectations by %

REGION	Kimberleys	Pilbara	Gascoyne	Central West	Lower & South West	Great Southern	Central Wheat Belt	Goldfields	Eucla		WA
Town Sample	Broome	Port Hedland	Carnarvon	Geraldton	Perth	Albany	Merredin	Kalgoorlie	Eucla	Giles	Overall
Summer	-25%	-68%	-94%	80%	81%	46%	60%	-32%	12%	12%	-15%
Autumn	-14%	-43%	284%	-23%	35%	-2%	-57%	4%	11%	-89%	2%
Winter	-91%	-17%	-25%	-23%	-5%	2%	-23%	-32%	11%	-96%	-13%
Spring	-74%	-100%	-63%	-3%	3%	6%	119%	34%	37%	62%	21%
Overall	-27%	-53%	49%	-17%	11%	24%	-3%	-5%	13%	-28%	-2%

2026 WA - Estimated Seasonal Rainfall Trend Variation by 25%

REGION	Kimberleys	Pilbara	Gascoyne	Central West	Lower & South West	Great Southern	Central Wheat Belt	Goldfields	Eucla		WA
Town Sample	Broome	Port Hedland	Carnarvon	Geraldton	Perth	Albany	Merredin	Kalgoorlie	Eucla	Giles	Overall
Summer	av	drr	drr	wtr	wtr	wtr	wtr	drr	av	av	av
Autumn	av	drr	wtr	av	wtr	av	drr	av	av	drr	av
Winter	drr	av	av	av	av	av	av	drr	av	drr	av
Spring	drr	drr	drr	av	av	av	wtr	wtr	wtr	wtr	av
Overall	drr	drr	av	av	wtr	wtr	av	wtr	wtr	av	wtr

2026 WA - Estimated Seasonal Maximum Temperature Average and Trend

Region	Kimberleys	Pilbara	Gascoyne	Central West	Lower & South West	Great Southern	Central Wheat Belt	Goldfields	Eucla		WA
Town Sample	Broome	Port Hedland	Carnarvon	Geraldton	Perth	Albany	Merredin	Kalgoorlie	Eucla	Giles	Overall
Summer Est	33.6	37.0	30.4	31.1	31.5	23.2	32.9	32.9	26.5	36.4	31.5
Trend	av	wmr	clr	av	av	clr	wmr	av	clr	wmr	clr
Autumn Est	33.6	35.9	29.7	29.1	26.9	22.2	27.6	27.0	25.9	31.6	28.9
Trend	av	wmr	clr	av	a-c	clr	wmr	wmr	a-w	wmr	av
Winter Est	29.9	28.6	24.4	21.4	19.3	16.5	18.2	18.5	19.0	22.0	21.8
Trend	av	a-c	clr	clr	a-c	clr	wmr	wmr	wmr	wmr	a-w
Spring Est	33.6	34.6	26.2	24.3	22.4	17.9	22.7	23.9	23.6	30.6	26.0
Trend	a-w	wmr	clr	clr	clr	clr	wmr	a-c	clr	wmr	clr

2026 WA - Estimated Seasonal Minimum Temperature Average and Trend

Region	Kimberleys	Pilbara	Gascoyne	Central West	Lower & South West	Great Southern	Central Wheat Belt	Goldfields	Eucla		WA
Town Sample	Broome	Port Hedland	Carnarvon	Geraldton	Perth	Albany	Merredin	Kalgoorlie	Eucla	Giles	Overall
Summer Est	26.7	25.9	22.1	18.9	17.6	14.0	17.2	17.8	14.7	22.7	19.7
Trend	av	a-w	av	a-w	wmr	av	av	av	clr	av	av
Autumn Est	23.8	22.4	19.8	17.2	13.6	12.4	13.0	14.1	12.3	18.0	16.7
Trend	wmr	wmr	wmr	wmr	av	a-w	av	wmr	a-c	wmr	a-w
Winter Est	13.7	13.2	10.1	10.4	7.6	7.0	6.4	6.6	4.7	7.9	8.8
Trend	a-c	av	clr	a-w	a-c	a-c	av	a-w	clr	av	av
Spring Est	20.2	18.6	15.0	10.7	10.6	9.3	9.0	10.9	9.4	16.3	13.0
Trend	clr	av	clr	a-c	av	av	a-c	av	clr	av	a-c

Monthly Breakdowns

2026 — WESTERN AUSTRALIA - Estimated Rainfall Expectations (mm)

REGION	Kimberleys	Pilbara	Gascoyne	Central West	Lower & South West	Great Southern	Central Wheat Belt	Goldfields	Eucla		WA
Town Sample	Broome	Port Hedland	Carnarvon	Geraldton	Perth	Albany	Merredin	Kalgoorlie	Eucla	Giles	Overall
January	81	6	1	0	0	30	2	4	1	0	125
February	258	48	2	23	48	16	30	28	7	45	503
March	113	22	59	17	2	18	2	62	26	4	324
April	19	33	151	18	150	77	19	8	38	5	519
May	0	1	19	46	40	99	17	9	24	0	256
June	0	35	40	47	151	134	17	2	60	1	486
July	1	0	19	69	119	142	59	7	18	0	434
August	2	0	22	65	113	113	30	39	13	0	397
September	0	0	2	16	29	82	24	21	11	18	202
October	2	0	1	28	55	65	58	15	6	0	231
November	1	0	3	15	61	84	48	31	66	77	386
December	1	1	10	5	38	221	11	37	42	69	436
Overall	478	146	329	347	806	1080	317	263	312	221	4299

2026 — WESTERN AUSTRALIA - Estimated Rainfall Variation from the Average (mm)

REGION	Kimberleys	Pilbara	Gascoyne	Central West	Lower & South West	Great Southern	Central Wheat Belt	Goldfields	Eucla		WA
Town Sample	Broome	Port Hedland	Carnarvon	Geraldton	Perth	Albany	Merredin	Kalgoorlie	Eucla	Giles	Overall
January	-109	-49	-8	-5	-13	4	-13	-22	-15	-32	-261
February	56	-48	-15	12	30	-6	14	-4	-12	6	34
March	3	-31	43	-2	-16	-21	-20	32	3	-44	-53
April	-8	12	135	-2	115	20	-4	-13	12	-9	257
May	-16	-22	-9	-19	-49	-4	-24	-17	-7	-20	-186
June	-15	12	-6	-36	11	6	-31	-22	30	-14	-66
July	-9	-14	-25	-19	-30	8	9	-16	-7	-13	-114
August	0	-5	4	1	-2	-6	-9	15	-14	-8	-24
September	-1	-2	-5	-19	-48	-17	-2	8	-11	7	-90
October	2	-0	-3	12	19	-2	39	-1	-13	-14	40
November	-9	-1	-2	4	33	32	33	10	46	44	191
December	-75	-16	-2	-0	30	196	-3	15	24	13	182
Overall	-181	-164	108	-72	81	210	-10	-14	37	-86	-91

2026 — WESTERN AUSTRALIA - Estimated Rainfall Trend Variation by 10mm

REGION	Kimberleys	Pilbara	Gascoyne	Central West	Lower & South West	Great Southern	Central Wheat Belt	Goldfields	Eucla		WA
Town Sample	Broome	Port Hedland	Carnarvon	Geraldton	Perth	Albany	Merredin	Kalgoorlie	Eucla	Giles	Overall
January	drr	drr	av	av	drr	av	drr	drr	drr	drr	drr
February	wtr	drr	drr	wtr	wtr	av	wtr	av	drr	av	wtr
March	av	drr	wtr	av	drr	drr	drr	wtr	av	drr	drr
April	av	wtr	wtr	av	wtr	wtr	av	drr	wtr	av	wtr
May	drr	drr	av	drr	drr	av	drr	drr	av	drr	drr
June	drr	wtr	av	drr	wtr	av	drr	drr	wtr	drr	drr
July	av	drr	drr	drr	drr	av	av	drr	av	drr	drr
August	av	av	av	av	av	av	av	wtr	drr	av	av
September	av	av	av	drr	drr	drr	av	av	drr	av	drr
October	av	av	av	wtr	wtr	av	wtr	av	drr	drr	wtr
November	av	av	av	av	wtr	wtr	wtr	wtr	wtr	wtr	wtr
December	drr	drr	av	av	wtr	wtr	av	wtr	wtr	wtr	wtr
Overall	drr	drr	wtr	av	av	wtr	av	av	av	av	av

2026 — WESTERN AUSTRALIA - Estimated Rainfall Expectations by %

REGION	Kimberleys	Pilbara	Gascoyne	Central West	Lower & South West	Great Southern	Central Wheat Belt	Goldfields	Eucla		WA
Town Sample	Broome	Port Hedland	Carnarvon	Geraldton	Perth	Albany	Merredin	Kalgoorlie	Eucla	Giles	Overall
January	-57%	-89%	-93%	-100%	-100%	14%	-89%	-84%	-94%	-100%	-68%
February	28%	-50%	-90%	113%	170%	-28%	89%	-11%	-63%	15%	7%
March	3%	-59%	276%	-12%	-88%	-53%	-93%	106%	15%	-92%	-14%
April	-30%	53%	853%	-10%	327%	35%	-19%	-60%	46%	-62%	98%
May	-100%	-96%	-31%	-29%	-55%	-3%	-59%	-64%	-22%	-99%	-42%
June	-100%	50%	-14%	-44%	8%	5%	-64%	-92%	99%	-92%	-12%
July	-92%	-100%	-56%	-21%	-20%	6%	19%	-71%	-27%	-98%	-21%
August	7%	-93%	22%	2%	-1%	-5%	-24%	62%	-52%	-100%	-6%
September	-100%	-100%	-70%	-54%	-63%	-17%	-7%	56%	-50%	56%	-31%
October	300%	-100%	-70%	77%	54%	-3%	210%	-7%	-68%	-100%	21%
November	-94%	-100%	-43%	42%	116%	60%	219%	51%	230%	132%	98%
December	-99%	-95%	-13%	-5%	387%	786%	-18%	65%	126%	22%	71%
Overall	-27%	-53%	49%	-17%	11%	24%	-3%	-5%	13%	-28%	-2%

2026 WESTERN AUSTRALIA - Estimated Rainfall Trend Variation by 25%

REGION	Kimberleys	Pilbara	Gascoyne	Central West	Lower & South West	Great Southern	Central Wheat Belt	Goldfields	Eucla		WA
Town Sample	Broome	Port Hedland	Carnarvon	Geraldton	Perth	Albany	Merredin	Kalgoorlie	Eucla	Giles	Overall
January	drr	drr	drr	drr	drr	av	drr	drr	drr	drr	drr
February	wtr	drr	drr	wtr	wtr	drr	wtr	av	drr	av	av
March	av	drr	wtr	av	drr	drr	drr	wtr	av	drr	av
April	drr	wtr	wtr	av	wtr	wtr	av	drr	wtr	drr	wtr
May	drr	drr	drr	drr	drr	av	drr	drr	av	drr	drr
June	drr	wtr	av	drr	av	av	drr	drr	wtr	drr	av
July	drr	drr	drr	av	av	av	av	drr	drr	drr	av
August	av	drr	av	av	av	av	av	wtr	drr	drr	av
September	drr	drr	drr	drr	drr	av	av	wtr	drr	wtr	drr
October	wtr	drr	drr	wtr	wtr	av	wtr	av	drr	drr	av
November	drr	drr	drr	wtr	wtr	wtr	wtr	wtr	wtr	wtr	wtr
December	drr	drr	av	av	wtr	wtr	av	wtr	wtr	av	wtr
Overall	drr	drr	av	av	wtr	wtr	av	wtr	wtr	av	wtr

2026 WESTERN AUSTRALIA - Estimated Average Maximum Temperature

Region	Kimberleys	Pilbara	Gascoyne	Central West	Lower & South West	Great Southern	Central Wheat Belt	Goldfields	Eucla		WA
Town Sample	Broome	Port Hedland	Carnarvon	Geraldton	Perth	Albany	Merredin	Kalgoorlie	Eucla	Giles	Overall
January	34	37	31	33	34	24	36	35	27	38	**33**
February	33	37	31	32	34	25	34	33	27	37	**32**
March	33	37	32	31	31	24	33	31	30	36	**32**
April	34	36	30	30	27	23	27	26	25	31	**29**
May	34	34	27	26	23	20	23	24	23	28	**26**
June	31	29	26	23	21	18	21	21	21	23	**23**
July	30	28	23	20	19	16	17	18	19	22	**21**
August	30	28	24	20	19	16	17	17	17	21	**21**
September	33	32	25	22	20	17	20	21	21	26	**24**
October	34	37	27	25	23	18	23	25	24	32	**27**
November	34	35	27	26	24	19	25	26	25	34	**27**
December	34	36	28	26	26	21	29	30	24	34	**29**
Overall	**33**	**34**	**28**	**26**	**25**	**20**	**25**	**26**	**24**	**30**	**27**

2026 WESTERN AUSTRALIA - Estimated Average Minimum Temperature

Region	Kimberleys	Pilbara	Gascoyne	Central West	Lower & South West	Great Southern	Central Wheat Belt	Goldfields	Eucla		WA
Town Sample	Broome	Port Hedland	Carnarvon	Geraldton	Perth	Albany	Merredin	Kalgoorlie	Eucla	Giles	Overall
January	25	27	23	20	18	14	19	19	15	23	**20**
February	28	27	24	22	20	16	19	19	16	24	**21**
March	26	25	23	19	17	15	17	18	16	23	**20**
April	23	23	21	18	14	12	13	14	12	18	**17**
May	23	19	16	14	10	11	8	10	10	14	**13**
June	16	15	12	13	10	8	9	9	7	10	**11**
July	12	13	9	10	7	7	5	6	4	8	**8**
August	14	12	9	8	6	6	5	5	3	6	**7**
September	16	16	12	9	8	8	7	9	5	12	**10**
October	21	20	15	10	11	9	9	11	9	17	**13**
November	24	21	18	13	13	12	12	13	13	20	**16**
December	27	23	19	15	15	12	14	16	13	21	**17**
Overall	21	20	17	14	12	11	11	12	10	16	**15**

2026	WESTERN AUSTRALIA - Estimated Degree of Variation from the Maximum Average									
Region	Kimberleys	Pilbara	Gascoyne	Central West	Lower & South West	Great Southern	Central Wheat Belt	Goldfields	Eucla		WA
Town Sample	Broome	Port Hedland	Carnarvon	Geraldton	Perth	Albany	Merredin	Kalgoorlie	Eucla	Giles	Overall
January	0.7	0.8	-0.2	1.5	2.0	-0.7	2.2	1.6	1.4	0.8	1.0
February	-0.2	0.9	-1.7	-0.3	1.7	-0.2	0.5	1.1	0.8	0.5	0.3
March	-1.4	0.4	0.1	0.3	1.5	0.1	2.6	1.1	4.6	2.6	1.2
April	0.1	0.7	0.7	2.3	1.3	0.8	1.1	0.9	1.4	1.1	1.0
May	2.0	3.7	1.2	1.9	0.7	0.8	2.3	3.6	1.8	4.0	2.2
June	1.5	1.6	2.2	2.6	1.8	1.3	3.1	3.0	2.1	2.7	2.2
July	0.5	0.9	0.7	1.0	0.6	0.0	1.0	1.1	0.6	1.6	0.8
August	-0.7	-0.9	1.2	0.2	0.0	-0.4	-0.9	-1.7	-1.7	-1.6	-0.6
September	1.2	0.0	0.5	-0.1	0.1	-0.4	-0.9	-1.7	0.0	-2.0	-0.3
October	1.0	1.4	0.7	0.7	-0.2	-0.9	-2.1	-0.6	1.2	0.5	0.2
November	-0.1	-1.3	-1.0	-1.4	-1.8	-1.9	-3.5	-3.4	1.0	-0.6	-1.4
December	0.0	-1.2	-1.6	-3.3	-3.2	-2.4	-2.7	-2.6	-0.5	-1.8	-1.9
Overall	0.4	0.6	0.2	0.4	0.4	-0.3	0.2	0.2	1.1	0.6	0.4

2026	WESTERN AUSTRALIA - Estimated Degree of Variation from the Minimum Average									
Region	Kimberleys	Pilbara	Gascoyne	Central West	Lower & South West	Great Southern	Central Wheat Belt	Goldfields	Eucla		WA
Town Sample	Broome	Port Hedland	Carnarvon	Geraldton	Perth	Albany	Merredin	Kalgoorlie	Eucla	Giles	Overall
January	-0.9	1.0	0.3	1.3	1.3	0.1	0.9	0.7	-1.9	-0.2	0.3
February	2.1	1.4	0.4	2.4	2.9	1.9	1.0	1.1	-1.2	0.7	1.3
March	0.6	0.6	1.2	1.5	1.0	1.6	1.2	2.1	-0.5	2.0	1.1
April	0.1	1.8	1.3	2.9	0.9	-0.2	0.6	1.1	-1.7	1.3	0.8
May	4.2	1.5	0.7	1.1	-0.4	0.9	-0.8	1.7	-0.8	2.3	1.0
June	0.5	1.0	-0.5	1.9	1.2	0.0	2.2	2.7	-1.3	2.2	1.0
July	-2.0	0.2	-1.8	0.6	-1.5	-0.4	-0.2	0.4	-2.9	0.6	-0.7
August	-1.3	-1.4	-2.3	-0.5	-1.9	-1.7	-0.4	-0.3	-4.5	-2.7	-1.7
September	-2.3	0.1	-1.6	-0.6	-0.5	-0.5	-0.1	0.5	-3.7	-1.1	-1.0
October	-2.0	1.0	-1.3	-1.0	0.3	-0.6	-1.1	-0.6	-2.0	-0.3	-0.7
November	-1.4	-0.7	-1.2	-0.6	-0.2	0.9	-1.7	-0.8	-0.2	0.1	-0.6
December	0.1	-1.4	-1.7	-1.6	-0.5	-0.1	-1.8	-1.2	-2.3	-1.0	-1.1
Overall	-0.2	0.4	-0.5	0.6	0.2	0.2	0.0	0.6	-1.9	0.3	0.0

2026	WESTERN AUSTRALIA - Estimated Maximum Temperature Trend									
Region	Kimberleys	Pilbara	Gascoyne	Central West	Lower & South West	Great Southern	Central Wheat Belt	Goldfields	Eucla		WA
Town Sample	Broome	Port Hedland	Carnarvon	Geraldton	Perth	Albany	Merredin	Kalgoorlie	Eucla	Giles	Overall
January	a-w	a-w	av	wmr	wmr	a-c	wmr	wmr	wmr	a-w	a-w
February	av	a-w	clr	av	wmr	av	av	wmr	a-w	av	av
March	clr	av	av	av	wmr	av	wmr	wmr	wmr	wmr	wmr
April	av	a-w	a-w	wmr	wmr	a-w	wmr	a-w	wmr	wmr	a-w
May	wmr	wmr	wmr	wmr	a-w	a-w	wmr	wmr	wmr	wmr	wmr
June	wmr	wmr	wmr	wmr	wmr	wmr	wmr	wmr	wmr	wmr	wmr
July	av	a-w	a-w	a-w	a-w	av	a-w	wmr	a-w	wmr	a-w
August	a-c	a-c	wmr	av	av	av	a-c	clr	clr	clr	a-c
September	wmr	av	av	av	av	av	a-c	clr	av	clr	av
October	a-w	wmr	a-w	a-w	av	a-c	clr	a-c	wmr	av	av
November	av	clr	a-c	clr	clr	clr	clr	clr	a-w	a-c	clr
December	av	clr	clr	clr	clr	clr	clr	clr	av	clr	clr
Overall	av	a-w	av	av	av	av	av	av	wmr	a-w	av

2026	WESTERN AUSTRALIA - Estimated Minimum Temperature Trend									
Region	Kimberleys	Pilbara	Gascoyne	Central West	Lower & South West	Great Southern	Central Wheat Belt	Goldfields	Eucla		WA
Town Sample	Broome	Port Hedland	Carnarvon	Geraldton	Perth	Albany	Merredin	Kalgoorlie	Eucla	Giles	Overall
January	a-c	a-w	av	wmr	wmr	av	a-w	a-w	clr	av	av
February	wmr	wmr	av	wmr	wmr	wmr	a-w	wmr	clr	a-w	wmr
March	a-w	a-w	wmr	wmr	a-w	wmr	wmr	wmr	av	wmr	wmr
April	av	wmr	wmr	wmr	a-w	av	a-w	wmr	clr	wmr	a-w
May	wmr	wmr	a-w	wmr	av	a-w	a-c	wmr	a-c	wmr	a-w
June	av	a-w	av	wmr	wmr	av	wmr	wmr	clr	wmr	a-w
July	clr	av	clr	a-w	clr	av	av	av	clr	a-w	a-c
August	clr	clr	clr	av	clr	clr	av	av	clr	clr	clr
September	clr	av	clr	a-c	av	av	av	av	clr	clr	a-c
October	clr	a-w	clr	a-c	av	a-c	clr	a-c	clr	av	a-c
November	clr	a-c	clr	a-c	av	a-w	clr	a-c	av	av	a-c
December	av	clr	clr	clr	av	av	clr	clr	clr	a-c	clr
Overall	av	av	av	a-w	av	av	av	a-w	clr	av	av

MAPS SECTION
2026

ESTIMATED ONE MONTH / PAGE RAIN MAPS 2026

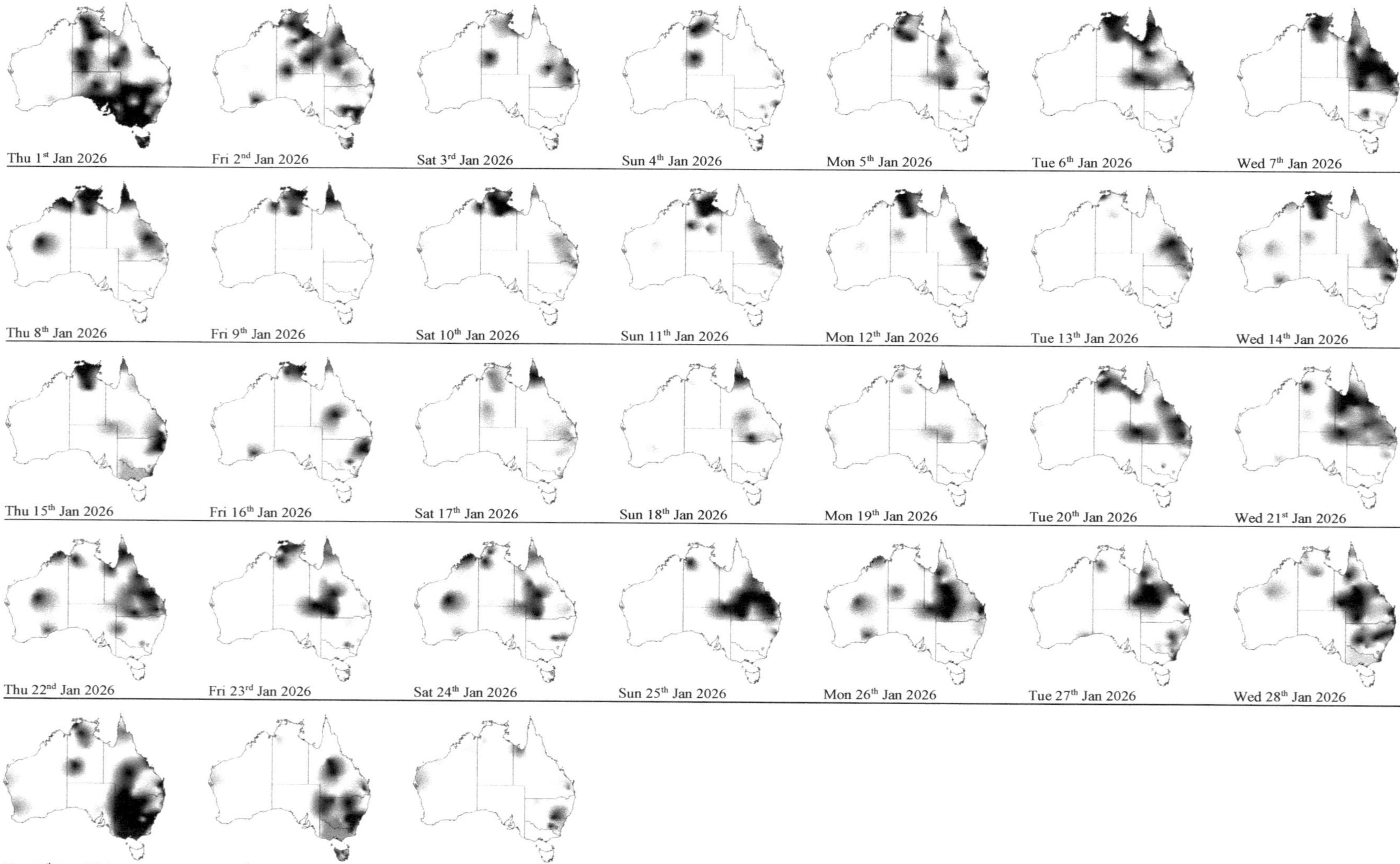

Thu 1st Jan 2026 | Fri 2nd Jan 2026 | Sat 3rd Jan 2026 | Sun 4th Jan 2026 | Mon 5th Jan 2026 | Tue 6th Jan 2026 | Wed 7th Jan 2026

Thu 8th Jan 2026 | Fri 9th Jan 2026 | Sat 10th Jan 2026 | Sun 11th Jan 2026 | Mon 12th Jan 2026 | Tue 13th Jan 2026 | Wed 14th Jan 2026

Thu 15th Jan 2026 | Fri 16th Jan 2026 | Sat 17th Jan 2026 | Sun 18th Jan 2026 | Mon 19th Jan 2026 | Tue 20th Jan 2026 | Wed 21st Jan 2026

Thu 22nd Jan 2026 | Fri 23rd Jan 2026 | Sat 24th Jan 2026 | Sun 25th Jan 2026 | Mon 26th Jan 2026 | Tue 27th Jan 2026 | Wed 28th Jan 2026

Thu 29th Jan 2026 | Fri 30th Jan 2026 | Sat 31st Jan 2026

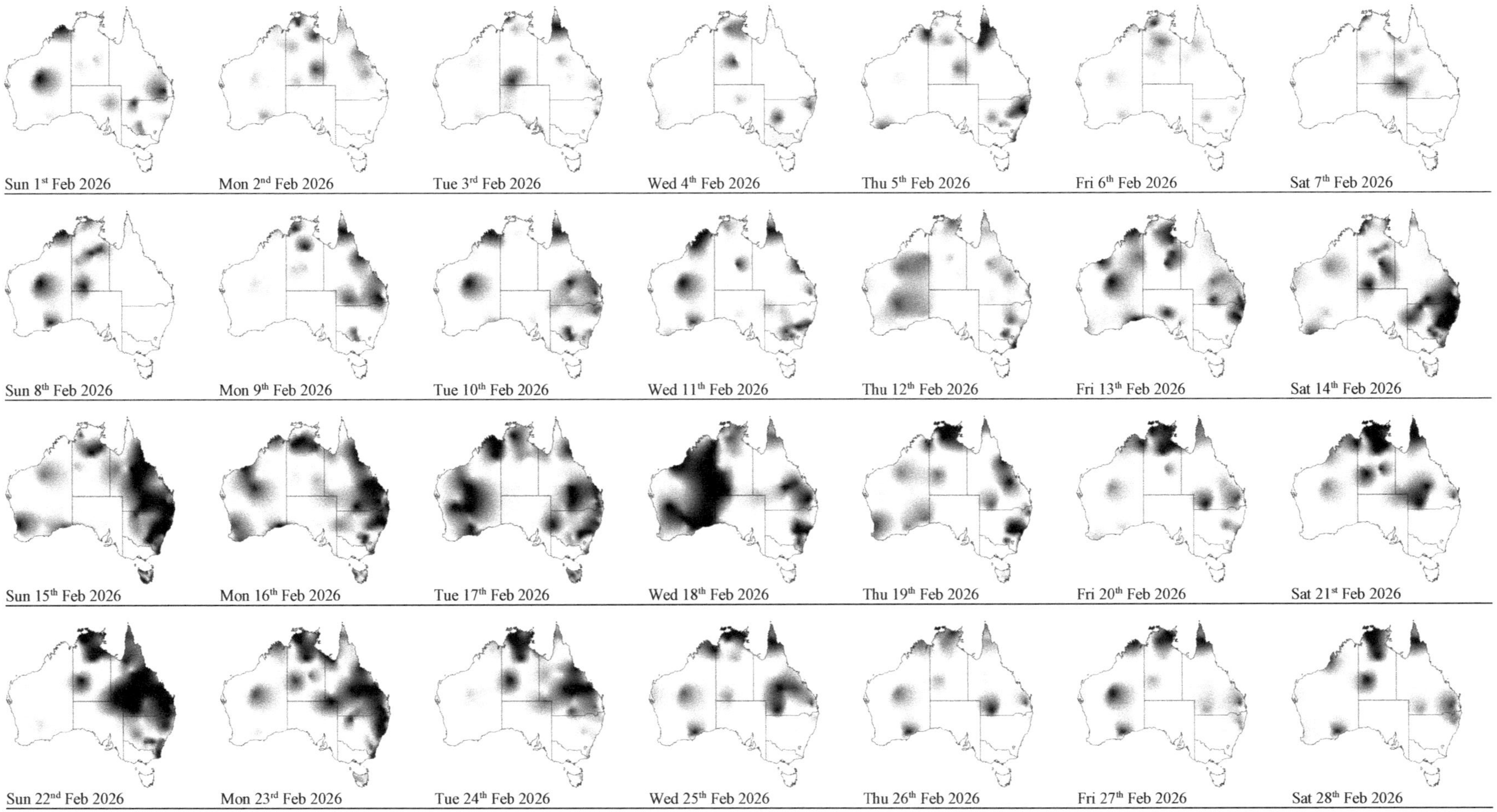
Sun 1st Feb 2026
Mon 2nd Feb 2026
Tue 3rd Feb 2026
Wed 4th Feb 2026
Thu 5th Feb 2026
Fri 6th Feb 2026
Sat 7th Feb 2026
Sun 8th Feb 2026
Mon 9th Feb 2026
Tue 10th Feb 2026
Wed 11th Feb 2026
Thu 12th Feb 2026
Fri 13th Feb 2026
Sat 14th Feb 2026
Sun 15th Feb 2026
Mon 16th Feb 2026
Tue 17th Feb 2026
Wed 18th Feb 2026
Thu 19th Feb 2026
Fri 20th Feb 2026
Sat 21st Feb 2026
Sun 22nd Feb 2026
Mon 23rd Feb 2026
Tue 24th Feb 2026
Wed 25th Feb 2026
Thu 26th Feb 2026
Fri 27th Feb 2026
Sat 28th Feb 2026

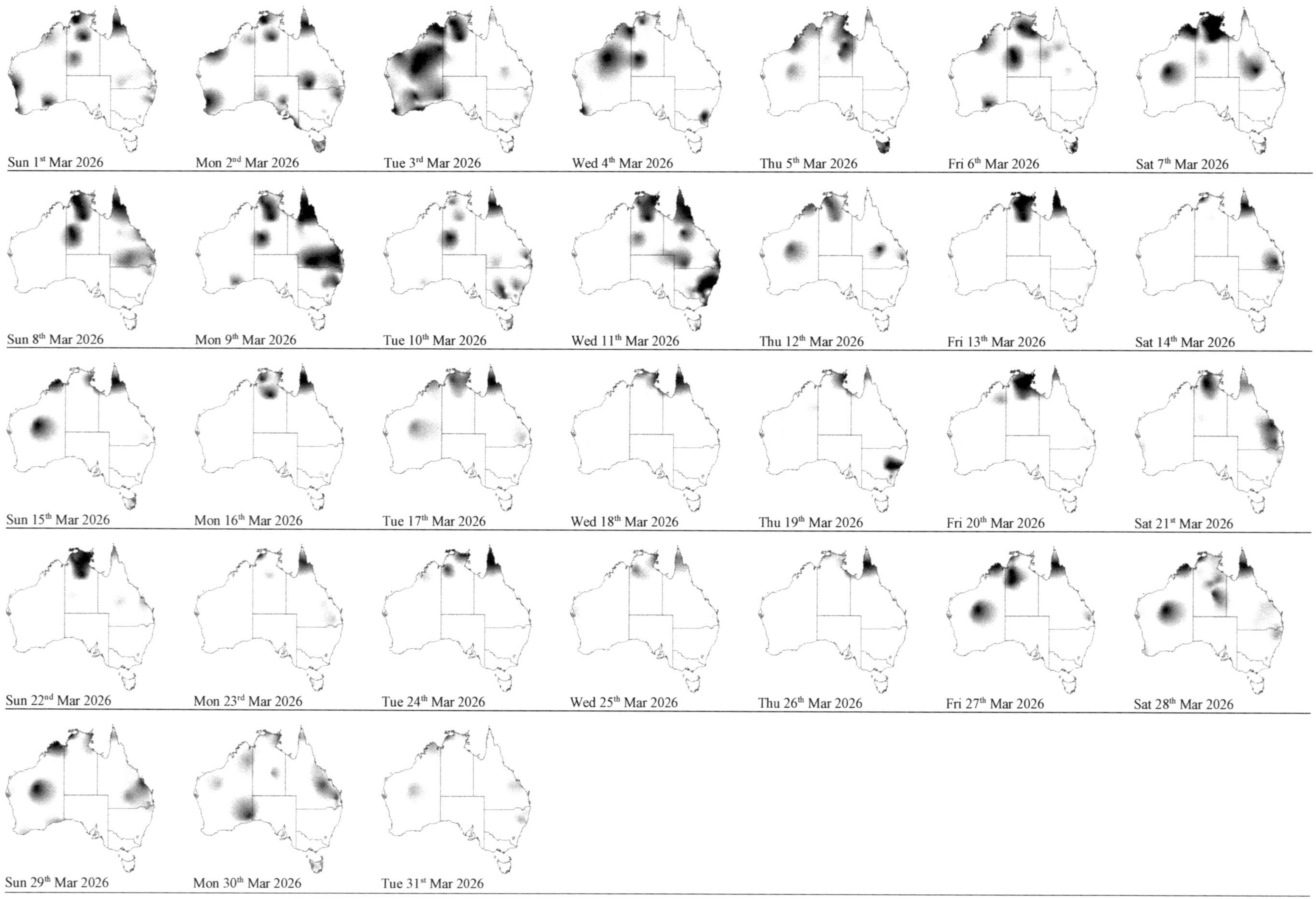
Sun 1st Mar 2026
Mon 2nd Mar 2026
Tue 3rd Mar 2026
Wed 4th Mar 2026
Thu 5th Mar 2026
Fri 6th Mar 2026
Sat 7th Mar 2026
Sun 8th Mar 2026
Mon 9th Mar 2026
Tue 10th Mar 2026
Wed 11th Mar 2026
Thu 12th Mar 2026
Fri 13th Mar 2026
Sat 14th Mar 2026
Sun 15th Mar 2026
Mon 16th Mar 2026
Tue 17th Mar 2026
Wed 18th Mar 2026
Thu 19th Mar 2026
Fri 20th Mar 2026
Sat 21st Mar 2026
Sun 22nd Mar 2026
Mon 23rd Mar 2026
Tue 24th Mar 2026
Wed 25th Mar 2026
Thu 26th Mar 2026
Fri 27th Mar 2026
Sat 28th Mar 2026
Sun 29th Mar 2026
Mon 30th Mar 2026
Tue 31st Mar 2026

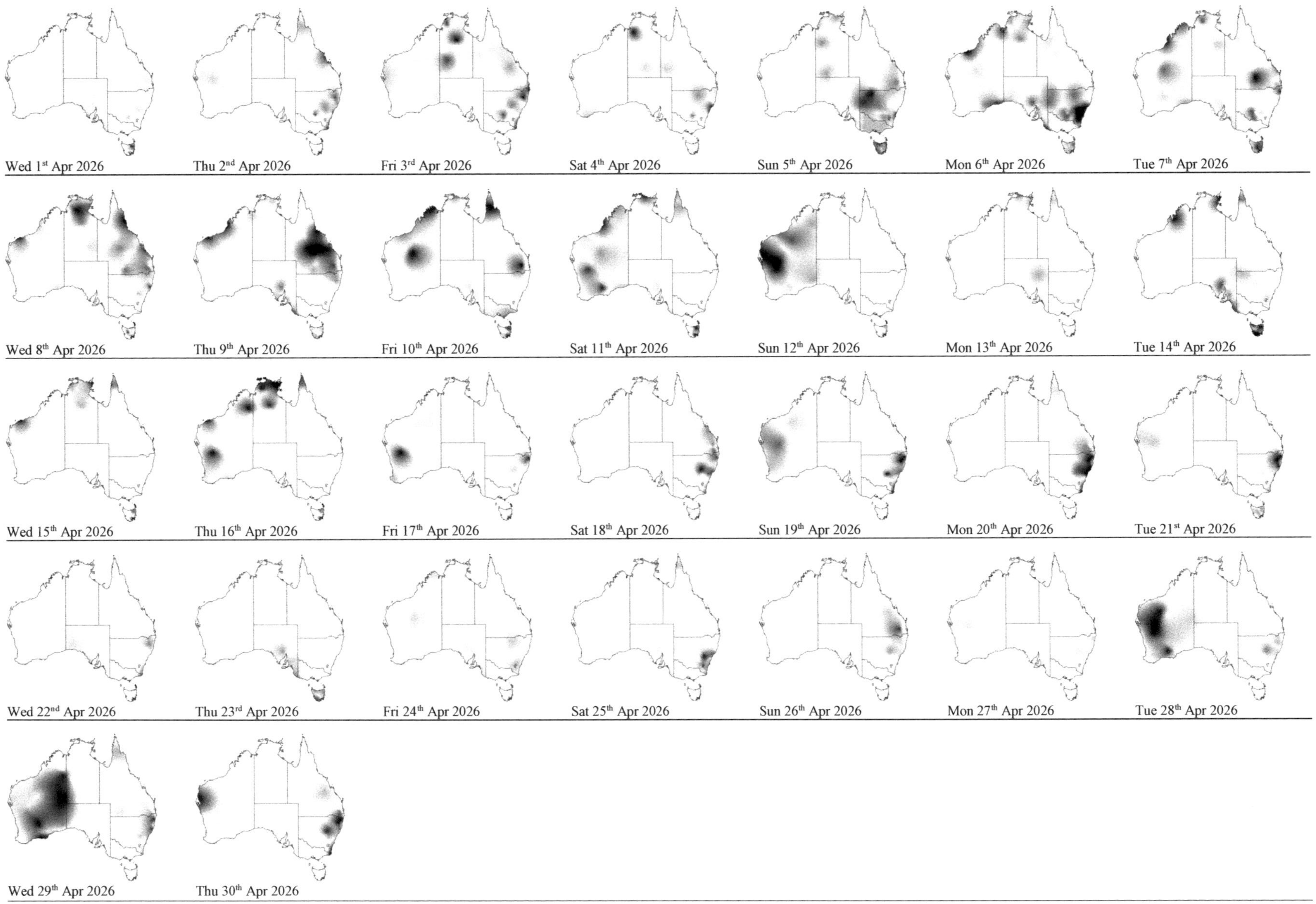
Wed 1st Apr 2026
Thu 2nd Apr 2026
Fri 3rd Apr 2026
Sat 4th Apr 2026
Sun 5th Apr 2026
Mon 6th Apr 2026
Tue 7th Apr 2026
Wed 8th Apr 2026
Thu 9th Apr 2026
Fri 10th Apr 2026
Sat 11th Apr 2026
Sun 12th Apr 2026
Mon 13th Apr 2026
Tue 14th Apr 2026
Wed 15th Apr 2026
Thu 16th Apr 2026
Fri 17th Apr 2026
Sat 18th Apr 2026
Sun 19th Apr 2026
Mon 20th Apr 2026
Tue 21st Apr 2026
Wed 22nd Apr 2026
Thu 23rd Apr 2026
Fri 24th Apr 2026
Sat 25th Apr 2026
Sun 26th Apr 2026
Mon 27th Apr 2026
Tue 28th Apr 2026
Wed 29th Apr 2026
Thu 30th Apr 2026

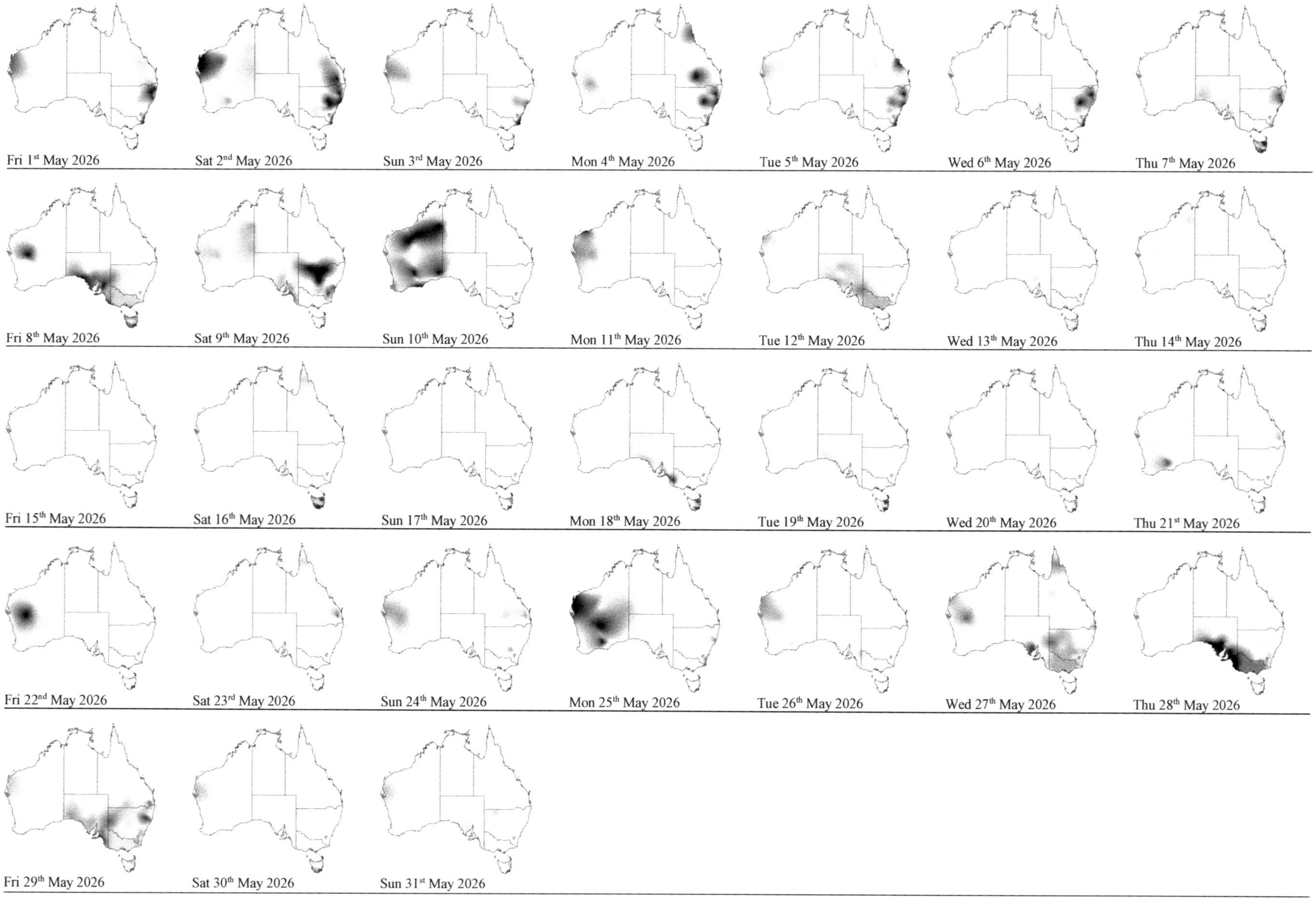
Fri 1st May 2026
Sat 2nd May 2026
Sun 3rd May 2026
Mon 4th May 2026
Tue 5th May 2026
Wed 6th May 2026
Thu 7th May 2026
Fri 8th May 2026
Sat 9th May 2026
Sun 10th May 2026
Mon 11th May 2026
Tue 12th May 2026
Wed 13th May 2026
Thu 14th May 2026
Fri 15th May 2026
Sat 16th May 2026
Sun 17th May 2026
Mon 18th May 2026
Tue 19th May 2026
Wed 20th May 2026
Thu 21st May 2026
Fri 22nd May 2026
Sat 23rd May 2026
Sun 24th May 2026
Mon 25th May 2026
Tue 26th May 2026
Wed 27th May 2026
Thu 28th May 2026
Fri 29th May 2026
Sat 30th May 2026
Sun 31st May 2026

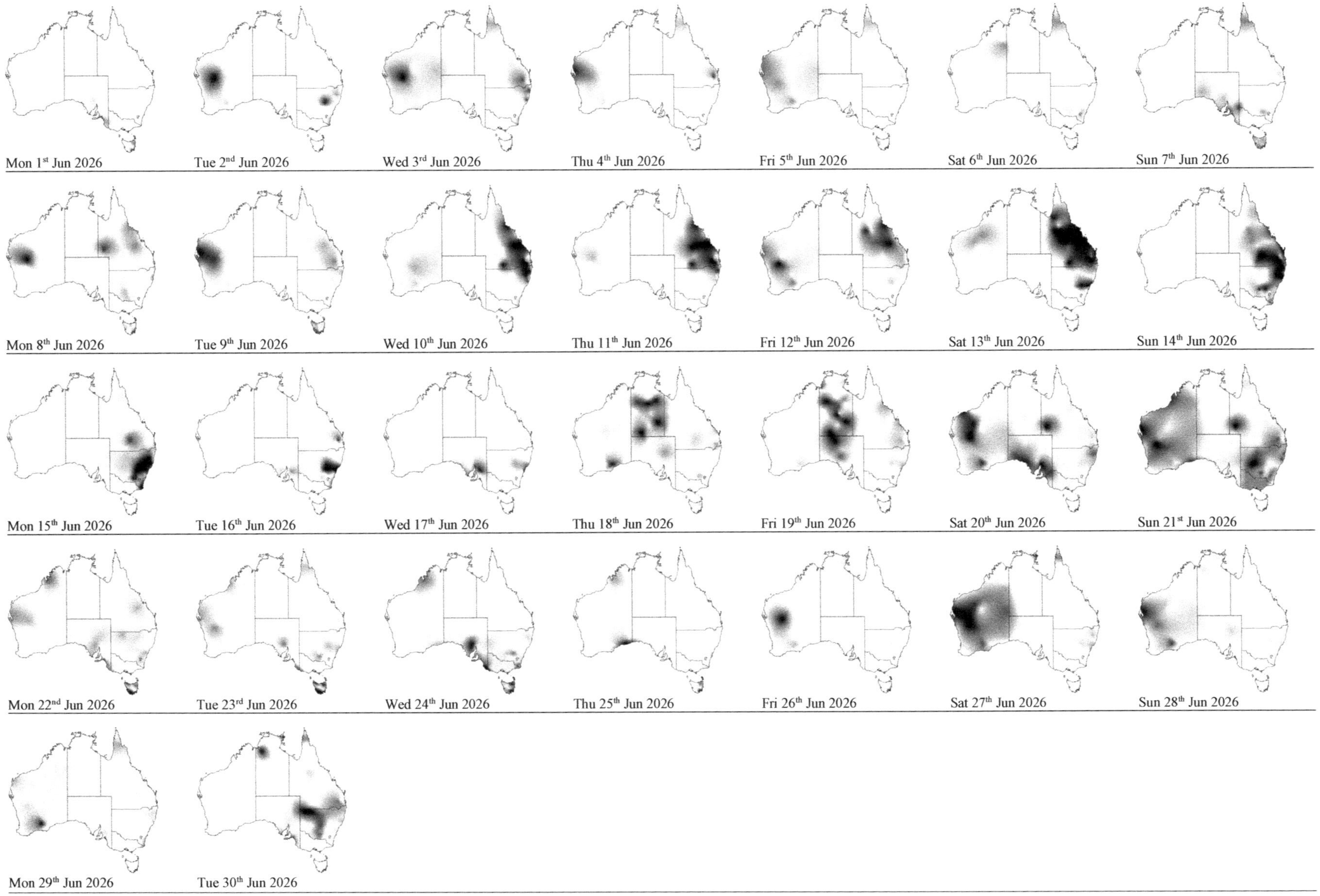
Mon 1st Jun 2026
Tue 2nd Jun 2026
Wed 3rd Jun 2026
Thu 4th Jun 2026
Fri 5th Jun 2026
Sat 6th Jun 2026
Sun 7th Jun 2026
Mon 8th Jun 2026
Tue 9th Jun 2026
Wed 10th Jun 2026
Thu 11th Jun 2026
Fri 12th Jun 2026
Sat 13th Jun 2026
Sun 14th Jun 2026
Mon 15th Jun 2026
Tue 16th Jun 2026
Wed 17th Jun 2026
Thu 18th Jun 2026
Fri 19th Jun 2026
Sat 20th Jun 2026
Sun 21st Jun 2026
Mon 22nd Jun 2026
Tue 23rd Jun 2026
Wed 24th Jun 2026
Thu 25th Jun 2026
Fri 26th Jun 2026
Sat 27th Jun 2026
Sun 28th Jun 2026
Mon 29th Jun 2026
Tue 30th Jun 2026

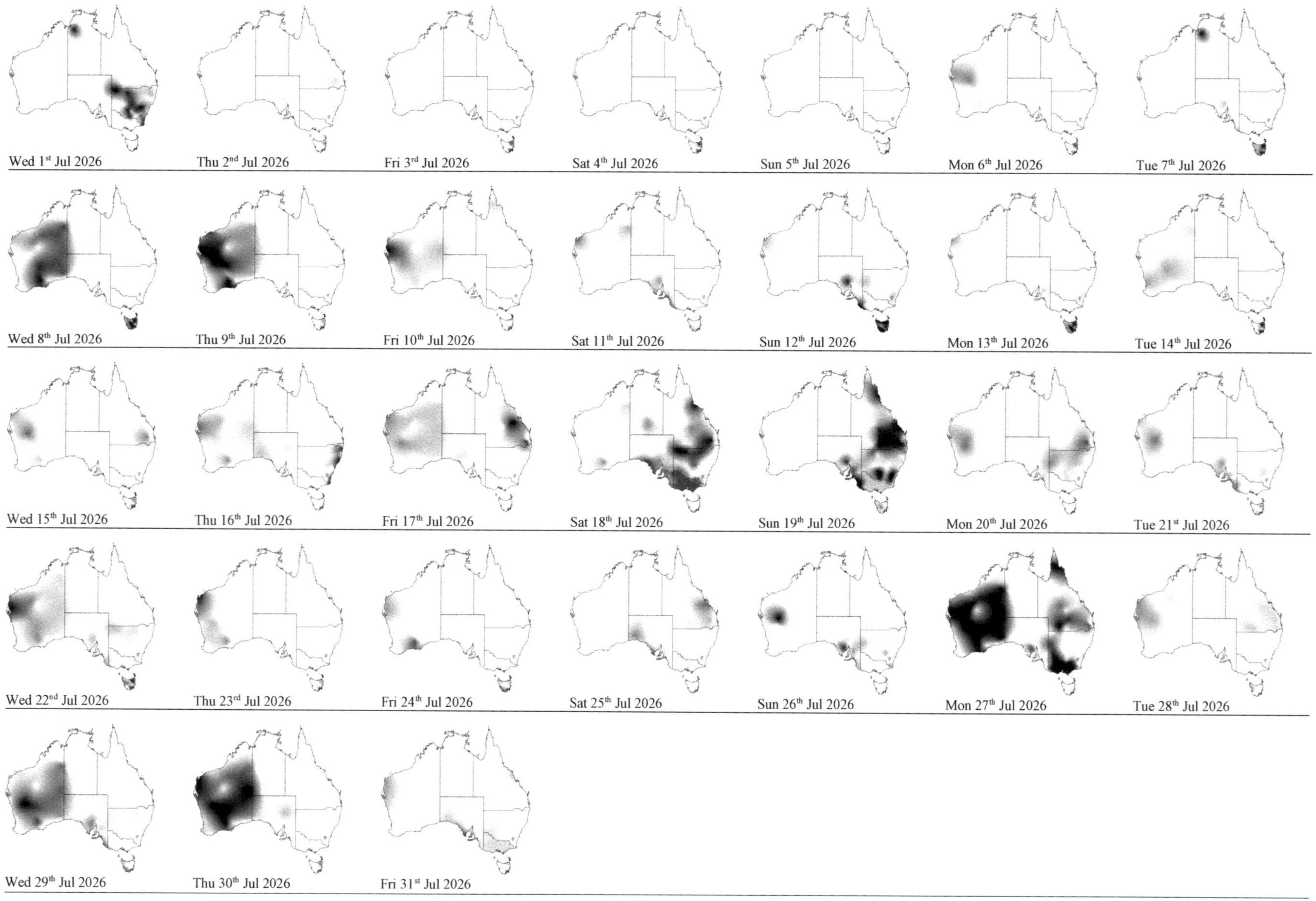
Wed 1st Jul 2026
Thu 2nd Jul 2026
Fri 3rd Jul 2026
Sat 4th Jul 2026
Sun 5th Jul 2026
Mon 6th Jul 2026
Tue 7th Jul 2026
Wed 8th Jul 2026
Thu 9th Jul 2026
Fri 10th Jul 2026
Sat 11th Jul 2026
Sun 12th Jul 2026
Mon 13th Jul 2026
Tue 14th Jul 2026
Wed 15th Jul 2026
Thu 16th Jul 2026
Fri 17th Jul 2026
Sat 18th Jul 2026
Sun 19th Jul 2026
Mon 20th Jul 2026
Tue 21st Jul 2026
Wed 22nd Jul 2026
Thu 23rd Jul 2026
Fri 24th Jul 2026
Sat 25th Jul 2026
Sun 26th Jul 2026
Mon 27th Jul 2026
Tue 28th Jul 2026
Wed 29th Jul 2026
Thu 30th Jul 2026
Fri 31st Jul 2026

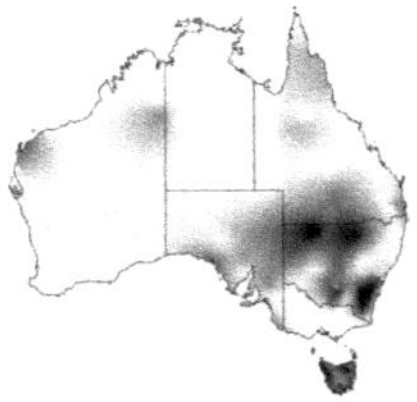
Sat 1st Aug 2026

Sun 2nd Aug 2026
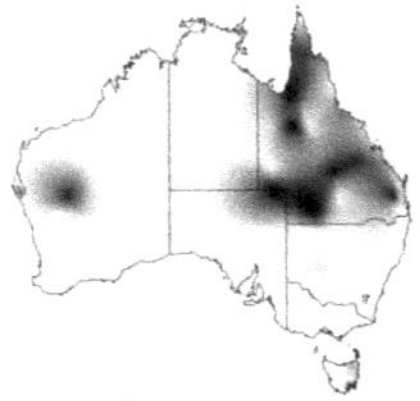
Mon 3rd Aug 2026
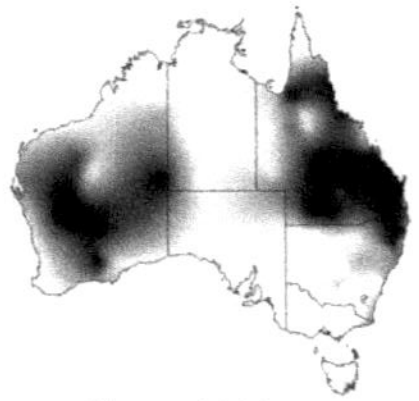
Tue 4th Aug 2026

Wed 5th Aug 2026
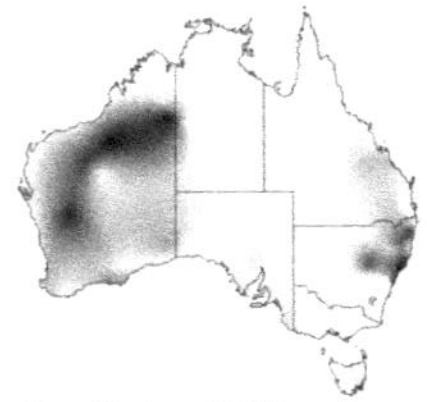
Thu 6th Aug 2026

Fri 7th Aug 2026
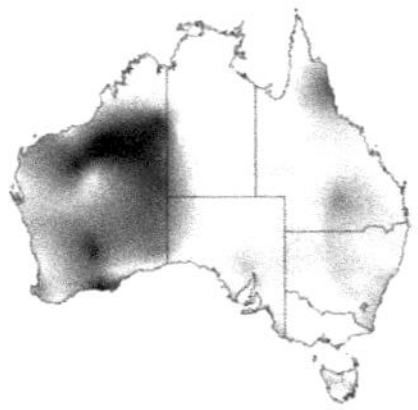
Sat 8th Aug 2026
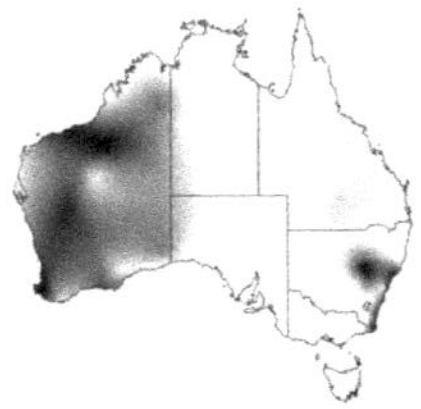
Sun 9th Aug 2026
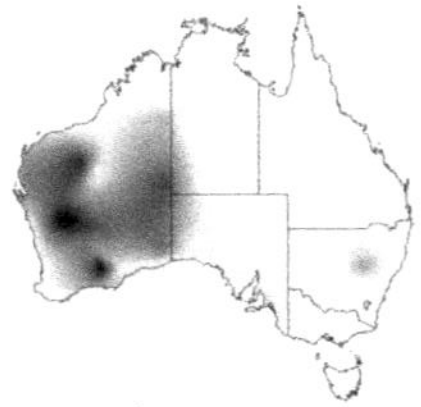
Mon 10th Aug 2026

Tue 11th Aug 2026

Wed 12th Aug 2026
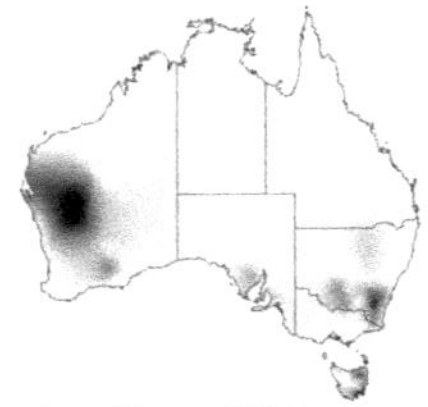
Thu 13th Aug 2026

Fri 14th Aug 2026

Sat 15th Aug 2026

Sun 16th Aug 2026

Mon 17th Aug 2026
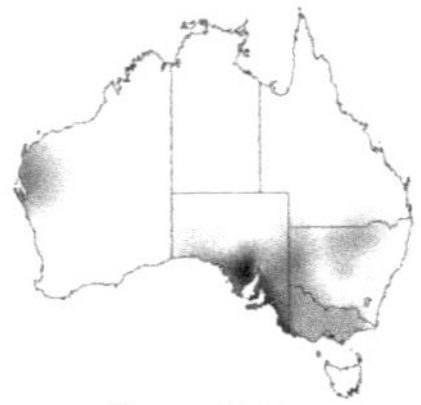
Tue 18th Aug 2026

Wed 19th Aug 2026

Thu 20th Aug 2026
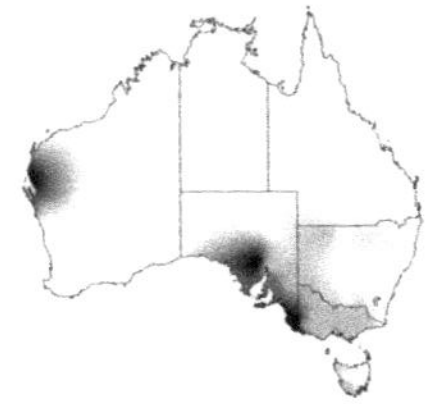
Fri 21st Aug 2026

Sat 22nd Aug 2026

Sun 23rd Aug 2026

Mon 24th Aug 2026

Tue 25th Aug 2026

Wed 26th Aug 2026

Thu 27th Aug 2026

Fri 28th Aug 2026

Sat 29th Aug 2026

Sun 30th Aug 2026

Mon 31st Aug 2026

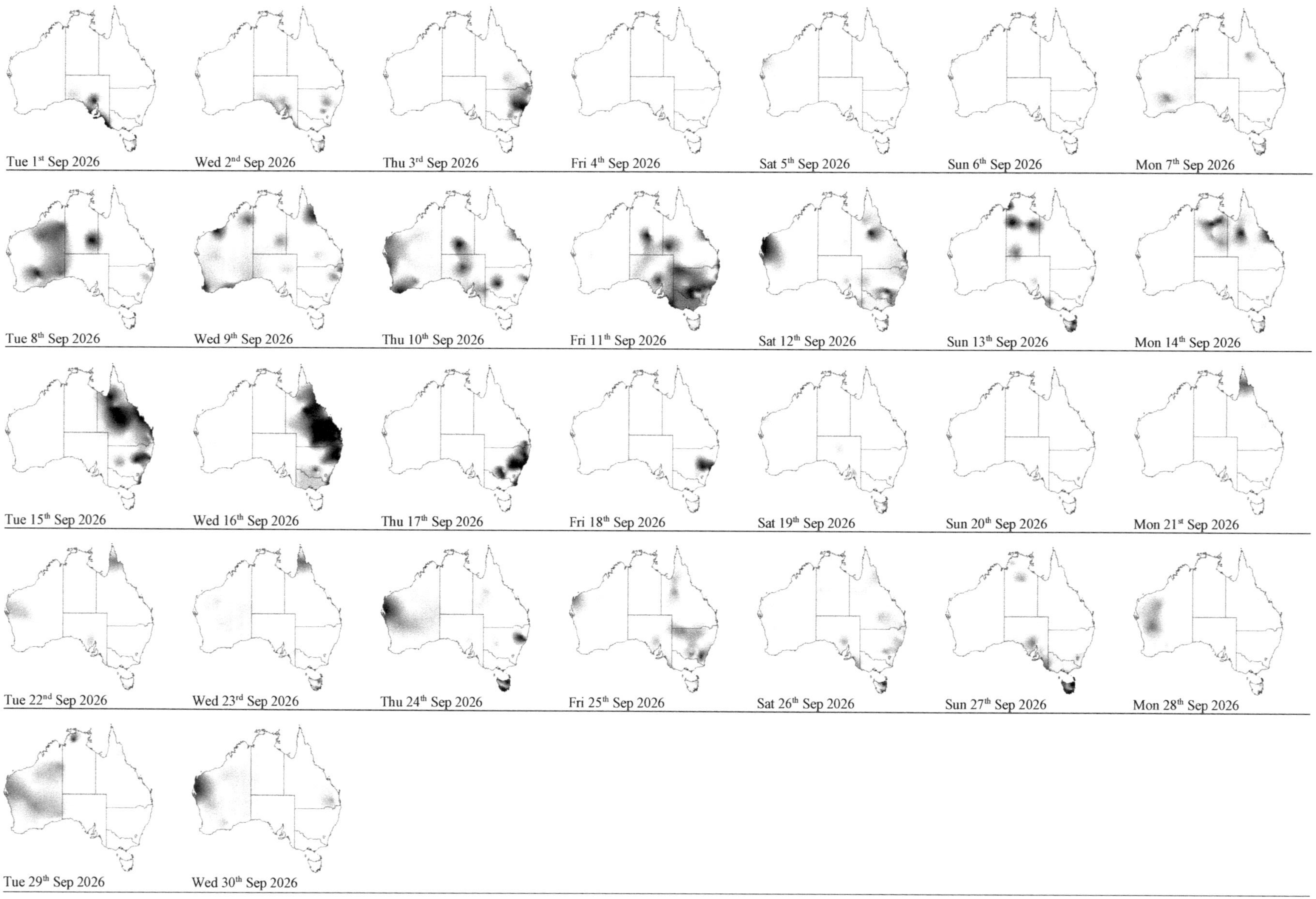
Tue 1st Sep 2026
Wed 2nd Sep 2026
Thu 3rd Sep 2026
Fri 4th Sep 2026
Sat 5th Sep 2026
Sun 6th Sep 2026
Mon 7th Sep 2026
Tue 8th Sep 2026
Wed 9th Sep 2026
Thu 10th Sep 2026
Fri 11th Sep 2026
Sat 12th Sep 2026
Sun 13th Sep 2026
Mon 14th Sep 2026
Tue 15th Sep 2026
Wed 16th Sep 2026
Thu 17th Sep 2026
Fri 18th Sep 2026
Sat 19th Sep 2026
Sun 20th Sep 2026
Mon 21st Sep 2026
Tue 22nd Sep 2026
Wed 23rd Sep 2026
Thu 24th Sep 2026
Fri 25th Sep 2026
Sat 26th Sep 2026
Sun 27th Sep 2026
Mon 28th Sep 2026
Tue 29th Sep 2026
Wed 30th Sep 2026

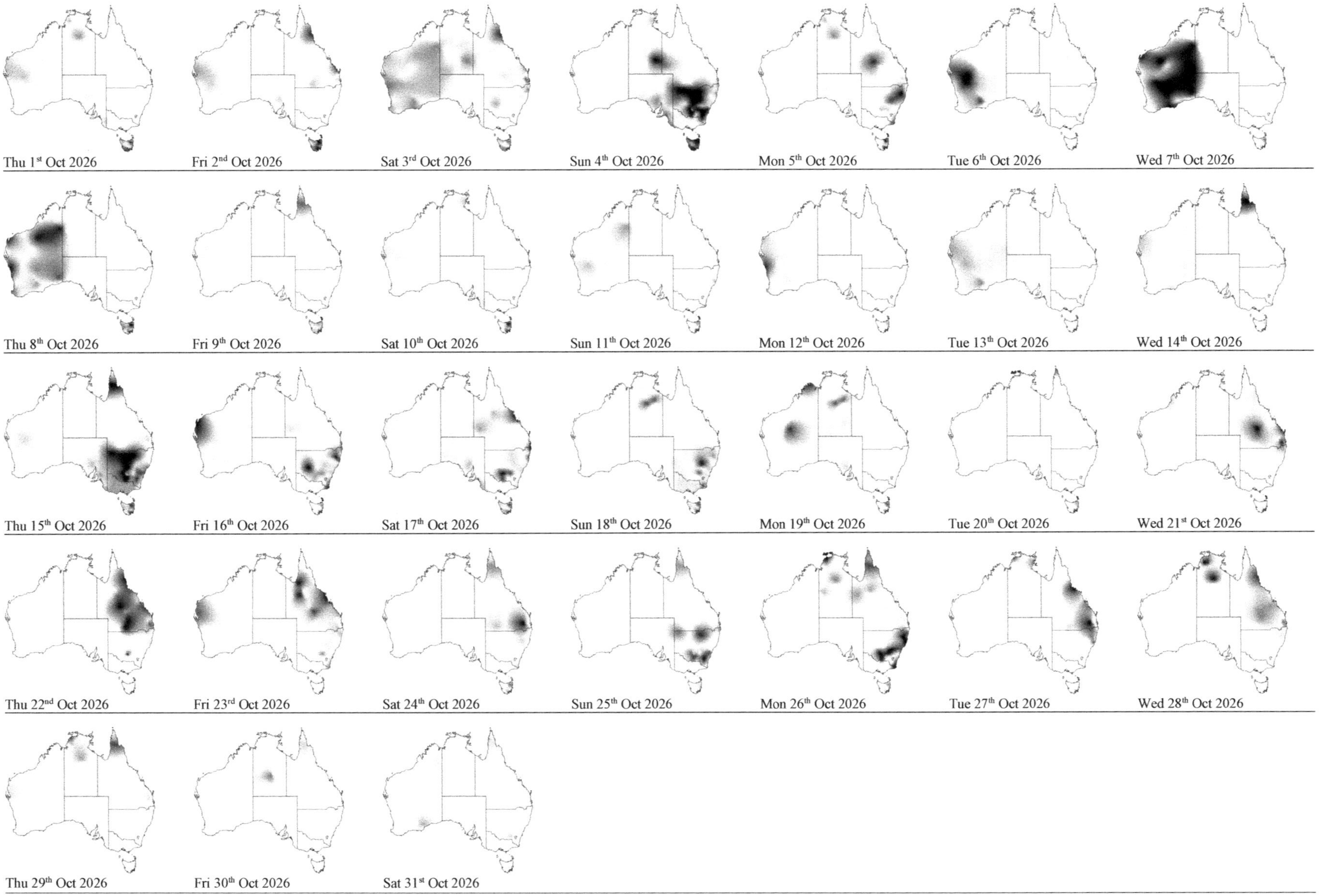
Thu 1st Oct 2026
Fri 2nd Oct 2026
Sat 3rd Oct 2026
Sun 4th Oct 2026
Mon 5th Oct 2026
Tue 6th Oct 2026
Wed 7th Oct 2026
Thu 8th Oct 2026
Fri 9th Oct 2026
Sat 10th Oct 2026
Sun 11th Oct 2026
Mon 12th Oct 2026
Tue 13th Oct 2026
Wed 14th Oct 2026
Thu 15th Oct 2026
Fri 16th Oct 2026
Sat 17th Oct 2026
Sun 18th Oct 2026
Mon 19th Oct 2026
Tue 20th Oct 2026
Wed 21st Oct 2026
Thu 22nd Oct 2026
Fri 23rd Oct 2026
Sat 24th Oct 2026
Sun 25th Oct 2026
Mon 26th Oct 2026
Tue 27th Oct 2026
Wed 28th Oct 2026
Thu 29th Oct 2026
Fri 30th Oct 2026
Sat 31st Oct 2026

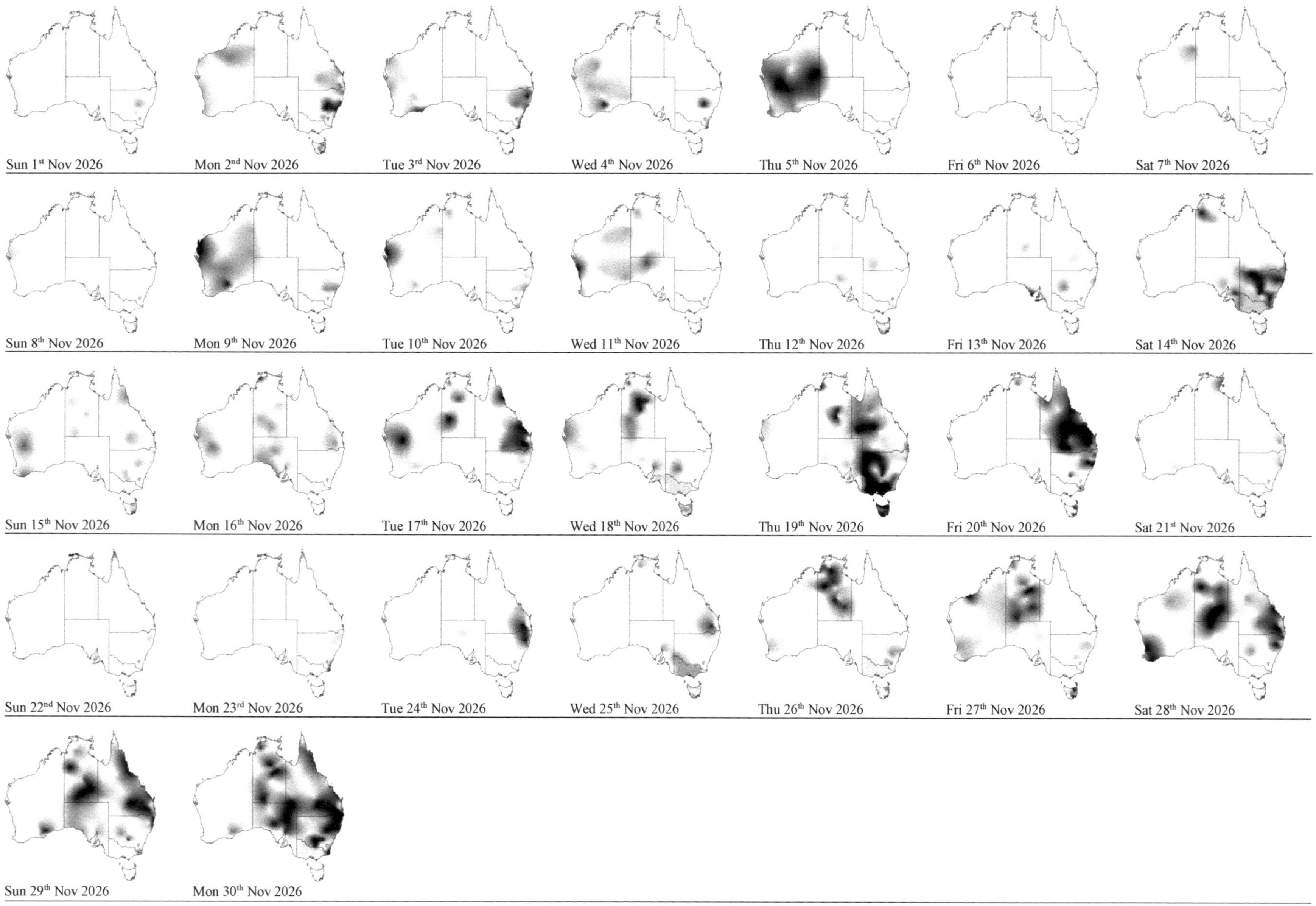

Sun 1st Nov 2026 | Mon 2nd Nov 2026 | Tue 3rd Nov 2026 | Wed 4th Nov 2026 | Thu 5th Nov 2026 | Fri 6th Nov 2026 | Sat 7th Nov 2026

Sun 8th Nov 2026 | Mon 9th Nov 2026 | Tue 10th Nov 2026 | Wed 11th Nov 2026 | Thu 12th Nov 2026 | Fri 13th Nov 2026 | Sat 14th Nov 2026

Sun 15th Nov 2026 | Mon 16th Nov 2026 | Tue 17th Nov 2026 | Wed 18th Nov 2026 | Thu 19th Nov 2026 | Fri 20th Nov 2026 | Sat 21st Nov 2026

Sun 22nd Nov 2026 | Mon 23rd Nov 2026 | Tue 24th Nov 2026 | Wed 25th Nov 2026 | Thu 26th Nov 2026 | Fri 27th Nov 2026 | Sat 28th Nov 2026

Sun 29th Nov 2026 | Mon 30th Nov 2026

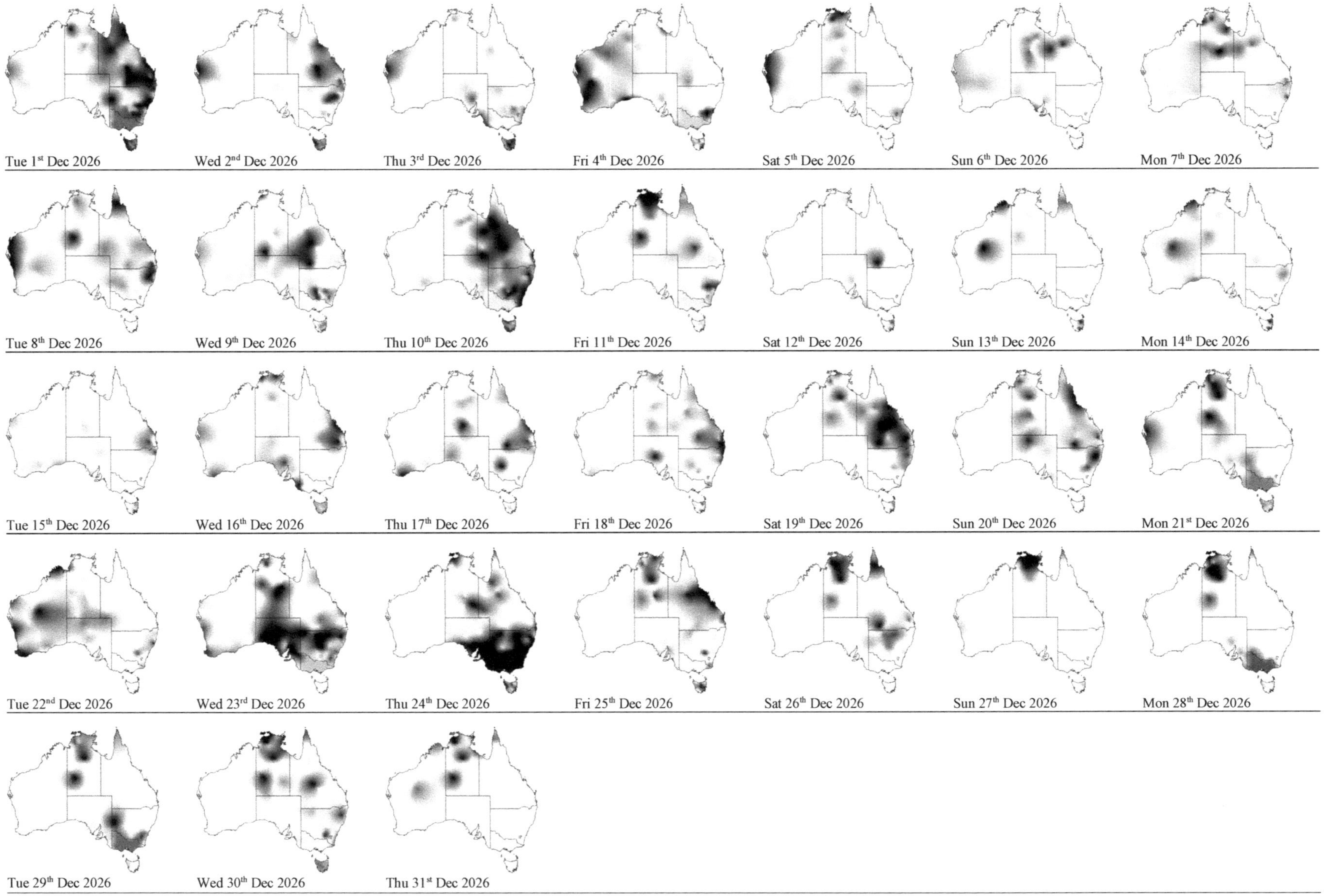
Tue 1st Dec 2026
Wed 2nd Dec 2026
Thu 3rd Dec 2026
Fri 4th Dec 2026
Sat 5th Dec 2026
Sun 6th Dec 2026
Mon 7th Dec 2026
Tue 8th Dec 2026
Wed 9th Dec 2026
Thu 10th Dec 2026
Fri 11th Dec 2026
Sat 12th Dec 2026
Sun 13th Dec 2026
Mon 14th Dec 2026
Tue 15th Dec 2026
Wed 16th Dec 2026
Thu 17th Dec 2026
Fri 18th Dec 2026
Sat 19th Dec 2026
Sun 20th Dec 2026
Mon 21st Dec 2026
Tue 22nd Dec 2026
Wed 23rd Dec 2026
Thu 24th Dec 2026
Fri 25th Dec 2026
Sat 26th Dec 2026
Sun 27th Dec 2026
Mon 28th Dec 2026
Tue 29th Dec 2026
Wed 30th Dec 2026
Thu 31st Dec 2026

ESTIMATED WEEKLY RAIN SUMMARY MAPS 2026

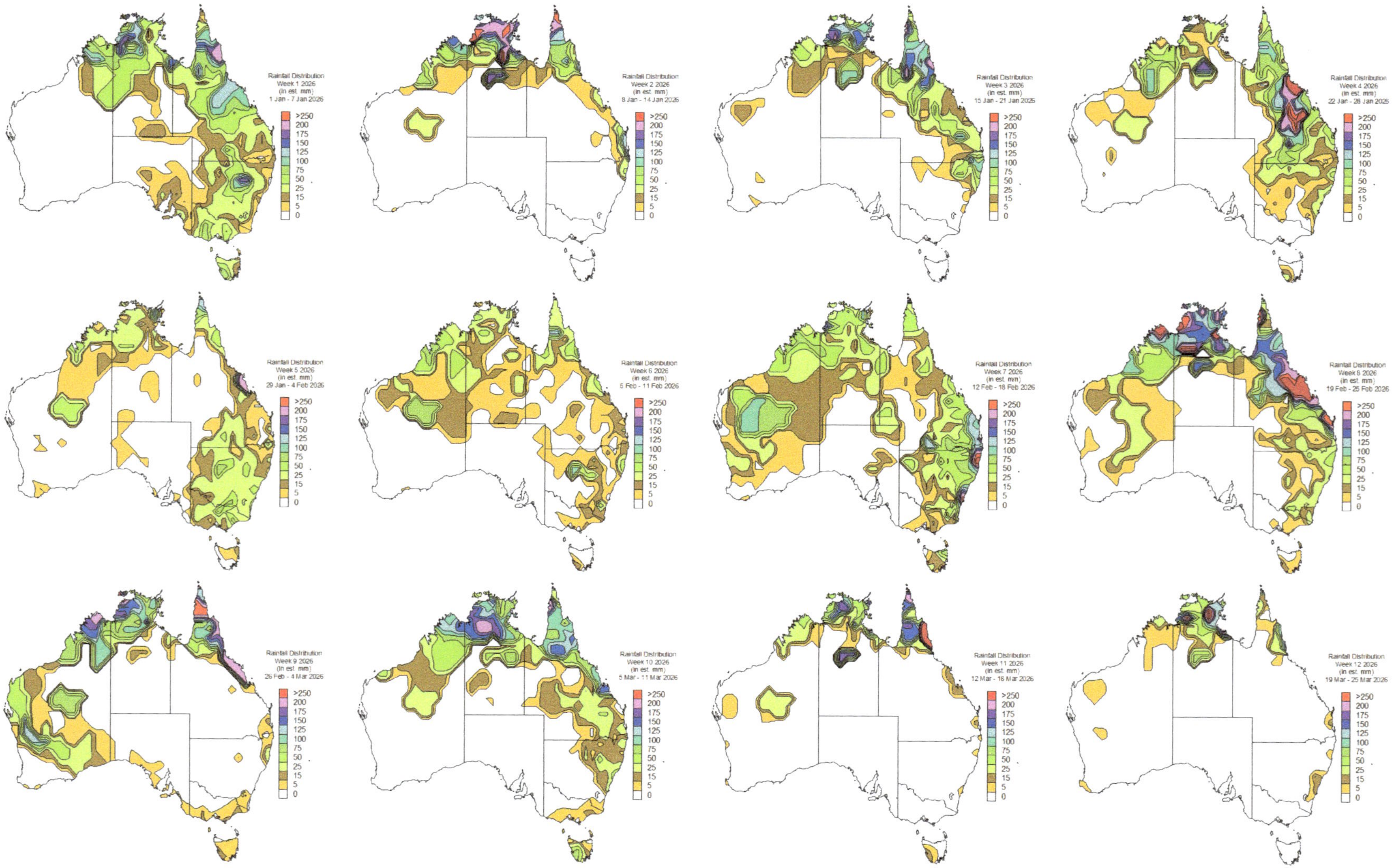

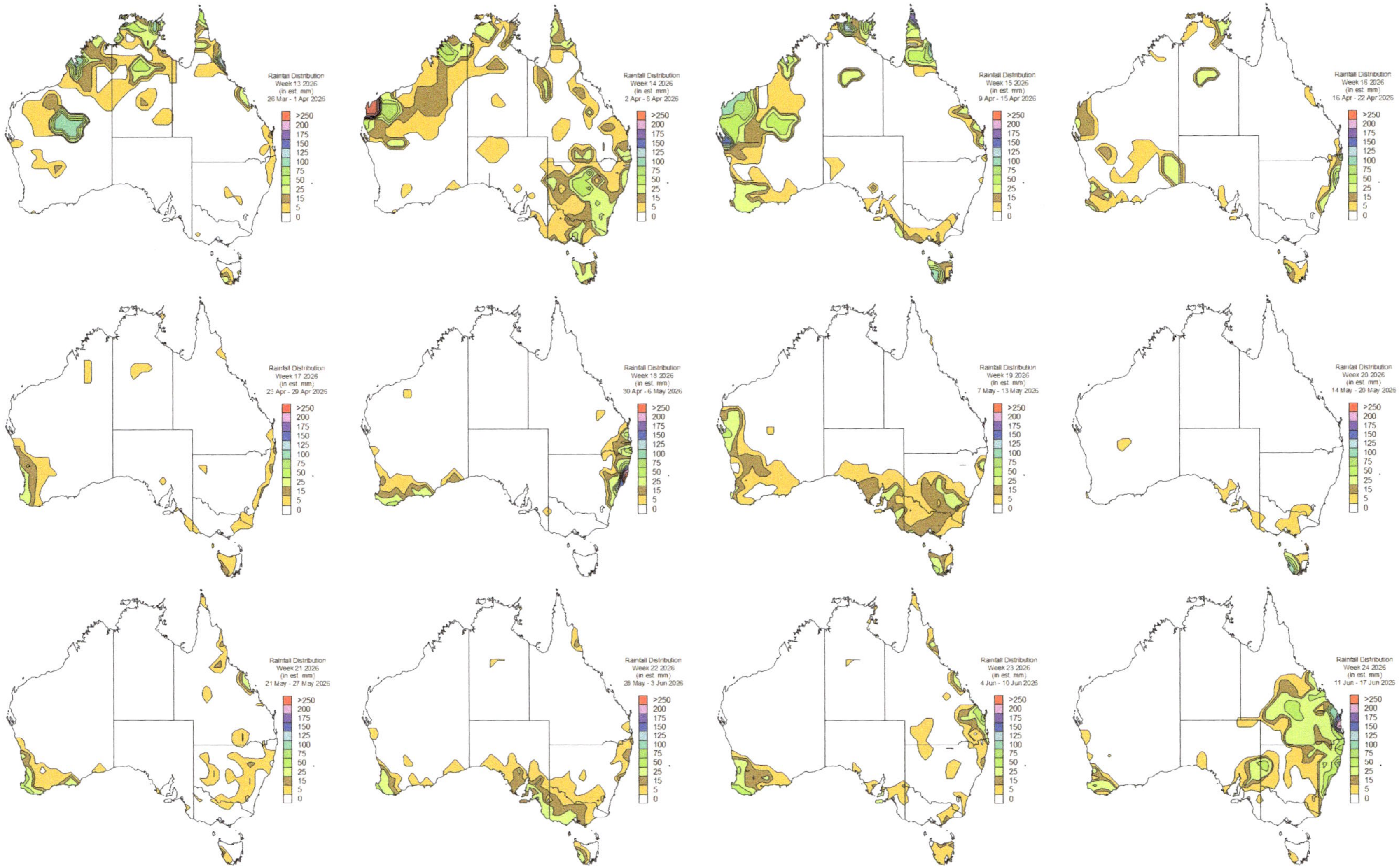
Rainfall Distribution
Week 13 2026
(in est. mm)
26 Mar - 1 Apr 2026
Rainfall Distribution
Week 14 2026
(in est. mm)
2 Apr - 8 Apr 2026
Rainfall Distribution
Week 15 2026
(in est. mm)
9 Apr - 15 Apr 2026
Rainfall Distribution
Week 16 2026
(in est. mm)
16 Apr - 22 Apr 2026
Rainfall Distribution
Week 17 2026
(in est. mm)
23 Apr - 29 Apr 2026
Rainfall Distribution
Week 18 2026
(in est. mm)
30 Apr - 6 May 2026
Rainfall Distribution
Week 19 2026
(in est. mm)
7 May - 13 May 2026
Rainfall Distribution
Week 20 2026
(in est. mm)
14 May - 20 May 2026
Rainfall Distribution
Week 21 2026
(in est. mm)
21 May - 27 May 2026
Rainfall Distribution
Week 22 2026
(in est. mm)
28 May - 3 Jun 2026
Rainfall Distribution
Week 23 2026
(in est. mm)
4 Jun - 10 Jun 2026
Rainfall Distribution
Week 24 2026
(in est. mm)
11 Jun - 17 Jun 2026
>250
200
175
150
125
100
75
50
25
15
5
0

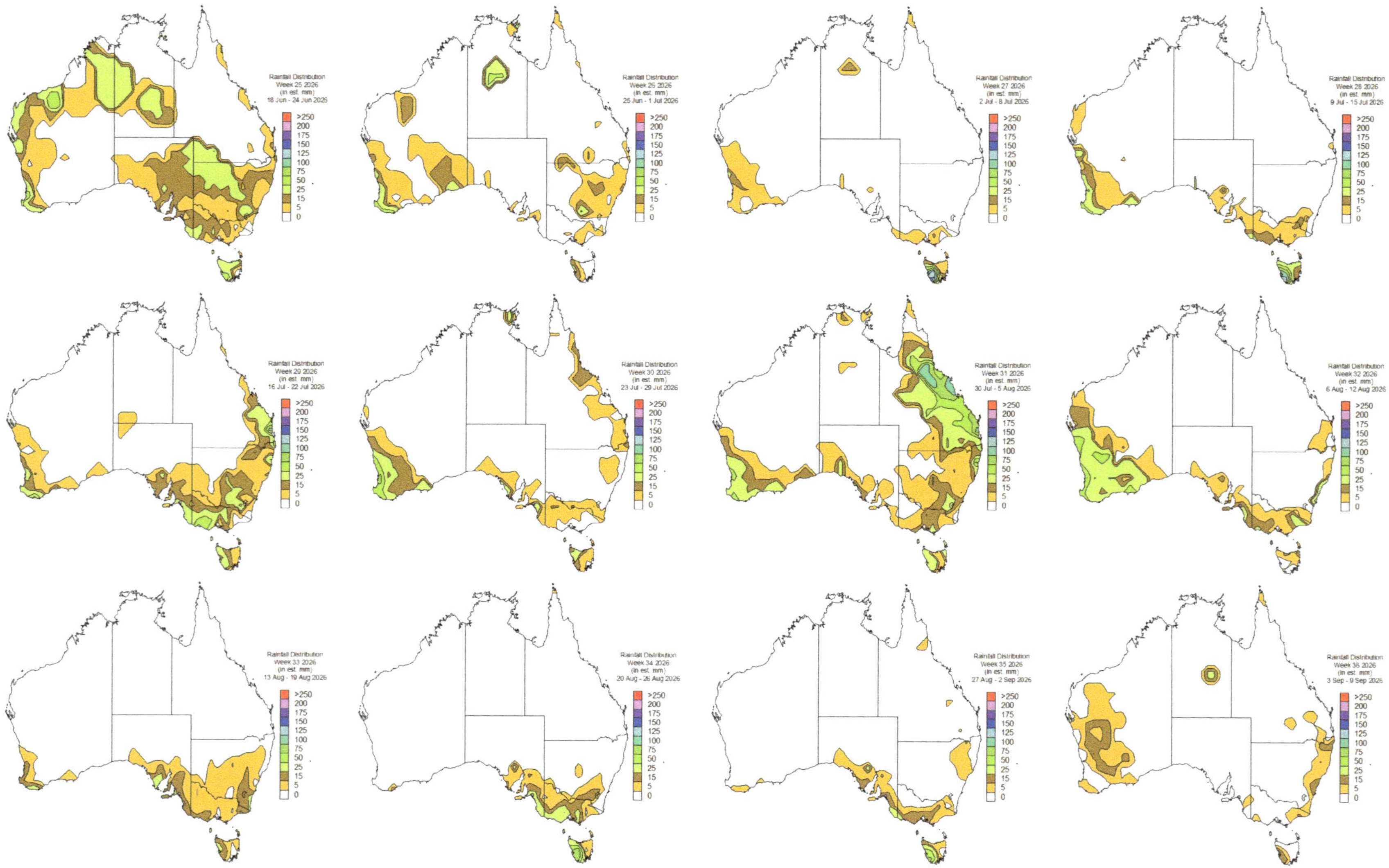
Rainfall Distribution
Week 25 2026
(in est. mm)
18 Jun - 24 Jun 2026
>250
200
175
150
125
100
75
50
25
15
5
0
Rainfall Distribution
Week 25 2026
(in est. mm)
25 Jun - 1 Jul 2026
Rainfall Distribution
Week 27 2026
(in est. mm)
2 Jul - 8 Jul 2026
Rainfall Distribution
Week 28 2026
(in est. mm)
9 Jul - 15 Jul 2026
Rainfall Distribution
Week 29 2026
(in est. mm)
16 Jul - 22 Jul 2026
Rainfall Distribution
Week 30 2026
(in est. mm)
23 Jul - 29 Jul 2026
Rainfall Distribution
Week 31 2026
(in est. mm)
30 Jul - 5 Aug 2026
Rainfall Distribution
Week 32 2026
(in est. mm)
6 Aug - 12 Aug 2026
Rainfall Distribution
Week 33 2026
(in est. mm)
13 Aug - 19 Aug 2026
Rainfall Distribution
Week 34 2026
(in est. mm)
20 Aug - 26 Aug 2026
Rainfall Distribution
Week 35 2026
(in est. mm)
27 Aug - 2 Sep 2026
Rainfall Distribution
Week 36 2026
(in est. mm)
3 Sep - 9 Sep 2026

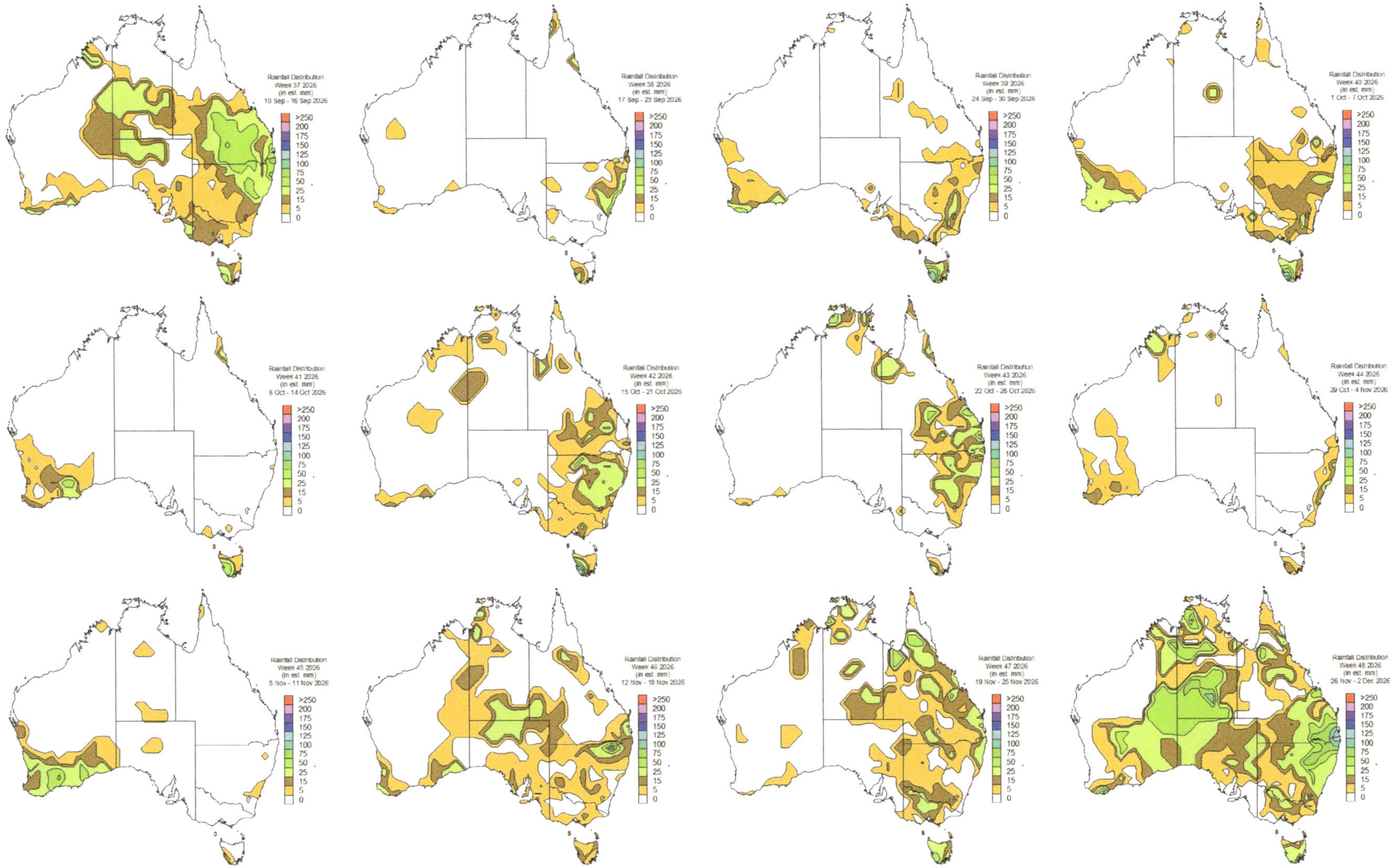
Rainfall Distribution
Week 37 2026
(in est. mm)
10 Sep - 16 Sep 2026
>250
200
175
150
125
100
75
50
25
15
5
0
Rainfall Distribution
Week 38 2026
(in est. mm)
17 Sep - 23 Sep 2026
>250
200
175
150
125
100
75
50
25
15
5
0
Rainfall Distribution
Week 39 2026
(in est. mm)
24 Sep - 30 Sep 2026
>250
200
175
150
125
100
75
50
25
15
5
0
Rainfall Distribution
Week 40 2026
(in est. mm)
1 Oct - 7 Oct 2026
>250
200
175
150
125
100
75
50
25
15
5
0
Rainfall Distribution
Week 41 2026
(in est. mm)
8 Oct - 14 Oct 2026
>250
200
175
150
125
100
75
50
25
15
5
0
Rainfall Distribution
Week 42 2026
(in est. mm)
15 Oct - 21 Oct 2026
>250
200
175
150
125
100
75
50
25
15
5
0
Rainfall Distribution
Week 43 2026
(in est. mm)
22 Oct - 28 Oct 2026
>250
200
175
150
125
100
75
50
25
15
5
0
Rainfall Distribution
Week 44 2026
(in est. mm)
29 Oct - 4 Nov 2026
>250
200
175
150
125
100
75
50
25
15
5
0
Rainfall Distribution
Week 45 2026
(in est. mm)
5 Nov - 11 Nov 2026
>250
200
175
150
125
100
75
50
25
15
5
0
Rainfall Distribution
Week 46 2026
(in est. mm)
12 Nov - 18 Nov 2026
>250
200
175
150
125
100
75
50
25
15
5
0
Rainfall Distribution
Week 47 2026
(in est. mm)
19 Nov - 25 Nov 2026
>250
200
175
150
125
100
75
50
25
15
5
0
Rainfall Distribution
Week 48 2026
(in est. mm)
26 Nov - 2 Dec 2026
>250
200
175
150
125
100
75
50
25
15
5
0

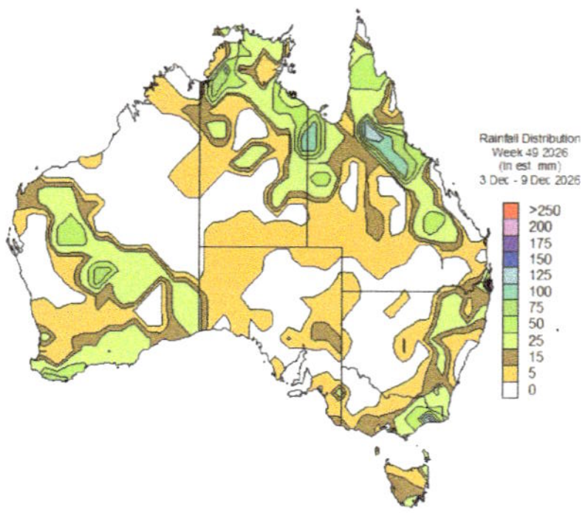
Rainfall Distribution
Week 49 2026
(in est. mm)
3 Dec - 9 Dec 2026
>250
200
175
150
125
100
75
50
25
15
5
0

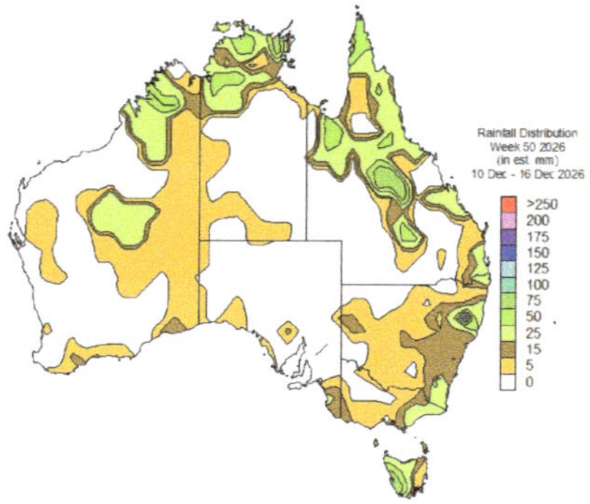
Rainfall Distribution
Week 50 2026
(in est. mm)
10 Dec - 16 Dec 2026
>250
200
175
150
125
100
75
50
25
15
5
0

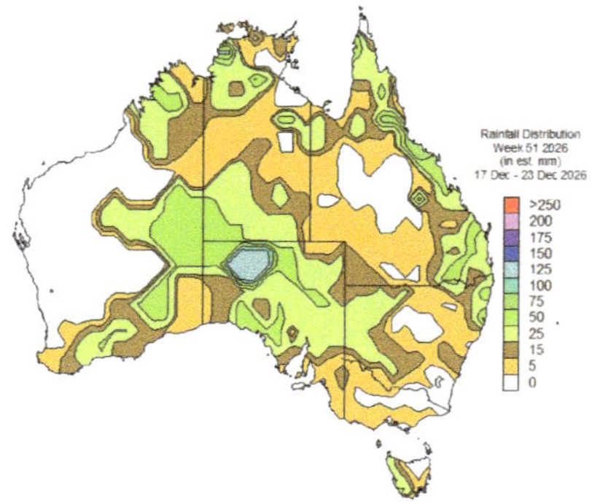
Rainfall Distribution
Week 51 2026
(in est. mm)
17 Dec - 23 Dec 2026
>250
200
175
150
125
100
75
50
25
15
5
0

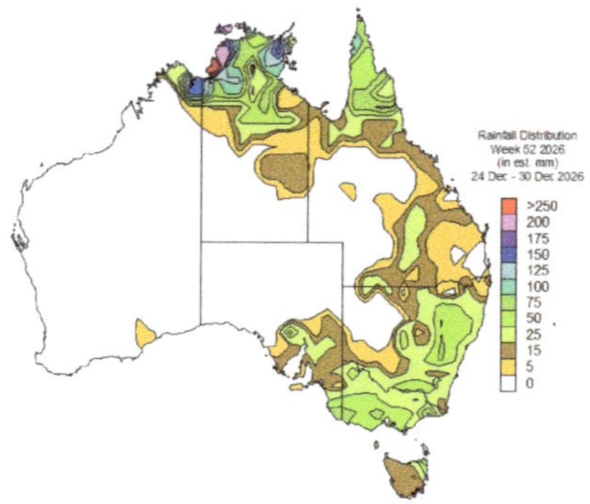
Rainfall Distribution
Week 52 2026
(in est. mm)
24 Dec - 30 Dec 2026
>250
200
175
150
125
100
75
50
25
15
5
0

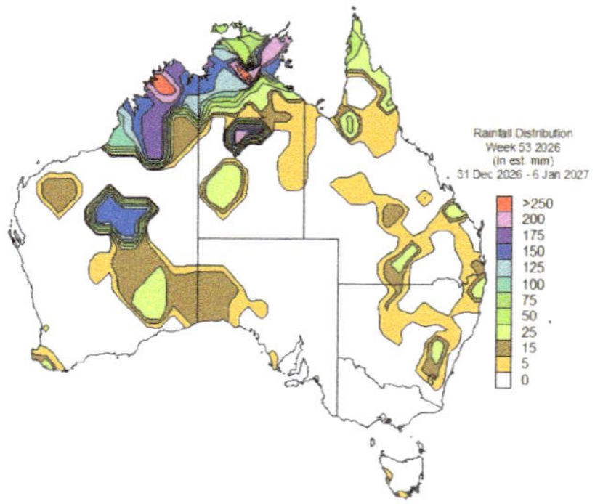
Rainfall Distribution
Week 53 2026
(in est. mm)
31 Dec 2026 - 6 Jan 2027
>250
200
175
150
125
100
75
50
25
15
5
0

ESTIMATED SEASONAL RAIN SUMMARY MAPS 2026

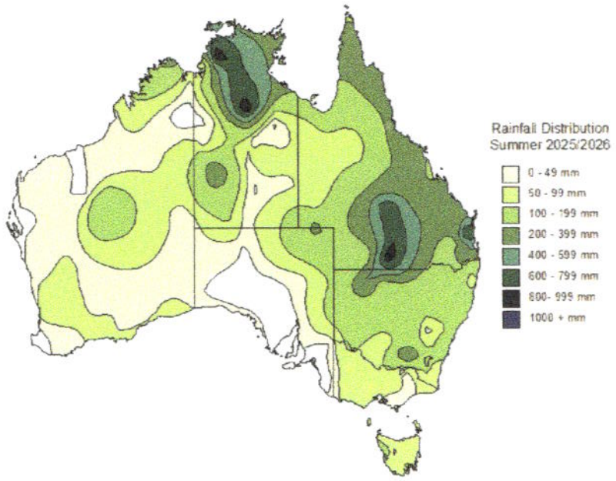

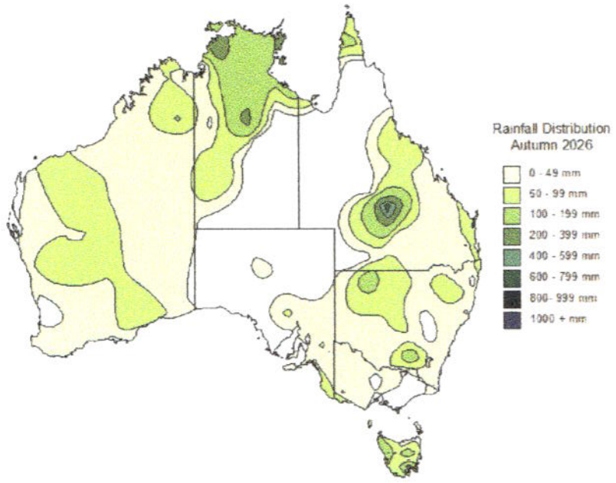

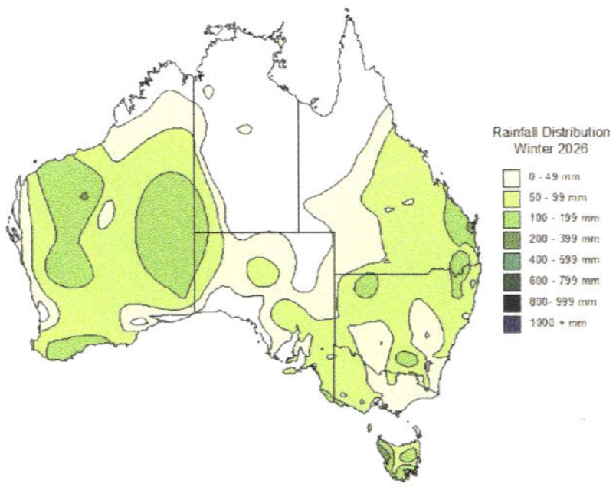

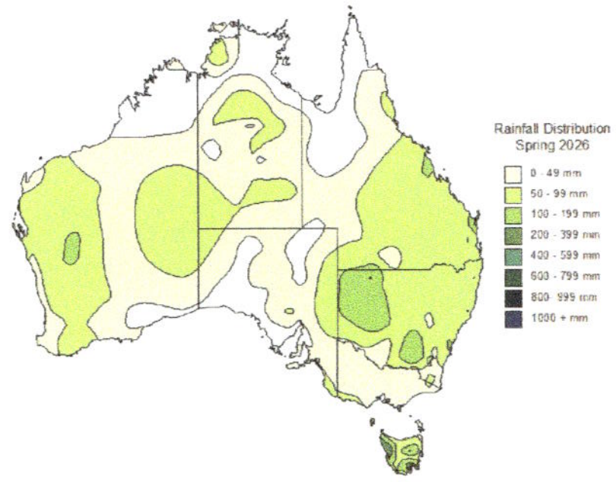

ESTIMATED MAXIMUM AND MINIMUM TEMPERATURE ANOMALIES 2026

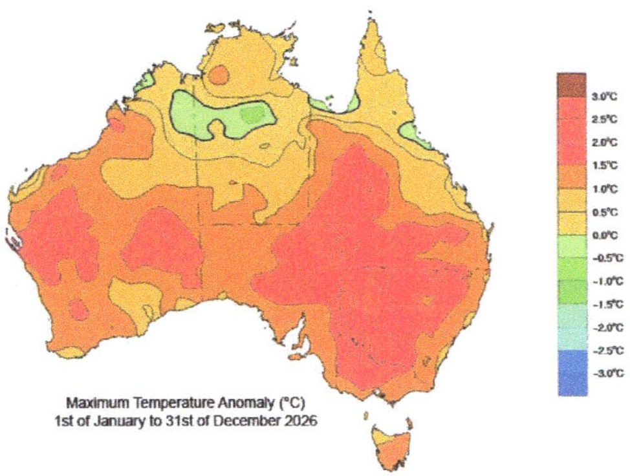

Maximum Temperature Anomaly (°C)
1st of January to 31st of December 2026

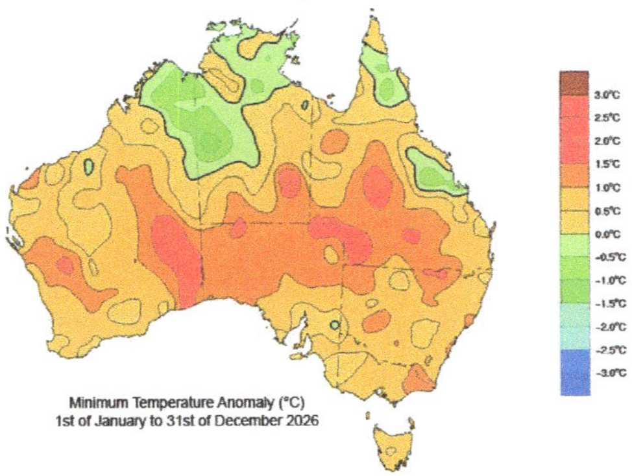

Minimum Temperature Anomaly (°C)
1st of January to 31st of December 2026

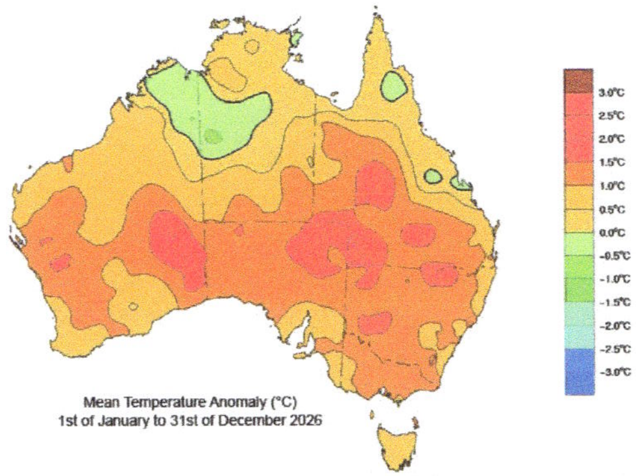

Mean Temperature Anomaly (°C)
1st of January to 31st of December 2026

NOTES ON WEATHER AND CLIMATE FOR AUSTRALIA

Much rain for NSW comes across the sea from the east but can emanate from all south and west southerly systems, whereas flooding that comes because of a laden Murray-Darling is a flow-on effect from QLD rain systems well to the north. Sometimes QLD rain may come from monsoonal systems and cyclonic developments to the north and across from the northwest moving southeast, and west from the Coral Sea. The monsoon can bring cool systems to the top of Australia, and in so doing, drain the heat from the interior. When the monsoon slackens, the interior dries out and can flood when first rains return, because the ground is baked hard.

Whether it is a wetter season or not is governed by regular cycles. These cycles in turn are determined by the Moon. The sun, moon and planets, acting in combinations of position and angles, to each other and to earth, are mostly responsible for the cycles. Then it is changes in ocean currents that determine weather. The orbiting of planets affects the tide of radiation from the Sun. Jupiter, Saturn, Neptune, Uranus and Mercury are of particular significance for long term forecasting. The gravitational orbits of the planets combine to find an average gravitational centre which many assume to be the Sun, when in fact it is a moving point called the Solar System Barycentre (SSB) that extends one sun's radius on either side of the sun and crosses the solar center about once every 30-40 years. There is therefore a tide of the ebb and flow of solar radiation that occurs on a cycle of around 72 years, with a half-cycle of 30-36 years.

JUPITER AND SATURN

When the planet Jupiter draws the Sun outwards from its normal line, Sunspot maxima occurs, and when the Sun is drawn inwards the minima is the result. When the gas giants Jupiter and Saturn pass across the line of the Sun's advance (Solar Apex), Sunspot minimum is very close to that date. Jupiter, Saturn, Uranus and Neptune, in that order affect the Sun the most, with a period of 178 years, roughly 10 lunar cycles. On the whole, the 11-12-year Sunspot period is the same as the orbit of Jupiter, and any variations are due to the varying positions of the other three great planets.

With each passage of Jupiter, either just before or just after, there is usually a major drought in Australia. In the prolonged drought of 1862, Jupiter and Saturn lined up on one side of the Sun in that year. Both straddled the Sun in 1871, were on one side of the Sun 1900-2, then on opposite sides of the Sun in 1931. Jupiter/Saturn together go through this cycle every 35-36 years, and the other two planets, Uranus and Neptune, are of much lesser importance. Jup/Sat were together again on one side of the Sun in 2020, the year of sunspot minimum.

The Sun is also on a 7deg tilt to earth, which must be added or subtracted from Earth's own tilt of 23deg. The difference between spring and autumn is because in our spring, southern Sunspots are turned more away from Earth's southern hemisphere and are more directed towards us in our autumn.

In 1982, Saturn/Jupiter were on the same side of the Sun and brought a severe drought to the east of Australia. In 2001 Jupiter crossed the vertex, bringing further drought. When Jupiter crosses the path of the Sun there is a drought just beforehand. Jupiter was crossing the Sun in 2007, when world-wide droughts brought a food crisis, sparking riots and the Murray Darling had the lowest inflows on record. Why? The solar cycle again.

No sunspots can mean little or no rain. Large spots can mean much rain. Flood years are around the time of maximum sunspots (2014). Jupiter's next path-cross was in 2019. If Jupiter and Saturn are together on one side of the Sun, this also can mean drought, and for some places was 2020/1. 19 years, the lunar Metonic Cycle, is approximately the period between the conjunctions of Jupiter and Saturn. That is to say, between the times when they are nearest to one another as they revolve round the Sun. The Bruckner cycle of 35½ years is the period required for three revolutions of Jupiter and 38 years for the interval between three consecutive conjunctions of Saturn and Jupiter.

A cycle of about 76 years shows rather conspicuously in the records of the Nile annual flood levels, but this also is a multiple of the Jupiter-Saturn conjunction period 19 x 4.76, and is slightly in excess of six revolutions of Jupiter's 71¼ years. After Jupiter crosses the Sun's path, sunspots drop which previously may have shown great activity.

Jupiter has been at high peaks in 1766-69, 1779-81, 1790-93, 1802-4 (drought Jup+Sat), 1814-17 (drought), 1827-30 (drought), 1838-40 (Jup+Sat drought), 1850-52, 1862-64, 1873-76, 1885-88, 1887-1900, 1909-12, 1921-23, 1933-35, 1945-47, 1957-59, 1968-71, 1980-83, 1992-95, 2004-6, 2016-19 (all drought years). Low Jupiter peaks have been 1774, 1786, 1798, 1809, 1821, 1833, 1845, 1857, 1881, 1904, 1928, 1939-40, 1952, 1964, 1976, 1987, 1999, 2011 and 2023.

Jupiter/Saturn peaks and conjunctions have been blamed for low yields and economic panic. Examples have been 1873, 1884, 1893, 1900-2, 1907, 1910, 1958-60, 1987, 1990, 1999, and the recession of 2011-12. The worst have been 1901 and 1959. The next is 2016-2023 as solar minimum creates potential for worldwide droughts through low rainfall and low agricultural yields in 2020. Closer perigees in minimum declination years typically bring longer lasting extreme events.

In 2020 Jupiter and Saturn were in "conjunction". That means they are occupying the same area of sky as seen from earth. When you looked up at them, they appeared very close together. Conjunction correlates with drought years. By 2023-5 these two gas giant planets will be at right angles to each other, called a "square", and the wetter years will begin. Finally, in 2030 they are in "opposition", being 180° apart, and typically this will be a year of floods. Half the cycle is then over.

The last previous year of equivalent conjunction was 2001. In 2005 was the last square of these planets, and their last equivalent opposition time was 2012, which was a wet year for eastern Australia. Following that, the last previous square was in 2015. Since 2015 it has been noticeably getting drier.

The order is:
J/S conjunction=drought years (2001, 2020),
J/S square=coming out of drought years, wetter (2005, 2025)
J/S opposition=flood years (2012, 2030)
J/S square=getting drier, approaching the next drought (2015, 2036).

Jupiter and Saturn together control the grand cycle of drought/flood. Jupiter orbits the Sun every 12 years, and Saturn every 29. Jupiter and Saturn are by far the largest planets in the cosmos, and together have an effect on the Sun, such that every 35-38 years the SSB (solar system barycenter) sees an event-repeat. Thus 2036 may be the repeat of some things that last occurred in 2001. And 2021 may see repeats of 1985. By looking at the known history of Australia we can spot trends that have resulted from this Jupiter/Saturn/Sun tidal model. The trends we see are the alternating 30–36-year SSB turnarounds, caused by the planets affecting the sun, creating the cyclic effects on the continent's climate.

During the 1860s-1890s Australia was wet and cold. The 1900s - 1930s became relatively hot and dry. The Great Federation Drought happened within this time, and the heat, warmer than today's temperature averages, that are ingrained in American history as the Dustbowl years. Arctic ice melted dramatically in the 1920s.

The 1940s-1970s again were cooler and wetter, starting around 1960, so much so that in the later 1970s an impending Ice Age was considered to be near. International media forecasted that glaciers would be spreading over Britain and the USA within the following decade.

1970s-1998 saw conditions warm up. Glaciers receded. The media suggested that coastal towns around the world would soon have to be evacuated as the poles would rapidly melt down and sea levels would rise. 2000-2030 brings the next big change. Around 2007/8 was the first signs of a cooler and wetter phase. There were droughts at first, but overall, after 2020, Australia can prepare for a decade that more resembles past cycles of milder years. That is, the next decade should overall be cooler and wetter than the last decade.

SOLAR CYCLES TO 2070

It is the Sun's heat that causes evaporation which brings rain. The amount of heat that the Sun puts out determines the amount of evaporation, which must then fall within about 7 days, near to the water source. Coastal locations are generally wetter than places that are some distance inland. Without a source of nearby water, a place quickly dries out. The *amount* of rain falling is due to the amount of solar radiation which causes evaporation. Therefore, rain amounts depend on the sunspot cycle, varying over 11 and/or 22-24 years. The amount of rain falling is also dependent on the extent of the air tide, which can be likened to volume of water in each sea tide. As the solar cycle of radiation is cyclic then so too are rain intensities.

Sea tide timing is predictable, subject to positions of Moon and Sun. The volume of water flowing into an estuary depends on wind direction subject to shape and size of surrounding hills, wind forces increased or confused by surface currents, and less visible factors like proximity of underwater eruptions, fissures, earthquakes, and time of the day due of temperature gradients. The amount of rain is influenced by Sun's warmth for evaporation and then how much cold air comes to the layers of the atmosphere containing clouds.

Sometimes colder air is at a higher or lower altitude than rain-filled clouds, and expected rains must wait for the marrying of the two factors. Until then a cloud may look like rain but heat from the ground is sufficient to buoy up the rain-laden cloud and prevent rain falling. Analyzing sunspot cycles is at least better than having nothing at all, by looking to see which previous years correspond to same points of the cycle. We are currently experiencing the beginnings of Cycle #25.

Solar, or sunspot cycles determine annual temperatures and amounts of annual rain. Moon cycles determine timing of weather events. Long-term, planet Earth is not getting hotter. The accepted figure of a warming of 0.7 of 1 degree over the past 100 years is merely the same as would be expected due to natural variation. That warming, as we continue to leave the last Ice age, has amounted to only 1°C over the past 8000 years. Ice ages typically last 50-60,000 years and the interglacial, the period between ice ages, lasts about 10,000 years.

It means that ice ages are the *natural* state of earth, comprising 80% of our known geological history. Earth spends most of its time frozen up. We are now about 11,000 years since the last ice age and at end of this current interglacial and preparing to enter the next ice age. But it will be gradual, losing about a degree every 5,000 years. We are global cooling - there have never been back-to-back interglacial stages.

The sun's radiation comes and goes in cycles that we have noticed correlating to periods of black spots on the sun. They have been called sunspot cycles and although observed and documented for hundreds of years by the Chinese, it is only since the 1700s that the sunspot cycles have been given agreed-upon numbers. Sunspot Cycle 23 began in 1998, peaked around 2000/1 (solar maximum), and died off (solar minimum) from 2007-2009. After a 2–3-year period of relative inactivity, Sunspot Cycle 24 began around midway through 2010, and peaked around 2013 and died in 2019/20. Around 2021 Cycle #25 began.

Sunspot variance is largely dependent on the orbits of the gas giants, Jupiter and Saturn. Whenever these two are on the same side as the sun or exactly opposite to each other with the sun between them, the tide of electromagnetism coming from the sun undergoes a surge. Sunspot cycles are about 11 years and 22 years, but there is much variance, such that the 11-year one can span 9-14 years. The SSB (solar system barycentre) correlates to this, and 12 and 24-year magnetic cycles are also common. The minimum to minimum of Cycle #23 was from 1996 to 2009, about 13.5 years. The last time that happened was immediately prior to a period called the 'Dalton Minimum' which coincided with a few decades of unusual cold on earth, from 1790 to 1830.

Within that period was 1816, in the northern hemisphere called the year without a summer, and the time before that ushered in the more extended 'Maunder Minimum' that spanned 1645 to 1715, coinciding with the depths of the "Little Ice Age" from the 16th century to the mid-19th century. It is why some current geologists and astronomers are suggesting we are entering a more extended Maunder-type grand minimum.

Just as Sun and Moon, either both on one side of Earth (new Moon) or on opposite sides (full Moon) cause internal stress on Earth to raise tides in land, sea and air, by the same process the largest of our solar system's planetary bodies on one side of the Sun or on opposite sides cause exaggerations in Sun-tides manifesting as Sunspots/solar wind.

Right now we are well into Sunspot cycle #25. Cycle #23 began in 1998, peaked around 2000/1 (solar maximum), and died off (solar minimum) from 2007-2009. After a 2-3 year period of relative inactivity Cycle #24 began around midway through 2010, and peaked in 2014. It bottomed out in 2020-21. Cycle #25 began around 2022, peaks in 2026 and bottoms out in 2032.

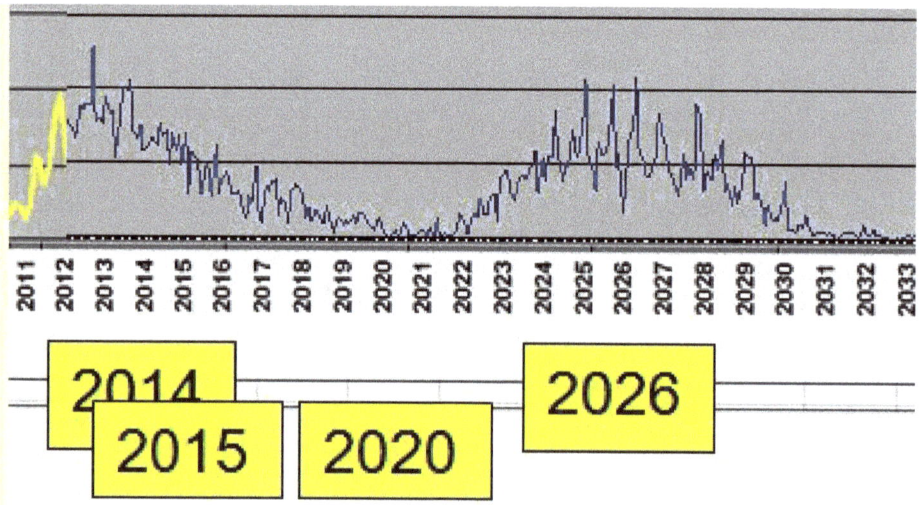

Just after a solar maximum (last one, 2015), there is usually a run of wetter years. The last solar minimum of 1996 was followed by the late 1990s which were wet. So were the late 1970s after the previous solar minimum of 1975. There is also typically more rain between peak and minimum years than between solar minima and maxima. 2014 and 2015 weather trends were always going to resemble the later 1970s and the mid-1990s for event timings, and the mid-1950s for intensity and rain amounts.

2016 was notable for its mild wet winter. 2023 is a near match to 1962/3. By combining lunar and solar influences, temperature (solar) may lead to amounts of rainfall (due to evaporation rates), whereas timing of severe weather events may follow strict lunar cycles. Note that rainy years are mainly on the rising side of the cycle, whereas dry years (including El Nino) tend to occur more at the peaks and troughs.

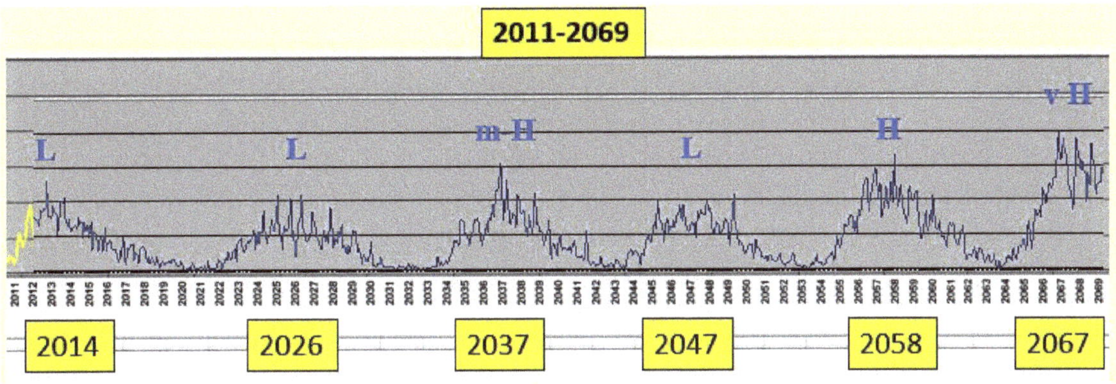

Looking at the trend in sunspot cycles the next 50 years may resemble the first 50 years of the 20th century. If so, it means the next 3 or 4 solar cycles may be weaker than the last 3 or 4, with low maximum sunspot numbers and long periods with no sunspots at all, perhaps initiating a 30-to-50-year decline in global temperatures. 2007 may have been the start of this decline and global temperatures may bottom out around 2030, perhaps even staying cooler until regaining higher sunspots around 2070. In the initial stages, heat stored in the oceans may keep things warm over the next decade, then polar outbreaks in both hemispheres may increase, with snow extending anew to lower latitudes and we may witness more heavy rains as increased amounts of polar air mix with moist equatorial systems. Just after a solar minimum (2020), there is usually a run of wetter years in Australia. Solar minimum years bring lower temperatures. Just after the solar minimum of 1996, the late 1990s were wet years. So were the 1970s after the previous solar minimum of 1975. And so, too, for 2021-2024 followed by 2029-2031.

Immediate weather is modified by cycles of solar activity (position and number of occurring Sunspots), and the cyclic position of the Moon relative to the earth. Solar influences determine major droughts. Sunspots are dark blotches on the part of the Sun corresponding to the tropical zone of the Earth, and are cyclonic on the Sun. Like on earth, where cyclones are very rarely seen on the equator, Sunspots are seldom seen on the equator of the Sun. Sunspots have a fixed swaying motion whereas cyclones on earth move, either north or south away from the equator in a parabolic fashion.

Sunspots have a cycle of 9-14 years, averaging about 11 years or 23 years. After solar minimum, 4- or 5-years elapse until Sunspot maximum, with spots nearer the Sun's equator. Then they gradually decline in number although still near to the Sun's equator, until minimum is reached. This decline is about 7 years. Intensity has a connection to the Sunspot cycle. The next solar maximum is 2025-6.

Very long term, the next 3 or 4 solar cycles are likely to be weaker than the last 3 or 4, with low maximum sunspot numbers and long periods with no sunspots at all, perhaps initiating a 40 to 50-year incremental decline in global temperatures. 2007 may have been the start of this decline and global temperatures may bottom out around 2030, perhaps even staying cooler until late in the century. In the initial stages, heat stored in the oceans and occasional El Ninos may keep things warm over the next decade, but then polar outbreaks in both hemispheres may conceivably increase, with snow extending anew to lower latitudes. We may eventually witness more heavy rains, as increased amounts of polar air mix with moist equatorial systems.

As events unfold around 2022/23, even in temperate latitudes there may be more cyclonic storms. This is because contrary to popular belief, warmer periods have less and weaker hurricanes while cooler periods have more and stronger hurricanes. Further into the cold period, post 2030 or so, after the oceans have lost much of their heat, the difference between polar cold and ocean warmth may be less, which means rainfall may reduce.

EL NINO VS LA NINA

Normally, cold water from Antarctica flows up the left side of South America. Upon reaching the equator, it is referred to as the (Humboldt) current, which cannot turn right because the Americas are in the way, so which deflects to the left and trundles on towards Australia. Due to the interface between surface currents and air, winds blowing to the west are the result. It is the norm.

After about three years of this, the ocean rises to 62cm higher at the Australian end, like the slow sloshing of your bath, and, just like your bath, a point is reached where it wants to flow the other way to re-stabilize the sea level. The mysterious reverse-direction phenomenon became known "the El Nino" condition. Stagnation spreads and sometime later the Gulf Stream stalls.

It takes an average 3-4 years of normal blowing, and 1-2 years of this reverse sea-level stabilizing phenomenon. The normal winds, blowing west along the equatorial band, were previously known as the 'trade winds' and were used by sail-driven trading ships for centuries. But every five years the winds underwent a lull, which the mariners avoided, in which the winds blew the other way. The Peruvians knew about this and kept 15,000 years of records. It was only identified as a changing cycle by Westerners in 1983. They called it the Peruvian name because warmth in the water went eastwards all the way from the top of Australia to Peru. It was a brief but significant phenomenon. It was found to mainly affect Australia, and so Alice Springs, at the very centre geographically of the continent, was selected.

El Nino is still called off on, at the end of any given year, and refers, in hindsight, to the condition of the dry season in Alice Springs from April until October. In time it was realised that an exotic name to the in-between years to balance El Nino was needed. Hence "La Nina" was born.

It was assumed that La Nina would adequately describe the wet seasons in Australia, between October through March. However, weather is not neatly divided into two systems. An anticyclone is not equal and opposite to a cyclone. And second, La Nina already had a name, the trade winds. So, they overlaid the old name, with the hypothetical notion of the new La Nina, as if they had discovered a whole new system. You don't hear about trade winds in the Pacific anymore, but the area still contains strong winds blowing westwards, picking up moist and warm temperatures and delivering these to the east coasts of Australia and NZ. Only they now blow namelessly.

To repeat, El Nino is only identified by looking back, at the end of each year, and newcomer La Nina was the same. For this reason, there cannot be any such *coming* thing as an El Nino summer. It is a description in hindsight, which has to run for 6 consecutive months to receive its El Nino status. Likewise, La Nina is descriptive of bad weather that *was*. So why do the forecasters indulge in misinformation? For example, although it is still summertime, they have claimed that " La Niña continues to bring its contrasting wet-and-dry flavors to the north and south". Are they not aware that six months does not describe January?

Whether at any stage we are in El Nino or La Nina is measured between Darwin and Tahiti on something called the ENSO meter. The last time it registered in the El Nino range was in 2020. The last time the needle swung anywhere near La Nina was March 2022. Then it was over, and the needle has drifted back to its present near-neutral level, where it should stay at least through to 2024. Anybody can check this out. The next El Nino is expected around 2025. And always, El Nino and La Nina are over before they are announced. The forecasters, in 2023, named La Niña for the February floods in the east of Australia, but the severe weather event was entirely due to the closest lunar perigee for over 1000 years. ENSO is not a function of perigee. Whenever rain occurs, when they default to La Niña as the reason, they misinform, but the ENSO needle does NOT lie.

(21st October 2024)

La Nina possibly brings a theoretical symmetry to El Nino. Where El Nino is used to scare everyone to fear a dry, hot period, La Nina is the parallel scare for a wet season. This is where the climate change movement has been described as too simplistic. Using La Nina, scientists switch the concept of climate in everybody's minds, ignoring what the climate actually does. Now, in weather, everything is to be feared. La Niña was intended to be an extended event. Its arrival must be ongoing for six months, not describing a wet afternoon, or heat showers due to the high humidity. When the normal west-blowing trade winds are returning, it has no name, but that is not unusual or alarming. El Nino, on the other hand, portends drought which is part of some new normal in the globalist mindset which says eastern states are facing a hot, dry, global-warming-forever future. But La Nina means 6 months of wet. It can't be both.

For 2020 the ENSO reading was slightly positive, meaning a potential El Nino was developing for the first half of this year (up to and including July). Accordingly, as is typical for a dry summer season, in the early part of 2020, NT received well below average monsoon activity. El Nino tends to put the brake on cyclones. After July, the situation slipped backwards, and ENSO became slightly negative, indicating the potential return of mild La Nina conditions, and the resumption of the approach toward monsoon activity. The result is that the following two years would prove to be wetter than the past two. In 2021 the monsoon affected the Top End, but in 2022 both WA and NT were affected.

After two years the monsoon slowed down again, with 2024 again a light monsoon year, ceasing with El Nino-type conditions of 2025/2026, both years of little or no monsoonal activity. In 2027 only QLD will be affected, but the wet seasons may again pick up, with a strong La Nina for 2029 and monsoon severely affecting the whole top half of the continent. That year, destructive floods may occur and 2030 should be very wet for Australia,

Jump ahead to the next El Nino or positive ENSO time for 2032-2034, then a series of neutral years until the next short El Nino of 2038, followed by the La Nina years to 2041. 2031 – 2037 are expected to be monsoonal, although 2032-2034 should be less active than the remainder. It is to be noted that 2034 is the year of minimum 18d lunar declination (lunar minor standstill, last one being 2015), but monsoons happen mostly around the in-between solar min-max years, and 2034 is the start of these. 1998 was an equivalent 18d and min-max solar year and was very wet.

This also brings the 35-year Bruckner Cycle into focus because it is almost identical to the 35.4-year tidal period. In 1958 maximum Sunspots occurred. This culminated in Jupiter, Saturn and Uranus peaks. There was an extensive drought in many parts of the world. The last time this came close to happening was 1901, the year of the Great Federation Drought.

Solar cycles peak when earth-Moon distance is on the decrease. Solar peaks are closer to La Nina than El Nino, which has some bearing on the phenomena of ocean currents and therefore sea surface temperatures. Strongest El Ninos form either just after solar minima or on the rising side of Sunspot cycles.

Taking note of averages of perigees over a year yield how in one particular year the Moon may be averagely closer to earth. These are when perigees occur as the declination over the equator sees the Moon rising due east, in the 8.85-year cycle that sees the Moon swapping earth hemispheres and spending around 2 years in each hemisphere and 2 years around the equator. The closest averagely shorter earth-Moon distance apart years over the past century or so have been 1905, 1922, 1940, 1958, 1976, 1993, 2011, and 2029. On or just after averagely closer perigee years, El Nino has occurred on 1905/6, 1923-6, 1940-2, 1957-8, 1976-7, 1994-5 and 2014-15. It is expected again in 2032.

These closest perigee-years happen typically when the Moon is at or near minimum declination (minor standstill). It is around these years that the Moon's inter-latitudinal range is narrower and the speed of changing hemispheres in the course of monthly declination is slower. Although we lack long-term data we can assume such years tend more towards the slackening of sea currents due to a more sluggish lunar lateral pull, and therefore more tendency towards El Nino systems.

El Nino conditions kicked in about July 2015. The next lunar declination minimum was 2015. The next solar minimum began about 2015 and peaked in 2020. From 1957 to 1998 spans 40 years and 5 strong, 5 moderate and 4 weak El Ninos. There is an oscillation alert every 2.5 years. If we take the 5 strong and 5 moderate events, we get an average of significant El Nino-type cycle of 4-5 years, which is a multiple of the lunar declination cycle.

THE MOON

The Moon is a planet like earth, also orbiting the Sun, such that Earth holds the Moon to its own orbit, so earth and Moon orbit the Sun together. The Moon is tied to both earth and the Sun, and alternately absorbs and shields solar electro-charges, according to phase. By this process the Moon controls the timing of weather events. The Sun is the engine, and the Moon is the driver. The bottom of the air and the surface of the sea are joined. Things that are joined are part of a joined system. Both do indeed comprise one interactive and connected interface, covering a surface area of 73% of the earth. Lunar forces begin with the Earth (or Land) tide, in which the land surface rises vertically around about 20cm per day in NZ, 50cm per day in Australia, and 55cm on the equator, towards the moon, and recedes again in regular daily tidal rhythm, and which in some locations and times produce earthquakes. The Land Tide supports the tide of the ocean which then gives rise (and fall) to a daily Air (or atmospheric) Tide. All tides are repeatable, measurable, and therefore predictable.

Matching the timing of the air-tide to oceanic tide-tables at a location yields a repeating frequency of weather extremes. Nobody questions the predictability of sea tide-tables, yet we have been conditioned to believe the weather is not subject to cycles but is somehow a random process. The more time spent outdoors, the more it will be realized that nothing in nature is random but governed by repeating cycles. The rotation of the earth and planets, the recurrence of day and night, summer and winter, animal, bird, insect and fish breeding, feeding and migration habits, and the cycles of the growing seasons can all be predicted, not exactly, but to a degree of usefulness. There is no logical or scientific reason why weather should be the sole exception.

The monthly cycle that we can all see, from New moon to Full and back to New again is called phase. That includes the intermediate phases Waxing Crescent, First Quarter, Waxing Gibbous, Waning Gibbous (Disseminating), Last Quarter, and Waning Crescent (Balsamic). For weather, gardening, and fishing, some only use this cycle. It will work in a general sense. For weather, a change to full or new or quarter moons very often brings a weather pattern change. If raining before a phase change it may stop on the day of changeover, but if dry before and rain is about, it may choose the day of the phase to manifest. A circle around the moon indicates rain expected soon. An old mariners' saying is that the full moon 'eats' clouds. Maximum temperature of the day will correspond with tides; and at the beach, temperature is likely to be highest when the tide is lowest. For gardening, plant above ground with attention to leafy things when the moon is climbing up in the sky in the daytime, which is before and up to full moon. After the full moon, the moon is descending in the daytime because it is rising on the other side of the earth, so give attention to root things, e.g. bulbs, carrots, potatoes. For fishing, the new and full moon tides are the best, around 12 noon and midnight, and the quarter moons are good around dawn and dusk.

The second cycle to consider is when the moon changes hemispheres because of the tilt of the earth. When it is in the northern hemisphere the tidal variation is greater because of greater gravitational pull, and higher tides mean more water filling a bay which means more fish competing for food in the faster currents, so bite- chances may be better, but also the water table rises in the land, and as well the sap tide in the trees. This means more insects appear, also more birds. It is why the Canadian Rockies Hunting calendar is identical to the Maori Fishing Calendar. When the moon is in the northern hemisphere in winter Ireland can get very cold winds, dragged southwards from the N Pole. Northernmost full moons bring northerly wind systems and snowfalls. It is why if the winter full moon rises further in the north it will be a colder month, and if the summer full moon rises far in the south, it will be unusually warm. So, when the cycle of the moon changing hemispheres coincides with the full moon cycle, such that the winter full moon is also when the moon is furthest north, one may plan for snow. The moon traverses different parts of the sky as it orbits the earth in the course of a month. The constellation (or *sign*) is the road map of the declination cycle.

In folklore functions were assigned to signs. Pisces moon was considered best for fishing, hence the fish zodiac symbol. Cancer full moons were said to bring more snow. In gardening, the constellations favor types of plant activity. Those of a leafy, flowery, fruity, or rootsy nature are said to have water (Scorpio, Pisces, Cancer), air (Libra, Aquarius, Gemini), fire (Leo, Sagittarius, Aries) or earth (Capricorn, Taurus, Virgo) signs. Air and fire signs are usually considered barren, and earth and water signs fertile. In weather, the constellations provide the template for changes in barometric pressure.

The Perigee-Apogee, or Apsidal Line comes about because the moon's orbit around the earth is not a circle but an elliptical or egg shape. The point in the month closest to earth is called Perigee and the furthest from earth for the month, the Apogee. At perigee the moon is about 350,000 km from earth and at apogee it is 400, 000 km from earth. This is a varying difference of about 50,000 km; and even that varies by about 10,000 kms over 12 months of the year. The gravitational forces that apply between earth and moon are increased at perigee and decrease at apogee and account for extremes in tidal movement particularly when coinciding with either a new moon or a full moon. In perigee the weather is more turbulent and strong winds are common. These may develop into gales or hurricanes and inflict property damage. Fishermen use apogee because fish are found in greater numbers in deeper or calmer water at that time. On perigee the tides are bigger, and more fish come closer in, but they don't like the breakers with their bigger swells, so surfcasting isn't the best that day. But estuaries, where fish hide from the turbulence of the shoreline are excellent for fishing during perigees.

Just as the moon affects the oceans tides, so it affects the sap-tide in a plant's structure. Each plant, seed, and tree, and organism in the garden has a tidal flow or magnetic force that we see daily in the ocean's tidal flow. At perigee (and a couple of days either side) you would avoid planting seeds, as germination can be poor and consequent growth of the seedling may be leggy and weak. But it is an opportune time to maximize on feeding and fertilizing as the nutrient is more easily distributed in the soil. Everything is 'going for it', so to speak. Perigee favors fungi and they can become prolific over a three-day period. Perigee is a good time to put fermentation brews to work. Some gather kelp or seaweed currently because the higher tides bring seaweed onto beaches. At apogee the magnetic force is weaker and the tidal pull also weaker, gentler and more consistent, so the ground is more settled for putting in things that need a calm environment to get going. While some plants may grow rapidly, others might run to seed quickly if planted in apogee.

All cycles should be considered. If the moon is descending (declination towards the north) good nourishment is coming to roots, especially if the moon was also waning, a good time for pruning as branches don't bleed so much if cut. If, on the other hand, the moon is ascending (declination towards south) and a waxing moon is taking energy upwards away from the ground, tending to draw water upwards, modulating root-ward flow, then this is a good time to sow.

Those are the three main Moon cycles, phase, declination and perigee. One can be seen easily, but the other two are visible but not noticed unless careful attention is paid. The least easily noticeable cycle is perigee-apogee. This closest point to Earth, the perigee, occurs over a different latitude/longitude each time down on earth. The Moon in perigee exerts increased gravitational pull-on fluids such as air and water (and land, influencing earthquakes), depending on the season, and by generating more turbulence, can cause weather to worsen.

Because the earth rotates beneath the Moon once every 24 hours, unsettled conditions can happen in many countries on the same day. Up to three days before or after a perigee it is common to see the month's most rain, particularly in autumn and winter. Clear skies might return shortly afterwards as the Moon recedes from the earth. The perigee is responsible for the king-tides, which are the highest coastal tides in any month.

For half the year perigee is accompanied by Full moon, then by new moon during the other half. e.g. in 2016, new moon perigees began the year, changed to full moon perigees in August. In 2017 January and February were full moon perigees, new moon March until Sept, and then back to perigeal full moons for October, November, and December. But for 2019, full moon perigees were from January to May, and the new moon perigees took over to November. For 2021, most months, from February-August, were full moon perigee.

Perigees vary in closeness. In 2014 the closest perigee was in August, with #2 closest in January. September had the closest perigee in 2015, perigee #2 in February, and perigee #3 in March. Consequently, the closer 2014-15 summer moon was set to fuel increased cyclone and monsoonal activity and some above normal temperatures. But perigees were weak in December 2015 through January 2016, and cyclones were weak. In 2016 the closest perigee was in November and tied to a full moon, and 2016-17 was a more intense cyclone season. 2016 was also an averagely closer moon across all months, whereas overall, 2017 was an averagely further away moon, but 2018 again started to come closer, with 2019 continuing this trend. Where 2020 was the 8th closest since 1905, 2021 is not amongst the closest perigee years. As for individual months, January 2023 was the closest perigee since 685AD. It will not be as close again until 2368. Perigees are on an 8.85-year cycle, traversing between both earth's hemispheres. When perigees are descending towards the southern hemispheres, cyclones are more numerous. In 2015 and 2016 perigees were at or near occurring over the equator and in 2017 perigees were between the equator and the northern declination. By winter of 2016 perigee was past the equator heading north. In the first week of January 2018, the closest perigee for that year coincided with northern declination. Perigee was again level with the equator by November 2019. It passed that point and again entered the southern hemisphere by the following summer, and 2021 was expected to intensify. In years when the Moon is averagely closer to earth monsoons and cyclones also tend to be more destructive. The moon is typically averagely closer to earth during La Nina years. Strong La Ninas can signal wetter seasons. During El Nino the moon is often further away, causing a weakening of currents and a slowing of air flows. It is typically when monsoonal activity and cyclones may reduce. After a neutral phase, a developing La Nina took effect around July of 2016.

In 2011 the Moon was averagely much closer to earth, being (since 1900) in the top four closest to earth years. It was, in fact, the averagely closest to earth for 19 years. We know how destructive 2011 was, with the Japanese tsunami, Cyclone Yasi and Christchurch earthquakes. By 2013 the moon was further away but 2015 saw the moon drift slightly back closer. 2013 and 2014 have been neutral years, when monsoons have been late and cyclones fairly weak. 2016 was another slightly closer-perigee year, but 2019 not quite as close. 2021 then saw cyclone activity coming back to life.

General trends are:

- Closer perigees = more cyclones (1997, 2002, 2007, 2016)
- Far perigees = less cyclones (1983, 1991, 2000, 2009, 2013, 2017)
- Maximum declination = less cyclones (1987, 2006, 2025)
- Minimum declination = more cyclones (1978, 1997, 2015, 2034)

In the 124 years that lie between 1900-2024, in order of closeness of perigee (moon-to-earth distance averaged over the year), 2015 ranked a high 26th. 2015 was both minimum declination and year of closer perigee = more cyclones. 2016 ranked 29th, so not quite as close as in 2015 but similar in closeness to 1997 (27th), 2002 (28th), 1954 (30th), and 1985 (31st). 2017 ranks well down the list - at 109th closest since 1900. It was on par with 1942 (108th) and 1960 (110th). So how did these years fare for cyclone intensities?

The 1997–98 South Pacific cyclone season was one of the most active and longest South Pacific tropical seasons then on record. The 2002–03 South Pacific cyclone season was the most active tropical cyclone season since 1997-98 (with ten tropical cyclones occurring within the South Pacific basin). 1954 saw extreme rain and flooding on the Gold Coast and the 1985–86 Australian region cyclone season was also a well above average season. Going to the other extreme, the 124th closest perigee occurred in 1999/2000, which was a quiet season for cyclones, as was 1990/91 which sits at 123rd closest perigee year. Quiet too, was 1938 (122nd) and 2009/10 (115th).

2015 saw the Moon drift slightly back closer. 2013 (114th) and 2014 (62nd) were years when monsoons were late and cyclones fairly weak. Comparing 2017 with 1942 and 1960, 1942 saw nine cyclones; three in the northeast and six affecting WA. In 1960, there were again three in the northeast, but five affected WA, with more activity from a further four in the north Indian Ocean, west of Australia. In 2016-17 were three developments in the northeast, but more on the western side of the country.

2019 was a light year for cyclones. 2019, in perigee-closeness, sits at 52nd closest, between 1988 (51st) and 2001 (53rd). The 1987-88 Australian region cyclone season was the least active tropical cyclone season since 1977–78. And looking at the 2001-02 cyclone season, it was a slightly below average tropical cyclone season. But 2020 is the 8th closest perigee. In comparison, the 7th closest was 1962, and the 9th closest was 1949. 2021 is the 95th closest. It was on a par with 1916 (94th closest) and 1982 (96th closest), the latter seeing an ex-tropical cyclone. 2022 is the 90th closest. There was flooding in Qld akin to 1965 in March 2021. There also will be flooding in 2023 as described for the states in earlier pages.

SUMMARY

Wettest years: between Solar max - Solar min (21/33) and mid-pt (23d) or lunar min (18d) (La Nina): 2021, 2023, 2024.
Driest years: at Solar max and Solar min (22/33) and about mid-pt (23d) or lunar max (28d) (El Nino).: 2020, 2025-2028
Warmest years: Solar max 2025-7
Coolest years: Solar min: 2021, 2031-3
Wetter years appear around minimum lunar declination (18deg), e.g. 1978, 1997 and 2015.
Drier years appear around maximum lunar declination (28deg), e.g. 1970, 1987 and 2005, 2020.
More El Ninos come on closer perigee years.
More La Ninas come on far away perigee years.
El Ninos cause least cyclones.
Monsoons weaken in El Nino years.
Monsoons stronger in La Nina years.
La Nina years are 2x wetter than El Nino years.
El Nino years are 2x drier than La Nina years.

2020	Close
2021	
2022	
2023	Close
2024	Close
2025	Close
2026	
2027	
2028	Close
2029	Close
2030	Close

Perigee (Closest)			Apogee (Furtherst)		
Year	Distance	Order	Year	Distance	Order
2020	361609	8th	2029	405615	5th
2024	361668	12th	2020	405546	18th
2029	361911	27th	2023	405499	27th
2028	361994	30th	2025	405490	36th
2025	362520	54th	2028	405460	49th
2030	362814	76th	2024	405428	60th
2023	362834	81st	2021	405356	88th
2022	362929	90th	2026	405302	107th
2021	363002	95th	2022	405294	108th
2026	363184	107th	2030	405293	110th
2027	363339	120th			

This means that for closest average annual perigees, relative to each other, e.g. 2020 being the 8/130, (between 1900-2030), and the years 2024 and 2025 will have more significant perigeal events than will 2026 and 2027. 2028-30 will again see destructive weather events around perigee times.

MOON SYMBOLS

1st Q	Moon in first-quarter phase.
3rd Q	Moon in third-quarter phase.
^	Northern declination Moon at furthest point north for month.
A	Apogee Moon furthest from earth for the month.
F	Full moon.
N	New moon.
P	Perigee Moon at closest point to the earth out of the 27 day cycle.
P(n)	Priority closest e.g. P3 means third closest for year.
V	Southern declination Moon at furthest point south for the month.
XhN	Moon crossing the equator heading north.
XhS	Moon crossing the equator heading south.

2025

```
              Perigee                                Apogee
-----------------------------------   -----------------------------------
Jan  7 23:36 370171 km     F-5d22h    Jan 21  4:56 404298 km     F+7d 6h
Feb  2  2:44 367456 km     N+3d14h    Feb 18  1:12 404881 km     F+5d11h
Mar  1 21:20 361966 km     N+1d20h    Mar 17 16:38 405752 km     F+3d 9h
Mar 30  5:27 358126 km     N+  18h    Apr 13 22:49 406294 km +   F+  22h
Apr 27 16:16 357118 km -   N-   3h    May 11  0:50 406244 km +   F-1d16h
May 26  1:38 359022 km     N-1d 1h    Jun  7 10:44 405551 km     F-3d21h
Jun 23  4:44 363176 km     N-2d 5h    Jul  5  2:30 404626 km     F-5d18h
Jul 20 13:54 368046 km     N-4d 5h    Aug  1 20:38 404163 km     F-7d11h
Aug 14 18:02 369286 km     F+5d10h    Aug 29 15:35 404551 km     N+6d 9h
Sep 10 12:11 364780 km     F+2d18h    Sep 26  9:47 405551 km     N+4d13h
Oct  8 12:37 359818 km     F+1d 8h    Oct 23 23:32 406444 km -   N+2d11h
Nov  5 22:30 356832 km ++  F+   9h    Nov 20  2:49 406692 km --  N-   3h
Dec  4 11:07 356961 km +   F-  12h    Dec 17  6:11 406322 km -   N-2d19h
```

2026

```
              Perigee                                Apogee
-----------------------------------   -----------------------------------
Jan  1 21:45 360347 km     F-1d12h    Jan 13 20:49 405436 km     N-4d23h
Jan 29 21:54 365877 km     F-3d 0h    Feb 10 16:54 404576 km     N-6d19h
Feb 24 23:20 370131 km     F-6d12h    Mar 10 13:44 404384 km     F+7d 2h
Mar 22 11:41 366856 km     N+3d10h    Apr  7  8:33 404973 km     F+5d 6h
Apr 19  6:58 361630 km     N+1d19h    May  4 22:32 405841 km     F+3d 5h
May 17 13:49 358073 km     N+  17h    Jun  1  4:34 406368 km +   F+  19h
Jun 14 23:19 357195 km     N-   3h    Jun 28  7:12 406266 km +   F-1d16h
Jul 13  7:51 359110 km     N-1d 1h    Jul 25 16:46 405548 km     F-3d21h
Aug 10 11:20 363287 km     N-2d 6h    Aug 22  8:22 404642 km     F-5d19h
Sep  6 20:27 368254 km     N-4d 7h    Sep 19  3:02 404216 km     F-7d13h
Oct  1 20:42 369336 km     F+5d 3h    Oct 16 22:57 404638 km     N+6d 7h
Oct 28 18:02 364410 km     F+2d13h    Nov 13 17:51 405618 km     N+4d10h
Nov 25 20:59 359346 km     F+1d 6h    Dec 11  6:47 406420 km --  N+2d 5h
Dec 24  8:31 356649 km ++  F+   7h
```

Disclaimer

The forecasts in this work are best-endeavour opinions of the authors and associates. No guarantee or promise of certainty is claimed. The authors cannot be held accountable for decisions made on the basis of information in these pages.

Thanks

I am hugely grateful to the supportive team of Glenys and Victoria for their many hours of work, project design, research and layout, without which the almanac would not be possible. Thank you all so much. Also Vicki in dispatch. We also acknowledge the NOAA for ongoing permission to adapt archival MSL maps, and the BoM for historical archival data.

Ken Ring

www.ingramcontent.com/pod-product-compliance
Lightning Source LLC
LaVergne TN
LVHW071345210926
848486LV00003B/231

9781067044534